Problems and Solutions in
Banach Spaces, Hilbert Spaces, Fourier
Transform, Wavelets, Generalized
Functions and Quantum Mechanics

Problems and Solutions in Banach Spaces, Hilbert Spaces, Fourier Transform, Wavelets, Generalized Functions and Quantum Mechanics

Willi-Hans Steeb
University of Johannesburg, South Africa

Wolfgang Mathis
Leibniz Universität Hannover, Germany

World Scientific

NEW JERSEY · LONDON · SINGAPORE · BEIJING · SHANGHAI · HONG KONG · TAIPEI · CHENNAI · TOKYO

Published by

World Scientific Publishing Co. Pte. Ltd.

5 Toh Tuck Link, Singapore 596224

USA office: 27 Warren Street, Suite 401-402, Hackensack, NJ 07601

UK office: 57 Shelton Street, Covent Garden, London WC2H 9HE

Library of Congress Cataloging-in-Publication Data
Names: Steeb, W.-H., author. | Mathis, Wolfgang, author.
Title: Problems and solutions in Banach spaces, Hilbert spaces, Fourier transform,
    wavelets, generalized functions and quantum mechanics /
    Willi-Hans Steeb (University of Johannesburg, South Africa),
    Wolfgang Mathis (Leibniz Universität Hannover, Germany).
Description: New Jersey : World Scientific, [2023] | Includes bibliographical references and index.
Identifiers: LCCN 2022026777 | ISBN 9789811245725 (hardcover) |
    ISBN 9789811246838 (paperback) | ISBN 9789811245732 (ebook for institutions) |
    ISBN 9789811245749 (ebook for individuals)
Subjects: LCSH: Functional analysis. | Quantum theory.
Classification: LCC QC174.52.F8 S74 2023 | DDC 530.1201/5157--dc23/eng20220829
LC record available at https://lccn.loc.gov/2022026777

British Library Cataloguing-in-Publication Data
A catalogue record for this book is available from the British Library.

For any available supplementary material, please visit
https://www.worldscientific.com/worldscibooks/10.1142/12513#t=suppl

Printed in Singapore

Preface

The purpose of this book is to supply a collection of problems in functional analysis and quantum mechanics together with their detailed solutions which will prove to be valuable to graduate students as well as to research workers in these fields. All the important concepts and definitions are provided either in the introduction or the problems. The functional analysis part covers normed spaces, Banach spaces, Hilbert spaces and generalized functions. The discrete and continuous Fourier transforms are also covered as well as wavelets. Some chapters also include programming problems. Supplementary problems in each chapter are also provided. Students can learn important principles and strategies required for problem solving.

In chapter 1 we consider normed and Banach spaces. Cauchy sequences are discussed. The dual space of a normed linear space is introduced. Attention is also given to Euclidean spaces. Chapter 2 is devoted to Hilbert spaces. A complete pre-Hilbert space is a Hilbert space. The pre-Hilbert space defines a scalar product. The properties of the scalar product are studied and inequalities in the Hilbert spaces are covered. Finite dimensional Hilbert spaces are introduced in chapter 3, namely $\mathbb{R}^n$, $\mathbb{C}^n$ with the corresponding scalar product and the Hilbert space $M(n, \mathbb{C})$ of $n \times n$ matrices over $\mathbb{C}$. Hermitian matrices, unitary matrices and projection matrices acting as operators in these Hilbert spaces are discussed. Chapter 4 concentrates on the Hilbert space of square integrable functions. Self-adjoint and unitary operators are also included. The Hilbert space $\ell_2(\mathbb{I})$ of sequences with $\mathbb{I}$ a countable index set is treated in chapter 5. The link between the Hilbert spaces $L_2(\Omega)$ and $\ell_2(\mathbb{I})$ is emphasized. Problems and solutions are provided for both discrete and continuous Fourier transforms in chapter 6 together with some programs. Wavelet theory is a form of mathematical transformation, similar to the Fourier transform that it takes a signal in time domain and represents it in frequency domain. Chapter 7 covers both discrete and continuous wavelets. The Hilbert space of holomorphic entire functions and the Bargmann transform are introduced in chapter 8. Generalized functions with emphasizes on the Banach-Gelfand triple are studied in chapter 9. The Fourier transform in the sense of generalized function is also introduced. The Dirac delta function plays a central role also in connection with the completeness relation. Chapter 10 provides applications of Hilbert space theory and generalized functions to quantum theory. Spin systems, Bose systems and Fermi system are covered as well as spin coherent states and coherent states.

Any useful suggestions and comments are welcome.

The International School for Scientific Computing (ISSC) provides certificate courses for this subject. Please contact the first author if you want to do this course.

e-mail address of the first author:

`steebwilli@gmail.com`

Contents

Symbol Index

$:=$	is defined as
$\in$	belongs to (a set)
$\notin$	does not belong to (a set)
$\cap$	intersection of sets
$\cup$	union of sets
$T \subset S$	subset T of set S
$S \cap T$	the intersection of the sets S and T
$S \cup T$	the union of the sets S and T
$\emptyset$	empty set
$\mathbb{N}$	set of natural numbers
$\mathbb{N}_0$	set of natural numbers including 0
$\mathbb{Z}$	set of integers
$\mathbb{Q}$	set of rational numbers
$\mathbb{R}$	set of real numbers
$\mathbb{R}^+$	set of nonnegative real numbers
$\mathbb{C}$	set of complex numbers
i	$\sqrt{-1}$ imaginary unit
$\Re(z)$	real part of the complex number z
$\Im(z)$	imaginary part of the complex number z
$|z|$	modulus of complex number z
	$|x + iy| = (x^2 + y^2)^{1/2}, \ x, y \in \mathbb{R}$
$\mathbb{R}^n$	n-dimensional Euclidean space
	space of column vectors with n real components
$\mathbb{C}^n$	n-dimensional complex linear space
	space of column vectors with n complex components
$\mathcal{B}$	Banach space
$\mathcal{H}$	Hilbert space
$f(S)$	image of set S under mapping f
$f \circ g$	composition of two mappings $(f \circ g)(x) = f(g(x))$
$f * g$	convolution of f and g
$\mathbf{x}$	column vector in $\mathbb{C}^n$
$\mathbf{x}^T$	transpose of $\mathbf{x}$ (row vector)
$\mathbf{0}$	zero (column) vector
$\|\cdot\|$	norm

$\mathbf{x} \cdot \mathbf{y} \equiv \mathbf{x}^* \mathbf{y}$	scalar product (inner product) in $\mathbb{C}^n$
$\mathbf{x} \times \mathbf{y}$	vector product in $\mathbb{R}^3$
A, B, C	$m \times n$ matrices
$M(n, \mathbb{R}), M(n, \mathbb{C})$	vector space of $n \times n$ matrices with real or complex elements
$\det(A)$	determinant of a square matrix A
$\mathrm{tr}(A)$	trace of a square matrix A
$\mathrm{rank}(A)$	rank of matrix A
A^T	transpose of matrix A
$\overline{A}$	conjugate of matrix A
A^*	conjugate transpose of matrix A
$A^\dagger$	conjugate transpose of operator A (physics)
A^{-1}	inverse of square matrix A (if it exists)
I_n	$n \times n$ unit matrix
I	unit operator
0_n	$n \times n$ zero matrix
AB	matrix product of $m \times n$ matrix A and $n \times p$ matrix B
$A \bullet B$	Hadamard product (entry-wise product) of $m \times n$ matrices A and B
$[A, B] := AB - BA$	commutator for square matrices A and B
$[A, B]_+ := AB + BA$	anticommutator for square matrices A and B
$A \otimes B$	Kronecker product of matrices A and B
$T_1 \otimes T_2$	Tensor product of tensors T_1 and T_2
$A \oplus B$	Direct sum of matrices A and B
$\delta_{j,k}$	Kronecker delta with $\delta_{j,k} = 1$ for $j = k$ and $\delta_{j,k} = 0$ for $j \neq k$
δ	delta function
Θ	Heaviside's function
λ	eigenvalue
ϵ	real parameter
t	time variable
$\hat{H}$	Hamilton operator
$S(\mathbb{R})$	Schwartz space
$L_p(D)$	the space of pth order Lebesgue integrable functions $f : D \to \mathbb{R},\ p \in [1, \infty)$
$L_p(D, \mu)$	the space of pth order μ integrable functions using the Lebesgue-Stieltjes integral
$L_\infty(D)$	the space of essentially bounded functions $f : D \to \mathbb{R}$
$\hbar$	reduced Planck constant

Chapter 1

Normed Spaces and Banach Spaces

1.1 Introduction

Let V be a vector space (sometimes called linear space) over the field $\mathbb{C}$ of complex number or over the field $\mathbb{R}$ of real numbers. Then the vector space V is called a normed linear space if for every $v \in V$ there is a map $V \mapsto \mathbb{R}$ $\|v\|$ called the *norm* such that the following conditions are satisfied

(i) $\|v\| \geq 0$
(ii) $\|v\| = 0$ if and only if $v = 0$
(iii) $\|cv\| = |c|\,\|v\|$ for every scalar $c \in \mathbb{C}$
(iv) $\|v + w\| \leq \|v\| + \|w\| \quad w \in V$.

The mapping $d(\cdot, \cdot) : V \times V \to \mathbb{R}$ with $d(v, w) := \|v - w\|$ is called norm-induced *metric*.

Let V, W be two normed spaces. Consider the map $T : V \to W$. Then T is *linear* if

$$T(\alpha x + \beta y) = \alpha T(x) + \beta T(y)$$

for every $x, y \in V$ and $\alpha, \beta \in \mathbb{C}$. The map T is *injective* if

$$T(x) = T(y) \text{ implies } x = y.$$

The *kernel* or *null space* of the map T is

$$\ker(T) := \{\, x \in V \, : \, T(x) = 0 \,\}.$$

The *range* of T is

$$\operatorname{range}(T) := \{\, T(x) \, : \, x \in V \,\}.$$

The *rank* of T is the vector space dimension of its range
$$\text{rank}(T) = \dim(\text{range}(T)).$$
The map T is finite-rank if $\text{range}(T)$ is finite dimensional. The map T is *surjective* if $\text{range}(T) = W$. The map T is a *bijection* if it is both injective and surjective.

Let V be a normed vector space and let $\{f_n\}_{n\in\mathbb{N}}$ be a sequence of elements of the normed vector space V.
(i) The sequence $\{f_n\}_{n\in\mathbb{N}}$ converges to $f \in V$ if $\lim_{n\to\infty} \|f - f_n\| = 0$, i.e. if
$$\forall \epsilon > 0, \ \exists N > 0, \ \forall n \geq N \ \|f - f_n\| < \epsilon.$$
One writes $\lim_{n\to\infty} f_n = f$.

(ii) A sequence $\{f_n\}_{n\in\mathbb{N}}$ is a *Cauchy sequence* if
$$\forall \epsilon > 0, \ \exists N > 0, \ \forall m, n \geq N, \ \|f_m - f_n\| < \epsilon.$$
Any convergent sequence in a normed vector space V is a Cauchy sequence. However not all Cauchy sequences are convergent.

A normed vector space V where all Cauchy sequences are convergent is said to be *complete*. A complete normed vector space is called a *Banach space* ([4],[8],[15],[16],[43],[50],[59],[60],[63],[65],[84]). Every normed space has a completion which is a Banach space.

A Banach space $\mathcal{B}$ is *separable* if it contains a dense set S, that is $\overline{S} = \mathcal{B}$, which is countable. Let $\mathcal{B}$ be a Banach space with the underlying field $\mathbb{R}$ or $\mathbb{C}$. The continuous dual space is the space of continuous linear maps from $\mathcal{B}$ into $\mathbb{R}$ or $\mathbb{C}$ (also called *linear functions*). Now $\mathbb{R}$ and $\mathbb{C}$ are Banach spaces itself with the absolute value as norm. The dual space of a Banach space is a Banach space.

The n-dimensional *Euclidean space* with metric tensor field
$$g = \sum_{j=1}^{n} (dx_j \otimes dx_j)$$
is a Banach space. Let $\mathbb{I}$ be a countable index set. Examples are $\mathbb{N}$, $\mathbb{N}_0$, $\mathbb{Z}$, $\mathbb{N} \times \mathbb{N}$, $\mathbb{Z} \times \mathbb{Z}$ etc. The following are Banach spaces
$$\ell_p(\mathbb{I}) := \left\{ (v_n)_{n\in\mathbb{I}} : \sum_{n\in\mathbb{I}} |v_n|^p < \infty \right\}, \ \ \|(v_n)\|_p = \left(\sum_{n\in\mathbb{I}} |a_n|^p \right)^{1/p}, \ 1 \leq p < \infty$$

$$\ell_\infty(\mathbb{I}) := \left\{ (v_n)_{n\in\mathbb{I}} \; : \; \sup_{n\in\mathbb{I}} |(v_n)|^p < \infty \right\}, \quad \|(v_n)\|_\infty = \left(\sum_{n\in\mathbb{I}} |v_n|^p \right)^{1/p}.$$

The underlying field are the complex numbers $\mathbb{C}$ or the real numbers $\mathbb{R}$.

The Banach spaces $\ell_p(\mathbb{N}_0)$ $(1 \le p < \infty)$ are separable, i.e. each has a countable dense subset. The Banach space $\ell_\infty(\mathbb{N}_0)$ is not separable.

Let (V, d) be a metric space. A *contraction* of V (sometimes called contraction mapping) is a function $f : V \to V$ that satisfies:

$$\text{For all} \quad v, w \in V \; : \; d(f(v), f(w)) \le \gamma d(v, w)$$

with some real number $0 < \gamma < 1$.

Let V be a normed linear space. Its *dual space* V' is the set of all continuous linear functionals with domain V, i.e.

$$V' := \{ T : V \to \mathbb{C} \; : \; T \text{ is continuous and linear} \}.$$

The dual space V' is a Banach space even if V is only a normed space.

One calls a triple of vector spaces $(\mathcal{B}, \mathcal{H}, \mathcal{B}')$ a *Banach-Gelfand triple* if $(\mathcal{B}, \|.\|_\mathcal{B})$ is a Banach space, which is dense in some Hilbert space $\mathcal{H}$, and which in turn is contained in $\mathcal{B}'$ the *dual* of $\mathcal{B}$. An example of a Banach Gelfand triple is $(\ell_1(\mathbb{Z}), \ell_2(\mathbb{Z}), \ell_\infty(\mathbb{Z}))$.

In the Banach space $S(\mathbb{R}^d)$ with $f \in S(\mathbb{R}^d)$ the classical *Poisson formula* can be written as

$$\sum_{k\in\mathbb{Z}^d} \hat{f}(k) = \sum_{n\in\mathbb{Z}^d} f(n)$$

where the sums are absolutely convergent on both sides.

Let $\mathcal{B}_1$ and $\mathcal{B}_2$ be Banach spaces. A linear transformation $T : \mathcal{B}_1 \to \mathcal{B}_2$ is *bounded* if and only if

$$\exists C_T, \quad \forall \mathbf{v} \in \mathcal{B}_1, \quad \|Y\mathbf{v}\|_{\mathcal{B}_2} \le C_T \|\mathbf{v}\|_{\mathcal{B}_1}.$$

If the linear transformation $T : \mathcal{B}_1 \to \mathcal{B}_2$ is a bounded linear transformation, then the norm is given by

$$\|T\|_{\mathcal{B}_1 \to \mathcal{B}_2} := \sup_{\mathbf{v}\in\mathcal{B}_1, \mathbf{v}\neq 0} \frac{\|T\mathbf{v}\|_{\mathcal{B}_2}}{\|\mathbf{v}\|_{\mathcal{B}_1}}.$$

Let $\mathbf{v} \in \mathbb{C}^n$. Then

$$\|\mathbf{v}\|_p = \left(\sum_{j=1}^{n} |v_j|^p \right)^{1/p}$$

for $1 \leq p < \infty$ defines a norm and for $p = \infty$

$$\|\mathbf{v}\|_\infty = \max_{1 \leq j \leq n} |v_j|$$

defines a norm. For the vector space $\mathbb{R}^n$ norms are

$$\|\mathbf{v}\|_1 = \sum_{j=1}^{n} |v_j|, \quad \|\mathbf{v}\|_2 = \left(\sum_{j=1}^{n} |v_j|^2 \right)^{1/2}, \quad \|\mathbf{v}\|_\infty = \sup_{j=1,2,\ldots,n} |v_j|$$

with $\mathbf{v} \in \mathbb{R}^n$.

Let $\mathcal{B}_1, \mathcal{B}_2$ be two Banach spaces. A bounded linear operator $T : \mathcal{B}_1 \to \mathcal{B}_2$ between two Banach spaces is defined to be the linear map for which the norm

$$\|T\| = \sup\{ \|Tv\| : \|v\| \leq 1 \}$$

is finite. The set of all operators from $\mathcal{B}$ to itself is an algebra. The multiplication is defined as

$$(T_1 T_2)(v) := T_1(T_2(v)).$$

Let $T \in \mathbb{R}^{n \times n}$. Then

$$\|T\|_\infty = \max_{1 \leq j \leq n} \sum_{k=1}^{n} |T_{jk}|$$

defines a norm.

Let $T : \mathcal{B} \to \mathcal{B}$ be a bounded linear operator on a Banach space $\mathcal{B}$ and I be the identity operator. If $\|I - T\| < 1$, then T has a unique bounded linear inverse T^{-1}.

The space $L_1(\mathbb{R})$ is the set of all functions over $\mathbb{R}$ with

$$\|f(x)\|_1 := \int_{\mathbb{R}} |f(x)| dx < \infty.$$

For all $f, g, h \in L_1(\mathbb{R})$ the *convolution product*

$$h(s) = (f \star g)(s) = \int_{-\infty}^{\infty} f(x) g(s - x) dx$$

exists. The vector space $L_\infty([a,b])$ consists of all bounded functions

$$\|f(x)\|_\infty := \inf\{c_f \geq 0, \ |f(x)| < c_f \ \text{for all} \ x \in [a,b]\}.$$

One has

$$C([a,b]) \subset L_\infty([a,b]) \subset \cdots \subset L_p([a,b]) \subset \cdots \subset L_2([a,b]) \subset L_1([a,b])$$

with $[a,b] \subset \mathbb{R}$ bounded.

The *Riemann-Lebesgue lemma* states that

$$\lim_{|n|\to\infty} \int_{-\pi}^{+\pi} f(x)e^{-inx}\,dx = 0$$

for all $f \in L_1([-\pi,\pi])$.

Let $S \subset \mathbb{R}^n$. The *diameter* of S denoted by $|S|$ is defined as

$$|S| := \sup_{\mathbf{x,y}\in S}\{\|\mathbf{x}-\mathbf{y}\|\}$$

where $\|\mathbf{x}-\mathbf{y}\|$ is the Euclidean norm of the vector $\mathbf{x}-\mathbf{y} \in \mathbb{R}^n$.

Let (V,d) be a complete metric space and $f : V \to V$. The *Lipschitz constant* of f is defined by

$$\text{Lip}(f) := \sup_{x\neq y} \frac{d(f(x),f(y))}{d(x,y)}.$$

Therefore if $\text{Lip}(f) = \lambda$, then $d(f(x),f(y)) \leq \lambda d(x,y)$ for all $x,y \in V$.

A complete separable metric space is called a Polish space.

1.2   Solved Problems

Problem 1. Two *Cauchy sequences* $\{x_k\}$ and $\{y_k\}$ are said to be equivalent if for all $\epsilon > 0$, there is a $k(\epsilon)$ such that for all $j \geq k(\epsilon)$ we have $d(x_j, y_j) < \epsilon$. One writes $\{x_k\} \sim \{y_k\}$. Obviously, $\sim$ is an equivalence relationship. Show that equivalent Cauchy sequences have the same limit.

Solution 1. Suppose $\{x_k\} \sim \{y_k\}$ and
$$\lim_{k \to \infty} x_k = x.$$
Given $\epsilon > 0$, there is $k_1(\epsilon)$ such that $k \geq k_1(\epsilon)$ implies $d(x_k, x) < \epsilon/2$. There is also $k_2(\epsilon)$ such that $k \geq k_2(\epsilon)$ implies $d(x_k, y_k) < \epsilon/2$. Therefore, letting
$$k(\epsilon) = \max\{k_1(\epsilon),\, k_2(\epsilon)\}$$
$k \geq k(\epsilon)$ implies
$$d(x, y_k) \leq d(x, x_k) + d(x_k, y_k) < \frac{\epsilon}{2} + \frac{\epsilon}{2} = \epsilon$$
which states that $y_k \to x$ for $k \to \infty$.

Problem 2. A sequence $\{f_n\}$ $(n \in \mathbb{N})$ of elements in a normed space E is called a *Cauchy sequence* if, for every $\epsilon > 0$, there exists a number M_ϵ, such that $\|f_p - f_q\| < \epsilon$ for $p, q > M_\epsilon$. Consider the Banach space $\mathbb{R}$. Show that
$$s_n = \sum_{j=1}^{n} \frac{1}{(j-1)!}, \qquad n \geq 1$$
is a Cauchy sequence.

Solution 2. Without loss of generality we can assume that $p > q$. First we must prove for $q \geq 4$ the inequality
$$\sum_{n=q}^{p-1} \frac{1}{n!} < \frac{p-q}{pq}.$$
We do this by induction over p. For $p = q + 1$ we have
$$\frac{1}{q!} < \frac{1}{(q+1)q}$$
for $q \geq 4$. Assume that $(p = q + k)$
$$\sum_{n=q}^{q+k-1} \frac{1}{n!} < \frac{k}{(q+k)q}.$$

It follows that $(p = q + k + 1)$

$$\sum_{n=q}^{q+k} \frac{1}{n!} < \frac{k}{(q+k)q} + \frac{1}{(q+k)!} < \frac{k+1}{q(q+k+1)}.$$

Now

$$\|s_p - s_q\| = \left| \sum_{j=1}^{p} \frac{1}{(j-1)!} - \sum_{j=1}^{q} \frac{1}{(j-1)!} \right| = \left| \sum_{j=q}^{p-1} \frac{1}{j!} \right| < \left| \frac{p-q}{pq} \right|.$$

Thus for $m_\epsilon > \max\{4, \frac{2}{\epsilon}\}$ and $p, q > m_\epsilon$ we have

$$\left| \frac{p-q}{pq} \right| < \frac{p}{pq} + \frac{q}{pq} = \frac{1}{q} + \frac{1}{p} < \frac{1}{m_\epsilon} + \frac{1}{m_\epsilon} < \epsilon.$$

Problem 3. Let $\mathcal{B}$ a Banach space. Show that each Cauchy sequence in $\mathcal{B}$ is bounded, that is $\sup_{n \in \mathbb{N}} \|x_n\| < \infty$.

Solution 3. If $\{x_n\}$ is a Cauchy sequence an $N \in \mathbb{N}$ can be chosen such that $\|x_n - x_N\| \leq 1$ for all $n \geq N$. Now, let $C := \max\{\|x_j\| : j = 1, \ldots, N\} + 1$ and it follows that $\|x_n\| \leq C$ for all $n \in \mathbb{N}$.

Problem 4. Consider the sequence $\{x_k\}$, $k = 1, 2, \ldots$ in $\mathbb{R}$ defined by $x_k = 1/k^2$ for all $k = 1, 2, \ldots$. Show that this sequence is a *Cauchy sequence*.

Solution 4. Let $\epsilon > 0$. Let $k(\epsilon)$ be any integer k that satisfies $k^2 \epsilon > 2$. For $m, \ell \geq k(\epsilon)$, we have

$$d(x_m, x_\ell) = \left| \frac{1}{m^2} - \frac{1}{\ell^2} \right| \leq \left| \frac{1}{m^2} + \frac{1}{\ell^2} \right| \leq \left| \frac{1}{(k(\epsilon))^2} + \frac{1}{(k(\epsilon))^2} \right| = \frac{2}{(k(\epsilon))^2}$$

and this last term is less than ϵ by choice of $k(\epsilon)$.

Problem 5. Let f, g be elements in a normed space. Show that

$$\|f - g\| \geq |\, \|f\| - \|g\| \,|.$$

Solution 5. From $\|f - g\| + \|g\| \geq \|f\|$ we obtain $\|f - g\| \geq \|f\| - \|g\|$. On the other hand

$$\|f - g\| = |-1| \, \|g - f\| \geq \|g\| - \|f\|.$$

Problem 6. The sequence space consists of the set of all (bounded or unbounded) sequences of complex numbers $x = (x_1, x_2, \dots)$. Thus we have a vector space. Can we define a metric in this vector space which is not implied by a norm?

Solution 6. Yes. An example is

$$d(x, y) = \sum_{j=1}^{\infty} \frac{1}{2^j} \frac{|x_j - y_j|}{1 + |x_j - y_j|}.$$

This metric cannot be obtained from a norm.

Problem 7. Let P be the set of prime numbers. We define the set

$$S := \{ (p, q) \ : \ p, q \in P \ \ p \le q \}.$$

Show that $d((p_1, q_1), (p_2, q_2)) := |p_1 q_1 - p_2 q_2|$ defines a metric.

Solution 7. The *triangle inequality* is proved as follows

$$
\begin{aligned}
d((p_1, q_1), (p_2, q_2)) = |p_1 q_1 - p_2 q_2| &= |(p_1 q_1 - p_3 q_3) + (p_3 q_3 - p_2 q_2)| \\
&\le |p_1 q_1 - p_3 q_3| + |p_3 q_3 - p_2 q_2| \\
&= d((p_1, q_1), (p_3, q_3)) + d((p_3, q_3), (p_2, q_2)).
\end{aligned}
$$

Problem 8. Let $(X_1, \|\cdot\|_1)$ and $(X_2, \|\cdot\|_2)$ be two normed spaces. Show that the product vector spaces $X = X_1 \times X_2$ is also a normed vector space if we define

$$\|x\| := \max(\|x_1\|_1, \|x_2\|_2)$$

with $x = (x_1, x_2)$.

Solution 8. Let $x = (x_1, x_2)$. We have

$$\|x\| = 0 \Leftrightarrow \|x_1\|_1 = \|x\|_2 = 0 \Leftrightarrow x = (0, 0) = 0.$$

Let $x_1, y_1 \in X_1$, $x_2, y_2 \in X_2$ and $x = (x_1, x_2)$, $y = (y_1, y_2)$. Then

$$
\begin{aligned}
\|x + y\| &= \max(\|x_1 + y_1\|_1, \|x_2 + y_2\|_2) \\
&\le \max(\|x_1\|_1 + \|y_1\|_1, \|x_2\|_2 + \|y_2\|_2) \\
&\le \max(\|x_1\|_1, \|x_2\|_2) + \max(\|y_1\|_1, \|y_2\|_2) \\
&= \|x\| + \|y\|.
\end{aligned}
$$

Problem 9. Find

$$\|A\| := \max_{\mathbf{v} \in \mathbb{R}^3, \|\mathbf{v}\|=1} \|A\mathbf{v}\|$$

where $\| \cdot \|$ is the Euclidean norm in $\mathbb{R}^3$ and A is the 3×3 matrix

$$A = \begin{pmatrix} 1 & 0 & 1 \\ 1 & 0 & 1 \\ 1 & 0 & 1 \end{pmatrix}.$$

Apply two different methods.

Solution 9. The matrix norm given above of the matrix A is the square root of the principal component for the matrix $A^T A$, where ^{T} denotes the transpose. This is equivalent to the *spectral norm*

$$\|A\| = \sqrt{\lambda_{max}}$$

where λ_{max} is the largest eigenvalue of $A^T A$. Now

$$A^T A = \begin{pmatrix} 3 & 0 & 3 \\ 0 & 0 & 0 \\ 3 & 0 & 3 \end{pmatrix}$$

and the largest eigenvalue is 6. Hence $\|A\| = \sqrt{6}$. With the *Lagrange multiplier method* we have to maximize the function

$$f(\mathbf{x}) = \mathbf{x}^T A^T A \mathbf{x} = (x_1 \ x_2 \ x_3) \begin{pmatrix} 3 & 0 & 3 \\ 0 & 0 & 0 \\ 3 & 0 & 3 \end{pmatrix} \begin{pmatrix} x_1 \\ x_2 \\ x_3 \end{pmatrix} = 3x_1^2 + 6x_1 x_3 + 3x_3^2$$

subject to the constraint $x_1^2 + x_2^2 + x_3^2 = 1$. Thus

$$L(\mathbf{x}) = f(\mathbf{x}) + \mu(1 - x_1^2 - x_2^2 - x_3^2).$$

From $\partial L/\partial x_1 = 0$, $\partial L/\partial x_2 = 0$, $\partial L/\partial x_3 = 0$ we obtain $x_2 = 0$ and

$$3x_1 + 3x_3 - \mu x_1 = 0, \quad 3x_3 + 3x_1 - \mu x_3 = 0$$

together with the constraint. This provides $x_1 = x_3 = 1/\sqrt{2}$ and

$$f(1/\sqrt{2}, 0, 1/\sqrt{2}) = 6.$$

Hence the norm is $\sqrt{6}$.

Problem 10. Let V be the linear space of functions $f : C \to \mathbb{C}$, where $C \subset \mathbb{C}$ is the unit circle in the complex plane $\mathbb{C}$.
(i) A norm can be defined by

$$\|f\|^2 := \frac{1}{2\pi i} \oint_C |f(z)|^2 \frac{dz}{z} < \infty.$$

Let $f(z) = z$. Find $\|f\|^2$.
(ii) An inner product can be defined via

$$\langle f, g \rangle := \frac{1}{2\pi i} \oint_C f(z) \bar{g}(z) \frac{dz}{z}$$

which implies a norm $\|f\|^2 = \langle f, f \rangle$. Let $f(z) = z$, $g(z) = z^2$. Find $\langle f, g \rangle$.

Solution 10. (i) With $z = re^{i\phi}$ and $dz = re^{i\phi} i d\phi$ we obtain

$$\|f(z)\| = \frac{1}{2\pi i} \int_0^{2\pi} i d\phi = 1.$$

(ii) We have $f(z) \bar{g}(z) = r^3 e^{-i\phi}$. Then

$$\frac{i}{2\pi i} \int_0^{2\pi} e^{-i\phi} d\phi = 0.$$

So the functions are orthogonal to each other.

Problem 11. Let V be the complex linear space of all polynomials $v \in V$ with

$$v(\tau) = c_0 + c_1 \tau + \cdots + c_n \tau^n$$

where $\tau \in \mathbb{C}$, $n \in \mathbb{N}$ and $c_j \in \mathbb{C}$ ($j \in \mathbb{N}_0$). Let V be equipped with the norm

$$\|v\| := |c_0| + |c_1| + |c_2| + \cdots + |c_n|.$$

Furthermore, let W be the subspace of V consisting of all polynomials w with $w(0) = 0$, i.e. $c_0 = 0$.
(i) Define the *differentiation operator* $D : V \to W$ as

$$(Dv)(\tau) := c_1 + 2c_2 \tau + \cdots + nc_{n-1} \tau^{n-1}$$

for all $v \in V$. Show that D is not a topological *isomorphism* with respect to the norm.
(ii) Define the operator $T : V \to W$ as

$$(Vw)(\tau) := c_1 + c_2 \tau + \cdots + c_n \tau^{n-1}$$

for all $w \in W$. Show that T is a topological isomorphism with respect to the norm. Therefore it is possible to show using T that W and V are topological isomorphic.

(iii) Let $\ell_1(\mathbb{N})$ be the normed space consisting of all sequences $u = \{u_1, u_2, u_3, \dots\}$ ($u_j \in \mathbb{C}$ for all $j \in \mathbb{N}$) such that

$$\|u\|_1 := \sum_{j=1}^{\infty} |u_j| < \infty.$$

Let S be the normed linear subspace of $\ell_1(\mathbb{N})$ consisting of all sequences where only a finite number of coefficients are nonzero. Show that the map $\psi : V \to S$ with

$$\psi(c_0 + c_1\tau + \cdots + c_n\tau^n) := \{c_0, c_1, \dots, c_n, 0, 0, \dots\}$$

is an isometric isomorphism and therefore V and S are isometrically isomorphic.

Solution 11. (i) In order to show that D is not a topological isomorphism it is sufficient to show that D is not continuous (or bounded). We consider a sequence of unit vectors w_N ($\|w_N\| = 1$) with

$$w_N := \frac{1}{N}(\tau + \tau^2 + \cdots + \tau^N)$$

for $N = 1, 2, \dots$. Applying the differentiation operator D we have

$$\|D(w_N)\| = \left\| \frac{1}{N}(1 + 2\tau + \cdots + N\tau^{N-1}) \right\| = \frac{1}{2}(N+1) \to \infty \text{ as } N \to \infty$$

such that the differentiation operator D is unbounded. The differentiation operator D is an injective operator and it maps W onto V, that is if $v(\tau) = b_0 + b_1\tau + \cdots + b_n\tau^n$ is an arbitrary element of V, then we have

$$D\left(b_0\tau + \frac{b_1}{2}\tau^2 + \cdots + \frac{b_n}{n+1}\tau^{n+1} \right) = v(\tau).$$

The range of D is the entire V and it follows that D^{-1} exists, i.e.

$$(D^{-1}v)(\tau) = D^{-1}(c_0 + c_1\tau + \cdots + c_n\tau^n) = c_0\tau + \frac{c_1}{2}\tau^2 + \cdots + \frac{c_n}{n+1}\tau^{n+1}$$

and we have $\|D^{-1}(v)\| \le \|v\|$ such that D^{-1} is continuous. D is not a topological isomorphism and therefore it is not possible to prove with D that V and W are topological isomorphic.

(ii) Applying the operator T to an arbitrary $v = c_1\tau + \cdots + c_n\tau^{n-1}$ we obtain

$$T(v) = c_1 + c_2\tau + \cdots + c_n\tau^{n-1}$$

with $\|T(v)\| = \sum_{j=1} |c_j|$ and therefore T is bounded. Moreover, the inverse T^{-1} exists with

$$T^{-1}(b_0 + b_1\tau + \cdots + b_n\tau^n) = b_1\tau + b_2\tau^2 + \cdots + b_n\tau^{n+1}$$

and is bounded. T is continuous and its inverse exists and it is a topological isomorphism as well as an isometric isomorphism. Hence W and V are isometrically isomorphic.

(iii) If $v(\tau) = c_0 + c_1\tau + \cdots + c_n\tau^n$ we find

$$\|\psi(v(\tau))\| = \|(c_0, c_1, \ldots, c_n, 0, 0, \ldots)\|_1 = \|v(\tau)\|.$$

The map ψ is bounded and therefore continuous. Furthermore ψ is injective and the inverse ψ^{-1} exists

$$\psi^{-1}((c_0, c_1, \ldots, c_n, 0, 0, \ldots)) = c_0 + c_1\tau + \cdots + c_n\tau^n$$

and is continuous. Hence V and S are isometrically isomorphic.

Problem 12. Consider $\ell_p(\mathbb{N})$ with $1 < p < \infty$, all sequences $\xi := (\xi_n)_{n \in \mathbb{N}}$ with coefficients $\xi_n \in \mathbb{C}$ and the norm

$$\|\xi\|_p := \left(\sum_{n \in \mathbb{N}} |\xi_n|^p \right)^{1/p} < \infty.$$

Show that all $\ell_p(\mathbb{N})$ are *reflexive*.

Solution 12. Since $(\ell_p(\mathbb{N}))^* = \ell_q(\mathbb{N})$ with $(1/p) + (1/q) = 1$ the corresponding isometry is even and the canonical embedding is surjective.

Problem 13. Let $\ell_\infty(\mathbb{N})$ be all the sequences $\{\xi_n\}$ with coefficients ξ_n and the norm

$$\|\xi\|_\infty := \sup_{n \in \mathbb{N}} |\xi_n| < \infty.$$

Show that $\ell_\infty(\mathbb{N})$ is not separable and not reflexive.

Solution 13. Consider the set S of binary sequences, that is

$$x_I \in S \leftrightarrow x_i \text{ with } I \subset \mathbb{N}, \ (x_I)_n = \begin{cases} 1 \text{ for } n \in I \\ 0 \text{ for } n \notin I \end{cases}$$

where $d_\infty = 1$ whenever $I \neq J$. Hence the set M is uncountable. For this purpose let

$$\tilde{B} := \left\{ B = \left(x_I, \frac{1}{2} \right) : x_I \in S \right\}$$

is an uncountably infinite collection of disjoint open balls in $\ell_\infty(\mathbb{N})$. Assume that S be any dense subset $\ell_\infty(\mathbb{N})$ then each ball of $\widetilde{B}$ must contain at least one element of S. Therefore S must be uncountable infinite and $\ell_\infty(\mathbb{N})$ is not separable. Using the proposition that every closed subspace of a reflexive space reflexive we find that the Banach space $\ell_\infty(\mathbb{N})$ is not reflexive since $\ell_\infty(\mathbb{N})$ contains a non-reflexive closed subspace $C(\mathbb{N})$.

Problem 14. Show that if the dual space V^* of a normed linear space V is separable, then V is separable as well.

Solution 14. From the separability of the dual space V^* follows that the dual unit sphere contains a countable dense subset $\{\, f_n \, : \, n \in \mathbb{N} \,\}$ with $\|f_n\| = 1$ for all n. Choose a sequence $\{\, x_n \,\}_{n \in \mathbb{N}}$ in the unit sphere ∂V_1 of V with $|f_n(x_n)| > 1/2$ for all $n \in \mathbb{N}$. Define the linear span $F := \langle x_n \, : \, n \in \mathbb{N} \rangle$ with rational coefficients and its closure $Y := \overline{F}$ which is separable. To prove that $Y = V$ assume that Y is not all of V. Then by the *Hahn-Banach theorem* there is an f in V^* with $\|f\| = 1$ and $f(y) = 0$ for all $y \in Y$. Then

$$\frac{1}{2} = \frac{\|f_n\|}{2} \le |f_n(x_n) - f(x_n)| \le \|f_n - f\|$$

in contraction to be assume of denseness of $\{f_n \, : \, n \in \mathbb{N}\}$ on the unit sphere of V^*. Therefore $Y = V$.

Problem 15. Show that the normed linear space $C_0(\mathbb{N})$

$$C_0(\mathbb{N}) := \{\, \{\xi_n\}_{n \in \mathbb{N}} \, : \, \xi_n \in \mathbb{C} \ (n \in \mathbb{N}), \ \lim_{n \to \infty} \xi_n = 0 \,\}$$

with the norm $\| \cdot \|_\infty$ is not isomorphic to $\ell_\infty(\mathbb{N})$ and $(\ell_\infty(\mathbb{N})^*)$.

Solution 15. The dual space of $C_0(\mathbb{N})$ is $\ell_1(\mathbb{N})$ which is separable. The dual space of $\ell_1(\mathbb{N})$ is $\ell_\infty(\mathbb{N})$ which is non-separable. Consequently $C_0(\mathbb{N})$ and $\ell_1(\mathbb{N})$ cannot be isomorphic. Since $C_0(\mathbb{N})$ is separable and $\ell_\infty(\mathbb{N})$ is non-separable and the dual space of $\ell_1(\mathbb{N})$ is $\ell_1(\mathbb{N})$ which is separable they cannot be isomorphic. We know that

$$C_0(\mathbb{N})^{**} = (\ell_1(\mathbb{N}))^* = \ell_\infty(\mathbb{N}).$$

The Banach space $\ell_\infty(\mathbb{N})$ is non-separable, but $C_0(\mathbb{N})$ is separable. Hence the canonical map J from $C_0(\mathbb{N})$ to $C_0(\mathbb{N})^{**} = \ell_\infty(\mathbb{N})$ must not be surjective and therefore $C_0(\mathbb{N})$ cannot be reflexive.

Problem 16. Show that the space of all Riemann square-integrable functions on the unit interval $[0, 1]$ is not complete.

Solution 16. Let $n \in \mathbb{N}$ and $m \in \mathbb{N}$. Consider the oscillating function

$$f_n(t) = \lim_{m \to \infty} (\cos(\pi n! t))^{2m} = \begin{cases} 1 \text{ if } & t = \frac{k}{n!}, \ k \text{ integer} \\ 0 \text{ otherwise} \end{cases}$$

The function f_n is Riemann integrable and the integral on the unit interval $[0, 1]$ is equal to 0. Now $f_n(t) \to f(t)$ with

$$f(t) = \begin{cases} 1 \text{ if } t \text{ is rational} \\ 0 \text{ if } t \text{ is irrational} \end{cases}$$

The function f is not Riemann integrable. The Lebesgue integral has to be used which is equal to 1.

Problem 17. Show that the normed vector space $\ell_\infty(\mathbb{N}_0)$ is a Banach space. The norm in $\ell_\infty(\mathbb{N}_0)$ is given as

$$\|x\|_\infty := \sup_{j \in \mathbb{N}_0} |x(j)|.$$

Solution 17. Let $\{x_n\}_{n=1}^\infty$ be a Cauchy sequence of elements in $\ell_\infty(\mathbb{N}_0)$. It follows that for each $\epsilon > 0$ there exists an N such that for all $n, m \geq N$ we have

$$\|x_n - x_m\|_\infty < \epsilon.$$

Therefore $|x_n(j) - x_m(j)| < \epsilon$ for all $j \in \mathbb{N}_0$. Hence the sequence $\{x_n(j)\}_{n=1}^\infty$ is a Cauchy sequence of real numbers and converges to some value $x(j)$, i.e.

$$\lim_{n \to \infty} x_n(j) = x(j)$$

exists. If we set $\epsilon = 1$, then for all $n, m \geq N$ we have

$$\|x_n - x_m\|_\infty < 1$$

and $\|x_n\| < 1 + \|x_m\|$ for $n, m \geq N$. Now we fix m. Therefore for all $n \geq N$ we obtain

$$|x_n(j)| \leq \|x_n\|_\infty < 1 + \|x_m\|_\infty$$

holds uniformly in j. Hence $x \in \ell_\infty(\mathbb{N}_0)$. Next we show that x_n converges to x in norm. Now

$$|x_n(j) - x_m(j)| < \epsilon \quad \text{for all } j \in \mathbb{N}_0 \text{ and } n, m \geq N.$$

If $n \to \infty$, then $x_n(j) \to x(j)$ and so

$$|x_n(j) - x_m(j)| < \epsilon \quad \text{for all } j \in \mathbb{N}_0 \ \ m \geq N.$$

Since this is true for all $j \in \mathbb{N}_0$ we conclude that $\|x - x_m\|_\infty < \epsilon$ for all $m \geq N$ and the sequence x_m converges to $x \in \ell_\infty(\mathbb{N}_0)$.

Problem 18. Let $\mathcal{B}$ be a Banach space and $f, g \in \mathcal{B}$. The *triangle inequality* is given by

$$|\, \|f\| - \|g\| \,| \leq \|f + g\| \leq \|f\| + \|g\|.$$

Let $\mathcal{B} = C([0,1])$ with the norm

$$\|f\| := \sup_{x \in [0,1]} |f(x)|$$

and $f(x) = x$, $g(x) = 1$. Find $\|f\|$, $\|g\|$, $\|f + g\|$, $\|f\| + \|g\|$. Discuss.

Solution 18. We have $\|f\| = 1$, $\|g\| = 1$, $\|f\| + \|g\| = 2$, $\|f + g\| = 2$. Hence the inequality takes the form $0 < 2 = 2$.

Problem 19. Consider the vector space of all continuous functions f on the interval $[-1, 1]$ and the L_1 norm

$$\|f\| = \int_{-1}^{+1} |f(x)| dx.$$

Show that this space is not complete.

Solution 19. Let $n \in \mathbb{N}$. Consider the continuous function $f_n : [-1, 1] \to [-1, 1]$

$$f_n(x) = \begin{cases} -1 & \text{for } x \in [-1, -1/n] \\ nx & \text{for } x \in [-1/n, 1/n] \\ 1 & \text{for } \quad x \in [1/n, 1] \end{cases}$$

Then $f_n \in C[-1, 1]$. However the limit for $n \to \infty$ is not continuous. One obtains

$$f(x) = \begin{cases} -1 & \text{for } x \in [-1, 0] \\ 0 & \text{for } \quad x = 0 \\ 1 & \text{for } x \in (0, 1] \end{cases}$$

and $\|f - f_n\|_{L_1[-1,1]} = 1/n$ and thus

$$\lim_{n \to \infty} \|f - f_n\|_{L_1[-1,1]} = 0.$$

Problem 20. Let $\mathbf{x} = \{x_n\}_{n=0}^{\infty}$ and define the Banach space

$$\ell_\infty(\mathbb{N}_0) := \{\, \mathbf{x} : x_n \in \mathbb{R},\ \forall n \in \mathbb{N}_0 \text{ and } \|\mathbf{x}\|_\infty = \sup_{n \in \mathbb{N}} |x_n| < \infty \,\}.$$

Define the map $T : \ell_\infty(\mathbb{N}_0) \mapsto \ell_\infty(\mathbb{N}_0)$ by

$$(T\mathbf{x})_0 = x_0 \text{ and } (T\mathbf{x})_n = x_n - x_{n-1}\ n \in \mathbb{N}.$$

Let $\mathbf{e} = (1, 1, 1, \dots)^T \in \ell_\infty(\mathbb{N}_0)$. Show that the equation $T\mathbf{x} = \mathbf{e}$ has no solution in the Banach space $\ell_\infty(\mathbb{N}_0)$.

Solution 20. The equation $T\mathbf{x} = \mathbf{e}$ implies that $x_0 = 1$. Now $x_1 - x_0 = 1$ with $x_1 = 2$. Utilizing induction on n we obtain $x_n = n + 1$. Thus the equation has no solution in the Banach space $\ell_\infty(\mathbb{N}_0)$.

Problem 21. Consider a Banach space $\mathcal{B}$ with norm $\|\cdot\|$. Let $g \in \mathcal{B}$ and consider the equation $f = g + Tf$ with T a linear bounded transformation with $\|T\| < 1$. Show that

$$f_n := \sum_{j=0}^{n} T^j g$$

is a *Cauchy sequence*.

Solution 21. Let $\mu, \nu \in \mathbb{N}$ and $\nu > \mu$. Then

$$\|f_\nu - f_\mu\| = \Big\| \sum_{j=\mu+1}^{\nu} T^j g \Big\| \leq \sum_{j=\mu+1}^{\nu} \|T^j g\|$$

$$\leq \sum_{j=\mu+1}^{\nu} \|T^j\| \cdot \|g\| \leq \|g\| \sum_{j=\mu+1}^{\nu} \|T\|^j$$

$$= \|g\| \cdot \|T\|^{\mu+1} \sum_{j=0}^{\nu=\mu-1} \|T\|^j \leq \|g\| \cdot \|T\|^{\mu+1} \sum_{j=0}^{\infty} \|T\|^j$$

$$= \|g\| \cdot \|T\|^{\mu+1} \cdot \frac{1}{1 - \|T\|}.$$

Since $\|T\| < 1$ we find that $1/(1 - \|T\|)$ is a finite number as well as $\|g\|$. Thus if μ is arbitrary large then $\|f_\nu - f_\mu\|$ is arbitrary small. Since a Banach space is complete we have $f_n \to f$, where $f \in \mathcal{B}$, $Tf_n \to Tf$ and $f = g + Tf$.

Problem 22. Consider the Banach space $L_1(\mathbb{R})$. Let

$$\hat{f}(\omega) = \int_{\mathbb{R}} f(t) e^{-i\omega t} dt.$$

Show that $\hat{f}$ is a continuous function.

Solution 22. We have

$$\hat{f}(\omega + \gamma) - \hat{f}(\omega) = \int_{\mathbb{R}} f(t)(e^{-i(\omega+\gamma)t} - e^{-i\omega t}) dt$$

$$= \int_{\mathbb{R}} f(t) e^{-i\omega t} (e^{-i\gamma t} - 1) dt.$$

The integrand is bounded from above by $2|f(t)|$ almost everywhere. Since $f \in L_1(\mathbb{R})$ the function $2|f(t)|$ is integrable. Utilizing the dominated convergence theorem provides

$$\lim_{\gamma \to 0} \hat{f}(\omega + \gamma) = \hat{f}(\omega).$$

Hence $\hat{f}$ is continuous.

Problem 23. Consider the vector space $C([0,1])$ of continuous functions. We define the *triangle function*

$$\Lambda(x) := \begin{cases} 2x & 0 \le x \le 1/2 \\ 2 - 2x & 1/2 < x \le 1 \end{cases}.$$

Let $\Lambda_0(x) := x$ and

$$\Lambda_n(x) := \Lambda(2^j x - k)$$

where $j = 0, 1, 2, \ldots$, $n = 2^j + k$ and $0 \le k < 2^j$. The functions

$$\{ 1, \Lambda_0, \Lambda_1, \ldots \}$$

are the *Schauder basis* for the vector space $C([0,1])$. Let $f \in C([0,1])$. Then

$$f(x) = a + bx + \sum_{n=1}^{\infty} c_n \Lambda_n(x).$$

Find the Schauder coefficients a, b, c_n.

Solution 23. (i) We have $f(0) = a$ and $f(1) = a + b$. For $f_1(x) = f(x) - a - bx$ we have

$$c_1 = f_1(1/2) = f(1/2) - \frac{1}{2}(f(0) + f(1)).$$

For $n = 2^j + k$, $0 \le k < 2^j$ we have

$$c_n = f\left(\frac{k+1/2}{2^j}\right) - \frac{1}{2}\left(f\left(\frac{k}{2^j}\right) + f\left(\frac{k+1}{2^j}\right)\right).$$

Problem 24. Consider the Euclidean space $\mathbb{E}^2$. The equation for the *lemniscate* is given by the curve $r^2 = a^2\cos(2\theta)$ with $a > 0$ of dimension length and $a = 1\,cm$. Note that $\cos(2\theta) \equiv \cos^2(\theta) - \sin^2(\theta)$. Consider the point $(p_1 = \frac{3}{2}cm, p_2 = 0)$ in the Euclidean space $\mathbb{E}^2$. Find the shortest distance from this point to the lemniscate. Apply the *Lagrange multiplier* method.

Solution 24. Using *polar coordinates* $x(r,\theta) = r\cos(\theta)$, $y(r,\theta) = r\sin(\theta)$ we obtain the constraint

$$(x^2 + y^2)(x^2 + y^2) = a^2(x^2 - y^2).$$

The square of the distance between two points (x_1, y_1) and (x_2, y_2) in the Euclidean plane is given by

$$d^2 = (x_1 - x_2)^2 + (y_1 - y_2)^2.$$

With $x_1 = p_1$, $y_1 = p_2 = 0$ we obtain

$$d^2 = (p_1 - x_2)^2 + y_2^2 = p_1^2 - 2p_1 x_2 + x_2^2 + y_2^2.$$

The constraint is $x_2^4 + y_2^4 + 2x_2^2 y_2^2 - a^2 x_2^2 + a^2 y_2^2 = 0$. Thus we have to minimize

$$f(x_2, y_2) = d^2 + \lambda(x_2^4 + y_2^4 + 2x_2^2 y_2^2 - a^2 x_2^2 + a^2 y_2^2).$$

From $\partial f/\partial x_2 = 0$, $\partial f/\partial y_2 = 0$ we obtain the two coupled equations

$$-2p_1 + 2x_2 + \lambda(4x_2^3 + 4x_2 y_2^2 - 2a^2 x_2) = 0$$
$$2y_2 + \lambda(4y_2^3 + 4x_2 y_2 + 2a^2 y_2) = 0.$$

From the second equation we obtain $y_2 = 0$. Inserting this solution into the first equation we obtain

$$-2p_1 + 2x_2 + \lambda(4x_2^3 - 2a^2 x_2) = 0$$

together with the constraint $x_2^4 = a^2 x_2^2$. The constraint admits the solutions $x_2 = 0$, $x_2 = a$, $x_2 = -a$. Owing to $a > 0$ we have $x_2 = a$. Then with $p_1 = \frac{3}{2}\,cm$ we have

$$d^2 = (p_1 - x_2)^2 = \frac{1}{4}\,cm^2.$$

Problem 25. Consider the d-dimensional Euclidean space $\mathbb{E}^d$ which is a Banach space. Let

$$\mathbf{x}_0, \ \mathbf{x}_1, \ \ldots, \ \mathbf{x}_n$$

be pairwise different $n+1$ points in the d-dimensional Euclidean space with $d \geq n$ and d_{jk}^2 be the square of the distance between $\mathbf{x}_j$ and $\mathbf{x}_k$

$$d_{jk}^2 = \|\mathbf{x}_j - \mathbf{x}_k\|^2, \quad j, k = 0, 1, \ldots, n.$$

These $n+1$ points can be considered as vertices of an n-dimensional *simplex*. So we can determine the volume V_n of this simplex. The square of the volume V_n^2 is given by the *Cayley-Menger determinant*

$$V_n^2 = \frac{(-1)^{n+1}}{(n!)^2 2^n} \begin{vmatrix} 0 & d_{01}^2 & d_{02}^2 & \cdots & d_{0n}^2 & 1 \\ d_{01}^2 & 0 & d_{12}^2 & \cdots & d_{1n}^2 & 1 \\ d_{02}^2 & d_{12}^2 & 0 & \cdots & d_{2n}^2 & 1 \\ \cdots & \vdots & \vdots & \ddots & \vdots & \vdots \\ d_{0n}^2 & d_{1n}^2 & d_{2n}^2 & \cdots & 0 & 1 \\ 1 & 1 & 1 & \cdots & 1 & 0 \end{vmatrix}.$$

(i) Consider the two-dimensional Euclidean space $\mathbb{E}^2$ and the three points

$$\mathbf{x}_0 = \begin{pmatrix} 0 \\ 0 \end{pmatrix}, \quad \mathbf{x}_1 = \begin{pmatrix} 2 \\ 0 \end{pmatrix}, \quad \mathbf{x}_2 = \begin{pmatrix} 0 \\ 1 \end{pmatrix}.$$

Find the volume V_2.
(ii) Let

$$M = \{\, (x, y) : 0 \leq x \leq 2, \ 0 \leq y \leq 1, \ x + 2y \leq 2 \,\}.$$

Find

$$\int_M dx\,dy.$$

Solution 25. (i) We have $d_{01}^2 = 4$, $d_{02}^2 = 1$, $d_{12}^2 = 5$. It follows that

$$V_2^2 = \frac{-1}{4 \cdot 4} \begin{vmatrix} 0 & d_{01}^2 & d_{02}^2 & 1 \\ d_{01}^2 & 0 & d_{12}^2 & 1 \\ d_{02}^2 & d_{12}^2 & 0 & 1 \\ 1 & 1 & 1 & 0 \end{vmatrix} = -\frac{1}{16} \begin{vmatrix} 0 & 4 & 1 & 1 \\ 4 & 0 & 5 & 1 \\ 1 & 5 & 0 & 1 \\ 1 & 1 & 1 & 0 \end{vmatrix} = 1.$$

(ii) We have

$$\int_M dx\,dy = \int_{y=0}^{y=1} \left(\int_{x=0}^{x=2-2y} dx \right) dy = \int_{y=0}^{y=1} (2 - 2y)dy = 1.$$

Problem 26. Let (M, d) be a complete metric space (for example a Banach space) and let $f : M \to M$ be a mapping such that

$$d(f^{(m)}(x), f^{(m)}(y)) \leq kd(x, y), \qquad \forall x, y \in M$$

for some $m \geq 1$, where $0 \leq k < 1$ is a constant. Show that the map f has a unique fixed point in M.

Solution 26. It follows from the contraction mapping principle that $f^{(m)}$ has a unique fixed point $x^* \in M$. Thus

$$x^* = f^{(m)}(x^*)$$

implies that

$$f(x^*) = f(f^{(m)}(x^*)) = f^{(m)}(f(x^*)).$$

Thus $f(x^*)$ is a fixed point of $f^{(m)}$ and therefore by uniqueness of such fixed points $x^* = f(x^*)$.

1.3   Supplementary Problems

Problem 1. (i) Let $b > a$. Show that the vector space $C([a, b])$ of all continuous (real or complex valued) functions on the interval $[a, b]$ with the norm

$$\|f\| := \max_{x \in [a,b]} |f(x)|$$

is a Banach space.

(ii) Let $k \in \mathbb{N}_0$, $a, b \in \mathbb{R}$ with $b > a$. Show that

$$C^{(k)}([a, b]), \qquad \|f\|_{C^{(k)}([a,b])} = \sum_{j=0}^{k} \sup_{x \in [a,b]} |d^j f(x)/dx^j|$$

is a Banach space.

Problem 2. (i) Let $n \geq 1$. Consider the continuous function

$$f_n(t) = \begin{cases} 0 & \text{for} & 0 \leq t < 1/2 - 1/n \\ 1/2 + \frac{n}{2}(t - 1/2) & \text{for} & 1/2 - 1/n \leq t \leq 1/2 + 1/n \\ 1 & \text{for} & 1/2 + 1/n \leq t \leq 1. \end{cases}$$

Show that the sequence $\{ f_n(t) \}$ is not a *Cauchy sequence* for the uniform norm, but with any of the L^p norms ($1 \leq p < \infty$) it is a Cauchy sequence.

(ii) Let $g_n : [-1, 1] \to [-1, 1]$ be defined by

$$g_n(x) = \begin{cases} 1 & -1 \leq x \leq 0 \\ \sqrt{1 - nx} & 0 \leq x \leq 1/n \\ 0 & 1/n \leq x \leq 1 \end{cases}$$

Show that $g_n \in L_2([-1, 1])$. Show that f_n is a Cauchy sequence.

Problem 3. The Banach space $\ell_1(\mathbb{N})$ consists of all sequences $\xi := (\xi_j)_{j \in \mathbb{N}}$ with $\xi_j \in \mathbb{C}$ and the norm

$$\|\xi\|_1 := \sum_{j \in \mathbb{N}} |\xi_j| < \infty.$$

Show that the Banach space $\ell_1(\mathbb{N})$ is separable, but not reflexive.

Problem 4. Let $C^1_{(2)}(\mathbb{R})$ be the linear space of the once continuously differentiable functions that vanish at infinity together with their first derivative and which are square integrable. Show that $C^1_{(2)}(\mathbb{R})$ is dense in the Hilbert space $L_2(\mathbb{R})$.

Problem 5. Show that for bounded operators A, B in a Banach space

$$\|AB\| \leq \|A\| \cdot \|B\|.$$

Problem 6. The *metric tensor field* of a sphere with radius R in the Euclidean space $\mathbb{E}^3$ is given by

$$g = R^2(d\theta \otimes d\theta + \sin^2(\theta)d\phi \otimes d\phi)$$

where $\theta \in [0, \pi]$ and $\phi \in [-\pi, \pi]$. Show that the *geodesic distance d* between the points p, p' on the sphere of radius R is given by

$$d(p, p') = R \arccos(\cos(\theta)\cos(\theta') + \sin(\theta)\sin(\theta')\cos(\phi - \phi')).$$

For example the distance from the north pole $(\theta, \phi) = (0, 0)$ to the south pole $(\theta, \phi) = (\pi, 0)$ is

$$d = R \arccos(-1) = \pi R.$$

Note that $\arccos(0) = \pi/2$, $\arccos(1) = 0$, $\arccos(-1) = \pi$.

Problem 7. Consider the Euclidean space $\mathbb{E}^3$ and *spherical coordinates*. Show that the expansion of the inverse distance r^{-1} between two points in $\mathbb{E}^3$ with spherical coordinates (r_1, θ_1, ϕ_1) and (r_2, θ_2, ϕ_2) in terms of Legendre polynomials $P_\ell(\cos(\theta_{12}))$ is given (*Laplace expansion*) by

$$r^{-1} = r_>^{-1} \sum_{\ell=0}^{\infty} \left(\frac{r_<}{r_>}\right)^\ell P_\ell(\cos(\theta_{12}))$$

where

$$r_< := \min(r_1, r_2), \qquad r_> := \max(r_1, r_2)$$

$$\cos(\theta_{12}) = \cos(\theta_1)\cos(\theta_2) + \sin(\theta_1)\cos(\theta_2)\cos(\phi_1 - \phi_2).$$

Problem 8. Consider the map $T : [0, 1] \to [0, 1]$, $T(x) = 2x$ mod 1. Show that the *Frobenius-Perron operator* of the map (T, μ), where μ denotes the Lebesgue measure is given by $V : L_1([0, 1]) \to L_1([0, 1])$ is given by

$$Vf(x) = \frac{1}{2}\sum_{j=0}^{1} f\left(\frac{1}{2}(x+j)\right) \equiv \frac{1}{2}f\left(\frac{1}{2}x\right) + \frac{1}{2}f\left(\frac{1}{2}(x+1)\right)$$

where $f \in L_1([0,1])$.

Problem 9. Let $s, t \in \mathbb{C}^n$ and ρ be an $n \times n$ density matrix, i.e. ρ is positive definite and $\mathrm{tr}(\rho) = 1$. Show that

$$\left| \sum_{j,k=1}^{n} \rho_{jk} s_j t_k \right| \leq \max_{1 \leq j,k \leq n} \{ |s_j| \cdot |t_k| \}.$$

Problem 10. Let T be a linear bounded invertible operator acting on a Banach space $\mathcal{H}$. Let $n \in \mathbb{Z}$ and suppose that $\|T^n\| < c$ for all $n \in \mathbb{Z}$. Show that there exists a linear bounded operator S such that STS^{-1} is unitary. Show that $\|S\| < c$ and $\|S^{-1}\| < c$.

Problem 11. Let $d \geq 1$. Consider the Banach space $\mathbb{R}^d$ and $\mathbf{x} \in \mathbb{R}^d$. Show that

$$I_d = \int_{\mathbb{R}^d} \frac{dx_1 dx_2 \ldots dx_d}{(1 + \|\mathbf{x}\|)^{(d+1)/2}} = \frac{\pi^{(d+1)/2}}{\Gamma((d+1)/2)}.$$

For $d = 1$ we have $I_1 = \pi$.

Problem 12. Let $T : \mathcal{B} \to \mathcal{B}$ be a bounded linear operator on a complex Banach space $\mathcal{B}$. Show that the limit (called the *spectral radius* of T)

$$\lim_{n \to \infty} \|T^n\|^{1/n}$$

exists. Note that $r_T \leq \|T\|$, where r_T is the spectral radius. Find the spectral radius of the 2×2 matrix

$$T = \begin{pmatrix} 0 & 1 \\ 0 & 0 \end{pmatrix}.$$

Find the spectral radius of $T \otimes T$, where $\otimes$ denotes the Kronecker product.

Problem 13. Let $T > 0$. Consider the continuous function $f : [0, T] \to \mathbb{R}$. Assume that

$$\int_0^T x^n f(x) dx = 0, \quad n = 1, 2, \ldots.$$

Show that $f(x) = 0$ for $0 \leq t \leq T$.

Problem 14. Let $b > a$.

(i) Consider the linear space $V = C([a, b])$. Show that

$$\|f\| = \int_a^b |f(x)| dx$$

defines a norm for V.

(ii) Consider the linear space $V = C([a, b])$. Does

$$\left| \int_a^b f(x) dx \right|$$

define a norm for V?

(iii) Consider the linear space $V = C^1([a, b])$. Show that

$$\|f\| = \max(|f(x)| + |df(x)/dx|)$$

defines a norm on V.

(iv) Consider the linear space $V = C^1([a, b])$. Show that

$$\|f\| = \int_a^b (|f(x)| + |df(x)/dx|) dx$$

defines a norm on V.

Problem 15. Consider the Banach space $L_p(\mathbb{R}, dx)$ with $p \in [1, \infty]$. Show that the linear operator

$$(Tf)(x) := f(x + 1)$$

is an invertible isometry for all $p \in [1, \infty]$ on the Banach spaces $L_p(\mathbb{R}, dx)$.

Problem 16. Consider the Banach spaces $L_p([0, 1])$ ($p \in [1, \infty]$) and the map $T : L_p([0, 1]) \to L_p([0, 1])$ given by

$$(Tf)(x_1) := \int_0^1 K(x_1, x_2) f(x_2) dx_2$$

with

$$K(x_1, x_2) := \begin{cases} 1 & \text{for} \quad x_1 + x_2 \leq 1 \\ 0 & \text{otherwise} \end{cases}$$

Find $\|T\|_1$, $\|T\|_2$, $\|T\|_\infty$. For $p = 2$ we have a Hilbert space.

Problem 17. Let T be a linear operator on a Banach space satisfying $\|T\| < 1$. Show that $(I - T)$ is invertible. Show that

$$(I - T)^{-1} = \sum_{j=0}^\infty T^j$$

with the sum norm convergent. Apply it to

$$T = \begin{pmatrix} 0 & 1/2 \\ 1/2 & 0 \end{pmatrix}.$$

Problem 18. Let the underlying field be $\mathbb{R}$. The continuous dual of the Banach space $\ell_1(\mathbb{N}_0)$ naturally identified with $\ell_\infty(\mathbb{N}_0)$ which is the Banach space of bounded sequences with the underlying field $\mathbb{R}$. Let $\mathbf{x} \in \ell_1(\mathbb{N}_0)$ and $\mathbf{y} \in \ell_\infty(\mathbb{N}_0)$. Is

$$\sum_{j=0}^{\infty} x_j y_j < \infty$$

for all $\mathbf{x} \in \ell_1(\mathbb{N}_0)$ and $\mathbf{y} \in \ell_\infty(\mathbb{N}_0)$?

Problem 19. Let $n \in \mathbb{N}$. The *Lie product formula* in a unital Banach algebra tells us that

$$(e^{A/n}e^{B/n})^n \to e^{A+B} \quad \text{as } n \to \infty.$$

Let

$$A = \sigma_1 = \begin{pmatrix} 0 & 1 \\ 1 & 0 \end{pmatrix}, \quad B = \sigma_3 = \begin{pmatrix} 1 & 0 \\ 0 & -1 \end{pmatrix}.$$

With $A^2 = I_2$ and $B^2 = I_2$ we have

$$e^{A/n} = I_2 \left(1 + \frac{1}{n^2} \cdot \frac{1}{2!} + \frac{1}{n^4} \cdot \frac{1}{4!} + \cdots \right) + A \left(\frac{1}{n} \cdot \frac{1}{1!} + \frac{1}{n^3} \cdot \frac{1}{3!} + \frac{1}{n^5} \cdot \frac{1}{5!} + \cdots \right)$$

$$e^{B/n} = I_2 \left(1 + \frac{1}{n^2} \cdot \frac{1}{2!} + \frac{1}{n^4} \cdot \frac{1}{4!} + \cdots \right) + B \left(\frac{1}{n} \cdot \frac{1}{1!} + \frac{1}{n^3} \cdot \frac{1}{3!} + \frac{1}{n^5} \cdot \frac{1}{5!} + \cdots \right).$$

Find

$$e^{A/n}e^{B/n}, \quad e^{A+B}, \quad \|e^{A/n}e^{B/n} - e^{A+B}\|.$$

Chapter 2

Hilbert Spaces

2.1 Introduction

A *Hilbert space* ([3],[19],[22],[32],[59],[75],[80],[85]) is a set, $\mathcal{H}$ of elements, or vectors, $(f, g, h, \dots)$ which satisfies the following conditions (1)–(5).

(1) If f and g belong to $\mathcal{H}$, then there is a unique element of $\mathcal{H}$, denoted by $f + g$, the operation of addition (+) being invertible, commutative and associative.

(2) If c is a complex number, then for any f in $\mathcal{H}$, there is an element cf of $\mathcal{H}$; and the multiplication of vectors by complex numbers thereby defined satisfies the distributive conditions

$$c(f + g) = cf + cg, \qquad (c_1 + c_2)f = c_1 f + c_2 f.$$

(3) Hilbert spaces $\mathcal{H}$ possess a zero element, 0, characterized by the property that $0 + f = f$ for all vectors f in $\mathcal{H}$.

(4) For each pair of vectors f, g in $\mathcal{H}$, there is a complex number $\langle f, g \rangle$, termed the inner product or scalar product, of f with g, such that

$$\langle f, g \rangle = \overline{\langle g, f \rangle}$$

$$\langle f, g + h \rangle = \langle f, g \rangle + \langle f, h \rangle$$

$$\langle f, cg \rangle = \overline{c}\langle f, g \rangle.$$

As a consequence we obtain $\langle cf, g \rangle = c\langle f, g \rangle$. So we assume that an inner product is linear in the first variable and conjugate linear in the second variable. Furthermore we have $\langle f, f \rangle \geq 0$. Equality in the last formula

27

occurs only if $f = 0$. The scalar product defines the norm $\|f\| := \langle f, f \rangle^{1/2}$.

(5) If $\{f_n\}$ is a sequence in $\mathcal{H}$ satisfying the *Cauchy condition* that

$$\|f_m - f_n\| \to 0$$

as m and n tend independently to infinity, then there is a unique element f of $\mathcal{H}$ such that $\|f_n - f\| \to 0$ as $n \to \infty$.

Let $B = \{\phi_n : n \in \mathbb{I}\}$ be an orthonormal basis in the Hilbert space $\mathcal{H}$ and $\mathbb{I}$ is the countable index set. Then

$$(1) \qquad \langle \phi_j, \phi_k \rangle = \delta_{j,k}$$

$$(2) \qquad \bigwedge_{f \in \mathcal{H}} f = \sum_{j \in \mathbb{I}} \langle f, \phi_j \rangle \phi_j$$

$$(3) \qquad \bigwedge_{f,g \in \mathcal{H}} \langle f, g \rangle = \sum_{j \in \mathbb{I}} \overline{\langle f, \phi_j \rangle} \langle g, \phi_j \rangle$$

$$(4) \qquad \left(\bigwedge_{\phi_j \in B} \langle f, \phi_j \rangle = 0 \right) \Rightarrow f = 0$$

$$(5) \qquad \bigwedge_{f \in \mathcal{H}} \|f\|^2 = \sum_{j \in \mathbb{I}} |\langle f, \phi_j \rangle|^2 .$$

Let $f, g \in \mathcal{H}$. Then we have the inequalities

$$|\langle f, g \rangle| \leq \|f\| \cdot \|g\|, \qquad \|f + g\| \leq \|f\| + \|g\|$$

and the equality $\|f + g\|^2 + \|f - g\|^2 = 2(\|f\|^2 + \|g\|^2)$. Note that

$$\langle f + g, f + g \rangle = \langle f, f \rangle + \langle g, g \rangle + \langle f, g \rangle + \langle g, f \rangle \geq 0.$$

Let $f, g \in \mathcal{H}$. We call f and g *orthogonal* to each other if $\langle f, g \rangle = 0$.

Let $f, g \in \mathcal{H}$. The *polarization identity* is given by

$$4\langle f, g \rangle = \|f + g\|^2 - \|f - g\|^2 - i\|f - ig\|^2 + i\|f + ig\|^2.$$

For a complex normed linear space $(V, \| \cdot \|)$ where for all $f, g \in V$ the *polarization identity*

$$\|f + g\|^2 + \|f - g\|^2 = 2\|f\|^2 + 2\|g\|^2$$

is valid an inner product can be defined by

$$\langle f, g \rangle := \frac{1}{4}(\|f + g\|^2 - \|f - g\|^2 - i\|f - ig\|^2 + i\|f + ig\|).$$

Let $\mathbb{I}$ be a finite or countable (index) set. Then the Hilbert space $\ell_2(\mathbb{I})$ is defined to be the space of all functions $f : \mathbb{I} \to \mathbb{C}$ such that

$$\|f\|_2 := \sqrt{\sum_{j \in \mathbb{I}} |f(j)|^2} < \infty.$$

This norm comes from the inner product

$$\langle f, g \rangle := \sum_{j \in \mathbb{I}} f(j)\overline{g(j)}.$$

The sum is absolutely convergent for all $f, g \in \ell_2(\mathbb{I})$.

Let S be a subset of the Hilbert space $\mathcal{H}$. The subset S is dense in $\mathcal{H}$ if it contains a countable dense subset $\{f_1, f_2, \dots\}$. The subset S is dense in $\mathcal{H}$ if for every $f \in \mathcal{H}$ there exists a Cauchy sequence $\{f_j\}$ in S such that $f_j \to f$ as $j \to \infty$.

A topological space is called *separable* if it contains a countable but dense subset. Every separable Hilbert space $\mathcal{H}$ admits a Hilbert basis, which is a countable set $\{\phi_j\}_{j \in \mathcal{H}} \subset \mathcal{H}$ and has the properties:
(i) The basis vectors are orthonormal, i.e. $\langle \phi_j, \phi_k \rangle = \delta_{j,k}$.

(ii) The set is complete, i.e. for all $f \in \mathcal{H}$ we have $f = \sum_{j=1}^{\infty} \langle \phi_j, f \rangle \phi_j$.

Examples of separable Hilbert spaces are $L_2(\mathbb{R})$ and $\ell_2(\mathbb{Z})$.

We also use the so-called *Dirac notation*. Let $\mathcal{H}$ be a Hilbert space and $\mathcal{H}_*$ be the dual space endowed with a multiplication law of the form

$$(c, \phi) = \bar{c}\phi$$

where $c \in \mathbb{C}$ and $\phi \in \mathcal{H}$. The inner product can be viewed as a bilinear form (duality) $\langle \cdot | \cdot \rangle : \mathcal{H}_* \times \mathcal{H} \to \mathbb{C}$ such that the linear maps

$$\langle \phi| : \psi \to \langle \phi | \psi \rangle, \quad \langle \cdot | : \mathcal{H}_* \to \mathcal{H}'$$

$$|\psi\rangle : \phi \to \langle \phi | \psi \rangle, \quad |\cdot\rangle : \mathcal{H} \to \mathcal{H}'_*$$

where prime denotes the space of linear continuous functionals on the corresponding space, are monomorphisms. The vectors $\langle \phi|$ and $|\psi\rangle$ are called bra and ket vectors, respectively. The ket vector $|\phi\rangle$ is uniquely determined by a vector $\phi \in \mathcal{H}$, therefore we can write $|\phi\rangle \in \mathcal{H}$.

A *functional* f on a Hilbert space $\mathcal{H}$ is a continuous linear map $f : \mathcal{H} \to \mathbb{C}$.

Riesz-Fischer theorem. [62], [63] Let $\{ e_j \}_{j=0}^{\infty}$ be an orthonormal basis for a (real or complex) infinite dimensional Hilbert space $\mathcal{H}$. If $\{ c_j \}_{j=0}^{\infty}$ is a sequence of (real or complex numbers) such that $\sum_{j=0}^{\infty} |c_j|^2 < \infty$, then there is an $f \in \mathcal{H}$ such that

$$f = \sum_{j=0}^{\infty} c_j e_j \quad \text{with} \quad c_j = \langle f, e_j \rangle.$$

Every continuous linear functional T on a pre-Hilbert space V is of the form $T(f) = \langle f, g \rangle$ for a uniquely determined g in the completion $\bar{V}$ of the pre-Hilbert space V. Hence every continuous linear functional f on a Hilbert space $\mathcal{H}$ with scalar product $\langle \cdot, \cdot \rangle$ is of the form $f(x) = \langle x, y_f \rangle$ for a uniquely determined y_f in $\mathcal{H}$.

The *Riesz representation theorem* tells us that the continuous dual if a Hilbert space is again a Hilbert space which is anti-isomorphic to the original Hilbert space. The dual space $\mathcal{H}^*$ has a natural norm

$$|f|_{\mathcal{H}^*} = \sup_{v \in \mathcal{H} : \|v\| \leq 1} |f(v)|, \quad \text{for } f \in \mathcal{H}^*.$$

Furthermore $\|y_f\|_{\mathcal{H}} = |f|_{\mathcal{H}^*}$.

A subspace $\mathcal{K}$ of a Hilbert space $\mathcal{H}$ is a subset of vectors which themselves form a Hilbert space. If $\mathcal{K}$ is a subspace, then so too is the set $\mathcal{K}^{\perp}$ of vectors orthogonal to all those in $\mathcal{K}$. The subspace $\mathcal{K}^{\perp}$ is called the *orthogonal complement* of $\mathcal{K}$ in $\mathcal{H}$. Any vector f in $\mathcal{H}$ may be uniquely decomposed into components $f_{\mathcal{K}}$ and $f_{\mathcal{K}^{\perp}}$, lying in $\mathcal{K}$ and $\mathcal{K}^{\perp}$, respectively, i.e. $f = f_{\mathcal{K}} + f_{\mathcal{K}^{\perp}}$.

Lax-Milgram theorem. [44] Let $\mathcal{H}$ be a Hilbert space and $B(v, w)$ $(v, w \in \mathcal{H})$ be a complex-valued functional defined on the product Hilbert space $\mathcal{H} \times \mathcal{H}$ satisfying the conditions:

Sesqui-linearity, i.e. $(\alpha_1, \alpha_2, \beta_1, \beta_2 \in \mathbb{C})$

$$B(\alpha_1 v_1 + \alpha_2 v_2, w) = \alpha_1 B(v_1, w) + \alpha_2 B(v_2, w)$$
$$B(v, \beta_1 w_1 + \beta_2 w_2) = \bar{\beta}_1 B(v, w_1) + \bar{\beta}_2 B(v, w_2).$$

Boundedness, i.e., there exists a positive constant γ with

$$|B(v, w)| \leq \gamma \|v\| \cdot \|w\|.$$

Positivity, i.e., there exists a positive constant δ with $B(v, v) \geq \delta \|v\|^2$. Then there exists a uniquely determined bounded linear operator T with a bounded linear inverse T^{-1} such that $\langle v, w \rangle = B(v, Tw)$ whenever $v, w \in \mathcal{H}$ and $\|T\| \leq \delta^{-1}$, $\|T^{-1}\| \leq \gamma$.

Given a closed, convex, non-empty subset $\mathcal{K}$ of a Hilbert space $\mathcal{H}$ and an element $v \in \mathcal{H}$ not in $\mathcal{K}$, then there is a unique element $w \in \mathcal{K}$ closest to v.

A linear operator, T, in a Hilbert space, $\mathcal{H}$, is a linear transformation of a linear manifold, $\mathcal{D}(T)$ ($\subset \mathcal{H}$), into $\mathcal{H}$. The manifold $\mathcal{D}(T)$ is termed the domain of definition, or simply the *domain*, of T. The linear operator T is termed *bounded* if the set of numbers, $\|Tf\|$, is bounded as f runs through the normalized vectors in $\mathcal{D}(T)$. In this case, we define $\|T\|$, the *norm* of T, to be the supremum, i.e. the least upper bound, of $\|Tf\|$, as f runs through these normalized vectors, i.e.

$$\|T\| := \sup_{\|f\|=1} \|Tf\|.$$

Let $\mathcal{H} = \mathbb{C}^n$. Then all $n \times n$ matrices over $\mathbb{C}$ are bounded linear operators. It follows from this definition that

$$\|Tf\| \leq \|T\|\|f\| \quad \text{for all vectors } f \text{ in } \mathcal{D}(T).$$

If T is bounded, we may take $\mathcal{D}(T)$ to be $\mathcal{H}$, since, even if this domain is originally defined to be a proper subset of $\mathcal{H}$, we can always extend it to the whole of this space as follows. Since

$$\|Tf_m - Tf_n\| \leq \|T\|\|f_m - f_n\|$$

we conclude that the convergence of a sequence of vectors $\{f_n\}$ in $\mathcal{D}(T)$ implies that of $\{Tf_n\}$. Hence, we may extend the definition of T to $\overline{\mathcal{D}(T)}$, the closure of $\mathcal{D}(T)$, by defining

$$T(\lim_{n \to \infty} f_n) := \lim_{n \to \infty} Tf_n.$$

We may then extend T to the full Hilbert space $\mathcal{H}$, by defining it to be zero on $\overline{\mathcal{D}(T)}^\perp$, the orthogonal complement of $\overline{\mathcal{D}(T)}$.

Let T be a bounded, non-negative, self-adjoint operator acting in a Hilbert space $\mathcal{H}$. Then there exists a bounded operator Q such that $Q = Q^* \geq 0$ and $Q^2 = T$. The operator Q is the norm limit of a sequence of polynomials in T.

A self-adjoint operator H in a Hilbert space $\mathcal{H}$ admits a uniquely determined spectral resolution. The *Cayley transform* of the self-adjoint operator H

$$U_H = (H - iI)(H + iI)^{-1}$$

is unitary, where I is the identity operator.

Let $\mathcal{H}$ be a Hilbert space and $T : \mathcal{H} \xrightarrow{\cdot} \mathcal{H}$ be a bounded linear operator. The operator T is called *isometric* if the operator T leaves the scalar product invariant, i.e.

$$\langle Tv, Tw \rangle = \langle v, w \rangle \quad \text{for all} \quad v, w \in \mathcal{H}.$$

If $R(T) = \mathcal{H}$ (range of T), then a bounded isometric operator T is called a *unitary operator*. One has $U^*U = I$ and $U^{-1} = U^*$.

Let $U(\tau)$ with $\tau \in \mathbb{R}$ be a group of unitary operators of class (C_0) in a Hilbert space $\mathcal{H}$. Then the infinitesimal generator G of $U(\tau)$ is $i = \sqrt{-1}$ times a self-adjoint operator H.

Given a Hilbert space $\mathcal{H}$ and a Hilbert subspace $\mathcal{G}$ of $\mathcal{H}$. The Hilbert space *projection theorem* states that for every $f \in \mathcal{H}$, there exists a unique $g \in \mathcal{G}$ such that

$$(i) \ f - g \in \mathcal{G}^{\perp} \qquad (ii) \ \|f - g\| = \inf_{h \in \mathcal{G}} \|f - h\|$$

where the space $\mathcal{G}^{\perp}$ is defined by

$$\mathcal{G}^{\perp} := \{ k \in \mathcal{H} : \langle k, u \rangle = 0 \text{ for all } u \in \mathcal{G} \}.$$

If g is the minimizer of $\|f - h\|$ over all $h \in \mathcal{G}$, then it is true that $f - g \in \mathcal{G}^{\perp}$.

Let $\mathcal{H}$ be a Hilbert space and $L(\mathcal{H})$ be the set of all linear bounded operators on $\mathcal{H}$. An operator $T \in L(\mathcal{H})$ is said to be a compact operator if the image of each bounded set under T is relatively compact. For every compact self-adjoint operator T on a real or complex separable infinite Hilbert space $\mathcal{H}$, there exists a countable infinite orthonormal basis $\{ e_j \}_{j \in \mathbb{N}_0}$ of $\mathcal{H}$ consisting of eigenvectors of T with corresponding real eigenvalues $\{ \lambda_j \}$ such that $\lambda_j \to 0$ as $j \to \infty$. Let $\mathcal{H}$ be a separable Hilbert space and $T : \mathcal{H} \to \mathcal{H}$ be a non-zero self-adjoint compact operator. Then there exists at least one eigenvalue given by $\lambda \in \{ \pm \|T\| \}$. Let $\mathcal{H}_1, \mathcal{H}_2$ be Hilbert spaces. A linear operator $T : \mathcal{H}_1 \to \mathcal{H}_2$ is called *compact* if for every sequence of functions $\{f_n\}_{j\in\mathbb{N}_0}$ in the unit ball of $\mathcal{H}_1$ (i.e. $\|f_j\|_2 \le 1$ for all j) there is a subsequence $\{f_{j_k}\}_{k\in\mathbb{N}_0}$ converges in the Hilbert space $\mathcal{H}_2$. If the linear operator T is compact and self-adjoint, then the singular numbers are the absolute values of the eigenvalues λ_j.

2.2 Solved Problems

Problem 1. Consider a Hilbert space $\mathcal{H}$ with scalar product $\langle \cdot, \cdot \rangle$. The scalar product implies a norm via $\|f\|^2 := \langle f, f \rangle$, where $f \in \mathcal{H}$.
(i) Show that

$$\|f + g\|^2 + \|f - g\|^2 = 2(\|f\|^2 + \|g\|^2).$$

Start with

$$\|f + g\|^2 + \|f - g\|^2 \equiv \langle f + g, f + g \rangle + \langle f - g, f - g \rangle.$$

(ii) Assume that $\langle f, g \rangle = 0$, where $f, g \in \mathcal{H}$. Show that

$$\|f + g\|^2 = \|f\|^2 + \|g\|^2.$$

Start with $\|f + g\|^2 \equiv \langle f + g, f + g \rangle$.

Solution 1. (i) We have

$$
\begin{aligned}
\|f + g\|^2 + \|f - g\|^2 &= \langle f + g, f + g \rangle + \langle f - g, f - g \rangle \\
&= \langle f, f \rangle + \langle g, f \rangle + \langle f, g \rangle + \langle g, g \rangle + \langle f, f \rangle \\
&\quad - \langle f, g \rangle - \langle g, f \rangle + \langle g, g \rangle \\
&= 2\langle f, f \rangle + 2\langle g, g \rangle \\
&= 2(\|f\|^2 + \|g\|^2).
\end{aligned}
$$

(ii) Since $\langle f, g \rangle = 0$ we have $\overline{\langle g, f \rangle} = 0$. Therefore $\langle g, f \rangle = 0$. It follows that

$$
\begin{aligned}
\|f + g\|^2 = \langle f + g, f + g \rangle &= \langle f, f \rangle + \langle f, g \rangle + \langle g, f \rangle + \langle g, g \rangle \\
&= \langle f, f \rangle + \langle g, g \rangle \\
&= \|f\|^2 + \|g\|^2.
\end{aligned}
$$

Problem 2. Let $\{e_j\}_{j=0}^{\infty}$ be an orthonormal basis in a Hilbert space $\mathcal{H}$. Find $\|e_j - e_k\|^2$ with $j \neq k$.

Solution 2. With $\langle e_j, e_j \rangle = 1$ and $\langle e_j, e_k \rangle = 0$ for $j \neq k$ we have

$$\|e_j - e_k\|^2 = \langle e_j - e_k, e_j - e_k \rangle = \langle e_j, e_j \rangle - \langle e_j, e_k \rangle - \langle e_k, e_j \rangle + \langle e_k, e_k \rangle = 2.$$

Problem 3. Let $\mathcal{H}$ be a Hilbert space. Show that the map (scalar product)

$$\langle \cdot, \cdot \rangle :: \mathcal{H} \times \mathcal{H} \to \mathbb{C}$$

is continuous as function of two variables.

Solution 3. Let $u, v, u', v' \in \mathcal{H}$. Suppose that $\|u - u'\| < \epsilon$ and $\|v - v'\| < \epsilon$. Then we have the identity

$$\langle u, v \rangle - \langle u', v' \rangle \equiv \langle u - u', v \rangle + \langle u', v - v' \rangle.$$

Using the triangle inequality for the absolute value and then the Cauchy-Schwarz inequality we arrive at

$$|\langle u, v \rangle - \langle u', v' \rangle| \leq |\langle u - u', v \rangle| + |\langle u', v - v' \rangle|$$
$$\leq \|u - u'\| \cdot \|v\| + \|u'\| \cdot \|v - v'\|$$
$$< \epsilon(\|v\| + \|u'\|).$$

Thus the scalar product is continuous.

Problem 4. Let $f, g \in \mathcal{H}$. Assume that $\langle f, g \rangle = -i$. Find $\langle g, f \rangle$.

Solution 4. We have

$$-i = \langle f, g \rangle = \overline{\langle g, f \rangle} \quad \Rightarrow \quad \langle g, f \rangle = i.$$

Problem 5. Let $f, g \in \mathcal{H}$. Assume that $\langle f, g \rangle = \langle g, f \rangle$. What can we conclude?

Solution 5. Since $\langle f, g \rangle = \overline{\langle g, f \rangle}$ we have $\overline{\langle g, f \rangle} = \langle g, f \rangle$. Hence $\langle g, f \rangle \in \mathbb{R}$.

Problem 6. Show that

$$\langle f, g \rangle = \frac{1}{4}\|f + g\|^2 - \frac{1}{4}\|f - g\|^2$$

or

$$\langle f, g \rangle = \frac{1}{4}\|f + g\|^2 - \frac{1}{4}\|f - g\|^2 + \frac{i}{4}\|f + ig\|^2 - \frac{i}{4}\|f - ig\|^2$$

depending on whether we are dealing with a real or complex Hilbert space.

Solution 6. For a real Hilbert space we have

$$\|f + g\|^2 - \|f - g\|^2 = \langle f + g, f + g \rangle - \langle f - g, f - g \rangle$$
$$= \langle f, f \rangle + \langle f, g \rangle + \langle g, f \rangle + \langle g, g \rangle$$
$$- \langle f, f \rangle - \langle g, g \rangle + \langle f, g \rangle + \langle g, f \rangle$$
$$= 2\langle f, g \rangle + 2\langle g, f \rangle$$
$$= 4\langle f, g \rangle.$$

With $i^2 = -1$, $i(-i) = 1$ in the second case we have

$$\|f + g\|^2 - \|f - g\|^2 + i\|f + ig\| - i\|f - ig\| =$$

$$\langle f + g, f + g \rangle - \langle f - g, f - g \rangle + i\langle f + ig, f + ig \rangle - i\langle f - ig, f - ig \rangle =$$

$$\langle f, g \rangle + \langle g, f \rangle + \langle f, g \rangle + \langle g, f \rangle + \langle f, g \rangle - \langle g, f \rangle + \langle f, g \rangle - \langle g, f \rangle = 4\langle f, g \rangle.$$

Problem 7. Let $\mathcal{H}$ be a Hilbert space with scalar product $\langle \cdot, \cdot \rangle$ and $u, v \in \mathcal{H}$. Let $\|.\|$ be the norm induced by the scalar product, i.e. $\|u\|^2 = \langle u, u \rangle$. Show that (*Schwarz-Cauchy inequality*)

$$|\langle u, v \rangle| \leq \|u\| \cdot \|v\|.$$

Obviously for $u = 0$ or $v = 0$ the inequality is an equality. So we can assume that $u \neq 0$ and $v \neq 0$ in the following.

Solution 7. We have $\langle u - v, u - v \rangle \geq 0$. Thus

$$\langle u, u \rangle + \langle v, v \rangle - \langle u, v \rangle - \langle v, u \rangle \geq 0$$

or $\langle u, u \rangle + \langle v, v \rangle - \langle u, v \rangle - \overline{\langle u, v \rangle} \geq 0$. It follows that

$$\langle u, u \rangle + \langle v, v \rangle - 2\Re(\langle u, v \rangle) \geq 0$$

and therefore

$$\Re\langle u, v \rangle \leq \frac{1}{2}(\langle u, u \rangle + \langle v, v \rangle).$$

Let c be a positive real number. We set $u \to cu$, $v \to v/c$. Then

$$\Re(\langle u, v \rangle) \leq \frac{1}{2}(c^2 \langle u, u \rangle + \frac{1}{c^2} \langle v, v \rangle)$$

must be true for all positive real numbers c. Thus we find the minimum of the right-hand side with respect to c. We obtain

$$c^2 = \sqrt{\frac{\langle v, v \rangle}{\langle u, u \rangle}}.$$

Then we have

$$\Re(\langle u, v \rangle) \le \sqrt{\langle u, u \rangle} \sqrt{\langle v, v \rangle}.$$

Now let $u \to e^{-i\alpha} u$, $\alpha \in \mathbb{R}$. Then

$$\Re(e^{-i\alpha} \langle u, v \rangle) \le \sqrt{\langle u, u \rangle} \sqrt{\langle v, v \rangle}.$$

We have

$$\Re(e^{-i\alpha} \langle u, v \rangle) = \Re(\cos(\alpha) \langle u, v \rangle - i \sin(\alpha) \langle u, v \rangle)$$
$$= \cos(\alpha) \Re(\langle u, v \rangle) + \sin(\alpha) \Im(\langle u, v \rangle).$$

This expression has the maximum when

$$\sqrt{(\Re \langle u, v \rangle)^2 + (\Im(\langle u, v \rangle)^2)} = |\langle u, v \rangle|.$$

Therefore we obtain the Cauchy-Schwarz inequality.

Problem 8. Let $\mathcal{H}$ be a Hilbert space and $f, g \in \mathcal{H}$. Show that

$$((\|f\|^2 + \|g\|^2)^2 - 4|\langle f, g \rangle|^2)^{1/2} \le \|f - g\| \cdot \|f + g\|.$$

Solution 8. We have

$$((\|f\|^2 + \|g\|^2)^2 - 4|\langle f, g \rangle|^2)^{1/2} \le ((\|f\|^2 + \|g\|^2)^2 - 4(\Re(\langle f, g \rangle))^2)^{1/2}$$
$$= (\|f\|^2 + \|g\|^2 - 2\Re(\langle f, g \rangle))^{1/2}$$
$$\cdot (\|f\|^2 + \|g\|^2 + 2\Re(\langle f, g \rangle))^{1/2}$$
$$= \|f - g\| \cdot \|f + g\|.$$

Problem 9. Let $\mathcal{O} = \{ u_1, u_2, \dots \}$ be an orthonormal set in an infinite dimensional Hilbert space $\mathcal{H}$. Let $v \in \mathcal{H}$. Show that if

$$v = \sum_{j=1}^{\infty} c_j u_j$$

then

$$\|v\|^2 = \sum_{j=1}^{\infty} |c_j|^2.$$

Solution 9. We have

$$\langle v, v \rangle = \sum_{j=1}^{\infty} \sum_{k=1}^{\infty} c_j \bar{c}_k \langle u_j, u_k \rangle = \sum_{j=1}^{\infty} \sum_{k=1}^{\infty} c_j \bar{c}_k \delta_{jk} = \sum_{j=1}^{\infty} |c_j|^2.$$

Problem 10. (i) Let $\alpha \in \mathbb{C}$ and $u, v \in \mathcal{H}$. Find $\langle u + \alpha v, u + \alpha v \rangle$.
(ii) Let $u, v \in \mathcal{H}$ and $v \neq 0$. Show that $\langle u, v \rangle \langle v, u \rangle \leq \langle u, u \rangle \langle v, v \rangle$.

Solution 10. (i) We find

$$\langle u + \alpha v, u + \alpha v \rangle = \langle u, u \rangle + \alpha \langle v, u \rangle + \bar{\alpha} \langle u, v \rangle + |\alpha|^2 \langle v, v \rangle.$$

(ii) For $v = 0$ we have an equality. Thus we can assume that $\|v\| \neq 0$. Utilizing (i) we have

$$0 \leq \langle u, u \rangle + \bar{\alpha} \langle u, v \rangle + \alpha \langle v, u \rangle + |\alpha|^2 \langle v, v \rangle.$$

Setting $\alpha := -\overline{\langle u, v \rangle}/\|v\|^2$ we obtain

$$0 \leq \langle u, u \rangle - \frac{\langle u, v \rangle \overline{\langle u, v \rangle}}{\|v\|^2} - \frac{\langle u, v \rangle \langle v, u \rangle}{\|v\|^2} + \frac{\langle u, v \rangle \langle v, u \rangle \langle v, v \rangle}{\|v\|^4}.$$

Hence

$$2\langle u, v \rangle \overline{\langle u, v \rangle} \leq \|v\|^2 \langle u, u \rangle + \langle u, v \rangle \langle v, u \rangle$$

and finally $\langle u, v \rangle \langle v, u \rangle \leq \langle u, u \rangle \langle v, v \rangle$.

Problem 11. Let $f, g \in \mathcal{H}$. Use the *Cauchy-Schwarz inequality*

$$|\langle f, g \rangle|^2 \leq \langle f, f \rangle \langle g, g \rangle = \|f\|^2 \|g\|^2$$

to prove the *triangle inequality* $\|f + g\| \leq \|f\| + \|g\|$.

Solution 11. We have

$$\|f+g\|^2 = \langle f+g, f \rangle + \langle f+g, g \rangle \leq \|f+g\| \|f\| + \|f+g\| \|g\| = \|f+g\|(\|f\| + \|g\|).$$

Problem 12. Let Π be a nonzero *projection operator* in a Hilbert space $\mathcal{H}$. Show that $\|\Pi\| = 1$.

Solution 12. Let $f \in \mathcal{H}$ and $\Pi f \neq 0$. Then applying the *Cauchy-Schwarz inequality* provides

$$\|\Pi f\| = \frac{\langle \Pi f, \Pi f \rangle}{\|\Pi f\|} = \frac{\langle f, \Pi^2 f \rangle}{\|\Pi f\|} = \frac{\langle f, \Pi f \rangle}{\|\Pi f\|} \leq \|f\|.$$

It follows that $\|\Pi\| \leq 1$. If $\Pi \neq 0$, then there is an $f \in \mathcal{H}$ with $\Pi f \neq 0$ and

$$\|\Pi(\Pi f)\| = \|\Pi f\|$$

so that $\|\Pi\| \geq 1$. Thus the result $\|\Pi\| = 1$ follows.

Problem 13. Let $|0\rangle$, $|1\rangle$, $|2\rangle$, ... be an orthonormal basis in a Hilbert space.
(i) Show that $\Pi = |j\rangle\langle j|$ ($j \in \mathbb{N}_0$) is a projection operator.
(ii) Let $j \neq k$ ($j, k \in \mathbb{N}_0$). Show that $\Pi = |j\rangle\langle j| + |k\rangle\langle k|$ is a projection operator.

Solution 13. (i) We have $\Pi = |j\rangle\langle j| = \Pi^*$ and with $\langle j|j\rangle = 1$ we find

$$\Pi^2 = |j\rangle\langle j|j\rangle\langle j| = |j\rangle\langle j| = \Pi.$$

(ii) We have $\Pi = \Pi^*$ and with $\langle j|k\rangle = 0$ we obtain

$$\Pi^2 = (|j\rangle\langle j| + |k\rangle\langle k|)(|j\rangle\langle j| + |k\rangle\langle k|) = |j\rangle\langle j| + |k\rangle\langle k| = \Pi.$$

Problem 14. Let $|\psi\rangle$, $|s\rangle$, $|\phi\rangle$ be normalized states in a Hilbert space $\mathcal{H}$. Let U be a unitary operator, i.e. $U^{-1} = U^*$ in the Hilbert space $\mathcal{H} \otimes \mathcal{H}$ such that

$$U(|\psi\rangle \otimes |s\rangle) = |\psi\rangle \otimes |\psi\rangle, \quad U(|\phi\rangle \otimes |s\rangle) = |\phi\rangle \otimes |\phi\rangle.$$

Show that $\langle \phi|\psi\rangle = \langle \phi|\psi\rangle^2$. Find solutions to this equation.

Solution 14. Taking the scalar product of these two equations with $U^* = U^{-1}$ and $\langle s|s\rangle = 1$ we obtain

$$((\langle\psi| \otimes \langle s|)U^*U(|\phi\rangle \otimes |s\rangle) = ((\langle\psi| \otimes \langle\psi|)(|\phi\rangle \otimes |\phi\rangle))$$
$$((\langle\psi| \otimes \langle s|)(|\phi\rangle \otimes |s\rangle)) = \langle\psi|\phi\rangle\langle\psi|\phi\rangle$$
$$\langle\psi|\phi\rangle = \langle\psi|\phi\rangle^2.$$

The equation can be satisfied if $\langle\psi|\phi\rangle = 0$ ($|\psi\rangle$ and $|\phi\rangle$ are orthonormal to each other) or $\langle\psi|\phi\rangle = 1$, i.e. $|\psi\rangle = |\phi\rangle$.

Problem 15. A family, $\{ \psi_j \}_{j\in\mathbb{I}}$ of vectors in the Hilbert space $\mathcal{H}$ is called a *frame* if for any $f \in \mathcal{H}$ there exist two constants $k_1 > 0$ and $0 < k_2 < \infty$, such that

$$k_1\|f\|^2 \leq \sum_{j\in\mathbb{I}} |\langle\psi_j, f\rangle|^2 \leq k_2\|f\|^2$$

for all $f \in \mathcal{H}$. One calls the two constants k_1, k_2 the *frame bounds*. Consider the Hilbert space $\mathcal{H} = \mathbb{R}^2$ and the family of vectors

$$\left\{ \psi_0 = \begin{pmatrix} 1 \\ 1 \end{pmatrix}, \quad \psi_1 = \begin{pmatrix} 1 \\ -1 \end{pmatrix} \right\}.$$

Show that we have a *tight frame* i.e. $k_1 = k_2$.

Solution 15. Any vector u in the Hilbert space $\mathcal{H} = \mathbb{R}^2$ can be written as

$$u = c_1 \begin{pmatrix} 1 \\ 0 \end{pmatrix} + c_2 \begin{pmatrix} 0 \\ 1 \end{pmatrix}, \qquad c_1, c_2 \in \mathbb{R}.$$

Thus $k_1 \|u\|^2 = k_1(c_1^2 + c_2^2)$, $k_2 \|u\|^2 = k_2(c_1^2 + c_2^2)$. Now

$$|\langle \psi_0 | u \rangle|^2 + |\langle \psi_1 | u \rangle|^2 = (c_1 + c_2)^2 + (c_1 - c_2)^2 = 2(c_1^2 + c_2^2).$$

Thus

$$k_1(c_1^2 + c_2^2) = 2(c_1^2 + c_2^2) = k_2(c_1^2 + c_2^2)$$

with $k_1 = k_2 = 2$.

Problem 16. Let $T : X \to Y$ be a linear map between linear spaces (vector spaces) X, Y. The *null space* or *kernel* of the linear map T, denoted by $\ker(T)$, is the subset of X defined by

$$\ker(T) := \{ x \in X : Tx = 0 \}.$$

The *range* of T, denoted by $\mathrm{ran}(T)$, is the subset of Y defined by

$$\mathrm{ran}(T) := \{ y \in Y : \text{there exists } x \in X \text{ such that } Tx = y \}.$$

Let Π be a projection operator in a Hilbert space $\mathcal{H}$. Show that $\mathrm{ran}(\Pi)$ is closed and

$$\mathcal{H} = \mathrm{ran}(\Pi) \oplus \ker(\Pi)$$

is the orthogonal direct sum of $\mathrm{ran}(\Pi)$ and $\ker(\Pi)$.

Solution 16. Suppose that Π is a projection operator in a Hilbert space $\mathcal{H}$. Then we have

$$\mathcal{H} = \mathrm{ran}(\Pi) \oplus \ker(\Pi).$$

If $x = \Pi y \in \mathrm{ran}(\Pi)$ and $z \in \ker(\Pi)$, then the scalar product of x and z yields

$$\langle x, z \rangle = \langle \Pi y, z \rangle = \langle y, \Pi z \rangle = 0.$$

Thus ran(Π) $\perp$ ker(Π). Therefore we find that $\mathcal{H}$ is the orthogonal direct sum of ran(Π) and ker(Π). It follows that ran(Π) = (ker(Π))$^\perp$, so ran(Π) is closed.

Problem 17. Let $\mathcal{H}$ be an arbitrary Hilbert space with scalar product $\langle \cdot, \cdot \rangle$. Show that if φ is a bounded linear functional on the Hilbert space $\mathcal{H}$, then there is a unique vector $u \in \mathcal{H}$ such that

$$\varphi(x) = \langle u, x \rangle \qquad \text{for all } x \in \mathcal{H}.$$

Solution 17. If $\varphi = 0$, then $u = 0$. Let us assume that $\varphi \neq 0$. Then ker(φ) is a proper closed subspace of the Hilbert space $\mathcal{H}$. There is a nonzero vector $v \in \mathcal{H}$ such that $v \perp$ ker(φ). We define a linear map $\Pi : \mathcal{H} \to \mathcal{H}$ by

$$\Pi x := \frac{\varphi(x)}{\varphi(v)} v.$$

Obviously $\Pi^2 = \Pi$ and $\mathcal{H} = \text{ran}(\Pi) \oplus \text{ker}(\Pi)$. Moreover

$$\text{ran}(\Pi) = \{ \alpha v : \alpha \in \mathbb{C} \}, \qquad \text{ker}(\Pi) = \text{ker}(\varphi)$$

so that ran(Π) $\perp$ ker(Π). Hence Π is an orthogonal projection and the Hilbert space can be expressed as an orthogonal direct sum

$$\mathcal{H} = \{ \alpha v : \alpha \in \mathbb{C} \} \oplus \text{ker}(\varphi).$$

Consequently we can write the element x of the Hilbert space $\mathcal{H}$ as

$$x = \alpha v + n, \qquad \alpha \in \mathbb{C} \text{ and } n \in \text{ker}(\varphi).$$

Taking the scalar product of this equation with v yields

$$\alpha = \frac{\langle v, x \rangle}{\|v\|^2}.$$

Evaluating φ on $x = \alpha v + n$ we obtain $\varphi(x) = \alpha \varphi(v)$. Inserting α into this equation we obtain

$$\varphi(x) = \langle u, x \rangle, \qquad u = \frac{\overline{\varphi(v)}}{\|v\|^2} v.$$

Thus, every bounded linear functional is given by the scalar product with a fixed vector (*Riesz representation*). Now $\varphi_u(x) = \langle u, x \rangle$ defines a bounded linear functional on $\mathcal{H}$ for every $u \in \mathcal{H}$. To prove that there is a unique u in $\mathcal{H}$ associated with a given linear functional, suppose that $\varphi_{u_1} = \varphi_{u_2}$. Then $\varphi_{u_1}(u) = \varphi_{u_2}(u)$ when $u = u_1 - u_2$, which implies that $\|u_1 - u_2\|^2 = 0$.

Thus $u_1 = u_2$.

Problem 18. Let $\mathcal{H}$ be an arbitrary Hilbert space. A bounded linear operator $T : \mathcal{H} \to \mathcal{H}$ satisfies the *Fredholm alternative* if one of the following two alternatives holds:
(i) either $Tx = 0$, $T^*x = 0$ have only the zero solution, and the linear equations $Tx = y$, $T^*x = y$ have a unique solution $x \in \mathcal{H}$ for every $y \in \mathcal{H}$;
(ii) or $Tx = 0$, $T^*x = 0$ have nontrivial, finite-dimensional solution spaces of the same dimension, $Tx = y$ has a (nonunique) solution if and only if $y \perp u$ for every solution u of $T^*u = 0$, and $T^*x = y$ has a (nonunique) solution if and only if $y \perp u$ for every solution u of $Tu = 0$.

Give an example of a bounded linear operator that satisfies the Fredholm alternative.

Solution 18. Any linear operator $T : \mathbb{C}^n \to \mathbb{C}^n$ (i.e. $n \times n$ matrices over $\mathbb{C}$) associated with an $n \times n$ system of linear equations $Tx = y$ satisfies the Fredholm alternative. The range of A and T^* are closed because they are finite-dimensional. The rank of T^* is equal to the rank of T.

Problem 19. Let $\mathcal{H}$ be a Hilbert space and let $f : \mathcal{H} \to \mathcal{H}$ be a monotone mapping such that for some constant $\beta > 0$
$$\|f(u) - f(v)\| \le \beta \|u - v\| \qquad \forall u, v \in \mathcal{H}.$$
Show that for any $w \in \mathcal{H}$, the equation $u + f(u) = w$ has a unique solution u.

Solution 19. If $\beta < 1$, then the mapping $u \mapsto w - f(u)$ is a contraction mapping and the result follows from the contraction mapping principle. Consider now $\beta \ge 1$. We notice that for $\lambda \ne 0$, u is a solution of
$$u = (1 - \lambda)u - \lambda f(u) + \lambda w$$
if, and only if, u solves $u + f(u) = w$. Consider the mapping
$$f_\lambda(u) := (1 - \lambda)u - \lambda f(u) + \lambda w.$$
It follows that
$$f_\lambda(u) - f_\lambda(v) = (1 - \lambda)(u - v) - \lambda(f(u) - f(v)).$$
Using the properties of the scalar product in the Hilbert space, we find
$$\|f_\lambda(u) - f_\lambda(v)\|^2 \le \lambda^2 \beta^2 \|u - v\|^2 - 2\Re\lambda(1 - \lambda)\langle f(u) - f(v), u - v \rangle$$
$$+ (1 - \lambda)^2 \|u - v\|^2.$$

Therefore, if $0 < \lambda < 1$, the monotonicity of f implies that

$$\|f_\lambda(u) - f_\lambda(v)\|^2 \le (\lambda^2\beta^2 + (1 - \lambda)^2)\|u - v\|^2.$$

Choosing $\lambda = 1/(\beta^2 + 1)$ we obtain

$$\|f_\lambda(u) - f_\lambda(v)\|^2 \le (1 - \lambda)\|u - v\|^2.$$

Thus for $\lambda \to 1$ we have $\|f_1(u) - f_1(v)\| = 0$ and the result follows.

Problem 20. Consider the inner product space

$$C([a, b]) := \{\, f(x) \, : \, f \text{ is continuous on } x \in [a, b] \,\}$$

with the inner product

$$\langle f, g \rangle := \int_a^b f(x)g^*(x)dx.$$

This implies a norm

$$\langle f, f \rangle = \int_a^b f(x)f^*(x)dx = \|f\|^2.$$

Show that the space $C([a, b])$ is incomplete. This means find a *Cauchy sequence* in the space $C([a, b])$ which converges to an element which is not in the space $C([a, b])$.

Solution 20. Consider the sequence of continuous functions

$$g_k(x) = \frac{1}{2} + \frac{1}{\pi}\arctan(kx), \qquad -1 \le x \le 1$$

i.e. $a = -1$ and $b = 1$. We find

$$\lim_{k,p\to\infty} \|g_k - g_p\|^2 = \lim_{k,p\to\infty} \int_{-1}^{+1} (g_k(x) - g_p(x))^2 dx = 0.$$

In other words, the sequence is a *Cauchy sequence*. For fixed x we have

$$\lim_{k\to\infty} g_k(x) = g(x) = \begin{cases} 1 & 0 < x \le 1 \\ 1/2 & x = 0 \\ 0 & -1 \le x < 0 \end{cases}$$

which is a discontinuous function, i.e. $g \notin C([-1, 1])$. We say that the sequence of functions $\{\, g_k \, : \, k = 1, 2, \ldots \,\}$ converges pointwise to the function g. We also have

$$\lim_{k\to\infty} \|g - g_k\|^2 = \lim_{k\to\infty} \int_{-1}^{+1} (g(x) - g_k(x))^2 dx = 0.$$

Thus the limit of the Cauchy sequence does not lie in the inner product space $C([-1,1])$. Thus the normed space $C([a,b])$ is incomplete.

Problem 21. Let Z be the linear space of functions $f : C \to \mathbb{C}$, where C is the unit circle in $\mathbb{C}$ and a scalar product is given by

$$\langle f, g \rangle := \frac{1}{2\pi i} \oint_C f(z)\bar{g}(z)\frac{dz}{z}$$

where $f, g \in Z$. Let $f(z) = z^3$, $g(z) = z^{-2}$. Find the scalar product $\langle f, g \rangle$. The complex Hilbert space Z contains the orthonormal basis

$$\{ \dots, z^{-2}, z^{-1}, 1, z^{+1}, z^{+2}, \dots \}$$

Solution 21. We set $z = e^{i\phi}$. Then $z^3 = e^{3i\phi}$, $z^{-2} = e^{-2i\phi}$, $\overline{z^{-2}} = e^{2i\phi}$. Then

$$\langle f(z), g(z) \rangle = \frac{1}{2\pi i} \int_0^{2\pi} i e^{5i\phi} d\phi = 0.$$

Problem 22. Let $\{ \phi_n \}_{n \in \mathbb{Z}}$ be an orthonormal basis in a Hilbert space $\mathcal{H}$. Then any vector $f \in \mathcal{H}$ can be written as

$$f = \sum_{n \in \mathbb{Z}} \langle f, \phi_n \rangle \phi_n.$$

Now suppose that $\{ \psi_n \}_{n \in \mathbb{Z}}$ is also a basis for $\mathcal{H}$, but it is not orthonormal. Show that if we can find a so-called dual basis $\{ \chi_n \}_{n \in \mathbb{Z}}$ satisfying

$$\langle \psi_n, \chi_m \rangle = \delta(n - m)$$

then for any vector $f \in \mathcal{H}$, we have

$$f = \sum_{n \in \mathbb{Z}} \langle f, \chi_n \rangle \psi_n.$$

Here $\delta(n - m)$ denotes the Kronecker delta with $\delta(n - m) = 0$ if $n = m$ and 1 otherwise.

Solution 22. Any $f \in \mathcal{H}$ can be represented by means of a basis $\{\psi_n\}_{n \in \mathbb{Z}}$ as

$$f = \sum_{n \in \mathbb{Z}} \langle f, \psi_n \rangle \psi_n.$$

Especially for each χ_m we have

$$\chi_m = \sum_{n \in \mathbb{Z}} \langle \chi_m, \psi_n \rangle \psi_n = \sum_{n \in \mathbb{Z}} \delta_{m,n} \psi_n = \psi_m$$

and therefore the desired representation of f follows.

Problem 23. Let A be a linear bounded self-adjoint operator in a Hilbert space $\mathcal{H}$. Let $u, v \in \mathcal{H}$ and $\lambda \in \mathbb{C}$. Consider the equation

$$Au - \lambda u = v.$$

(i) Show that for λ nonreal (i.e. it has an imaginary part) v cannot vanish unless u vanishes.
(ii) Show that for λ nonreal we have

$$\|(A - \lambda I)^{-1} v\| \leq \frac{1}{|\Im(\lambda)|} \|v\|.$$

Solution 23. (i) From the equation by taking the scalar product with u we obtain

$$\langle u, Au \rangle - \lambda \langle u, u \rangle = \langle u, v \rangle.$$

Taking the imaginary part of this equation and taking into account that $\langle u|Au \rangle$ is real yields

$$-\Im(\lambda)\|u\|^2 = \Im(\langle u, v \rangle).$$

This shows that for λ nonreal, v cannot vanish unless u vanishes. Thus λ is not an eigenvalue of A, and $(A - \lambda I)$ has an inverse.
(ii) By the Cauchy-Schwarz inequality we have

$$|\Im(\lambda)| \, \|u\|^2 = \|\Im \langle u, v \rangle\| \leq |\langle u, v \rangle| \leq \|u\| \cdot \|v\|.$$

However $u = (A - \lambda I)^{-1} v$. Therefore

$$\|(A - \lambda I)^{-1} v\| \leq \frac{1}{|\Im(\lambda)|} \|v\|.$$

Problem 24. Let $d \geq 1$ and $|0\rangle, |1\rangle, \ldots, |d\rangle$ be an orthonormal basis in the Hilbert space $\mathbb{C}^{d+1}$. Define the linear operators $((d+1) \times (d+1)$ matrices)

$$\hat{b} = \sum_{j=1}^{d} \sqrt{j} |j-1\rangle \langle j|, \quad \hat{b}^\dagger = \sum_{k=1}^{d} \sqrt{k} |k\rangle \langle k-1|.$$

Are the operators hermitian? Are the operators unitary? Find the commutator $[\hat{b}, \hat{b}^\dagger]$. Is the commutator hermitian? Is the commutator unitary?

Solution 24. The operators $\hat{b}$, $\hat{b}^\dagger$ are neither hermitian nor unitary. Utilizing $\langle j|k \rangle = \delta_{j,k}$ we have

$$[\hat{b}, \hat{b}^\dagger] = \sum_{j=0}^{d-1} |j\rangle\langle j| - d|d\rangle\langle d| = I_{d+1} - (d+1)|d\rangle\langle d|$$

where we used the *completeness relation*

$$\sum_{j=0}^{d} |j\rangle\langle j| = I_{d+1}.$$

Problem 25. Show that if two bounded self-adjoint linear operators S and T on a Hilbert space $\mathcal{H}$ are positive semi-definite and commute ($ST = TS$), then their product ST is positive semi-definite. We have to show that $\langle STf, f \rangle \geq 0$ for all $f \in \mathcal{H}$.

Solution 25. If $S = 0$, then obviously $\langle STf, f \rangle = 0$. Assume that $S \neq 0$, i.e. $\|S\| > 0$. We set

$$S_1 := \frac{1}{\|S\|} S$$

and

$$S_{k+1} = S_k - S_k^2, \qquad k = 1, 2, \ldots$$

For $n = 1$ the inequality $0 \leq S_k \leq I$ holds. The assumption $0 \leq S$ implies $0 \leq S_1$, and $S_1 \leq I$ is obtained by an application of the inequality $\|Sf\| \leq \|S\|\|f\|$ with $f \in \mathcal{H}$. We have

$$\langle S_1 f, f \rangle = \frac{1}{\|S\|} \langle Sf, f \rangle \leq \frac{1}{\|S\|} \|Sf\|\|f\| \leq \|f\|^2 = \langle If, f \rangle.$$

Suppose that $0 \leq S_k \leq I$ holds for an $k = j$. This means $0 \leq S_j \leq I$. Thus $0 \leq I - S_j \leq I$. Then, since S_j is self-adjoint, for every $f \in \mathcal{H}$ and $g = S_j f$ we get

$$\langle S_j^2(I - S_j)f, f \rangle = \langle (I - S_j)S_j f, S_j f \rangle = \langle (I - S_j)g, g \rangle \geq 0.$$

This shows that $S_j^2(I - S_j) \geq 0$. Analogously $S_j(I - S_j)^2 \geq 0$. Adding these two equations yields

$$0 \leq S_j^2(I - S_j) + S_j(I - S_j)^2 = S_j - S_j^2 = S_{j+1}.$$

Hence $0 \leq S_{j+1}$. Now $S_{j+1} \leq I$ follows from $S_j^2 \geq 0$ and $I - S_j \geq 0$ by addition, i.e.

$$0 \leq I - S_j + S_j^2 = I - S_{j+1}.$$

Problem 26. Let $\mathcal{H}$ be a separable Hilbert space and $H : \mathcal{H} \to \mathcal{H}$ be a bounded self-adjoint operator. The norm of H is given by

$$\|H\| := \sup_{v \neq 0} \frac{|\langle v, Hv \rangle|}{\|v\|^2}.$$

Show that if there exists a non-zero element $u \in \mathcal{H}$ such that

$$\frac{|\langle Hu, u \rangle|}{\|u\|^2} = \|H\|$$

then u is an eigenvector of the operator H with $Hu = \lambda u$ and for the eigenvalue we have $\lambda \in \{\pm \|H\|\}$.

Solution 26. We set

$$C := \sup_{v \neq 0} \frac{|\langle v, Hv \rangle|}{\|v\|^2}.$$

Since $|\langle v, Hv \rangle| \leq \|v\| \cdot \|Hv\| \leq \|H\| \cdot \|v\|^2$ we find that $C \leq \|H\|$. Now let $v, u \in \mathcal{H}$. Then straightforward calculation provides

$$
\begin{aligned}
\langle v + u, H(v + u) \rangle - \langle v - u, H(v - u) \rangle &= \langle v, Hu \rangle + \langle u, Hv \rangle + \langle v, Hu \rangle + \langle u, Hv \rangle \\
&= 2(\langle v, Hu \rangle + \langle Hu, v \rangle) \\
&= 2(\langle v, Hu \rangle + \overline{\langle v, Hu \rangle}) \\
&= 4\Re \langle v, Hu \rangle.
\end{aligned}
$$

If the vectors $u, v \in \mathcal{H}$ are normalized we obtain

$$|\Re \langle v, Hu \rangle| \leq \frac{C}{4}(\|v + u\|^2 + \|v - u\|^2) = \frac{C}{4}(2\|v\|^2 + 2\|u\|^2) = C.$$

With $v \mapsto e^{i\theta} v$ ($\theta \in \mathbb{R}$) chosen such that $e^{i\theta} \langle v, Hu \rangle$ one obtains

$$|\langle v, Hu \rangle| \leq C \quad \text{for all } \|v\| = \|u\| = 1.$$

It follows that

$$\|H\| = \sup_{\|v\| = \|u\| = 1} |\langle v, Hu \rangle| \leq C.$$

Let u be a nonzero element of $\mathcal{H}$ and $\|H\| = |\langle Hu, u\rangle|/\|u\|^2$. Applying the *Cauchy-Schwarz inequality* it follows that

$$\|H\| = \frac{|\langle Hu, u\rangle|}{\|u\|^2} \leq \frac{\|Hu\|}{\|u\|^2} \leq \|H\|.$$

Hence $|\langle Hu, u\rangle| = \|Hu\| \cdot \|u\|$. Thus Hu and u are linearly dependent. This means $Hu = \lambda u$ for some λ (the eigenvalue). It follows that $|\lambda| = \|H\|$. Since the operator H is self-adjoint we have

$$\lambda\|u\|^2 = \langle \lambda u, u\rangle = \langle Hu, u\rangle = \langle u, Hu\rangle = \langle u, \lambda u\rangle = \overline{\lambda}\langle u, u\rangle.$$

It follows that $\lambda \in \mathbb{R}$ and hence $\lambda = +\|H\|$ or $\lambda = -\|H\|$.

Problem 27. Let $\mathcal{H}$ be a real Hilbert space with $\|x\| = \langle x, x\rangle^{1/2}$, where $x \in \mathcal{H}$ and $\langle x, x\rangle$ denotes the scalar product.
(i) Let $x \in \mathcal{H}$ and $f(x) = \|x\|$. Find the *Gateaux derivative*

$$\lim_{\epsilon \to 0} \frac{f(x + \epsilon h) - f(x)}{\epsilon}.$$

(ii) Let $g(x) = \|x\|^2 = \langle x, x\rangle$. Find the Gateaux derivative

$$\lim_{\epsilon \to 0} \frac{g(x + \epsilon h) - g(x)}{\epsilon}.$$

Solution 27. (i) We have

$$f(x + \epsilon h) - f(x) = \langle x + \epsilon h, x + \epsilon h\rangle^{1/2} - \langle x, x\rangle^{1/2}$$

$$= \frac{(\langle x + \epsilon h, x + \epsilon h\rangle^{1/2} - \langle x, x\rangle^{1/2})(\langle x + \epsilon h, x + \epsilon h\rangle^{1/2} + \langle x, x\rangle^{1/2})}{\langle x + \epsilon h, x + \epsilon h\rangle^{1/2} + \langle x, x\rangle^{1/2}}$$

$$= \frac{\langle x + \epsilon h, x + \epsilon h\rangle - \langle x, x\rangle}{\|x + \epsilon h\| + \|x\|}$$

$$= \frac{2\langle x, \epsilon h\rangle + \langle \epsilon h, \epsilon h\rangle}{\|x + \epsilon h\| + \|x\|}.$$

Then

$$\lim_{\epsilon \to 0} \frac{f(x + \epsilon h) - f(x)}{\epsilon} = \lim_{\epsilon \to 0} \frac{1}{\epsilon} \frac{2\langle x, \epsilon h\rangle + \langle \epsilon h, \epsilon h\rangle}{\|x + \epsilon h\| + \|x\|} = \lim_{\epsilon \to 0} \frac{2\langle x, h\rangle + \langle h, \epsilon h\rangle}{\|x + \epsilon h\| + \|x\|}$$

$$= \lim_{\epsilon \to 0} \frac{2\langle x, h\rangle}{\|x + \epsilon h\| + \|x\|} + \lim_{\epsilon \to 0} \frac{\langle h, \epsilon h\rangle}{\|x + \epsilon h\| + \|x\|}$$

$$= \lim_{\epsilon \to 0} \frac{2\langle x, h\rangle}{\|x + \epsilon h\| + \|x\|}$$

$$= \frac{\langle x, h\rangle}{\|x\|}.$$

(ii) We have

$$g(x + \epsilon h) = \|x + \epsilon h\|^2 = \langle x + \epsilon h, x + \epsilon h \rangle = \langle x, x \rangle + 2\langle x, \epsilon h \rangle + \epsilon^2 \langle h, h \rangle.$$

Then $g(x + \epsilon h) - g(x) = 2\epsilon\langle x, h \rangle + \epsilon^2\langle h, h \rangle$ and

$$\lim_{\epsilon \to 0} \frac{1}{\epsilon}(g(x + \epsilon h) - g(x)) = \lim_{\epsilon \to 0} \frac{1}{\epsilon}(2\epsilon\langle x, h \rangle + \epsilon^2\langle h, h \rangle) = 2\langle x, h \rangle.$$

Problem 28. Consider the Hilbert space $\mathbb{R}^3$. Show that the three vectors

$$\mathbf{v}_1 = \begin{pmatrix} \sin(\theta)\cos(\phi) \\ \sin(\theta)\sin(\phi) \\ \cos(\theta) \end{pmatrix}, \quad \mathbf{v}_2 = \begin{pmatrix} \cos(\theta)\cos(\phi) \\ \cos(\theta)\sin(\phi) \\ -\sin(\theta) \end{pmatrix}, \quad \mathbf{v}_3 = \begin{pmatrix} -\sin(\phi) \\ \cos(\phi) \\ 0 \end{pmatrix}$$

form an orthonormal basis in the Hilbert space $\mathbb{R}^3$.

Solution 28. With $\sin^2(\phi) + \cos^2(\phi) = 1$ we obtain

$$\|\mathbf{v}_1\| = \|\mathbf{v}_2\| = \|\mathbf{v}_3\| = 1.$$

For the scalar products we obtain $\mathbf{v}_1^T\mathbf{v}_2 = 0$, $\mathbf{v}_2^T\mathbf{v}_3 = 0$, $\mathbf{v}_3^T\mathbf{v}_1 = 0$. Hence we have an orthonormal basis in $\mathbb{R}^3$.

Problem 29. Let $\{f_n\}_{n \in \mathbb{N}_0}$ be a set of elements from a separable Hilbert space $\mathcal{H}$. The two statements are equivalent

(i) $\{f_n\}_{n \in \mathbb{N}_0}$ is complete in $\mathcal{H}$.

(ii) The only element $g \in \mathcal{H}$ which satisfies $\langle g, f_n \rangle = 0$ for every $n \in \mathbb{N}_0$ is $g = 0$.

This means each statement implies the other. Show that (ii) follows from (i).

Solution 29. Suppose that $\{f_n\}_{n \in \mathbb{N}_0}$ is complete. Now assume that $g \in \mathcal{H}$ satisfies $\langle g, f_n \rangle = 0$ for every $n \in \mathbb{N}_0$. Choose any $\epsilon > 0$. Then by definition of completeness we can find an $N(\epsilon) > 0$ and expansion coefficients $c_0, c_1, \ldots, c_{N-1}$ such that the element $h = \sum_{n=0}^{N-1} c_n f_n$ lies within ϵ of g, i.e. $\|g - h\| < \epsilon$, where

$$\langle g, h \rangle = \langle g, \sum_{n=0}^{N-1} c_n f_n \rangle = \sum_{n=0}^{N-1} \bar{c}_n \langle g, f_n \rangle = 0.$$

Therefore applying the Cauchy-Schwarz inequality we have

$$\|g\|^2 = \langle g, g \rangle = \langle g, g - h \rangle + \langle g, h \rangle = \langle g, g - h \rangle \leq \|g\| \, \|g - h\| \leq \epsilon \|g\|.$$

It follows that $\|g\| \leq \epsilon$. This is true for every $\epsilon > 0$. Hence $\|g\| = 0$ and therefore $g = 0$.

Problem 30. Consider the set $G = \{+1, -1, +i, -i\}$ and multiplication. This provide us with a finite commutative *group* which is a cyclic subgroup of $U(1)$. Let $g \in G$. Given a Hilbert space $\mathcal{H}$ with scalar product $\langle v, w \rangle$ and the implied norm $\|v\| = \sqrt{\langle v, v \rangle}$ ($v, w \in \mathcal{H}$). Show that

$$\langle v, v \rangle = \frac{1}{4} \sum_{g \in G} g \|v + gw\|^2.$$

Solution 30. We have

$$2\Re(\langle v, w \rangle) = \|v + w\|^2 - \|v\|^2 - \|w\|^2$$

and

$$-2\Re(\langle v, w \rangle) = \|v - w\|^2 - \|v\|^2 - \|w\|^2.$$

Subtracting these two equations provides the identity

$$4\Re(\langle v, w \rangle) = \|v + w\|^2 - \|v - w\|^2.$$

Replacing $w \in \mathcal{H}$ by $iv \in \mathcal{H}$ we obtain

$$4\Im(\langle v, w \rangle) = \|v + iw\|^2 - \|v - iw\|^2.$$

It follows that

$$\langle v, v \rangle = \frac{1}{4} \sum_{g \in G} g \|v + gw\|^2.$$

Problem 31. Given a Hilbert space $\mathcal{H}$ with scalar product $\langle v, w \rangle$ ($v, w \in \mathcal{H}$) and corresponding norm $\|v\| = \sqrt{\langle v, v \rangle}$. Show that

$$|\langle v, w \rangle - \langle v', w' \rangle| \leq \|w\| \cdot \|v - v'\| + \|v\| \cdot \|w - w'\| + \|v - v'\| \cdot \|w - w'\|.$$

Solution 31. We have

$$|\langle v, w \rangle - \langle v', w' \rangle| = |\langle v - v', w \rangle + \langle v', w - w' \rangle|$$
$$\leq \|w\| \cdot \|v - v'\| + \|v'\| \cdot \|w - w'\|$$
$$\leq \|w\| \cdot \|v - v'\| + (\|v\| + \|v - v'\|)\|w - w'\|$$
$$= \|w\| \cdot \|v - v'\| + \|v\| \cdot \|w - w'\| + \|v - v'\| \cdot \|w - w'\|.$$

Problem 32. Consider the Hilbert space $L_2([0, a])$ with $a > 0$. Let $Tf(x) := xf(x)$. Show that $\mathcal{D}(T) = L_2([0, a])$ and $\|T\| = a$.

Solution 32. With $f \in L_2([0, a])$ we have $xf(x) \in L_2([0, a])$. Then

$$\|Tf\|_2 = \left(\int_0^a |Tf(x)|^2 dx \right)^{1/2} = \left(\int_0^a |xf(x)|^2 dx \right)^{1/2}$$
$$\leq \left(\max_{x \in [0,a]} (x) \right) \left(\int_0^a |f(x)|^2 dx \right)^{1/2}$$
$$= a\|f\|_2.$$

Hence $\|T\| \leq a$. By considering the action of T on functions that are supported in a small interval about point $x = a$, where x has its maximum, it can be seen that $\|T\| = a$.

2.3   Supplementary Problems

Problem 1. Let $f, g \in \mathcal{H}$.
(i) Show that $\langle f, g \rangle \langle g, f \rangle \in \mathbb{R}$.
(ii) Find all solutions to the equations $\langle f, g \rangle \langle g, f \rangle = 1$.
(iii) Find all solutions to the equations $\langle f, g \rangle \langle g, f \rangle = i$.
(iv) Show that $\|f\| \cdot \|g\| \geq \langle f, g \rangle \langle g, f \rangle$.
(v) Show that

$$\langle f, f \rangle \langle g, g \rangle \geq \frac{1}{4} (\langle f, g \rangle + \langle g, f \rangle)^2.$$

Problem 2. Let u, v_1, v_2 be elements of a Hilbert space. Show that

$$2\|u - v_1\|^2 + 2\|u - v_2\|^2 = \left\|2\left(u - \frac{v_1 + v_2}{2}\right)\right\|^2 + \|v_1 - v_2\|^2.$$

Problem 3. Consider a Hilbert space $\mathcal{H}$ and $\| \cdot \|$ be the norm implied by the scalar product. Let $\mathbf{u}, \mathbf{v} \in \mathcal{H}$.
(i) Show that $\|\mathbf{u} - \mathbf{v}\| + \|\mathbf{v}\| \geq \|\mathbf{u}\|$.
(ii) Show that $\langle \mathbf{u}, \mathbf{v} \rangle + \langle \mathbf{v}, \mathbf{u} \rangle \leq 2\|\mathbf{u}\| \cdot \|\mathbf{v}\|$.

Problem 4. Given a Hilbert space $\mathcal{H}$ with scalar product $\langle v, w \rangle$ $(v, w \in \mathcal{H})$ and corresponding norm $\|v\| = \sqrt{\langle v, v \rangle}$. Let $\epsilon \in \mathbb{R}$. Show that

$$\|u - (v + \epsilon w)\|^2 = \|u - v\|^2 - 2\epsilon \Re(\langle u - v, w \rangle) + \epsilon^2 \|w\|^2 \geq \|u - v\|^2 - 2\epsilon(\langle u - y, w \rangle).$$

Problem 5. Let $\mathcal{H}$ be a Hilbert space and $|u\rangle, |v\rangle, |w\rangle \in \mathcal{H}$ with

$$\langle u | u \rangle = \langle v | v \rangle = \langle w | w \rangle = 1$$

i.e. the states are normalized. Show that

$$|\langle u | u \rangle \langle v | w \rangle - \langle u | v \rangle \langle u | w \rangle| \leq (\langle u | u \rangle^2 - |\langle u | v \rangle|^2)(\langle u | u \rangle^2 - |\langle u | w \rangle|^2).$$

Problem 6. Consider a Hilbert space $\mathcal{H}$ with scalar product $\langle \cdot, \cdot \rangle$. Let $\mathbf{u}$, $\mathbf{v}$, $\mathbf{w}$ be elements of the Hilbert space with $\|\mathbf{u}\| = \|\mathbf{v}\| = \|\mathbf{w}\| = 1$. Then

$$|\langle \mathbf{u}, \mathbf{v} \rangle| \leq 1, \quad |\langle \mathbf{u}, \mathbf{w} \rangle| \leq 1, \quad |\langle \mathbf{v}, \mathbf{w} \rangle| \leq 1.$$

Show that $\sqrt{1 - |\langle \mathbf{u}, \mathbf{v} \rangle|^2} \le \sqrt{1 - |\langle \mathbf{u}, \mathbf{w} \rangle|^2} + \sqrt{1 - |\langle \mathbf{w}, \mathbf{v} \rangle|^2}$.

Problem 7. Let $\mathcal{H}$ be a Hilbert space and $f \in \mathcal{H}$. Let $\{ e_k : k \in \mathbb{I} \}$ ($\mathbb{I}$ countable index set) be an orthonormal sequence in the Hilbert space $\mathcal{H}$. Then (*Bessel inequality*)

$$\sum_{k \in \mathbb{I}} |\langle f, e_k \rangle| \le \|f\|.$$

Consider the Hilbert space $\mathbb{R}^3$ and

$$f = \frac{1}{\sqrt{3}} \begin{pmatrix} 1 \\ 1 \\ 1 \end{pmatrix}, \quad e_1 = \frac{1}{\sqrt{2}} \begin{pmatrix} 1 \\ 0 \\ 1 \end{pmatrix}, \quad e_2 = \frac{1}{\sqrt{2}} \begin{pmatrix} 1 \\ 0 \\ -1 \end{pmatrix}.$$

Find the left-hand side and right-hand side of the inequality.

Problem 8. Let $\tau = (1 + \sqrt{5})/2$ be the *golden mean number*. Do the column vectors in the 6×6 matrix

$$\begin{pmatrix} \tau & 0 & 1 & -1 & \tau & 0 \\ 1 & \tau & 0 & 0 & -1 & \tau \\ 0 & 1 & \tau & \tau & 0 & -1 \\ 1 & 0 & -\tau & \tau & 1 & 0 \\ -\tau & 1 & 0 & 0 & \tau & 1 \\ 0 & -\tau & 1 & 1 & 0 & \tau \end{pmatrix}$$

form a basis in the Hilbert space $\mathcal{H} = \mathbb{C}^6$? Find the determinant of the matrix.

Problem 9. The projection operators Π and Q of $\mathcal{H}$ onto the sub Hilbert spaces $\mathcal{M}$ and $\mathcal{N}$, respectively, are given by

$$\Pi f(x) := \frac{1}{2}(f(x) + f(-x)), \qquad Q f(x) := \frac{1}{2}(f(x) - f(-x)).$$

Show that for any function f we have $f(x) = \Pi f(x) + Q f(x)$. Notice that $I - \Pi = Q$, where I is the identity operator. An example is

$$g(x) = e^{-k|x|} \cos(kx), \qquad h(x) = e^{-k|x|} \sin(kx), \quad k > 0$$

where g is an odd function and h is an even function.

Problem 10. Consider the sequence

$$f_n(x) = \sin(nx), \quad n = 1, 2, \ldots$$

in the Hilbert space $L_2([0,1])$. Show that the sequence does not tend to a limit in the sense of *strong convergence*. Show that the sequence tends to 0 in the sense of *weak convergence*.

Problem 11. Show that a bounded operator H in a Hilbert space $\mathcal{H}$ is self-adjoint if and only if the bounded operators

$$U(\tau) = \sum_{j=0}^{\infty} \frac{(iH\tau)^j}{j!}$$

are unitary for all $\tau \in \mathbb{R}$.

Problem 12. Let $\mathcal{H}$ be a separable Hilbert space and T be a compact self-adjoint operator $T : \mathcal{H} \to \mathcal{H}$. Show that there exists a complete orthonormal set $\{v_j\}_{j=1}^{\infty}$ in $\mathcal{H}$ and real numbers λ_j (note that T is self-adjoint) which converge to 0 as $j \to \infty$ such that

$$Tv_j = \lambda_j v_j$$

for all $j = 1, 2, \ldots$. Show that each non-zero eigenvalue of T has finite multiplicity. Apply it to the compact self-adjoint operator $T : \ell_2(\mathbb{N}) \to \ell_2(\mathbb{N})$

$$T = \begin{pmatrix} 0 & 1 & 0 & 0 & 0 & \cdots \\ 1 & 0 & 1/2! & 0 & 0 & \cdots \\ 0 & 1/2! & 0 & 1/3! & 0 & \cdots \\ 0 & 0 & 1/3! & 0 & 1/4! & \cdots \\ 0 & 0 & 0 & 1/4! & 0 & \cdots \\ \vdots & \vdots & \vdots & \vdots & & \ddots \end{pmatrix}.$$

Problem 13. Let $\mathcal{H}_1$, $\mathcal{H}_2$ be Hilbert spaces with $\langle \cdot, \cdot \rangle_{\mathcal{H}_1}$ and $\langle \cdot, \cdot \rangle_{\mathcal{H}_2}$ the inner product for the two Hilbert spaces. An *isomorphism* between the two Hilbert spaces is a map $T : \mathcal{H}_1 \to \mathcal{H}_2$ such that (i) T is a linear map; (ii) T is surjective; (iii) for all $f, g \in \mathcal{H}_1$ one has

$$\langle f, g \rangle_{\mathbb{H}_1} = \langle Tf, Tg \rangle_{\mathbb{H}_2}.$$

Show that the Hilbert space $L_2([0,1])$ and $\ell_2(\mathbb{Z})$ are isomorphic. A basis in the Hilbert space $L_2([0,1])$ is given by

$$B = \{ \exp(2\pi i n) \; : \; n \in \mathbb{Z} \}.$$

Problem 14. Let $\mathcal{H}$ be a Hilbert space and I be the identity operator in this Hilbert space. A unitary operator U acting on the tensor product space $\mathcal{H} \otimes \mathcal{H}$ is called *multiplicative* if it satisfies the *pentagon equation*

$$U_{23}U_{12} = U_{12}U_{13}U_{23}$$

where $U_{12} := U \otimes I$, $U_{23} := I \otimes U$, $U_{13} := (I \otimes F)(U \otimes I)(I \otimes F)$. The operators act on the Hilbert space $\mathcal{H} \otimes \mathcal{H} \otimes \mathcal{H}$. The *flip operator* is defined by

$$F(v \otimes w) := w \otimes v, \quad w, v \in \mathcal{H}.$$

Let $\mathcal{H} = \mathbb{R}^2$, I the 2×2 identity matrix and the flip operator is

$$F = \begin{pmatrix} 1 & 0 & 0 & 0 \\ 0 & 0 & 1 & 0 \\ 0 & 1 & 0 & 0 \\ 0 & 0 & 0 & 1 \end{pmatrix}.$$

Find solutions of $V_{23}V_{12} = V_{12}V_{13}V_{23}$.

Problem 15. Consider the Hilbert space $\mathcal{H} = L_2(\mathbb{R})$. Let

$$(Tf)(x) := f(-x), \quad f \in \mathcal{H}.$$

Find the spectrum of T.

Problem 16. Let T be a linear operator in a Hilbert space $\mathcal{H}$ such that $\langle Tf, f \rangle = 0$ for every $f \in \mathcal{H}$. Show that $T = 0$.

Problem 17. Let H be a self-adjoint operator in a Hilbert space $\mathcal{H}$ and B be a bounded operator in $\mathcal{H}$. Show that B^*HB is self-adjoint.

Problem 18. Let T be a bounded linear operator on a Hilbert space $\mathcal{H}$. One defines the operator $|T|$ as

$$|T| := (T^*T)^{1/2}.$$

Show that there exists a linear operator C with $\|C\| \leq 1$ such that

$$|T| = C^*T \quad \Leftrightarrow \quad T = C|T|.$$

Apply it to the 2×2 *nonnormal matrix* (i.e. $TT^* \neq T^*T$)

$$T = \begin{pmatrix} 0 & 2 \\ 0 & 0 \end{pmatrix}.$$

Problem 19. Consider the Hilbert space $L_2([-\pi, \pi])$. Show that the functions

$$\phi_n(x) = \begin{cases} (2\pi)^{-1/2} & \text{for } n = 0 \\ \pi^{-1/2}\cos(nx) & \text{for } n \leq -1 \\ \pi^{-1/2}\sin((n-1/2)x) & \text{for } n \geq 1 \end{cases}$$

with $n \in \mathbb{Z}$ form an orthonormal basis in the Hilbert space $L_2([-\pi, \pi])$.

Problem 20. Let T be a self-adjoint operator. Show that if T admits the inverse T^{-1}, then T^{-1} is also a self-adjoint operator.

Problem 21. Let H be a bounded self-adjoint operator. Then

$$\|H\| = \sup_{\|v\| \leq 1} |\langle Hv, v \rangle| \qquad v \in \mathcal{H}.$$

Show that $|\langle Hv, v \rangle| \leq \|Hv\| \cdot \|v\|$.

Problem 22. Consider the following theorem. Let $\mathcal{H}$ be a Hilbert space and $S \subset \mathcal{H}$. Then for any $v \in \mathcal{H}$ there exists a unique $w \in S$ such that

$$\|v - w\| = d(v, S) = \inf_{u \in S} \|v - u\|.$$

Furthermore if S is a vector subspace of $\mathcal{H}$, then the vector w may be characterized as the unique point in S such that $(v - w) \perp S$. Visualize the theorem for the Hilbert space $\mathbb{R}^2$ and $S = \{(x_1, x_2) : x_1^2 + x_2^2 \leq 1\}$.

Problem 23. Consider the Banach-Gelfand triple $S(\mathbb{R}) \subset L_2(\mathbb{R}) \subset S'(\mathbb{R})$. Functions in the space $S(\mathbb{R})$ are in $L_2(\mathbb{R})$. $S(\mathbb{R})$ is dense in $L_2(\mathbb{R})$. Hence we have an expansion

$$f(x) = \sum_{n=0}^{\infty} c_n \phi_n(x)$$

for any $f \in L_2(\mathbb{R})$, where

$$\phi_n(x) := \pi^{-1/4} 2^{-n/2} (n!)^{-1/2} e^{x^2/2} \frac{d^n}{dx^n} e^{-x^2}, \quad n = 0, 1, 2, \ldots.$$

They form an orthonormal basis in $L_2(\mathbb{R})$. Let $f \in S(\mathbb{R})$ and

$$c_n := \int_{\mathbb{R}} \phi_n(x) f(x) dx.$$

Show that for any $m \in \mathbb{N}$ we have

$$\sum_{n=0}^{\infty} |c_n|^2 (n+1)^m < \infty.$$

Chapter 3

Finite Dimensional Hilbert Spaces

3.1 Introduction

We consider the Hilbert space $\mathbb{C}^n$ with the scalar product ($\mathbf{v}, \mathbf{w} \in \mathbb{C}^n$)

$$\langle \mathbf{v}, \mathbf{w} \rangle := \mathbf{v}^* \mathbf{w} = \sum_{j=1}^{n} \overline{v}_j w_j$$

and the Hilbert space $M(n, \mathbb{C})$ for $n \times n$ matrices over $\mathbb{C}$ with the scalar product

$$\langle A, B \rangle := \operatorname{tr}(AB^*).$$

The *standard basis* in $\mathbb{C}^n$ is given by

$$\begin{pmatrix} 1 \\ 0 \\ 0 \\ \vdots \\ 0 \end{pmatrix}, \begin{pmatrix} 0 \\ 1 \\ 0 \\ \vdots \\ 0 \end{pmatrix}, \dots, \begin{pmatrix} 0 \\ 0 \\ 0 \\ \vdots \\ 1 \end{pmatrix}.$$

In the Hilbert space $\mathbb{C}^2$ an orthonormal basis is given by

$$\begin{pmatrix} e^{i\phi} \cos(\theta) \\ \sin(\theta) \end{pmatrix}, \begin{pmatrix} e^{i\phi} \sin(\theta) \\ -\cos(\theta) \end{pmatrix}.$$

In the Hilbert space $\mathbb{C}^2$ the spin-$\frac{1}{2}$ coherent state is given by

$$\frac{1}{\sqrt{1+|z|^2}} \begin{pmatrix} 1 \\ z \end{pmatrix}$$

with $z \in \mathbb{C}$. In the Hilbert space $\mathbb{C}^3$ an often utilized orthonormal basis is

$$\frac{1}{\sqrt{2}} \begin{pmatrix} 1 \\ 0 \\ 1 \end{pmatrix}, \begin{pmatrix} 0 \\ 1 \\ 0 \end{pmatrix}, \frac{1}{\sqrt{2}} \begin{pmatrix} 1 \\ 0 \\ -1 \end{pmatrix}.$$

57

In the Hilbert space $\mathbb{C}^4$ an important orthonormal basis is given by the *Bell states*

$$\frac{1}{\sqrt{2}}\begin{pmatrix}1\\0\\0\\1\end{pmatrix}, \quad \frac{1}{\sqrt{2}}\begin{pmatrix}0\\1\\1\\0\end{pmatrix}, \quad \frac{1}{\sqrt{2}}\begin{pmatrix}0\\1\\-1\\0\end{pmatrix}, \quad \frac{1}{\sqrt{2}}\begin{pmatrix}1\\0\\0\\-1\end{pmatrix}.$$

Another useful orthonormal basis in $\mathbb{C}^4$ is

$$\frac{1}{2}\begin{pmatrix}-1\\1\\1\\1\end{pmatrix}, \quad \frac{1}{2}\begin{pmatrix}1\\-1\\1\\1\end{pmatrix}, \quad \frac{1}{2}\begin{pmatrix}1\\1\\-1\\1\end{pmatrix}, \quad \frac{1}{2}\begin{pmatrix}1\\1\\1\\-1\end{pmatrix}.$$

The standard basis in the Hilbert space $M(n, \mathbb{C})$ are the *elementary* $n \times n$ *matrices* (E_{jk}) with the matrix (E_{jk}) containing a 1 at the position (entry) (j, k) and 0 otherwise. We have

$$(E_{jk})(E_{\ell m}) = \delta_{k,\ell}(E_{jm})$$

and for the commutator we find

$$[(E_{jk}), (E_{\ell m})] = \delta_{k,\ell}(E_{jm}) - \delta_{m,j}(E_{\ell k}).$$

Thus the standard basis in $M(2, \mathbb{C})$ is given by

$$\begin{pmatrix}1&0\\0&0\end{pmatrix}, \quad \begin{pmatrix}0&1\\0&0\end{pmatrix}, \quad \begin{pmatrix}0&0\\1&0\end{pmatrix}, \quad \begin{pmatrix}0&0\\0&1\end{pmatrix}.$$

For the Hilbert space of 2×2 matrices a basis is given by $\{\sigma_0 = I_2, \sigma_1, \sigma_2, \sigma_3\}$ with the *Pauli spin matrices*

$$\sigma_1 = \begin{pmatrix}0&1\\1&0\end{pmatrix}, \quad \sigma_2 = \begin{pmatrix}0&-i\\i&0\end{pmatrix}, \quad \sigma_3 = \begin{pmatrix}1&0\\0&-1\end{pmatrix}$$

and the 2×2 identity matrix $I_2 = \sigma_0$.

Utilizing the *Kronecker product* we can form bases in higher dimensional vector spaces. For example given the basis in $\mathbb{C}^2$

$$\frac{1}{\sqrt{2}}\begin{pmatrix}1\\1\end{pmatrix}, \quad \frac{1}{\sqrt{2}}\begin{pmatrix}1\\-1\end{pmatrix}$$

one obtains the orthonormal basis in $\mathbb{C}^4$

$$\frac{1}{\sqrt{2}}\begin{pmatrix}1\\1\end{pmatrix} \otimes \frac{1}{\sqrt{2}}\begin{pmatrix}1\\1\end{pmatrix}, \quad \frac{1}{\sqrt{2}}\begin{pmatrix}1\\1\end{pmatrix} \otimes \frac{1}{\sqrt{2}}\begin{pmatrix}1\\-1\end{pmatrix},$$

$$\frac{1}{\sqrt{2}}\begin{pmatrix} 1 \\ -1 \end{pmatrix} \otimes \frac{1}{\sqrt{2}}\begin{pmatrix} 1 \\ 1 \end{pmatrix}, \quad \frac{1}{\sqrt{2}}\begin{pmatrix} 1 \\ -1 \end{pmatrix} \otimes \frac{1}{\sqrt{2}}\begin{pmatrix} 1 \\ -1 \end{pmatrix}.$$

Utilizing the *Kronecker product* we can form bases in higher dimensional vector spaces. Given the Pauli spin matrices σ_0, σ_1, σ_2, σ_3 with $\sigma_0 = I_2$ we can form a basis in $M(4, \mathbb{C})$

$$\sigma_j \otimes \sigma_k, \quad j, k = 0, 1, 2, 3.$$

Gram-Schmidt technique: Given a basis $\{v_1, \ldots, v_n\}$ of a finite dimensional inner product space V one obtains an orthogonal basis $\{w_1, \ldots, w_n\}$ by setting $w_1 = v_1$ and

$$w_m = v_n - \sum_{j=1}^{m-1} \frac{\langle v_m, w_j \rangle}{\|w_j\|^2} w_j, \quad m = 2, 3, \ldots, n.$$

Setting $u_j = w_j / \|w_j\|$ $(j = 1, \ldots, n)$ we obtain an orthonormal basis for V, i.e. $\{u_1, \ldots, u_n\}$.

An $n \times n$ matrix H over $\mathbb{C}$ is called *hermitian* (sometimes also called *self-adjoint*) if $H = H^*$. The eigenvalues of a hermitian matrix are real.

An $n \times n$ matrix U over $\mathbb{C}$ is called *unitary* if

$$U^* = U^{-1}.$$

The eigenvalues of a unitary matrix are of the form $e^{i\phi}$, where $\phi \in \mathbb{R}$.

An $n \times n$ matrix Π is called a *projection matrix* if $\Pi^* = \Pi$, $\Pi^2 = \Pi$. The eigenvalues of a projection matrix are elements of the set $\{0, 1\}$. The $n \times n$ zero matrix and the $n \times n$ identity matrix are projection matrices. If Π is an $n \times n$ projection matrix, then $I_n - \Pi$ is a projection matrix. If $|\psi\rangle$ is a normalized vector in $\mathbb{C}^n$, then $|\psi\rangle\langle\psi|$ is a projection matrix.

A bounded operator K on the Hilbert space $\mathcal{H}$ is *self-adjoint* if and only if the bounded operators

$$U(\tau) := \sum_{n=0}^{\infty} \frac{(iK\tau)^n}{n!}$$

are unitary for all $\tau \in \mathbb{R}$. So it applies to $n \times n$ hermitian matrices acting in the Hilbert space $\mathbb{C}^n$.

Let $M(n, \mathbb{C})$ be the Hilbert space of $n \times n$ matrices with scalar product $\langle A, B \rangle = \mathrm{tr}(AB^*)$. Let $\lambda_1, \lambda_2, \ldots, \lambda_n$ be the eigenvalues of the $n \times n$ matrix A. The following statements are equivalent

(i) A is normal
(ii) A is unitarily diagonalizable
(iii)

$$\sum_{j=1}^{n} \sum_{k=1}^{n} |a_{j,k}|^2 = \sum_{j=1}^{n} |\lambda_j|^2$$

(iv) There is an orthonormal set of n eigenvectors of A.

Let $n \geq 1$. A *tensor* T can be represented as a multidimensional array

$$T = (t_{j_1, j_2, \ldots, j_n})$$

with $j_1 = 1, 2, \ldots, d_1$, $j_2 = 1, 2, \ldots, d_2$, $\ldots$, $j_n = 1, 2, \ldots, d_n$. Let

$$\{e_{j_1}^{(1)} : j_1 = 1, \ldots, d_1\}$$

be the standard basis in the Hilbert space $\mathbb{C}^{d_1}$,

$$\{e_{j_2}^{(2)} : j_2 = 1, \ldots, d_2\}$$

be the standard basis in the Hilbert space $\mathbb{C}^{d_2}$ etc. Then the tensor T can be written as vector $T_\mathbf{v}$ in the Hilbert space $\mathbb{C}^{d_1 \cdot d_2 \cdots d_n}$

$$T_\mathbf{v} = \sum_{j_1=1}^{d_1} \sum_{j_2=1}^{d_2} \cdots \sum_{j_n=1}^{d_n} t_{j_1, j_2, \ldots, j_n} e_{j_1}^{(1)} \otimes e_{j_2}^{(2)} \otimes \cdots \otimes e_{j_n}^{(n)}$$

where $\otimes$ denotes the Kronecker product. The *norm* of the vector $T_\mathbf{v}$ is given by

$$\|T_\mathbf{v}\| = \sqrt{\sum_{j_1=1}^{d_1} \sum_{j_2=1}^{d_2} \cdots \sum_{j_n=1}^{d_n} t_{j_1, j_2, \ldots, j_n} \bar{t}_{j_1, j_2, \ldots, j_n}}.$$

3.2   Solved Problems

Problem 1. Consider the Hilbert space $\mathbb{R}^4$ and the vectors

$$\mathbf{v}_1 = \begin{pmatrix} 1 \\ 0 \\ 0 \\ 0 \end{pmatrix}, \quad \tilde{\mathbf{v}}_2 = \begin{pmatrix} 1 \\ 1 \\ 0 \\ 0 \end{pmatrix}, \quad \mathbf{v}_3 = \begin{pmatrix} 1 \\ 1 \\ 1 \\ 0 \end{pmatrix}, \quad \mathbf{v}_4 = \begin{pmatrix} 1 \\ 1 \\ 1 \\ 1 \end{pmatrix}.$$

(i) Show that the vectors are linearly independent.
(ii) Use the *Gram-Schmidt technique* to find mutually orthogonal vectors.

Solution 1. Setting an arbitrary linear combination equal to the zero vector

$$a \begin{pmatrix} 1 \\ 0 \\ 0 \\ 0 \end{pmatrix} + b \begin{pmatrix} 1 \\ 1 \\ 0 \\ 0 \end{pmatrix} + c \begin{pmatrix} 1 \\ 1 \\ 1 \\ 0 \end{pmatrix} + d \begin{pmatrix} 1 \\ 1 \\ 1 \\ 1 \end{pmatrix} = \begin{pmatrix} 0 \\ 0 \\ 0 \\ 0 \end{pmatrix}$$

leads to the system of four linear equations

$$d = 0, \quad c + d = 0, \quad b + c + d = 0, \quad a + b + c + d = 0.$$

Thus $a = b = c = d = 0$. We find the orthonormal basis $\{\mathbf{w}_1, \mathbf{w}_2, \mathbf{w}_3, \mathbf{w}_4\}$

$$\mathbf{w}_1 = \mathbf{v}_1 = \mathbf{e}_1$$

$$\mathbf{w}_2 = \mathbf{v}_2 - \frac{\mathbf{w}_1^T \mathbf{v}_2}{\mathbf{w}_1^T \mathbf{w}_1} \mathbf{w}_1 = \mathbf{e}_2$$

$$\mathbf{w}_3 = \mathbf{v}_3 - \frac{\mathbf{w}_2^T \mathbf{v}_3}{\mathbf{w}_2^T \mathbf{w}_2} \mathbf{w}_2 - \frac{\mathbf{w}_1^T \mathbf{v}_3}{\mathbf{w}_1^T \mathbf{w}_1} \mathbf{w}_1 = \mathbf{e}_3$$

$$\mathbf{w}_4 = \mathbf{v}_4 - \frac{\mathbf{w}_3^T \mathbf{v}_4}{\mathbf{w}_3^T \mathbf{w}_3} \mathbf{w}_3 - \frac{\mathbf{w}_2^T \mathbf{v}_4}{\mathbf{w}_2^T \mathbf{w}_2} \mathbf{w}_2 - \frac{\mathbf{w}_1^T \mathbf{v}_4}{\mathbf{w}_1^T \mathbf{w}_1} \mathbf{w}_1 = \mathbf{e}_4$$

where $\mathbf{e}_1, \mathbf{e}_2, \mathbf{e}_3, \mathbf{e}_4$ is the standard basis.

Problem 2. Consider the Hilbert space $\mathbb{R}^4$. Show that the vectors

$$\mathbf{v}_1 = \frac{1}{\sqrt{2}} \begin{pmatrix} 1 \\ 0 \\ 0 \\ 1 \end{pmatrix}, \quad \mathbf{v}_2 = \frac{1}{\sqrt{2}} \begin{pmatrix} 1 \\ 0 \\ 0 \\ -1 \end{pmatrix}, \quad \mathbf{v}_3 = \frac{1}{\sqrt{2}} \begin{pmatrix} 0 \\ 1 \\ 1 \\ 0 \end{pmatrix}, \quad \mathbf{v}_4 = \frac{1}{\sqrt{2}} \begin{pmatrix} 0 \\ 1 \\ -1 \\ 0 \end{pmatrix}$$

are linearly independent. Show that the vectors form an orthonormal basis (*Bell basis*) in this Hilbert space.

Solution 2. We have to show that

$$
c_1 \frac{1}{\sqrt{2}} \begin{pmatrix} 1 \\ 0 \\ 0 \\ 1 \end{pmatrix} + c_2 \frac{1}{\sqrt{2}} \begin{pmatrix} 1 \\ 0 \\ 0 \\ -1 \end{pmatrix} + c_3 \frac{1}{\sqrt{2}} \begin{pmatrix} 0 \\ 1 \\ 1 \\ 0 \end{pmatrix} + c_4 \frac{1}{\sqrt{2}} \begin{pmatrix} 0 \\ 1 \\ -1 \\ 0 \end{pmatrix} = \begin{pmatrix} 0 \\ 0 \\ 0 \\ 0 \end{pmatrix}
$$

implies that $c_1 = c_2 = c_3 = c_4 = 0$. We obtain the four linear equations

$$
c_1 + c_2 = 0, \quad c_1 - c_2 = 0, \quad c_3 + c_4 = 0, \quad c_3 - c_4 = 0.
$$

The only solution is $c_1 = c_2 = c_3 = c_4 = 0$. We could also argue as follows: Since the four vectors are linearly independent, normalized and the scalar product between each pair is 0 we have an orthonormal basis. The vectors are normalized, i.e. $\|\mathbf{v}_1\| = \|\mathbf{v}_2\| = \|\mathbf{v}_3\| = \|\mathbf{v}_4\| = 1$. The vectors are pairwise orthogonal to each other, i.e.

$$
\mathbf{v}_1^* \mathbf{v}_2 = 0, \quad \mathbf{v}_1^* \mathbf{v}_3 = 0, \quad \mathbf{v}_1^* \mathbf{v}_4 = 0, \quad \mathbf{v}_2^* \mathbf{v}_3 = 0, \quad \mathbf{v}_2^* \mathbf{v}_4 = 0, \quad \mathbf{v}_3^* \mathbf{v}_4 = 0.
$$

Thus since $\dim(\mathbb{R}^4) = 4$ we have an orthonormal basis.

Problem 3. Let $|0\rangle$, $|1\rangle$ be an orthonormal basis in the Hilbert space $\mathbb{C}^2$. Consider the 2×2 matrices $A = |0\rangle\langle 1|$, $B = |1\rangle\langle 0|$. Find the *commutator* $[A, B]$ and the *anticommutator* $[A, B]_+$.

Solution 3. With $\langle 0|0\rangle = 1$, $\langle 1|1\rangle = 1$, $\langle 0|1\rangle = 0$, $\langle 1|0\rangle = 0$ we have

$$
[A, B] = |0\rangle\langle 1|1\rangle\langle 0| - |1\rangle\langle 0|0\rangle\langle 1| = |0\rangle\langle 0| - |1\rangle\langle 1|
$$

and

$$
[A, B]_+ = |0\rangle\langle 1|1\rangle\langle 0| + |1\rangle\langle 0|0\rangle\langle 1| = |0\rangle\langle 0| + |1\rangle\langle 1| = I_2
$$

where I_2 is the 2×2 identity matrix.

Problem 4. Consider the Hilbert space $\mathbb{R}^4$. Find all pairwise orthogonal vectors (column vectors) $\mathbf{x}_1, \dots, \mathbf{x}_p$, where the entries of the column vectors can only be $+1$ or -1. Calculate the matrix

$$
\sum_{i=1}^{p} \mathbf{x}_i \mathbf{x}_i^T
$$

and find the eigenvalues and eigenvectors of this matrix.

Solution 4. The integer p cannot exceed 4 since that would imply $\dim(\mathbb{R}^4) > 4$. We have

$$\mathbf{x}_1 = \begin{pmatrix} 1 \\ 1 \\ 1 \\ 1 \end{pmatrix}, \quad \mathbf{x}_2 = \begin{pmatrix} 1 \\ -1 \\ 1 \\ -1 \end{pmatrix}, \quad \mathbf{x}_3 = \begin{pmatrix} 1 \\ -1 \\ -1 \\ 1 \end{pmatrix}, \quad \mathbf{x}_4 = \begin{pmatrix} 1 \\ 1 \\ -1 \\ -1 \end{pmatrix}.$$

Thus

$$\sum_{i=1}^{4} \mathbf{x}_i \mathbf{x}_i^T = 4 I_4.$$

The eigenvalue is 4 with multiplicity 4. The eigenvectors are all $\mathbf{x} \in \mathbb{R}^4$.

Problem 5. (i) Consider the Hilbert space $\mathbb{R}^n$. Let $\mathbf{x}, \mathbf{y} \in \mathbb{R}^n$. Show that

$$\|\mathbf{x} + \mathbf{y}\|^2 + \|\mathbf{x} - \mathbf{y}\|^2 \equiv 2(\|\mathbf{x}\|^2 + \|\mathbf{y}\|^2).$$

Note that $\|\mathbf{x}\|^2 := \langle \mathbf{x}, \mathbf{x} \rangle$.
(ii) Consider the Hilbert space $\mathbb{R}^n$ and $u, v \in \mathbb{R}^n$. Show that

$$\langle u, v \rangle = \frac{1}{2}(\|u + v\|^2 - \|u\|^2 - \|v\|^2).$$

Solution 5. (i) We have

$$\begin{aligned}
\langle \mathbf{x} + \mathbf{y}, \mathbf{x} + \mathbf{y} \rangle + \langle \mathbf{x} - \mathbf{y}, \mathbf{x} - \mathbf{y} \rangle &\equiv \langle \mathbf{x}, \mathbf{x} \rangle + \langle \mathbf{y}, \mathbf{y} \rangle + \langle \mathbf{x}, \mathbf{y} \rangle + \langle \mathbf{y}, \mathbf{x} \rangle \\
&\quad + \langle \mathbf{x}, \mathbf{x} \rangle + \langle \mathbf{y}, \mathbf{y} \rangle - \langle \mathbf{x}, \mathbf{y} \rangle - \langle \mathbf{y}, \mathbf{x} \rangle \\
&\equiv 2\langle \mathbf{x}, \mathbf{x} \rangle + 2\langle \mathbf{y}, \mathbf{y} \rangle.
\end{aligned}$$

(ii) We have

$$\begin{aligned}
\|u + v\|^2 - \|u\|^2 - \|v\|^2 &= \langle u + v, u + v \rangle - \langle u, u \rangle - \langle v, v \rangle \\
&= \langle u, u \rangle + \langle u, v \rangle + \langle v, u \rangle + \langle v, v \rangle - \langle u, u \rangle - \langle v, v \rangle \\
&= \langle u, v \rangle + \langle v, u \rangle = 2\langle u, v \rangle.
\end{aligned}$$

Problem 6. (i) Consider the two orthonormal bases in the Hilbert space $\mathbb{C}^2$

$$|v_0\rangle = \begin{pmatrix} 1 \\ 0 \end{pmatrix}, \quad |v_1\rangle = \begin{pmatrix} 0 \\ 1 \end{pmatrix}, \qquad |w_0\rangle = \frac{1}{\sqrt{2}}\begin{pmatrix} 1 \\ 1 \end{pmatrix}, \quad |w_1\rangle = \frac{1}{\sqrt{2}}\begin{pmatrix} 1 \\ -1 \end{pmatrix}.$$

Show that $V = |w_0\rangle\langle v_0| + |w_1\rangle\langle v_1|$ is a 2×2 unitary matrix.
(ii) Consider the Hilbert space $\mathbb{C}^2$ and let $|0\rangle$, $|1\rangle$ be an orthonormal basis in $\mathbb{C}^2$. Let $\phi \in \mathbb{C}$. Is $V = |0\rangle\langle 1| + e^{i\phi}|1\rangle\langle 0|$ a unitary matrix?

Solution 6. (i) We obtain

$$V = \frac{1}{\sqrt{2}}\begin{pmatrix} 1 \\ 1 \end{pmatrix}(1\ 0) + \frac{1}{\sqrt{2}}\begin{pmatrix} 1 \\ -1 \end{pmatrix}(0\ 1) = \frac{1}{\sqrt{2}}\begin{pmatrix} 1 & 1 \\ 1 & -1 \end{pmatrix}.$$

Since $V^* = V^{-1}$ we have a unitary matrix. Furthermore $V = V^*$.
(ii) With $V^* = |1\rangle\langle 0| + e^{-i\phi}|0\rangle\langle 1|$ and $\langle 0|0\rangle = \langle 1|1\rangle = 1$, $\langle 1|0\rangle = \langle 0|1\rangle = 0$ we have

$$VV^* = |0\rangle\langle 0| + |1\rangle\langle 1| = I_2$$

where I_2 is the 2×2 unit matrix. Hence V is unitary.

Problem 7. (i) Let $d \geq 2$. Consider the Hilbert space $\mathbb{C}^d$ and let

$$|v_0\rangle, \ |v_1\rangle, \ \ldots, \ |v_{d-1}\rangle, \qquad |w_0\rangle, \ |w_1\rangle, \ \ldots, \ |w_{d-1}\rangle$$

be two orthonormal bases in $\mathbb{C}^d$. Show that

$$U = \sum_{j=0}^{d-1} |w_j\rangle\langle v_j|$$

is a *unitary matrix*.
(ii) Consider the Hilbert space $\mathbb{C}^n$. Let $\mathbf{u}_j$, $j = 1, 2, \ldots, n$, and $\mathbf{v}_j$, $j = 1, 2, \ldots, n$ be orthonormal bases in $\mathbb{C}^n$, where $\mathbf{u}_j$, $\mathbf{v}_j$ are considered as column vectors. Show that

$$U = \sum_{j=1}^{n} \mathbf{u}_j \mathbf{v}_j^*$$

is a unitary $n \times n$ matrix.

Solution 7. (i) From U we obtain

$$U^* = \sum_{k=0}^{d-1} |v_k\rangle\langle w_k|.$$

Then with $\langle v_j|v_k\rangle = \delta_{j,k}$ it follows that

$$UU^* = \sum_{j=0}^{d-1}\sum_{k=0}^{d-1} |w_j\rangle\langle v_j|v_k\rangle\langle w_k| = \sum_{j=0}^{d-1}\sum_{k=0}^{d-1} |w_j\rangle\delta_{j,k}\langle w_k| = \sum_{j=0}^{d-1} |w_j\rangle\langle w_j| = I_d.$$

(ii) We have

$$U^*U = \sum_{k=1}^{n}(\mathbf{v}_k\mathbf{u}_k^*)\sum_{j=1}^{n}(\mathbf{u}_j\mathbf{v}_j^*) = \sum_{k=1}^{n}\sum_{j=1}^{n}\mathbf{v}_k(\mathbf{u}_k^*\mathbf{u}_j)\mathbf{v}_j^*$$
$$= \sum_{k=1}^{n}\sum_{j=1}^{n}\mathbf{v}_k\delta_{k,j}\mathbf{v}_j^* = \sum_{j=1}^{n}\mathbf{v}_j\mathbf{v}_j^* = I_n.$$

Problem 8. (i) Let $d \geq 2$. Consider the Hilbert space $\mathbb{C}^d$ and the two orthonormal bases

$$|v_0\rangle, \ |v_1\rangle, \ \ldots, \ |v_{d-1}\rangle, \qquad |w_0\rangle, \ |w_1\rangle, \ \ldots, \ |w_{d-1}\rangle.$$

Let $\lambda_j \in \mathbb{C}$ $(j = 0, 1, \ldots, d-1)$. Then

$$V = \sum_{j=0}^{d-1}\lambda_j|v_j\rangle\langle v_j|, \qquad W = \sum_{k=0}^{d-1}\lambda_k|w_k\rangle\langle w_k|$$

are $d \times d$ matrices (*spectral representation*). Consider the $d \times d$ unitary matrix

$$U = \sum_{\ell=0}^{d-1}|w_\ell\rangle\langle v_\ell|.$$

Show that $UVU^* = W$.

(ii) Let A, B be two normal $d \times d$ matrices with the spectral representations

$$A = \sum_{j=0}^{d-1}\lambda_j|a_j\rangle\langle a_j|, \qquad B = \sum_{k=0}^{d-1}\lambda_k|b_k\rangle\langle b_k|$$

i.e. we assume that A and B have the same eigenvalues (counting multiplicity) and both $\{|a_0\rangle, \ldots, |a_{d-1}\rangle\}$, $\{|b_0\rangle, \ldots, |b_{d-1}\rangle\}$ form orthonormal bases in the Hilbert space $\mathbb{C}^d$. Then

$$U = \sum_{\ell=0}^{d-1}|b_\ell\rangle\langle a_\ell| \ \Rightarrow \ U^* = \sum_{m=0}^{d-1}|a_m\rangle\langle b_m|$$

is a unitary matrix. Show that $UAU^* = B$.

Solution 8. (i) We have

$$
UVU^* = \left(\sum_{\ell=0}^{d-1} |w_\ell\rangle\langle v_\ell| \right) \left(\sum_{j=0}^{d-1} \lambda_j |v_j\rangle\langle v_j| \right) \left(\sum_{m=0}^{d-1} |v_m\rangle\langle w_m| \right)
$$

$$
= \sum_{\ell=0}^{d-1}\sum_{j=0}^{d-1}\sum_{m=0}^{d-1} \lambda_j |w_\ell\rangle\langle v_\ell|v_j\rangle\langle v_j|v_m\rangle\langle w_m|
$$

$$
= \sum_{\ell=0}^{d-1}\sum_{j=0}^{d-1}\sum_{m=0}^{d-1} \lambda_j |w_\ell\rangle\delta_{\ell,j}\delta_{j,m}\langle w_m|
$$

$$
= \sum_{j=0}^{d-1} \lambda_j |w_j\rangle\langle w_j| = W.
$$

(ii) We have

$$
UAU^* = \sum_{\ell=0}^{d-1}\sum_{j=0}^{d-1}\sum_{m=0}^{d-1} \lambda_j |b_\ell\rangle\langle a_\ell|a_j\rangle\langle a_j|a_m\rangle\langle b_m|
$$

$$
= \sum_{\ell=0}^{d-1}\sum_{j=0}^{d-1}\sum_{m=0}^{d-1} \lambda_j |b_\ell\rangle\delta_{\ell,j}\delta_{j,m}\langle b_m| = \sum_{j=0}^{d-1} \lambda_j |b_j\rangle\langle b_j| = B.
$$

Problem 9. (i) Let $|0\rangle$, $|1\rangle$ be an orthonormal basis in the Hilbert space $\mathbb{C}^2$ and

$$
|\psi\rangle = \cos(\theta/2)|0\rangle + e^{i\phi}\sin(\theta/2)|1\rangle
$$

where $\theta, \phi \in \mathbb{R}$. Find $\langle\psi|\psi\rangle$. Find the *probability* $|\langle 0|\psi\rangle|^2$. Discuss $|\langle 0|\psi\rangle|^2$ as a function of θ. Assume that (standard basis)

$$
|0\rangle = \begin{pmatrix} 1 \\ 0 \end{pmatrix}, \qquad |1\rangle = \begin{pmatrix} 0 \\ 1 \end{pmatrix}.
$$

Find the 2×2 matrix $|\psi\rangle\langle\psi|$ and calculate the eigenvalues.
(ii) Consider the Hilbert space $\mathbb{C}^2$ and the vectors

$$
|0\rangle = \begin{pmatrix} i \\ i \end{pmatrix}, \qquad |1\rangle = \begin{pmatrix} 1 \\ -1 \end{pmatrix}.
$$

Normalize these vectors and then calculate the *probability* $|\langle 0|1\rangle|^2$.

Solution 9. (i) With $\langle 0|0\rangle = \langle 1|1\rangle = 1$ and $\langle 0|1\rangle = \langle 1|0\rangle = 0$ we obtain

$$
\langle\psi|\psi\rangle = \cos^2(\theta/2) + \sin^2(\theta/2) = 1.
$$

Since $\langle 0|1\rangle = 0$ and $\langle 0|0\rangle = 1$ we obtain $|\langle 0|\psi\rangle|^2 = \cos^2(\theta/2)$. Hence $|\langle 0|\psi\rangle|^2$ depends on θ. For $\theta = 0$ we have $|\langle 0|\psi\rangle|^2 = 1$. We obtain

$$|\psi\rangle\langle\psi| = \begin{pmatrix} 1 & 0 \\ 0 & 0 \end{pmatrix}.$$

Now $|\psi\rangle\langle\psi|$ is a *projection matrix*, i.e. $|\psi\rangle\langle\psi| = (|\psi\rangle\langle\psi|)^2 = |\psi\rangle\langle\psi|$ and $|\psi\rangle\langle\psi|$ is hermitian. The eigenvalues are $+1$ and 0.

(ii) The normalized vectors are

$$|0\rangle = \frac{1}{\sqrt{2}}\begin{pmatrix} i \\ i \end{pmatrix}, \quad |1\rangle = \frac{1}{\sqrt{2}}\begin{pmatrix} 1 \\ -1 \end{pmatrix}.$$

With $\langle 0| = \frac{1}{\sqrt{2}}\begin{pmatrix} -i & -i \end{pmatrix}$ we obtain the probability $|\langle 0|1\rangle|^2 = 0$.

Problem 10. Consider the Hilbert space $\mathbb{R}^2$. Given the vectors

$$\mathbf{u}_1 = \begin{pmatrix} 0 \\ 1 \end{pmatrix}, \quad \mathbf{u}_2 = \begin{pmatrix} \sqrt{3}/2 \\ -1/2 \end{pmatrix}, \quad \mathbf{u}_3 = \begin{pmatrix} -\sqrt{3}/2 \\ -1/2 \end{pmatrix}.$$

The three vectors $\mathbf{u}_1$, $\mathbf{u}_2$, $\mathbf{u}_3$ are at 120 degrees of each other and are normalized, i.e. $\|\mathbf{u}_j\| = 1$ for $j = 1, 2, 3$. Every given two-dimensional vector $\mathbf{v}$ can be written as $\mathbf{v} = c_1\mathbf{u}_1 + c_2\mathbf{u}_2 + c_3\mathbf{u}_3$, $c_1, c_2, c_3 \in \mathbb{R}$ in many different ways. Given the vector $\mathbf{v}$ minimize

$$\frac{1}{2}(c_1^2 + c_2^2 + c_3^2)$$

subject to the constraint $\mathbf{v} - c_1\mathbf{u}_1 - c_2\mathbf{u}_2 - c_3\mathbf{u}_3 = \mathbf{0}$.

Solution 10. The *Lagrange function* is

$$L(c_1, c_2, c_3, \lambda_1, \lambda_2) = \frac{1}{2}(c_1^2 + c_2^2 + c_3^2) + \lambda_1(v_1 - c_1 u_{1,1} - c_2 u_{2,1} - c_3 u_{3,1})$$
$$+ \lambda_2(v_2 - c_1 u_{1,2} - c_2 u_{2,2} - c_3 u_{3,2}).$$

Thus we have

$$\frac{\partial L}{\partial c_1} = 0 \Rightarrow c_1 - \lambda_1 u_{1,1} - \lambda_2 u_{1,2} = 0$$

$$\frac{\partial L}{\partial c_2} = 0 \Rightarrow c_2 - \lambda_1 u_{2,1} - \lambda_2 u_{2,2} = 0$$

$$\frac{\partial L}{\partial c_3} = 0 \Rightarrow c_3 - \lambda_1 u_{3,1} - \lambda_2 u_{3,2} = 0$$

together with the constraints

$$v_1 - c_1 u_{1,1} - c_2 u_{2,1} - c_3 u_{3,1} = 0, \quad v_2 - c_1 u_{1,2} - c_2 u_{2,2} - c_3 u_{3,2} = 0.$$

Eliminating c_1, c_2, c_3 from these equations using the first three equations yields

$$v_1 = \lambda_1(u_{1,1}^2 + u_{2,1}^2 + u_{3,1}^2) + \lambda_2(u_{1,1}u_{1,2} + u_{2,1}u_{2,2} + u_{3,1}u_{3,2})$$
$$v_2 = \lambda_1(u_{1,1}u_{1,2} + u_{2,1}u_{2,2} + u_{3,1}u_{3,2}) + \lambda_2(u_{1,2}^2 + u_{2,2}^2 + u_{3,2}^2).$$

Inserting these values of the vectors $\mathbf{u}_1$, $\mathbf{u}_2$, $\mathbf{u}_3$ provides

$$\lambda_1 = \frac{2}{3}v_1, \qquad \lambda_2 = \frac{2}{3}v_2.$$

Thus

$$c_1 = \frac{2}{3}v_2, \qquad c_2 = \frac{1}{\sqrt{3}}v_1 - \frac{1}{3}v_2, \qquad c_3 = -\frac{1}{\sqrt{3}}v_1 - \frac{1}{3}v_2$$

and

$$c_1^2 + c_2^2 + c_3^2 = \frac{2}{3}v_1^2 + \frac{2}{3}v_2^2.$$

Problem 11. Consider the Hilbert space $M(2, \mathbb{C})$ of all 2×2 matrices over $\mathbb{C}$ with the scalar product $\langle A, B \rangle := \text{tr}(AB^*)$, where $A, B \in M(2, \mathbb{C})$. Do the four 2×2 matrices

$$B_1 = \begin{pmatrix} 0 & 1 \\ 1 & 0 \end{pmatrix}, \quad B_2 = \begin{pmatrix} 0 & -i \\ i & 0 \end{pmatrix}, \quad B_3 = \begin{pmatrix} 1 & 0 \\ 0 & -1 \end{pmatrix}, \quad B_4 = \begin{pmatrix} -i & 0 \\ 0 & -i \end{pmatrix}$$

form a basis in this Hilbert space. Calculate the scalar products $\langle B_j, B_k \rangle = \text{tr}(B_j B_k^*)$.

Solution 11. For the scalar products we have

$$\langle B_1, B_1 \rangle = 2, \quad \langle B_1, B_2 \rangle = 0, \quad \langle B_1, B_3 \rangle = 0, \quad \langle B_1, B_4 \rangle = 0,$$

$$\langle B_2, B_2 \rangle = 2, \quad \langle B_2, B_3 \rangle = 0, \quad \langle B_2, B_4 \rangle = 0,$$

$$\langle B_3, B_3 \rangle = 2, \quad \langle B_3, B_4 \rangle = 0, \quad \langle B_4, B_4 \rangle = 2.$$

The matrices B_1, B_2, B_3, B_4 are nonzero. Thus this result indicates we have a basis.

Problem 12. Consider the Hilbert space $\mathcal{H} = M(2, \mathbb{C})$ of the 2×2 matrices over the complex numbers with the scalar product

$$\langle A, B \rangle := \text{tr}(AB^*), \qquad A, B \in \mathcal{H}.$$

Show that the rescaled *Pauli spin matrices* $\mu_j := \frac{1}{\sqrt{2}}\sigma_j$, $j = 1, 2, 3$

$$\mu_1 = \frac{1}{\sqrt{2}}\begin{pmatrix} 0 & 1 \\ 1 & 0 \end{pmatrix}, \quad \mu_2 = \frac{1}{\sqrt{2}}\begin{pmatrix} 0 & -i \\ i & 0 \end{pmatrix}, \quad \mu_3 = \frac{1}{\sqrt{2}}\begin{pmatrix} 1 & 0 \\ 0 & -1 \end{pmatrix}$$

plus the rescaled 2×2 identity matrix

$$\mu_0 = \frac{1}{\sqrt{2}}\begin{pmatrix} 1 & 0 \\ 0 & 1 \end{pmatrix}$$

form an orthonormal basis in the Hilbert space $\mathcal{H}$.

Solution 12. Let $c_0, c_1, c_2, c_3 \in \mathbb{C}$. Solving

$$c_0\mu_0 + c_1\mu_1 + c_2\mu_2 + c_3\mu_3 = \begin{pmatrix} 0 & 0 \\ 0 & 0 \end{pmatrix}$$

yields $c_0 = c_1 = c_2 = c_3 = 0$ as the only solution, i.e. the matrices are linearly independent. Since

$$\langle \mu_j, \mu_k \rangle = \begin{cases} 1 \text{ for } j = k \\ 0 \text{ for } j \neq k \end{cases}$$

and μ_j $(j = 0, 1, 2, 3)$ are nonzero matrices we have an orthonormal basis.

Problem 13. Consider the Hilbert space $\mathbb{R}^4$. Let A be a symmetric 4×4 matrix over $\mathbb{R}$. Assume that the eigenvalues are given by $\lambda_1 = 0$, $\lambda_2 = 1$, $\lambda_3 = 2$ and $\lambda_4 = 3$ with the corresponding normalized eigenfunctions

$$\mathbf{v}_1 = \frac{1}{\sqrt{2}}\begin{pmatrix} 1 \\ 0 \\ 0 \\ 1 \end{pmatrix}, \quad \mathbf{v}_2 = \frac{1}{\sqrt{2}}\begin{pmatrix} 1 \\ 0 \\ 0 \\ -1 \end{pmatrix}, \quad \mathbf{v}_3 = \frac{1}{\sqrt{2}}\begin{pmatrix} 0 \\ 1 \\ 1 \\ 0 \end{pmatrix}, \quad \mathbf{v}_4 = \frac{1}{\sqrt{2}}\begin{pmatrix} 0 \\ 1 \\ -1 \\ 0 \end{pmatrix}.$$

Find the matrix A by means of the *spectral theorem*.

Solution 13. We obtain the hermitian matrix

$$A = \sum_{j=1}^{4} \lambda_j \mathbf{v}_j \mathbf{v}_j^* = \begin{pmatrix} 1/2 & 0 & 0 & -1/2 \\ 0 & 5/2 & -1/2 & 0 \\ 0 & -1/2 & 5/2 & 0 \\ -1/2 & 0 & 0 & 1/2 \end{pmatrix}$$

with $\text{tr}(A) = 6 = \lambda_1 + \lambda_2 + \lambda_3 + \lambda_4$ and $\det(A) = 0$ since $\lambda_1 = 0$.

Problem 14. Let $d \geq 2$ and $|0\rangle, |1\rangle, \ldots, |d-1\rangle$ be an orthonormal basis in the Hilbert space $\mathbb{C}^d$. Assume that $\lambda_j \in \mathbb{C}$ $(j = 0, 1, \ldots, d-1)$. Is the $d \times d$ matrix

$$T = \sum_{j=0}^{d-1} \lambda_j |j\rangle\langle j|$$

normal?

Solution 14. The answer is yes. We have

$$T^* = \sum_{j=0}^{d-1} \overline{\lambda}_j |j\rangle\langle j|$$

and with $\langle j|k\rangle = \delta_{j,k}$ we obtain

$$TT^* = \sum_{j=0}^{d-1} \overline{\lambda}_j \lambda_j |j\rangle\langle j| = T^*T.$$

Hence the matrix is normal.

Problem 15. Let A, B be two $n \times n$ matrices over $\mathbb{C}$. We introduce the scalar product

$$\langle A, B \rangle := \frac{\mathrm{tr}(AB^*)}{\mathrm{tr}(I_n)} \equiv \frac{1}{n}\mathrm{tr}(AB^*).$$

This provides us with a Hilbert space.
The *Lie group* $SU(n)$ is defined by the complex $n \times n$ matrices U

$$SU(n) := \{ U : U^*U = UU^* = I_n , \det(U) = 1 \}.$$

The dimension is $n^2 - 1$. The Lie algebra $su(n)$ is defined by the $n \times n$ matrices X

$$su(n) := \{ X : X^* = -X , \mathrm{tr}(X) = 0 \}.$$

(i) Let $U \in SU(n)$. Calculate the scalar product $\langle U, U \rangle$.
(ii) Let A be an arbitrary complex $n \times n$ matrix. Let $U \in SU(n)$. Calculate $\langle UA, UA \rangle$.
(iii) Consider the Lie algebra $su(2)$. Provide a basis. The elements of the basis should be orthogonal to each other with respect to the scalar product given above. Calculate the commutators of these matrices.

Solution 15. (i) We have

$$\langle U, U \rangle = \frac{\text{tr}(UU^*)}{\text{tr}(I_n)} = \frac{\text{tr}(I_n)}{\text{tr}(I_n)} = 1.$$

(ii) Since $U^*U = I_n$ we find $\langle UA, UA \rangle = \langle A, U^*UA \rangle = \langle A, A \rangle$.

(iii) The dimension of the Lie algebra is 3. We find

$$X_1 = \begin{pmatrix} i & 0 \\ 0 & -i \end{pmatrix}, \quad X_1 = -X_1^*,$$

$$X_2 = \begin{pmatrix} 0 & i \\ i & 0 \end{pmatrix}, \quad X_2 = -X_2^*,$$

$$X_3 = \begin{pmatrix} 0 & 1 \\ -1 & 0 \end{pmatrix}, \quad X_3 = -X_3^*.$$

Problem 16. Consider the Hilbert space $M(4, \mathbb{C})$ of all 4×4 matrices over $\mathbb{C}$ with the scalar product $\langle A, B \rangle := \text{tr}(AB^*)$, where $A, B \in M(4, \mathbb{C})$. The γ-matrices are given by

$$\gamma_1 = \begin{pmatrix} 0 & 0 & 0 & -i \\ 0 & 0 & -i & 0 \\ 0 & i & 0 & 0 \\ i & 0 & 0 & 0 \end{pmatrix}, \quad \gamma_2 = \begin{pmatrix} 0 & 0 & 0 & -1 \\ 0 & 0 & 1 & 0 \\ 0 & 1 & 0 & 0 \\ -1 & 0 & 0 & 0 \end{pmatrix},$$

$$\gamma_3 = \begin{pmatrix} 0 & 0 & -i & 0 \\ 0 & 0 & 0 & i \\ i & 0 & 0 & 0 \\ 0 & -i & 0 & 0 \end{pmatrix}, \quad \gamma_4 = \begin{pmatrix} 1 & 0 & 0 & 0 \\ 0 & 1 & 0 & 0 \\ 0 & 0 & -1 & 0 \\ 0 & 0 & 0 & -1 \end{pmatrix}$$

and

$$\gamma_5 = \gamma_1 \gamma_2 \gamma_3 \gamma_4 = \begin{pmatrix} 0 & 0 & -1 & 0 \\ 0 & 0 & 0 & -1 \\ -1 & 0 & 0 & 0 \\ 0 & -1 & 0 & 0 \end{pmatrix}.$$

We define the six 4×4 matrices

$$\sigma_{jk} := \frac{i}{2}[\gamma_j, \gamma_k], \quad j < k$$

where $j = 1, 2, 3$, $k = 2, 3, 4$ and $[\cdot, \cdot]$ denotes the commutator.

(i) Calculate σ_{12}, σ_{13}, σ_{14}, σ_{23}, σ_{24}, σ_{34}.

(ii) Do the 16 matrices

$$I_4, \ \gamma_1, \ \gamma_2, \ \gamma_3, \ \gamma_4, \ \gamma_5, \ \gamma_5\gamma_1, \ \gamma_5\gamma_2, \ \gamma_5\gamma_3, \ \gamma_5\gamma_4, \ \sigma_{12}, \ \sigma_{13}, \ \sigma_{14}, \ \sigma_{23}, \ \sigma_{24}, \ \sigma_{34}$$

form a basis in the Hilbert space $M(4, \mathbb{C})$? If so is the basis orthogonal?

Solution 16. (i) We have

$$\sigma_{12} = \frac{i}{2}[\gamma_1, \gamma_2] = \text{diag}(-1, +1, -1, +1)$$

$$\sigma_{13} = \frac{i}{2}[\gamma_1, \gamma_3] = \begin{pmatrix} 0 & -i & 0 & 0 \\ i & 0 & 0 & 0 \\ 0 & 0 & 0 & -i \\ 0 & 0 & i & 0 \end{pmatrix}$$

$$\sigma_{14} = \frac{i}{2}[\gamma_1, \gamma_4] = \begin{pmatrix} 0 & 0 & 0 & -1 \\ 0 & 0 & -1 & 0 \\ 0 & -1 & 0 & 0 \\ -1 & 0 & 0 & 0 \end{pmatrix}$$

$$\sigma_{23} = \frac{i}{2}[\gamma_2, \gamma_3] = \begin{pmatrix} 0 & -1 & 0 & 0 \\ -1 & 0 & 0 & 0 \\ 0 & 0 & 0 & -1 \\ 0 & 0 & -1 & 0 \end{pmatrix}$$

$$\sigma_{24} = \frac{i}{2}[\gamma_2, \gamma_4] = \begin{pmatrix} 0 & 0 & 0 & i \\ i & 0 & -i & 0 \\ 0 & i & 0 & 0 \\ -i & 0 & 0 & 0 \end{pmatrix}$$

$$\sigma_{34} = \frac{i}{2}[\gamma_3, \gamma_4] = \begin{pmatrix} 0 & 0 & -1 & 0 \\ i & 0 & 0 & 1 \\ -1 & 0 & 0 & 0 \\ 0 & 1 & 0 & 0 \end{pmatrix}.$$

(ii) The 16 matrices are linearly independent since from the equation

$$c_0 I_4 + \sum_{j=1}^{5} c_j \gamma_j + \sum_{j=1}^{4} d_j \gamma_5 \gamma_j + \sum_{j<k}^{4} e_{jk} \sigma_{jk} = 0_4$$

it follows that all the coefficients c_j, d_j, e_{jk} are equal to 0. Calculating the scalar product of all possible pairs of matrices we find 0. For example

$$\langle \gamma_1, \gamma_2 \rangle = \text{tr}(\gamma_1 \gamma_2^*) = \text{tr} \begin{pmatrix} i & 0 & 0 & 0 \\ 0 & -i & 0 & 0 \\ 0 & 0 & i & 0 \\ 0 & 0 & 0 & -i \end{pmatrix} = 0.$$

Problem 17. Find the spectrum (eigenvalues and normalized eigenvectors) of matrix

$$A = \begin{pmatrix} 1 & 1 & 1 \\ 1 & 1 & 1 \\ 1 & 1 & 1 \end{pmatrix} \equiv \begin{pmatrix} 1 \\ 1 \\ 1 \end{pmatrix} \begin{pmatrix} 1 & 1 & 1 \end{pmatrix}.$$

Find $\|A\| := \sqrt{\langle A, A \rangle}$, where $\| \cdot \|$ denotes the norm.

Solution 17. The eigenvalues of A are $\lambda_1 = \lambda_2 = 0$, $\lambda_3 = 3$. For the eigenvectors we have the orthonormal basis in the Hilbert space $\mathbb{R}^3$

$$\mathbf{v}_1 = \frac{1}{\sqrt{2}} \begin{pmatrix} 1 \\ 0 \\ -1 \end{pmatrix}, \quad \mathbf{v}_2 = \frac{1}{2} \begin{pmatrix} 1 \\ -2 \\ 1 \end{pmatrix}, \quad \mathbf{v}_3 = \frac{1}{\sqrt{3}} \begin{pmatrix} 1 \\ 1 \\ 1 \end{pmatrix}.$$

We have $A = A^T$, $AA^T = 3A$ and $\text{tr}(A^T A) = 9$. Hence the norm of A is 3.

Problem 18. Let $\mathbf{u}_j$ $(j = 1, 2, \ldots, m)$ be an orthonormal basis in the Hilbert space $\mathbb{R}^m$, $\mathbf{v}_k$ $(k = 1, 2, \ldots, n)$ be an orthonormal basis in the Hilbert space $\mathbb{R}^n$ and $\otimes$ be the Kronecker product. Then $\mathbf{u}_j \otimes \mathbf{v}_k$ $(j = 1, 2, \ldots, m)$, $(k = 1, 2, \ldots, n)$ is an orthonormal basis in the Hilbert space $\mathbb{R}^{m+n}$. With

$$\mathbf{u} = \sum_{j=1}^{m} c_j \mathbf{u}_j, \qquad \mathbf{v} = \sum_{k=1}^{n} d_k \mathbf{v}_k$$

we have

$$\mathbf{u} \otimes \mathbf{v} = \sum_{j=1}^{m} \sum_{k=1}^{n} c_j d_k \mathbf{u}_j \otimes \mathbf{v}_k.$$

Any vector $\mathbf{w}$ in the Hilbert space $\mathbb{R}^{m \times n}$ can be written as

$$\mathbf{w} = \sum_{j=1}^{m} \sum_{k=1}^{n} t_{j,k} \mathbf{u}_j \otimes \mathbf{v}_k.$$

Let $m = n = 2$ and the orthonormal bases

$$\mathbf{u}_1 = \begin{pmatrix} 1 \\ 0 \end{pmatrix}, \quad \mathbf{u}_2 = \begin{pmatrix} 0 \\ 1 \end{pmatrix}, \quad \mathbf{v}_1 = \frac{1}{\sqrt{2}} \begin{pmatrix} 1 \\ 1 \end{pmatrix}, \quad \mathbf{v}_2 = \frac{1}{\sqrt{2}} \begin{pmatrix} 1 \\ -1 \end{pmatrix}.$$

Find the expansion of the normalized vector $\mathbf{w}$ in $\mathbb{R}^4$

$$\mathbf{w} = \frac{1}{2}\begin{pmatrix} 1 \\ 1 \\ 1 \\ -1 \end{pmatrix}.$$

Solution 18. We have the basis in $\mathbb{R}^4$

$$\mathbf{u}_1 \otimes \mathbf{v}_1 = \begin{pmatrix} 1 \\ 0 \end{pmatrix} \otimes \frac{1}{\sqrt{2}}\begin{pmatrix} 1 \\ 1 \end{pmatrix} = \frac{1}{\sqrt{2}}\begin{pmatrix} 1 \\ 1 \\ 0 \\ 0 \end{pmatrix},$$

$$\mathbf{u}_1 \otimes \mathbf{v}_2 = \begin{pmatrix} 1 \\ 0 \end{pmatrix} \otimes \frac{1}{\sqrt{2}}\begin{pmatrix} 1 \\ -1 \end{pmatrix} = \frac{1}{\sqrt{2}}\begin{pmatrix} 1 \\ -1 \\ 0 \\ 0 \end{pmatrix},$$

$$\mathbf{u}_2 \otimes \mathbf{v}_1 = \begin{pmatrix} 0 \\ 1 \end{pmatrix} \otimes \frac{1}{\sqrt{2}}\begin{pmatrix} 1 \\ 1 \end{pmatrix} = \frac{1}{\sqrt{2}}\begin{pmatrix} 0 \\ 0 \\ 1 \\ 1 \end{pmatrix},$$

$$\mathbf{u}_2 \otimes \mathbf{v}_2 = \begin{pmatrix} 0 \\ 1 \end{pmatrix} \otimes \frac{1}{\sqrt{2}}\begin{pmatrix} 1 \\ -1 \end{pmatrix} = \frac{1}{\sqrt{2}}\begin{pmatrix} 0 \\ 0 \\ 1 \\ -1 \end{pmatrix}$$

and

$$\mathbf{w} = \frac{1}{\sqrt{2}}(\mathbf{u}_1 \otimes \mathbf{v}_1 + \mathbf{u}_2 \otimes \mathbf{v}_2).$$

Problem 19. Consider the 3×3 matrices over the real numbers

$$A = \begin{pmatrix} 2 & 0 & 2 \\ 1 & 0 & 0 \\ 0 & 0 & 1 \end{pmatrix}.$$

(i) The matrix A can be considered as an element of the Hilbert space of the 3×3 matrices over the real numbers $M(3, \mathbb{R})$ with the scalar product

$$\langle B, C \rangle := \mathrm{tr}(BC^T).$$

Find the norm of A with respect to this scalar product.

(ii) On the other hand the matrix A can be considered as a linear operator acting in the Hilbert space $\mathbb{R}^3$. Find the norm

$$\|A\| := \sup_{\|\mathbf{x}\|=1} \|A\mathbf{x}\|, \qquad \mathbf{x} \in \mathbb{R}^3.$$

(iii) Find the eigenvalues of A and $A^T A$. Compare the result with (i) and (ii).

Solution 19. (i) We have

$$\mathrm{tr}(AA^T) = \mathrm{tr}\begin{pmatrix} 8 & 2 & 2 \\ 2 & 1 & 0 \\ 2 & 0 & 1 \end{pmatrix} = 10.$$

Thus $\|A\| = \sqrt{10}$.

(ii) Using the method of the *Lagrange multiplier* we start with

$$L(x_1, x_2, x_3) = (2x_1 + 2x_3)^2 + x_1^2 + x_3^2 + \lambda(x_1^2 + x_2^2 + x_3^2 - 1).$$

Thus

$$\frac{\partial L}{\partial x_1} = 0 = 8(x_1 + x_3) + 2(\lambda + 1)x_1$$

$$\frac{\partial L}{\partial x_2} = 0 = 2\lambda x_2$$

$$\frac{\partial L}{\partial x_3} = 0 = 8(x_1 + x_3) + 2(\lambda + 1)x_3$$

$$\frac{\partial L}{\partial \lambda} = 0 = x_1^2 + x_2^2 + x_3^2 - 1.$$

If $\lambda = 0$ we find $x_1 = x_3 = 0$ and $x_2 = \pm 1$. If $\lambda \neq 0$, then $x_2 = 0$ and $x_1 = \pm x_3$. Using the constraint we find the solutions

$$x_1 = \frac{1}{\sqrt{2}}, \qquad x_2 = 0, \qquad x_3 = \pm\frac{1}{\sqrt{2}}.$$

Evaluating $(2x_1 + 2x_3)^2 + x_1^2 + x_3^2$ for these solutions gives 1 and 9. The maximum is 9 and thus $\|A\| = 3$.

(iii) The eigenvalues of A are $0, 1, 2$. The eigenvalues of $A^T A$ are $0, 1, 9$. Thus $\|A\|$ from (ii) is the square root of the largest eigenvalue of $A^T A$, namely 3.

Problem 20. Consider the Hilbert space $M(2,\mathbb{C})$ of all 2×2 matrices over $\mathbb{C}$ with scalar product $\langle A, B \rangle := \text{tr}(AB^*)$, $A, B \in M(2,\mathbb{C})$. The *standard basis* is

$$E_{11} = \begin{pmatrix} 1 & 0 \\ 0 & 0 \end{pmatrix}, \quad E_{12} = \begin{pmatrix} 0 & 1 \\ 0 & 0 \end{pmatrix}, \quad E_{21} = \begin{pmatrix} 0 & 0 \\ 1 & 0 \end{pmatrix}, \quad E_{22} = \begin{pmatrix} 0 & 0 \\ 0 & 1 \end{pmatrix}.$$

A *mutually unbiased basis* is

$$\mu_0 = \frac{1}{\sqrt{2}}\sigma_0 = \frac{1}{\sqrt{2}}\begin{pmatrix} 1 & 0 \\ 0 & 1 \end{pmatrix}, \quad \mu_1 = \frac{1}{\sqrt{2}}\sigma_1 = \frac{1}{\sqrt{2}}\begin{pmatrix} 0 & 1 \\ 1 & 0 \end{pmatrix},$$

$$\mu_2 = \frac{1}{\sqrt{2}}\sigma_2 = \frac{1}{\sqrt{2}}\begin{pmatrix} 0 & -i \\ i & 0 \end{pmatrix}, \quad \mu_3 = \frac{1}{\sqrt{2}}\sigma_3 = \frac{1}{\sqrt{2}}\begin{pmatrix} 1 & 0 \\ 0 & -1 \end{pmatrix}.$$

(i) Express the *Hadamard matrix*

$$A = \frac{1}{\sqrt{2}}\begin{pmatrix} 1 & 1 \\ 1 & -1 \end{pmatrix}$$

with this mutually unbiased basis.

(ii) Express the *Bell matrix*

$$B = \frac{1}{\sqrt{2}}\begin{pmatrix} 1 & 0 & 0 & 1 \\ 0 & 1 & 1 & 0 \\ 0 & 1 & -1 & 0 \\ 1 & 0 & 0 & -1 \end{pmatrix}$$

with the basis (sixteen-dimensional) given by $\mu_j \otimes \mu_k$ and $j, k = 0, 1, 2, 3$.

Solution 20. (i) We have the expansion

$$A = \sum_{j=0}^{3} \langle A, \mu_j \rangle \mu_j$$

with $\langle A, \mu_0 \rangle = 0$, $\langle A, \mu_1 \rangle = 1$, $\langle A, \mu_2 \rangle = 0$, $\langle A, \mu_3 \rangle = 1$. Hence $A = \mu_1 + \mu_3$.

(ii) We have the expansion

$$B = \sum_{j=0}^{3}\sum_{k=0}^{3} \langle B, \mu_j \otimes \mu_k \rangle (\mu_j \otimes \mu_k).$$

The only nonzero expansion coefficients are

$$\langle B, \mu_3 \otimes \mu_0 \rangle = \sqrt{2}, \quad \langle B, \mu_1 \otimes \mu_1 \rangle = \sqrt{2}.$$

Hence $B = \sqrt{2}\mu_3 \otimes \mu_0 + \sqrt{2}\mu_1 \otimes \mu_1$.

Problem 21. (i) Consider the Hilbert space $\mathbb{C}^4$. Show that the matrices

$$\Pi_1 = \frac{1}{2}(I_2 \otimes I_2 + \sigma_1 \otimes \sigma_1), \qquad \Pi_2 = \frac{1}{2}(I_2 \otimes I_2 - \sigma_1 \otimes \sigma_1)$$

are projection matrices in $\mathbb{C}^4$.
(ii) Find $\Pi_1 \Pi_2$. Discuss.
(iii) Let e_1, e_2, e_3, e_4 be the standard basis in $\mathbb{C}^4$. Calculate

$$\Pi_1 e_j, \qquad \Pi_2 e_j, \qquad j = 1, 2, 3, 4$$

and show that we obtain 2 two-dimensional Hilbert spaces under these projections.

Solution 21. (i) We have $\Pi_1^* = \Pi_1$, $\Pi_2^* = \Pi_2$, $\Pi_1^2 = \Pi_1$, $\Pi_2^2 = \Pi_2$. Thus Π_1 and Π_2 are projection matrices.
(ii) We obtain $\Pi_1 \Pi_2 = 0_4$. Let $\mathbf{u}$ be an arbitrary vector in $\mathbb{C}^4$. Then the vectors $\Pi_1 \mathbf{u}$ and $\Pi_2 \mathbf{u}$ are perpendicular.
(iii) We obtain

$$\Pi_1 e_1 = \frac{1}{2}\begin{pmatrix} 1 \\ 0 \\ 0 \\ 1 \end{pmatrix}, \quad \Pi_1 e_2 = \frac{1}{2}\begin{pmatrix} 0 \\ 1 \\ 1 \\ 0 \end{pmatrix}, \quad \Pi_1 e_3 = \frac{1}{2}\begin{pmatrix} 0 \\ 1 \\ 1 \\ 0 \end{pmatrix}, \quad \Pi_1 e_4 = \frac{1}{2}\begin{pmatrix} 1 \\ 0 \\ 0 \\ 1 \end{pmatrix}$$

$$\Pi_2 e_1 = \frac{1}{2}\begin{pmatrix} 1 \\ 0 \\ 0 \\ -1 \end{pmatrix}, \quad \Pi_2 e_2 = \frac{1}{2}\begin{pmatrix} 0 \\ 1 \\ -1 \\ 0 \end{pmatrix}, \quad \Pi_2 e_3 = \frac{1}{2}\begin{pmatrix} 0 \\ -1 \\ 1 \\ 0 \end{pmatrix}, \quad \Pi_2 e_4 = \frac{1}{2}\begin{pmatrix} -1 \\ 0 \\ 0 \\ 1 \end{pmatrix}.$$

Thus Π_1 projects into a two-dimensional Hilbert space spanned by the normalized vectors

$$\left\{ \frac{1}{\sqrt{2}}\begin{pmatrix} 1 \\ 0 \\ 0 \\ 1 \end{pmatrix}, \quad \frac{1}{\sqrt{2}}\begin{pmatrix} 0 \\ 1 \\ 1 \\ 0 \end{pmatrix} \right\}.$$

The projection matrix Π_2 projects into a two-dimensional Hilbert space spanned by the normalized vectors

$$\left\{ \frac{1}{\sqrt{2}}\begin{pmatrix} 1 \\ 0 \\ 0 \\ -1 \end{pmatrix}, \quad \frac{1}{\sqrt{2}}\begin{pmatrix} 0 \\ 1 \\ -1 \\ 0 \end{pmatrix} \right\}.$$

The four vectors we find are the *Bell basis*.

Problem 22. Consider the Hilbert space $\mathbb{C}^3$ and let $|0\rangle$, $|1\rangle$, $|2\rangle$ be an orthonormal basis in $\mathbb{C}^3$ with $\langle 0|$, $\langle 1|$, $\langle 2|$ be the dual basis.
(i) Let T_1 be a 3×3 matrix defined by $T_1|0\rangle = |2\rangle, T_1|1\rangle = |0\rangle, T_1|2\rangle = |1\rangle$.
Find T_1.
(ii) Let T_2 be a 3×3 matrix defined by $T_2|0\rangle = |1\rangle, T_2|1\rangle = |2\rangle, T_2|2\rangle = |0\rangle$.
Find T_2.

Solution 22. (i) We obtain $T_1 = |0\rangle\langle 1| + |2\rangle\langle 0| + |1\rangle\langle 2|$. If $|0\rangle$, $|1\rangle$, $|2\rangle$ denotes the standard basis in $\mathbb{C}^3$, then T_1 takes the form of a permutation matrix

$$T_1 = \begin{pmatrix} 0 & 1 & 0 \\ 0 & 0 & 1 \\ 1 & 0 & 0 \end{pmatrix}.$$

(ii) We obtain $T_2 = |1\rangle\langle 0| + |2\rangle\langle 1| + |0\rangle\langle 2|$. If $|0\rangle$, $|1\rangle$, $|2\rangle$ denotes the standard basis in $\mathbb{C}^3$, then T_2 takes the form of a permutation matrix

$$T_2 = \begin{pmatrix} 0 & 0 & 1 \\ 1 & 0 & 0 \\ 0 & 1 & 0 \end{pmatrix}.$$

Problem 23. Let U_1, U_2 be $n \times n$ unitary matrices and Π_1, Π_2 be projection matrices with $\Pi_1\Pi_2 = 0_n$, $\Pi_1 + \Pi_2 = I_n$. Show that $U_1 \otimes \Pi_1 + U_2 \otimes \Pi_2$ is a unitary $n^2 \times n^2$ matrix.

Solution 23. With $U_1 U_1^* = I_n$, $U_2 U_2^* = I_n$ we have

$$(U_1 \otimes \Pi_1 + U_2 \otimes \Pi_2)(U_1^* \otimes \Pi_1 + U_2^* \otimes \Pi_2) = I_n \otimes \Pi_1 + U_1 U_2^* \otimes \Pi_1 \Pi_2$$
$$+ U_2 U_1^* \otimes \Pi_2 \Pi_1 + I_n \otimes \Pi_n$$
$$= I_n \otimes \Pi_1 + I_n \otimes \Pi_2$$
$$= I_n \otimes (\Pi_1 + \Pi_2)$$
$$= I_n \otimes I_n.$$

Problem 24. Consider the Hilbert space $\mathbb{R}^2$. The vectors

$$\left\{ \frac{1}{\sqrt{2}} \begin{pmatrix} 1 \\ 1 \end{pmatrix}, \ \frac{1}{\sqrt{2}} \begin{pmatrix} 1 \\ -1 \end{pmatrix} \right\}$$

form an orthonormal basis. Let $\otimes$ be the *Kronecker product*. The vectors

$$\frac{1}{\sqrt{2}}\begin{pmatrix}1\\1\end{pmatrix}\otimes\frac{1}{\sqrt{2}}\begin{pmatrix}1\\1\end{pmatrix},\qquad \frac{1}{\sqrt{2}}\begin{pmatrix}1\\1\end{pmatrix}\otimes\frac{1}{\sqrt{2}}\begin{pmatrix}1\\-1\end{pmatrix},$$

$$\frac{1}{\sqrt{2}}\begin{pmatrix}1\\-1\end{pmatrix}\otimes\frac{1}{\sqrt{2}}\begin{pmatrix}1\\1\end{pmatrix},\qquad \frac{1}{\sqrt{2}}\begin{pmatrix}1\\-1\end{pmatrix}\otimes\frac{1}{\sqrt{2}}\begin{pmatrix}1\\-1\end{pmatrix}$$

form an orthonormal basis in the Hilbert space $\mathbb{R}^4$. Consider the 4×4 matrix Q which is constructed from the four vectors given above, i.e. the columns of the 4×4 matrix are the four vectors. Find Q^T. Is Q invertible? If so find the inverse Q^{-1}. What is the use of the matrix Q?

Solution 24. With

$$\frac{1}{\sqrt{2}}\begin{pmatrix}1\\1\end{pmatrix}\otimes\frac{1}{\sqrt{2}}\begin{pmatrix}1\\1\end{pmatrix}=\frac{1}{2}\begin{pmatrix}1\\1\\1\\1\end{pmatrix},\qquad \frac{1}{\sqrt{2}}\begin{pmatrix}1\\1\end{pmatrix}\otimes\frac{1}{\sqrt{2}}\begin{pmatrix}1\\-1\end{pmatrix}=\frac{1}{2}\begin{pmatrix}1\\-1\\1\\-1\end{pmatrix},$$

$$\frac{1}{\sqrt{2}}\begin{pmatrix}1\\-1\end{pmatrix}\otimes\frac{1}{\sqrt{2}}\begin{pmatrix}1\\1\end{pmatrix}=\frac{1}{2}\begin{pmatrix}1\\1\\-1\\-1\end{pmatrix},\qquad \frac{1}{\sqrt{2}}\begin{pmatrix}1\\-1\end{pmatrix}\otimes\frac{1}{\sqrt{2}}\begin{pmatrix}1\\-1\end{pmatrix}=\frac{1}{2}\begin{pmatrix}1\\-1\\-1\\1\end{pmatrix}$$

we find the orthogonal matrix

$$Q=\frac{1}{2}\begin{pmatrix}1&1&1&1\\1&-1&1&-1\\1&1&-1&-1\\1&-1&-1&1\end{pmatrix}$$

with $Q=Q^T=Q^{-1}$. Note that

$$Q\frac{1}{2}\begin{pmatrix}1\\1\\1\\1\end{pmatrix}=\begin{pmatrix}1\\0\\0\\0\end{pmatrix}.$$

Problem 25. Let e_1, e_2 be the standard basis in the Hilbert space $\mathbb{C}^2$. Write the *GHZ-state*

$$|GHZ\rangle=\frac{1}{\sqrt{2}}(1\,0\,0\,0\,0\,0\,0\,1)^T$$

in the form

$$|GHZ\rangle = \sum_{j,k,\ell=1}^{2} t_{j,k,\ell} \mathbf{e}_j \otimes \mathbf{e}_k \otimes \mathbf{e}_\ell.$$

Solution 25. We have

$$|GHZ\rangle = t_{1,1,1}\mathbf{e}_1 \otimes \mathbf{e}_1 \otimes \mathbf{e}_1 + t_{2,2,2}\mathbf{e}_2 \otimes \mathbf{e}_2 \otimes \mathbf{e}_2$$

with $t_{1,1,1} = t_{2,2,2} = 1/\sqrt{2}$.

Problem 26. (i) Let $d \geq 2$ and $|0\rangle$, $|1\rangle$, ..., $|d-1\rangle$ be an orthonormal basis in the Hilbert space $\mathbb{C}^d$. Let μ_0, μ_1, ..., μ_{d-1} are non-negative real numbers with $\sum_{j=0}^{d-1} \mu_j = 1$. Show that

$$\rho = \sum_{j=0}^{d-1} \mu_j |j\rangle\langle j|$$

is a *density matrix*.
(ii) Let $d = 2$ and

$$|0\rangle = \frac{1}{\sqrt{2}}\begin{pmatrix} 1 \\ 1 \end{pmatrix}, \quad |1\rangle = \frac{1}{\sqrt{2}}\begin{pmatrix} 1 \\ -1 \end{pmatrix}$$

with $\mu_0 = 1/2$, $\mu_1 = 1/2$. Find ρ.
(iii) Let $d = 3$ and

$$|0\rangle = \frac{1}{\sqrt{2}}\begin{pmatrix} 1 \\ 0 \\ 1 \end{pmatrix}, \quad |1\rangle = \begin{pmatrix} 0 \\ 1 \\ 0 \end{pmatrix}, \quad |2\rangle = \frac{1}{\sqrt{2}}\begin{pmatrix} 1 \\ 0 \\ -1 \end{pmatrix}$$

with $\mu_0 = 1/4$, $\mu_1 = 1/2$, $\mu_2 = 1/4$. Find ρ.

Solution 26. (i) Since μ_j are real numbers and $(|j\rangle\langle j|)^* = |j\rangle\langle j|$ we have $\rho = \rho^*$. Furthermore $\text{tr}(\rho) = 1$ since $\langle j|k\rangle = \delta_{j,k}$.
(ii) Straightforward calculation yields

$$\rho = \begin{pmatrix} 1/2 & 0 \\ 0 & 1/2 \end{pmatrix}.$$

(iii) Straightforward calculation yields

$$\rho = \begin{pmatrix} 1/4 & 0 & 0 \\ 0 & 1/2 & 0 \\ 0 & 0 & 1/4 \end{pmatrix}.$$

Problem 27. Let $a, b > 0$. Consider the 4×4 hermitian matrix

$$H = \begin{pmatrix} a & 0 & 0 & b \\ 0 & a & 0 & b \\ 0 & 0 & a & b \\ b & b & b & a \end{pmatrix}$$

operating in the Hilbert space $\mathbb{C}^4$, where $a = \hbar\omega_1$, $b = \hbar\omega_2$. Find the eigenvalues and normalized eigenvectors of H.

Solution 27. We only have to consider the 4×4 matrix

$$S = \begin{pmatrix} 0 & 0 & 0 & 1 \\ 0 & 0 & 0 & 1 \\ 0 & 0 & 0 & 1 \\ 1 & 1 & 1 & 0 \end{pmatrix}.$$

The eigenvectors of H and S are the same and the eigenvalues E_j of H follow from the eigenvalues λ_j of S are $E_j = a + \lambda_j b$. The eigenvalues and normalized eigenvectors of S are

$$-\sqrt{3}, \quad 0 \ (2\times), \quad +\sqrt{3}$$

$$\frac{1}{\sqrt{6}}\begin{pmatrix} 1 \\ 1 \\ 1 \\ -\sqrt{3} \end{pmatrix}, \quad \frac{1}{\sqrt{2}}\begin{pmatrix} 1 \\ 0 \\ -1 \\ 0 \end{pmatrix}, \quad \frac{1}{\sqrt{2}}\begin{pmatrix} 0 \\ 1 \\ -1 \\ 0 \end{pmatrix}, \quad \frac{1}{\sqrt{6}}\begin{pmatrix} 1 \\ 1 \\ 1 \\ \sqrt{3} \end{pmatrix}.$$

The two eigenvectors for the eigenvalue 0 are linear independent, but not orthonormal. The two vectors are *entangled*. By linear combination (adding and subtracting) of the two vectors we obtain the orthonormal vectors

$$\frac{1}{2}\begin{pmatrix} 1 \\ 1 \\ -2 \\ 0 \end{pmatrix}, \quad \frac{1}{\sqrt{2}}\begin{pmatrix} 1 \\ -1 \\ 0 \\ 0 \end{pmatrix}.$$

The second vector is not entangled. Hence the four eigenvalues of H are

$$E_0 = a - \sqrt{3}b, \quad E_{1,2} = a, \quad E_3 = a + \sqrt{3}b.$$

Problem 28. Let $|0\rangle$, $|1\rangle$, $|2\rangle$, $|3\rangle$, $|4\rangle$, $|5\rangle$ be an orthonormal basis in the Hilbert space $\mathbb{C}^6$ and H be a 6×6 hermitian matrix. Assume that

$$H|0\rangle = a|0\rangle + b|1\rangle + b|5\rangle, \quad H|1\rangle = b|0\rangle + a|1\rangle + b|2\rangle$$
$$H|2\rangle = b|1\rangle + a|2\rangle + b|3\rangle, \quad H|3\rangle = b|2\rangle + a|3\rangle + b|4\rangle$$
$$H|4\rangle = b|3\rangle + a|4\rangle + b|5\rangle, \quad H|5\rangle = b|0\rangle + b|4\rangle + a|5\rangle$$

with $a = \hbar\omega_1$ and $b = \hbar\omega_2$.
(i) Find the matrix representation of H.
(ii) Find the eigenvalues and normalized eigenvectors of this matrix.

Solution 28. (i) We obtain the 6×6 hermitian matrix

$$H = \begin{pmatrix} a & b & 0 & 0 & 0 & b \\ b & a & b & 0 & 0 & 0 \\ 0 & b & a & b & 0 & 0 \\ 0 & 0 & b & a & b & 0 \\ 0 & 0 & 0 & b & a & b \\ b & 0 & 0 & 0 & b & a \end{pmatrix}.$$

(ii) For the eigenvalues and eigenvectors we only have to study the 6×6 matrix

$$S = \begin{pmatrix} 0 & 1 & 0 & 0 & 0 & 1 \\ 1 & 0 & 1 & 0 & 0 & 0 \\ 0 & 1 & 0 & 1 & 0 & 0 \\ 0 & 0 & 1 & 0 & 1 & 0 \\ 0 & 0 & 0 & 1 & 0 & 1 \\ 1 & 0 & 0 & 0 & 1 & 0 \end{pmatrix}.$$

The normalized eigenvectors of H and S are the same and the eigenvalues E_j of H we find from the eigenvalues of S as $E_j = a\lambda_j + b$. The six eigenvalues of S are

$$-1 \ (2\times), \quad 1 \ (2\times), \quad -2 \ (1\times), \quad 2 \ (1\times).$$

The six normalized eigenvectors are

$$\frac{1}{2}\begin{pmatrix} 1 \\ 0 \\ -1 \\ 1 \\ 0 \\ -1 \end{pmatrix}, \quad \frac{1}{2}\begin{pmatrix} 0 \\ 1 \\ -1 \\ 0 \\ 1 \\ -1 \end{pmatrix}, \quad \frac{1}{2}\begin{pmatrix} 1 \\ 0 \\ -1 \\ -1 \\ 0 \\ 1 \end{pmatrix}, \quad \frac{1}{2}\begin{pmatrix} 0 \\ 1 \\ 1 \\ 0 \\ -1 \\ -1 \end{pmatrix},$$

$$\frac{1}{\sqrt{6}} \begin{pmatrix} 1 \\ -1 \\ 1 \\ -1 \\ 1 \\ -1 \end{pmatrix}, \quad \frac{1}{\sqrt{6}} \begin{pmatrix} 1 \\ 1 \\ 1 \\ 1 \\ 1 \\ 1 \end{pmatrix}.$$

These eigenvectors form a basis in the Hilbert space $\mathbb{C}^6$, but not an orthonormal basis. Forming linear combinations of the first two eigenvectors and the third and fourth eigenvectors we obtain an orthonormal basis.

Problem 29. Consider the finite dimensional Hilbert spaces $\mathcal{H}_1$ and $\mathcal{H}_2$ with $\dim(\mathcal{H}_1) = d_1$ ($d_1 \geq 2$) and $\dim(\mathcal{H}_2) = d_2$ ($d_2 \geq 2$) with $d_1 \leq d_2$. Any state of the product Hilbert space $\mathcal{H}_1 \otimes \mathcal{H}_2$ having the form

$$\frac{1}{\sqrt{d_1}} \sum_{j=1}^{d_1} \sum_{k=1}^{d_2} C_{j,k} |j\rangle \otimes |k\rangle$$

where the $d_1 \times d_2$ matrix $C = (C_{j,k})$ satisfies $CC^* = I_{d_1}$ is maximally entangled. Here I_{d_1} is the $d_1 \times d_1$ unit matrix. Study the case $d_1 = 2$, $d_2 = 3$ and the normalized state in the Hilbert space $\mathbb{C}^6$

$$\frac{1}{\sqrt{2}} (1\,0\,0\,0\,0\,1)^T.$$

Solution 29. We have

$$\frac{1}{\sqrt{2}} \begin{pmatrix} 1 \\ 0 \\ 0 \\ 0 \\ 0 \\ 1 \end{pmatrix} = \frac{1}{\sqrt{2}} \left(\begin{pmatrix} 1 \\ 0 \end{pmatrix} \otimes \begin{pmatrix} 1 \\ 0 \\ 0 \end{pmatrix} + \begin{pmatrix} 0 \\ 1 \end{pmatrix} \otimes \begin{pmatrix} 0 \\ 0 \\ 1 \end{pmatrix} \right).$$

Hence $c_{11} = 1$, $c_{12} = 0$, $c_{13} = 0$, $c_{21} = 0$, $c_{22} = 0$, $c_{23} = 1$ and

$$\begin{pmatrix} 1 & 0 & 0 \\ 0 & 0 & 1 \end{pmatrix} \begin{pmatrix} 1 & 0 \\ 0 & 0 \\ 0 & 1 \end{pmatrix} = \begin{pmatrix} 1 & 0 \\ 0 & 1 \end{pmatrix}.$$

Thus the state is maximally entangled.

Problem 30. Consider the Hilbert spaces $\mathbb{C}^2$, $\mathbb{C}^3$ and $\mathbb{C}^6$. Let $v \in \mathbb{C}^2$ and $w \in \mathbb{C}^3$. Find the 6×6 permutation matrix P such that $P(v \otimes w) = (w \otimes v)$.

Solution 30. With

$$v \otimes w = \begin{pmatrix} v_1 w_1 \\ v_1 w_2 \\ v_1 w_3 \\ v_2 w_1 \\ v_2 w_2 \\ v_2 w_3 \end{pmatrix}, \qquad w \otimes v = \begin{pmatrix} w_1 v_1 \\ w_1 v_2 \\ w_2 v_1 \\ w_2 v_2 \\ w_3 v_1 \\ w_3 v_2 \end{pmatrix}$$

we obtain the permutation matrix

$$P = \begin{pmatrix} 1 & 0 & 0 & 0 & 0 & 0 \\ 0 & 0 & 0 & 1 & 0 & 0 \\ 0 & 1 & 0 & 0 & 0 & 0 \\ 0 & 0 & 0 & 0 & 1 & 0 \\ 0 & 0 & 1 & 0 & 0 & 0 \\ 0 & 0 & 0 & 0 & 0 & 1 \end{pmatrix} \equiv (1) \oplus \begin{pmatrix} 0 & 0 & 1 & 0 \\ 1 & 0 & 0 & 0 \\ 0 & 0 & 0 & 1 \\ 0 & 1 & 0 & 0 \end{pmatrix} \oplus (1)$$

where $\oplus$ denotes the direct sum.

Problem 31. Consider the Hilbert space $\mathcal{H} = \mathbb{C}^6$ and the normalized state

$$|\psi\rangle = \frac{1}{\sqrt{2}} \begin{pmatrix} 1 \\ 0 \\ 0 \\ 0 \\ 0 \\ -1 \end{pmatrix}.$$

The state cannot be written as a Kronecker product of a state in $\mathbb{C}^2$ and a state in $\mathbb{C}^3$ and also not as a Kronecker product of a state in $\mathbb{C}^3$ and a state in $\mathbb{C}^2$. The *density matrix* is given by

$$\rho = \frac{1}{2} |\psi\rangle\langle\psi| = \begin{pmatrix} 1 & 0 & 0 & 0 & 0 & -1 \\ 0 & 0 & 0 & 0 & 0 & 0 \\ 0 & 0 & 0 & 0 & 0 & 0 \\ 0 & 0 & 0 & 0 & 0 & 0 \\ 0 & 0 & 0 & 0 & 0 & 0 \\ -1 & 0 & 0 & 0 & 0 & 1 \end{pmatrix}.$$

Consider the Hilbert spaces $\mathcal{H}_1 = \mathbb{C}^2$ and $\mathcal{H}_2 = \mathbb{C}^3$. Then $\mathcal{H} = \mathcal{H}_1 \otimes \mathcal{H}_2$. Find the *partial traces* $\rho_2 = \text{tr}_1(\rho)$, $\rho_1 = \text{tr}_2(\rho)$.

Solution 31. To calculate $\rho_2 = \mathrm{tr}_1(\rho)$ we utilize the basis

$$\left\{ \begin{pmatrix} 1 \\ 0 \end{pmatrix} \otimes I_3, \ \begin{pmatrix} 0 \\ 1 \end{pmatrix} \otimes I_3 \right\}.$$

Then

$$\left(\begin{pmatrix} 1 \\ 0 \end{pmatrix} \otimes I_3 \right)^* = (1 \ 0) \otimes I_3, \quad \left(\begin{pmatrix} 0 \\ 1 \end{pmatrix} \otimes I_3 \right)^* (0 \ 1) \otimes I_3.$$

Hence

$$\rho_2 = \mathrm{tr}_1(\rho) = ((1 \ 0) \otimes I_3)\rho \left(\begin{pmatrix} 1 \\ 0 \end{pmatrix} \otimes I_3 \right) + ((0 \ 1) \otimes I_3)\rho \left(\begin{pmatrix} 0 \\ 1 \end{pmatrix} \otimes I_3 \right)$$

$$= \frac{1}{2} \begin{pmatrix} 1 & 0 & 0 \\ 0 & 0 & 0 \\ 0 & 0 & 1 \end{pmatrix}.$$

To calculate $\rho_1 = \mathrm{tr}_2(\rho)$ we utilize the basis

$$\left\{ \begin{pmatrix} 1 \\ 0 \\ 0 \end{pmatrix} \otimes I_2, \ \begin{pmatrix} 0 \\ 1 \\ 0 \end{pmatrix} \otimes I_2, \ \begin{pmatrix} 0 \\ 0 \\ 1 \end{pmatrix} \otimes I_2 \right\}.$$

Then

$$\left(\begin{pmatrix} 1 \\ 0 \\ 0 \end{pmatrix} \otimes I_2 \right)^* = (1 \ 0 \ 0) \otimes I_2,$$

$$\left(\begin{pmatrix} 0 \\ 1 \\ 0 \end{pmatrix} \otimes I_2 \right)^* = (0 \ 1 \ 0) \otimes I_2,$$

$$\left(\begin{pmatrix} 0 \\ 0 \\ 1 \end{pmatrix} \otimes I_2 \right)^* = (0 \ 0 \ 1) \otimes I_2.$$

Hence

$$\rho_1 = \mathrm{tr}_2(\rho) = \frac{1}{2} \begin{pmatrix} 1 & 0 \\ 0 & 1 \end{pmatrix}.$$

Problem 32. Let $|0\rangle$, $|1\rangle$, $|2\rangle$, $|3\rangle$ be an orthonormal basis in the Hilbert space $\mathbb{C}^4$.

(i) Show that $U = 1|0\rangle\langle0| - 1|1\rangle\langle1| + i|2\rangle\langle2| - i|3\rangle\langle3|$ is a unitary matrix.

(ii) Consider the case that the orthonormal basis is the standard basis.

(iii) Consider the case that the orthonormal basis is the *Bell basis*

$$|0\rangle = \frac{1}{\sqrt{2}}\begin{pmatrix}1\\0\\0\\1\end{pmatrix}, \quad |1\rangle = \frac{1}{\sqrt{2}}\begin{pmatrix}0\\1\\1\\0\end{pmatrix}, \quad |2\rangle = \frac{1}{\sqrt{2}}\begin{pmatrix}0\\1\\-1\\0\end{pmatrix}, \quad |3\rangle = \frac{1}{\sqrt{2}}\begin{pmatrix}1\\0\\0\\-1\end{pmatrix}.$$

Solution 32. (i) We have $U^* = 1|0\rangle\langle0| - 1|1\rangle\langle1| - i|2\rangle\langle2| + i|3\rangle\langle3|$. Then with $\langle j|k\rangle = \delta_{j,k}$ we obtain

$$UU^* = 1|0\rangle\langle0| + 1|1\rangle\langle1| + 1|2\rangle\langle2| + 1|3\rangle\langle3| = I_4.$$

(ii) For the standard basis we obtain the diagonal matrix

$$U = \begin{pmatrix}1 & 0 & 0 & 0\\0 & -1 & 0 & 0\\0 & 0 & i & 0\\0 & 0 & 0 & -i\end{pmatrix}.$$

(iii) Since

$$|0\rangle\langle0| = \frac{1}{2}\begin{pmatrix}1&0&0&1\\0&0&0&0\\0&0&0&0\\1&0&0&1\end{pmatrix}, \quad |1\rangle\langle1| = \frac{1}{2}\begin{pmatrix}0&0&0&0\\0&1&1&0\\0&1&1&0\\0&0&0&0\end{pmatrix},$$

$$|2\rangle\langle2| = \frac{1}{2}\begin{pmatrix}0&0&0&0\\0&1&-1&0\\0&-1&1&0\\0&0&0&0\end{pmatrix}, \quad |3\rangle\langle3| = \frac{1}{2}\begin{pmatrix}1&0&0&-1\\0&0&0&0\\0&0&0&0\\-1&0&0&1\end{pmatrix},$$

we find

$$U = \frac{1}{2}\begin{pmatrix}1-i & 0 & 0 & 1+i\\0 & 1+i & 1-i & 0\\0 & 1-i & 1+i & 0\\1+i & 0 & 0 & 1-i\end{pmatrix}.$$

Problem 33. Let $d \geq 2$ and $|0\rangle$, $|1\rangle$, $\ldots$, $|d-1\rangle$ be an orthonormal basis in the Hilbert space $\mathbb{C}^d$. Consider the $d \times d$ matrix

$$A = \sum_{j=0}^{d-1} \lambda_j |j\rangle\langle j|.$$

Assume that $\lambda_j \neq 0$ for $j = 0, 1, \ldots, d-1$. Find the inverse of A.

Solution 33. Obviously the inverse is given by

$$A^{-1} = \sum_{k=0}^{d-1} \frac{1}{\lambda_k} |k\rangle\langle k|$$

since $\langle j|k \rangle = \delta_{j,k}$ and $\sum_{j=0}^{d-1} |j\rangle\langle j| = I_d$ (*completeness relation*).

Problem 34. Consider the Hilbert space $\mathbb{C}^2$ and the unitary matrix ($\tau \in \mathbb{R}$)

$$U(\tau) = \begin{pmatrix} \cos(\tau) & i\sin(\tau) \\ i\sin(\tau) & \cos(\tau) \end{pmatrix} \quad \Rightarrow \quad U^*(\tau) = \begin{pmatrix} \cos(\tau) & -i\sin(\tau) \\ -i\sin(\tau) & \cos(\tau) \end{pmatrix}.$$

Note that $\det(U) = 1$ and thus $U(\tau)$ is an element of the Lie group $SU(2)$. Find the hermitian (self-adjoint) 2×2 matrix T such that

$$U(\tau) = \exp(i\tau T).$$

Solution 34. First we find the *spectral decomposition* of $U(\tau)$. The eigenvalues of $U(\tau)$ are $\lambda_1 = e^{i\tau}$, $\lambda_2 = e^{-i\tau}$ with the corresponding normalized eigenvectors

$$\mathbf{v}_1 = \frac{1}{\sqrt{2}} \begin{pmatrix} 1 \\ 1 \end{pmatrix}, \quad \mathbf{v}_2 = \frac{1}{\sqrt{2}} \begin{pmatrix} 1 \\ -1 \end{pmatrix}.$$

Hence the spectral decomposition of $U(\tau)$ is

$$U(\tau) = \lambda_1 \mathbf{v}_1 \mathbf{v}_1^* + \lambda_2 \mathbf{v}_2 \mathbf{v}_2^* = e^{i\tau} \frac{1}{2} \begin{pmatrix} 1 & 1 \\ 1 & 1 \end{pmatrix} + e^{-i\tau} \frac{1}{2} \begin{pmatrix} 1 & -1 \\ -1 & 1 \end{pmatrix}$$

$$= \frac{1}{2} \begin{pmatrix} e^{i\tau} + e^{-i\tau} & e^{i\tau} - e^{-i\tau} \\ e^{i\tau} - e^{-i\tau} & e^{i\tau} + e^{-i\tau} \end{pmatrix}.$$

With $\ln(e^{i\tau}) = i\tau$, $\ln(e^{-i\tau}) = -i\tau$ we obtain

$$i\tau T = i\tau \frac{1}{2} \begin{pmatrix} 1 & 1 \\ 1 & 1 \end{pmatrix} - i\tau \frac{1}{2} \begin{pmatrix} 1 & -1 \\ -1 & 1 \end{pmatrix} = \begin{pmatrix} 0 & i\tau \\ i\tau & 0 \end{pmatrix}.$$

Hence

$$T = \begin{pmatrix} 0 & 1 \\ 1 & 0 \end{pmatrix}$$

which is the first Pauli spin matrix σ_1.

Problem 35. (i) Let $\lambda_1, \lambda_2, \mu_1, \mu_2 \in \mathbb{C}$ and $\mathbf{v}_1, \mathbf{v}_2$ be an orthonormal basis in the Hilbert space $\mathbb{C}^2$. We define the 2×2 matrices

$$A = \lambda_1 \mathbf{v}_1 \mathbf{v}_1^* + \lambda_2 \mathbf{v}_2 \mathbf{v}_2^*, \quad B = \mu_1 \mathbf{v}_1 \mathbf{v}_1^* + \mu_2 \mathbf{v}_2 \mathbf{v}_2^*.$$

Find the commutator $[A, B]$. Find the conditions on $\lambda_1, \lambda_2, \mu_1, \mu_2$ such that $[A, B] = 0_2$.
(ii) Consider the $n \times n$ matrices A, B. Assume that $\mathbf{v}_j$ $(j = 1, \ldots, n)$ form an orthonormal basis in $\mathbb{C}^n$ and

$$A\mathbf{v}_j = \lambda_j \mathbf{v}_j, \quad B\mathbf{v}_j = \mu_j \mathbf{v}_j, \quad j = 1, \ldots, n.$$

Find the commutator $[A, B]$.

Solution 35. (i) With $\mathbf{v}_j^* \mathbf{v}_j = 1$ $(j = 1, 2)$ we obtain

$$[A, B] = (\lambda_1 \mu_2 - \lambda_2 \mu_1)\mathbf{v}_1 \mathbf{v}_1^* \mathbf{v}_2 \mathbf{v}_2^* + (\lambda_2 \mu_1 - \lambda_1 \mu_2)\mathbf{v}_2 \mathbf{v}_2^* \mathbf{v}_1 \mathbf{v}_1^*.$$

Since $\mathbf{v}_1^* \mathbf{v}_2 = 0$, $\mathbf{v}_2^* \mathbf{v}_1 = 0$ we obtain $[A, B] = 0_2$. This means that the result is independent of $\lambda_1, \lambda_2, \mu_1, \mu_2$. Note that $\Pi_1 = \mathbf{v}_1 \mathbf{v}_1^*$, $\Pi_2 = \mathbf{v}_2 \mathbf{v}_2^*$ are projection matrices.
(ii) Since

$$A = \sum_{j=1}^n \lambda_j \mathbf{v}_j \mathbf{v}_j^*, \quad B = \sum_{k=1}^n \mu_k \mathbf{v}_k \mathbf{v}_k^*$$

and $\mathbf{v}_j^* \mathbf{v}_k = \delta_{j,k}$ we obtain $[A, B] = 0_n$. Find the anticommutator $[A, B]_+$.

Problem 36. Let $n \geq 2$ and I_n be the $n \times n$ identity matrix and J_n be the counter diagonal identity matrix. For example for $n = 3$ we have

$$I_3 + J_3 = \begin{pmatrix} 1 & 0 & 0 \\ 0 & 1 & 0 \\ 0 & 0 & 1 \end{pmatrix} + \begin{pmatrix} 0 & 0 & 1 \\ 0 & 1 & 0 \\ 1 & 0 & 0 \end{pmatrix} = \begin{pmatrix} 1 & 0 & 1 \\ 0 & 2 & 0 \\ 1 & 0 & 1 \end{pmatrix}.$$

Show that 0 and 2 are the eigenvalues (with appropriate multiplicity) of $I_n + J_n$.

Solution 36. Let λ be an eigenvalue of $I_n + J_n$ with eigenvector $\mathbf{v}$, i.e.

$$(I_n + J_n)\mathbf{v} = \lambda\mathbf{v}.$$

Multiplying the eigenvalue equation with $I_n + J_n$ and utilizing $J_n^2 = I_n$ we obtain

$$2(I_n + J_n)\mathbf{v} = \lambda(I_n + J_n)\mathbf{v} \quad \Rightarrow \quad (I_n + J_n)\mathbf{v} = \frac{\lambda^2}{2}\mathbf{v}.$$

The solutions of the quadratic equation $\lambda^2/2 = \lambda$ admits the solutions $\lambda = 0$ and $\lambda = 2$.

Problem 37. Let $N \geq 2$. Consider the Hilbert space $\mathbb{C}^N$ and $\{|e_j\rangle\}$ $(j = 1,\ldots,N)$ be an orthonormal basis in $\mathbb{C}^N$. Let $|\psi\rangle$ be a normalized vector in $\mathbb{C}^N$. We define the complex numbers

$$c_j(|\psi\rangle) := \langle e_j|\psi\rangle, \quad j = 1,\ldots,N.$$

So we can construct the *metric tensor field*

$$g = \sum_{j=1}^{N}(d\bar{c}_j \otimes dc_j).$$

Let $N = 2$ and

$$|e_1\rangle = \begin{pmatrix} 1 \\ 0 \end{pmatrix}, \quad |e_2\rangle = \begin{pmatrix} 0 \\ 1 \end{pmatrix}, \quad |\psi\rangle = \begin{pmatrix} e^{i\phi}\cos(\theta) \\ \sin(\theta) \end{pmatrix}.$$

Find the metric tensor field g.

Solution 37. We have

$$c_1(|\psi\rangle) = \langle e_1|\psi\rangle = e^{i\phi}\cos(\theta), \qquad c_2(|\psi\rangle) = \langle e_2|\psi\rangle = \sin(\theta).$$

Then

$$dc_1 = ie^{i\phi}\cos(\theta)d\phi - e^{i\phi}\sin(\theta)d\theta \;\Rightarrow\; d\bar{c}_1 = -ie^{-i\phi}\cos(\theta)d\phi - e^{-i\phi}\sin(\theta)d\theta$$

$$dc_2 = \cos(\theta)d\theta \;\Rightarrow\; d\bar{c}_2 = \cos(\theta)d\theta.$$

Then

$$d\bar{c}_1 \otimes dc_1 = \cos^2(\theta)d\phi \otimes d\phi + \sin^2(\theta)d\theta \otimes d\theta$$
$$+ i\cos(\theta)\sin(\theta)d\phi \otimes d\theta - i\cos(\theta)\sin(\theta)d\theta \otimes d\phi$$
$$d\bar{c}_2 \otimes dc_2 = \cos^2(\theta)d\theta \otimes d\theta.$$

It follows that

$$g = \cos^2(\theta)d\phi \otimes d\phi + d\theta \otimes d\theta + i\cos(\theta)\sin(\theta)d\phi \otimes d\theta - i\sin(\theta)\cos(\theta)d\theta \otimes d\phi$$

with the corresponding hermitian matrix

$$\begin{pmatrix} \cos^2(\theta) & i\cos(\theta)\sin(\theta) \\ -i\cos(\theta)\sin(\theta) & 1 \end{pmatrix}.$$

The determinant is $\cos^4(\theta)$. Find the Killing vector fields for g.

Problem 38. Consider the Hilbert space $\mathbb{C}^2$ and the 2×2 matrix

$$M(\theta) = \begin{pmatrix} \sin(\theta) & \cos(\theta) \\ \cos(\theta) & -\sin(\theta) \end{pmatrix}.$$

Find the eigenvalues and normalized eigenvectors of $M(\theta)$.

Solution 38. The matrix is symmetric over $\mathbb{R}$. We have $\text{tr}(M(\theta)) = 0$, $\det(M(\theta)) = -1$. Hence the eigenvalues are $\lambda_1 = +1$ and $\lambda_2 = -1$ with the corresponding normalized eigenvectors

$$\mathbf{v}_1 = \frac{1}{\sqrt{2}} \begin{pmatrix} \sqrt{1 - \sin(\theta)} \\ \sqrt{1 + \sin(\theta)} \end{pmatrix}, \quad \mathbf{v}_2 = \frac{1}{\sqrt{2}} \begin{pmatrix} \sqrt{1 + \sin(\theta)} \\ -\sqrt{1 - \sin(\theta)} \end{pmatrix}.$$

Extend the problem to the matrix

$$\begin{pmatrix} \sin(\theta) & e^{i\phi} \cos(\theta) \\ e^{-i\phi} \cos(\theta) & -\sin(\theta) \end{pmatrix}.$$

This matrix contains the *Pauli spin matrices* σ_1, σ_2, σ_3 for $\theta = 0$, $\phi = 0$; $\theta = 0$, $\phi = 3\pi/2$; $\theta = \pi/2$, $\phi = 0$.

Problem 39. Consider the Hilbert space $\mathbb{C}^8$ and the standard basis e_1, e_2, ..., e_8 in $\mathbb{C}^8$. We form the three normalized vectors in $\mathbb{C}^8$

$$v_1 = \frac{1}{2}(e_1 + e_2 + e_3 + e_4), \quad v_2 = \frac{1}{2}(e_3 - e_4 + e_5 + e_6), \quad v_3 = \frac{1}{2}(e_5 - e_6 + e_7 + e_8).$$

Let $\mu_1, \mu_2, \mu_3 \neq 0$. We define the 8×8 matrix

$$T := \mu_1 v_1 v_1^T + \mu_2 v_2 v_2^T + \mu_3 v_3 v_3^T.$$

Find the eigenvalues and normalized eigenvectors of T.

Solution 39. We have $v_1^T v_2 = 0$, $v_1^T v_3 = 0$, $v_2^T v_3 = 0$. Hence μ_1, μ_2, μ_3 are eigenvalues T with the corresponding normalized eigenvectors v_1, v_2, v_3. The remaining five eigenvalues are equal to 0, i.e.

$$\mu_4 = \mu_5 = \mu_6 = \mu_7 = \mu_8 = 0.$$

What are the corresponding eigenvectors?

Problem 40. Consider the Hilbert space $\mathbb{C}^n$. Let $\alpha \in \mathbb{R}$ and U be a unitary and hermitian $n \times n$ matrix. Calculate the matrix $\exp(i\alpha U)$.

Solution 40. Since U is unitary and hermitian we have $U^* = U^{-1}$, $U = U^*$. It follows that $U = U^{-1}$ and $U^2 = I_n$. With

$$(i\alpha U)^2 = -\alpha^2 I_n, \quad (i\alpha U)^3 = -i\alpha^3 U, \quad (i\alpha U)^4 = \alpha^4 I_n,$$

$$(i\alpha U)^5 = i\alpha^5 U, \quad (i\alpha U)^6 = -\alpha I_n, \quad \ldots$$

we obtain

$$\exp(i\alpha U) = I_n \cos(\alpha) + iU \sin(\alpha).$$

Obviously this matrix is unitary again.

Problem 41. Consider the Hilbert space $\mathbb{R}^3$. Let P be a 3×3 permutation matrix. Can we conclude that $\Pi P \Pi$ is a normal matrix.

Solution 41. No not in general. Consider

$$\Pi = \begin{pmatrix} 0 & 0 & 0 \\ 0 & 1 & 0 \\ 0 & 0 & 1 \end{pmatrix}, \quad P = \begin{pmatrix} 0 & 0 & 1 \\ 1 & 0 & 0 \\ 0 & 1 & 0 \end{pmatrix} \Rightarrow \Pi P \Pi = \begin{pmatrix} 0 & 0 & 0 \\ 0 & 0 & 0 \\ 0 & 1 & 0 \end{pmatrix}.$$

This matrix is *nonnormal.* For the matrix $\Pi P \Pi$ to be normal one has to satisfy $\Pi P \Pi = \Pi P^* \Pi$ with $P^* = P^T$, where ^{T} denotes the transpose.

Problem 42. (i) Consider the Hilbert space $\mathbb{C}^n$. Let $T_P : \mathbb{C}^n \to \mathbb{C}^n$ be a linear map ($n \times n$ matrix) obeying

$$T_P \begin{pmatrix} x_1 \\ x_2 \\ \vdots \\ x_n \end{pmatrix} = \begin{pmatrix} x_{\sigma(1)} \\ x_{\sigma(2)} \\ \vdots \\ x_{\sigma(n)} \end{pmatrix}, \quad x_1, \ldots, x_n \in \mathbb{C}$$

where $\sigma : \{1, 2, \ldots, n\} \to \{1, 2, \ldots, n\}$ is a one-to-one map (*permutation*). Find the *permutation matrix* satisfying $P\mathbf{x} = T_P(\mathbf{x})$ for all $\mathbf{x} \in \mathbb{C}^n$. Illustrate the answer for the case

$$T_P \begin{pmatrix} x_1 \\ x_2 \\ x_3 \end{pmatrix} = \begin{pmatrix} x_3 \\ x_1 \\ x_2 \end{pmatrix}, \quad x_1, x_2, x_3 \in \mathbb{C}.$$

(ii) Consider the Hilbert space $M(n, \mathbb{R})$. Let P be an $n \times n$ permutation matrix. Find $\langle P, P \rangle$ and the induced norm $\|P\|$.

Solution 42. (i) Since T_P is linear on $\mathbb{C}^n$, it is completely defined by its action on the standard basis $\mathbf{e}_1, \mathbf{e}_2, \ldots, \mathbf{e}_n$ in $\mathbb{C}^n$, i.e.

$$P\mathbf{e}_1 = T_P \mathbf{e}_1 = T_P \begin{pmatrix} 1 \\ 0 \\ \vdots \\ 0 \end{pmatrix} = \mathbf{e}_{\sigma^{-1}(1)}$$

which provides the first column of the permutation matrix P. The remaining columns are determined in a similar way, namely

$$P = (\mathbf{e}_{\sigma^{-1}(1)} \; \mathbf{e}_{\sigma^{-1}(2)} \; \cdots \; \mathbf{e}_{\sigma^{-1}(n)}).$$

Now consider

$$T_P \begin{pmatrix} x_1 \\ x_2 \\ x_3 \end{pmatrix} = \begin{pmatrix} x_3 \\ x_1 \\ x_2 \end{pmatrix}$$

which provides $\sigma(1) = 3$, $\sigma(2) = 1$, $\sigma(3) = 2$. Hence we find that $\sigma^{-1} = 2$, $\sigma^{-1}(2) = 3$, $\sigma^{-1}(3) = 1$. Consequently the permutation matrix is

$$P = (\mathbf{e}_2 \; \mathbf{e}_3 \; \mathbf{e}_1) = \begin{pmatrix} 0 & 0 & 1 \\ 1 & 0 & 0 \\ 0 & 1 & 0 \end{pmatrix}.$$

(ii) Since $P^{-1} = P^T$ we have

$$\langle P, P \rangle = \text{tr}(PP^T) = \text{tr}(PP^{-1}) = \text{tr}(I_n) = n.$$

Then $\|P\| = \sqrt{n}$.

Problem 43. Consider the Hilbert spaces $M(n, \mathbb{C})$ and $\mathbb{C}^{2n}$. Show that the *vec-operator* defines an *isomorphism* from $M(n, \mathbb{C})$ onto $\mathbb{C}^{n^2}$.

Solution 43. Since addition and scalar multiplication in $M(n, \mathbb{C})$ and $\mathbb{C}^{n^2}$ are defined componentwise, it follows that the vec-operator is linear, i.e.

$$\text{vec}(\alpha A + \beta B) = \alpha \, \text{vec}(A) + \beta \, \text{vec}(B)$$

for all $A, B \in M(n, \mathbb{C})$ and all $\alpha, \beta \in \mathbb{C}$. Obviously the vec-operator is bijective. Finally we have to show that the vec-operator is compatible with the respective inner product. Let $A := (a_{jk})$, $B := (b_{jk})$ $(j, k = 1, 2, \ldots, n)$ be matrices in the Hilbert space $M(n, \mathbb{C})$. For the matrices we have

$$\langle A, B \rangle = \text{tr}(AB^*) = \text{tr}(B^*A) = \sum_{k=1}^{n} \sum_{j=1}^{n} \overline{b_{jk}} a_{jk}$$

where we used that the trace of the square matrix is the sum of the diagonal elements. For the vectors $\text{vec}(A)$, $\text{vec}(B)$ in $\mathbb{C}^{n^2}$ we have

$$\langle \text{vec}(A), \text{vec}(B) \rangle = \text{vec}(B)^* \text{vec}(A) = \sum_{k=1}^{n} \sum_{j=1}^{n} \overline{b_{jk}} a_{jk}.$$

Problem 44. The normalized vectors

$$\mathbf{v}_1 = \frac{1}{\sqrt{2}} \begin{pmatrix} 1 \\ 0 \\ 1 \end{pmatrix}, \quad \mathbf{v}_2 = \begin{pmatrix} 0 \\ 1 \\ 0 \end{pmatrix}, \quad \mathbf{v}_3 = \frac{1}{\sqrt{2}} \begin{pmatrix} 1 \\ 0 \\ -1 \end{pmatrix}$$

form an orthonormal basis in the Hilbert space $\mathbb{C}^3$.
(i) Find the unitary matrices U_{12}, U_{23}, U_{31} such that $U_{12}\mathbf{v}_1 = \mathbf{v}_2$, $U_{23}\mathbf{v}_2 = \mathbf{v}_3$, $U_{31}\mathbf{v}_3 = \mathbf{v}_1$.
(ii) Find the unitary matrix $U_{31}U_{23}U_{12}$.
(iii) Calculate (*spectral decomposition*)

$$V = \lambda_1 \mathbf{v}_1 \mathbf{v}_1^* + \lambda_2 \mathbf{v}_2 \mathbf{v}_2^* + \lambda_3 \mathbf{v}_3 \mathbf{v}_3^*$$

where the complex numbers λ_1, λ_2, λ_3 satisfy $\lambda_1\bar{\lambda}_1 = 1$, $\lambda_2\bar{\lambda}_2 = 1$, $\lambda_3\bar{\lambda}_3 = 1$.

Solution 44. (i) We have

$$\begin{pmatrix} 0 & 1 & 0 \\ 1/\sqrt{2} & 0 & 1/\sqrt{2} \\ 1/\sqrt{2} & 0 & -1/\sqrt{2} \end{pmatrix} \frac{1}{\sqrt{2}} \begin{pmatrix} 1 \\ 0 \\ 1 \end{pmatrix} = \begin{pmatrix} 0 \\ 1 \\ 0 \end{pmatrix},$$

$$\begin{pmatrix} 0 & 1/\sqrt{2} & 1/\sqrt{2} \\ 1 & 0 & 0 \\ 0 & -1/\sqrt{2} & 1/\sqrt{2} \end{pmatrix} \begin{pmatrix} 0 \\ 1 \\ 0 \end{pmatrix} = \frac{1}{\sqrt{2}} \begin{pmatrix} 1 \\ 0 \\ -1 \end{pmatrix},$$

$$\begin{pmatrix} 1 & 0 & 0 \\ 0 & 1 & 0 \\ 0 & 0 & -1 \end{pmatrix} \begin{pmatrix} 1/\sqrt{2} \\ 0 \\ -1/\sqrt{2} \end{pmatrix} = \begin{pmatrix} 1/\sqrt{2} \\ 0 \\ 1/\sqrt{2} \end{pmatrix}$$

i.e.

$$U_{12} = \begin{pmatrix} 0 & 1 & 0 \\ 1/\sqrt{2} & 0 & 1/\sqrt{2} \\ 1/\sqrt{2} & 0 & -1/\sqrt{2} \end{pmatrix}, \quad U_{23} = \begin{pmatrix} 0 & 1/\sqrt{2} & 1/\sqrt{2} \\ 1 & 0 & 0 \\ 0 & -1/\sqrt{2} & 1/\sqrt{2} \end{pmatrix},$$

$$U_{31} = \begin{pmatrix} 1 & 0 & 0 \\ 0 & 1 & 0 \\ 0 & 0 & -1 \end{pmatrix}.$$

(ii) Obviously we have $U_{31}U_{23}U_{12} = I_3$, where I_3 is the 3×3 identity matrix.
(iii) We obtain

$$V = \frac{1}{2} \begin{pmatrix} \lambda_1 + \lambda_3 & 0 & \lambda_2 - \lambda_3 \\ 0 & 2\lambda_2 & 0 \\ \lambda_1 - \lambda_3 & 0 & \lambda_1 + \lambda_3 \end{pmatrix} \Rightarrow V^* = \frac{1}{2} \begin{pmatrix} \bar{\lambda}_1 + \bar{\lambda}_3 & 0 & \bar{\lambda}_1 - \bar{\lambda}_3 \\ 0 & 2\bar{\lambda}_2 & 0 \\ \bar{\lambda}_1 - \bar{\lambda}_3 & 0 & \bar{\lambda}_1 + \bar{\lambda}_3 \end{pmatrix}$$

with $VV^* = I_3$.

Problem 45. (i) Do the three vectors

$$\mathbf{v}_1 = \begin{pmatrix} \cos(\theta) \\ \sin(\theta) \\ 0 \end{pmatrix}, \quad \mathbf{v}_2 = \begin{pmatrix} -\sin(\theta) \\ \cos(\theta) \\ 0 \end{pmatrix}, \quad \mathbf{v}_1 \times \mathbf{v}_2$$

form an orthonormal basis in the Hilbert space $\mathbb{R}^3$?

(ii) Consider the orthonormal basis

$$\mathbf{v}_1 = \frac{1}{\sqrt{2}} \begin{pmatrix} 1 \\ 0 \\ 1 \end{pmatrix}, \quad \mathbf{v}_2 = \begin{pmatrix} 0 \\ 1 \\ 0 \end{pmatrix}, \quad \mathbf{v}_3 = \frac{1}{\sqrt{2}} \begin{pmatrix} 1 \\ 0 \\ -1 \end{pmatrix}$$

in the Hilbert space $\mathbb{R}^3$. Do the three vectors in $\mathbb{R}^3$

$$\mathbf{v}_1 \times \mathbf{v}_2, \quad \mathbf{v}_2 \times \mathbf{v}_3, \quad \mathbf{v}_3 \times \mathbf{v}_1$$

form an orthonormal basis in the Hilbert space $\mathbb{R}^3$?

Solution 45. (i) We obtain

$$\mathbf{v}_1 \times \mathbf{v}_2 = \begin{pmatrix} 0 \\ 0 \\ 1 \end{pmatrix}$$

and $\mathbf{v}_1^T \mathbf{v}_2 = 0$, $\mathbf{v}_1^T \mathbf{v}_3 = 0$, $\mathbf{v}_2^T \mathbf{v}_3 = 0$. Furthermore all three vectors are normalized. Hence we have an orthonormal basis in $\mathbb{R}^3$.

(ii) We have

$$\mathbf{v}_1 \times \mathbf{v}_2 = \frac{1}{\sqrt{2}} \begin{pmatrix} -1 \\ 0 \\ 1 \end{pmatrix}, \quad \mathbf{v}_2 \times \mathbf{v}_3 = \frac{1}{\sqrt{2}} \begin{pmatrix} -1 \\ 0 \\ -1 \end{pmatrix}, \quad \mathbf{v}_3 \times \mathbf{v}_1 = \begin{pmatrix} 0 \\ 1 \\ 0 \end{pmatrix}.$$

Hence we have an orthonormal basis again.

Is this true in general? Given an (arbitrary) orthonormal basis $\mathbf{w}_1$, $\mathbf{w}_2$, $\mathbf{w}_3$ in $\mathbb{R}^3$. Do the vectors $\mathbf{w}_1 \times \mathbf{w}_2$, $\mathbf{w}_2 \times \mathbf{w}_3$, $\mathbf{w}_3 \times \mathbf{w}_1$ form an orthonormal basis in $\mathbb{R}^3$?

The vector product can also be written utilizing the exterior product $\wedge$ and the *Hodge duality operator* $\star$ with the underlying metric tensor field $g = dx_1 \otimes dx_1 + dx_2 \otimes dx_2 + dx_3 \otimes dx_3$. We have

$$\mathbf{v} \times \mathbf{w} = \begin{pmatrix} v_2 w_3 - v_3 w_2 \\ v_3 w_1 - v_1 w_3 \\ v_1 w_2 - v_2 w_1 \end{pmatrix} = \star(\mathbf{v} \wedge \mathbf{w}).$$

Study higher dimensional cases using the exterior product and the Hodge duality operator.

Problem 46. Let $|0\rangle$, $|1\rangle$ be the standard basis in the Hilbert space $\mathbb{C}^2$, i.e.

$$|0\rangle = \begin{pmatrix} 1 \\ 0 \end{pmatrix}, \quad |1\rangle = \begin{pmatrix} 0 \\ 1 \end{pmatrix}$$

$x_1, x_2, x_3 \in \{0, 1\}$ and $\oplus$ be the *XOR operation* with $0 \oplus 0 = 0$, $0 \oplus 1 = 1$, $1 \oplus 0 = 1$, $1 \oplus 1 = 0$.

(i) Consider the Hilbert space $\mathbb{C}^4$. Find the 4×4 permutation matrix P such that

$$P(|x_1\rangle \otimes |x_2\rangle) = |x_1\rangle \otimes |x_1 \oplus x_2\rangle.$$

(ii) Consider the Hilbert space $\mathbb{C}^8$. Find the 8×8 permutation matrix P such that

$$P(|x_1\rangle \otimes |x_2\rangle \otimes |x_3\rangle) = |x_1\rangle \otimes |x_1 \oplus x_2\rangle \otimes |x_1 \oplus x_2 \oplus x_3\rangle.$$

Solution 46. (i) We have

$$x_1 = 0, x_2 = 0, x_1 \oplus x_2 = 0 \Rightarrow |0\rangle \otimes |0\rangle \mapsto |0\rangle \otimes |0\rangle$$
$$x_1 = 0, x_2 = 1, x_1 \oplus x_2 = 1 \Rightarrow |0\rangle \otimes |1\rangle \mapsto |0\rangle \otimes |1\rangle$$
$$x_1 = 1, x_2 = 0, x_1 \oplus x_2 = 1 \Rightarrow |1\rangle \otimes |0\rangle \mapsto |1\rangle \otimes |1\rangle$$
$$x_1 = 1, x_2 = 1, x_1 \oplus x_2 = 0 \Rightarrow |1\rangle \otimes |1\rangle \mapsto |1\rangle \otimes |0\rangle.$$

Hence the permutation matrix is

$$P = \begin{pmatrix} 1 & 0 & 0 & 0 \\ 0 & 1 & 0 & 0 \\ 0 & 0 & 0 & 1 \\ 0 & 0 & 1 & 0 \end{pmatrix}.$$

(ii) We have

$$(x_1, x_2, x_3) = (0, 0, 0), x_1 \oplus x_2 = 0, x_1 \oplus x_2 \oplus x_3 = 0$$
$$\Rightarrow |0\rangle \otimes |0\rangle \otimes |0\rangle \mapsto |0\rangle \otimes |0\rangle \otimes |0\rangle$$
$$x_1 = 0, x_2 = 0, x_3 = 1, x_1 \oplus x_2 = 0, x_1 \oplus x_2 \oplus x_3 = 1$$
$$\Rightarrow |0\rangle \otimes |0\rangle \otimes |1\rangle \mapsto |0\rangle \otimes |0\rangle \otimes |1\rangle$$
$$x_1 = 0, x_2 = 1, x_3 = 0, x_1 \oplus x_2 = 1, x_1 \oplus x_2 \oplus x_3 = 1$$
$$\Rightarrow |0\rangle \otimes |1\rangle \otimes |1\rangle \mapsto |0\rangle \otimes |1\rangle \otimes |0\rangle$$
$$x_1 = 0, x_2 = 1, x_3 = 1, x_1 \oplus x_2 = 1, x_1 \oplus x_2 \oplus x_3 = 0$$
$$\Rightarrow |0\rangle \otimes |1\rangle \otimes |1\rangle \mapsto |0\rangle \otimes |1\rangle \otimes |0\rangle$$
$$x_1 = 1, x_2 = 0, x_3 = 0, x_1 \oplus x_2 = 1, x_1 \oplus x_2 \oplus x_3 = 1$$
$$\Rightarrow |1\rangle \otimes |0\rangle \otimes |0\rangle \mapsto |1\rangle \otimes |1\rangle \otimes |1\rangle$$
$$x_1 = 1, x_2 = 0, x_3 = 1, x_1 \oplus x_2 = 1, x_1 \oplus x_2 \oplus x_3 = 0$$
$$\Rightarrow |1\rangle \otimes |0\rangle \otimes |1\rangle \mapsto |1\rangle \otimes |1\rangle \otimes |0\rangle$$

$x_1 = 1, x_2 = 1, x_3 = 0, x_1 \oplus x_2 = 0, x_1 \oplus x_2 \oplus x_3 = 0$

$$\Rightarrow |1\rangle \otimes |1\rangle \otimes |0\rangle \mapsto |1\rangle \otimes |0\rangle \otimes |0\rangle$$

$x_1 = 1, x_2 = 1, x_3 = 1, x_1 \oplus x_2 = 0, x_1 \oplus x_2 \oplus x_3 = 1$

$$\Rightarrow |1\rangle \otimes |1\rangle \otimes |1\rangle \mapsto |1\rangle \otimes |0\rangle \otimes |1\rangle.$$

Hence the 8×8 permutation matrix is

$$P = \begin{pmatrix} 1 & 0 \\ 0 & 1 \end{pmatrix} \oplus \begin{pmatrix} 0 & 1 & 0 & 0 & 0 & 0 \\ 1 & 0 & 0 & 0 & 0 & 0 \\ 0 & 0 & 0 & 0 & 1 & 0 \\ 0 & 0 & 0 & 0 & 0 & 1 \\ 0 & 0 & 0 & 1 & 0 & 0 \\ 0 & 0 & 1 & 0 & 0 & 0 \end{pmatrix}$$

where $\oplus$ denotes the direct sum.

Problem 47. Consider the Hilbert space $\mathbb{C}^2$. Let $k > 0$.
(i) Find the eigenvalues and normalized eigenvectors of the 2×2 matrix

$$A(k, x) = \begin{pmatrix} \cos(kx) & (1/k)\sin(kx) \\ -k\sin(kx) & \cos(kx) \end{pmatrix}.$$

Note that the determinant is equal to $+1$. Do the eigenvectors form an orthonormal basis in $\mathbb{C}^2$? Is the matrix $A(k, x)$ normal?
(ii) Consider the Hilbert space $\mathbb{C}^4$. Find the eigenvalues and normalized eigenvectors of the 4×4 matrix $A(k, x) \otimes A(k, x)$. Do the eigenvectors form an orthonormal basis in $\mathbb{C}^2$?

Solution 47. (i) From $(\cos(kx) - \lambda)(\cos(kx) - \lambda) + \sin^2(kx) = 0$ we obtain

$$\lambda^2 - 2\lambda\cos(kx) = -1 \Rightarrow \lambda^2 - 2\lambda\cos(kx) + \cos^2(kx) = -\sin^2(kx).$$

Hence the eigenvalues are $\lambda_+ = e^{ikx}$, $\lambda_- = e^{-ikx}$ with the corresponding normalized eigenvectors

$$v_+ = \frac{1}{\sqrt{1 + k^2}} \begin{pmatrix} 1 \\ ik \end{pmatrix}, \qquad v_- = \frac{1}{\sqrt{1 + k^2}} \begin{pmatrix} i/k \\ 1 \end{pmatrix}.$$

The scalar product in the Hilbert space $\mathbb{C}^2$ between the vectors v_+ and v_- is given by

$$\langle v_+, v_- \rangle = i\frac{1 - k^2}{1 + k^2}$$

which is 0 only if $k = 1$. Testing $A^T(k, x)A(k, x) = A(k, x)A^T(k, x)$ we find that the matrix is normal only if $k = 1$.

(ii) The eigenvalues of $A(k,x) \otimes A(k,x)$ can be found from λ_+ and λ_- and are given by

$$\lambda_+^2, \quad \lambda_+\lambda_-, \quad \lambda_-\lambda_+, \quad \lambda_-^2$$

with the corresponding normalized eigenvectors

$$v_+ \otimes v_+, \quad v_+ \otimes v_-, \quad v_- \otimes v_+, \quad v_- \otimes v_-.$$

For example we have

$$\lambda_+\lambda_- = 1 \Rightarrow v_+ \otimes v_- = \frac{k}{1+k^2} \begin{pmatrix} i/k \\ 1 \\ -1 \\ ik \end{pmatrix}$$

$$\lambda_-\lambda_+ = 1 \Rightarrow v_- \otimes v_+ = \frac{k}{1+k^2} \begin{pmatrix} i/k \\ -1 \\ 1 \\ ik \end{pmatrix}.$$

Problem 48. The three spin matrices $S_1^{(s)}$, $S_2^{(s)}$, $S_3^{(s)}$ for spin $s = \frac{1}{2}$, $s = 1$, $s = \frac{3}{2}$, $s = 2$, ... satisfy the commutation relations

$$[S_1^{(s)}, S_2^{(s)}] = iS_3^{(s)}, \quad [S_2^{(s)}, S_3^{(s)}] = iS_1^{(s)}, \quad [S_3^{(s)}, S_1^{(s)}] = iS_2^{(s)}.$$

Let $z \in \mathbb{C}$. Show that

$$\exp(zS_2^{(s)})S_1^{(s)}\exp(-zS_2^{(s)}) = \cosh(z)S_1^{(s)} - i\sinh(z)S_3^{(s)}$$
$$\exp(zS_3^{(s)})S_2^{(s)}\exp(-zS_3^{(s)}) = \cosh(z)S_2^{(s)} - i\sinh(z)S_1^{(s)}$$
$$\exp(zS_1^{(s)})S_3^{(s)}\exp(-zS_1^{(s)}) = \cosh(z)S_3^{(s)} - i\sinh(z)S_2^{(s)}$$

or written in matrix form

$$\begin{pmatrix} \exp(zS_2^{(s)})S_1^{(s)}\exp(-zS_2^{(s)}) \\ \exp(zS_3^{(s)})S_2^{(s)}\exp(-zS_3^{(s)}) \\ \exp(zS_1^{(s)})S_3^{(s)}\exp(-zS_1^{(s)}) \end{pmatrix} = \begin{pmatrix} \cosh(z) & 0 & -i\sinh(z) \\ -i\sinh(z) & \cosh(z) & 0 \\ 0 & -i\sinh(z) & \cosh(z) \end{pmatrix} \begin{pmatrix} S_1^{(s)} \\ S_2^{(s)} \\ S_3^{(s)} \end{pmatrix}.$$

For $z = 0$ we have the 3×3 identity matrix.

Solution 48. We could utilize the Baker-Campbell-Hausdorff formula or setting

$$f_{12}(z) := \exp(zS_2^{(s)})S_1^{(s)}\exp(-zS_2^{(s)}), \quad f_{23}(z) := \exp(zS_3^{(s)})S_2^{(s)}\exp(-zS_3^{(s)}),$$

$$f_{31}(z) := \exp(zS_1^{(s)})S_3^{(s)}\exp(-zS_1^{(s)})$$

and then solve the matrix valued linear differential equation with the appropriate initial conditions.

Problem 49. Consider the 4×4 matrices

$$\Pi_+ = \begin{pmatrix} 1 & 0 & 0 & 0 \\ 0 & 1/2 & 1/2 & 0 \\ 0 & 1/2 & 1/2 & 0 \\ 0 & 0 & 0 & 1 \end{pmatrix}, \quad \Pi_- = \begin{pmatrix} 0 & 0 & 0 & 0 \\ 0 & 1/2 & -1/2 & 0 \\ 0 & -1/2 & 1/2 & 0 \\ 0 & 0 & 0 & 0 \end{pmatrix}.$$

Show that Π_+ and Π_- are projection matrices. Find $\Pi_+\Pi_-$ and $\Pi_+ + \Pi_-$. Discuss.

Solution 49. We have $\Pi_+ = \Pi_+^*$, $\Pi_+^2 = \Pi_+$ and $\Pi_- = \Pi_-^*$, $\Pi_-^2 =' \Pi_-$. Hence Π_+ and Π_- are projection matrices. We find $\Pi_+\Pi_- = 0_4$ and $\Pi_+ + \Pi_- = I_4$. We have

$$\Pi_+ \begin{pmatrix} x_1 \\ x_2 \\ x_3 \\ x_4 \end{pmatrix} = \begin{pmatrix} x_1 \\ (x_2 + x_3)/2 \\ (x_2 + x_3)/2 \\ x_4 \end{pmatrix}, \quad \Pi_- \begin{pmatrix} x_1 \\ x_2 \\ x_3 \\ x_4 \end{pmatrix} = \begin{pmatrix} 0 \\ (x_2 - x_3)/2 \\ (-x_2 + x_3)/2 \\ 0 \end{pmatrix}.$$

Thus Π_+ projects into a three-dimensional subspace and Π_- projects into a one-dimensional subspace.

Problem 50. Let $\mathcal{H}$ be the Hilbert space of $n \times n$ matrices over $\mathbb{C}$ with the scalar product $\langle A, B^* \rangle$.
(i) Consider the map $f : \mathcal{H} \to \mathcal{H}$, $f(A) = A^2$. Find the *Gateaux derivative*

$$\frac{d}{d\tau}(f(A + \tau B)) \Big|_{\tau=0}$$

(ii) Consider the map $f : \mathcal{H} \to \mathcal{H}$, $f(A) = A^3$. Find the Gateaux derivative

$$\frac{d}{d\tau}(f(A + \tau B)) \Big|_{\tau=0}$$

(iii) Consider the map $f : \mathcal{H} \to \mathcal{H} \otimes \mathcal{H}$, $f(A) = A \otimes A$. Find the Gateaux derivative

$$\frac{d}{d\tau}(f(A + \tau B)) \Big|_{\tau=0}$$

(iv) Consider the map $f : \mathcal{H} \to \mathcal{H} \otimes \mathcal{H}$, $f(A) = A \otimes I_n + I_n \otimes A$. Find the Gateaux derivative

$$\frac{d}{d\tau}(f(A + \tau B)) \Big|_{\tau=0}.$$

Solution 50. (i) We have

$$f(A + \tau B) = (A + \tau B)(A + \tau B) = A^2 + \tau AB + \tau BA + \tau^2 B^2.$$

Thus

$$\frac{d}{d\tau}f(A + \tau B)\bigg|_{\tau=0} = AB + BA + 2\tau B^2\big|_{\tau=0} = AB + BA \equiv [A, B]_+.$$

(ii) With

$$(A+\tau B)^3 \equiv A^3 + \tau(ABA+BAA+AAB)+\tau^2(BBA+ABB+BAB)+\tau^3 B^3.$$

Thus

$$\frac{d}{d\tau}f(A + \tau B)\bigg|_{\tau=0} = ABA + BAA + AAB = A[A, B]_+ + BBA.$$

(iii) We have

$$f(A + \tau B) = A \otimes A + \tau(A \otimes B) + \tau(B \otimes A) + \tau^2 B \otimes B.$$

Thus

$$\frac{d}{d\tau}f(A + \tau B)\bigg|_{\tau=0} = A \otimes B + B \otimes A.$$

(iv) We have

$$f(A+\tau B) = (A+\tau B)\otimes I_n + I_n\otimes(A+\tau B) \equiv A\otimes I_n + \tau B\otimes I_n + I_n\otimes A + \tau I_n\otimes B.$$

Thus

$$\frac{d}{d\tau}f(A + \tau B)\bigg|_{\tau=0} = B \otimes I_n + I_n \otimes B.$$

Problem 51. Let $|0\rangle$, $|1\rangle$ be an orthonormal basis in the Hilbert space $\mathbb{C}^2$. Then

$$U = \frac{1}{\sqrt{2}}(|0\rangle\langle 0| - |1\rangle\langle 1| + |0\rangle\langle 1| + |1\rangle\langle 0|)$$

is a 2×2 unitary matrix. For example for the standard basis we obtain the Hadamard matrix

$$U_H = \frac{1}{\sqrt{2}}\begin{pmatrix} 1 & 1 \\ 1 & -1 \end{pmatrix}.$$

Now $\rho = (|0\rangle \otimes |0\rangle)(\langle 0| \otimes \langle 0|)$ is a *density matrix* acting in the Hilbert space $\mathbb{C}^4$. Find the density matrix

$$\widetilde{\rho} = (U \otimes U)\rho(U^* \otimes U^*).$$

Solution 51. With $\langle 0|0 \rangle = \langle 1|1 \rangle = 1$ and $\langle 0|1 \rangle = \langle 1|0 \rangle = 0$ we obtain

$$\tilde{\rho} = \frac{1}{4}(|0\rangle\langle 0| \otimes |0\rangle\langle 0| + |0\rangle\langle 0| \otimes |0\rangle\langle 1| + |0\rangle\langle 0| \otimes |1\rangle\langle 0| + |0\rangle\langle 0| \otimes |1\rangle\langle 1| +$$
$$|0\rangle\langle 1| \otimes |0\rangle\langle 0| + |0\rangle\langle 1| \otimes |0\rangle\langle 1| + |0\rangle\langle 1| \otimes |1\rangle\langle 0| + |0\rangle\langle 1| \otimes |1\rangle\langle 1| +$$
$$|1\rangle\langle 0| \otimes |0\rangle\langle 0| + |1\rangle\langle 0| \otimes |0\rangle\langle 1| + |1\rangle\langle 0| \otimes |1\rangle\langle 0| + |1\rangle\langle 0| \otimes |1\rangle\langle 1| +$$
$$|1\rangle\langle 1| \otimes |0\rangle\langle 0| + |1\rangle\langle 1| \otimes |0\rangle\langle 1| + |1\rangle\langle 1| \otimes |1\rangle\langle 0| + |1\rangle\langle 1| \otimes |1\rangle\langle 1|).$$

For the standard basis we obtain

$$\tilde{\rho} = \frac{1}{4}\begin{pmatrix} 1 & 1 & 1 & 1 \\ 1 & 1 & 1 & 1 \\ 1 & 1 & 1 & 1 \\ 1 & 1 & 1 & 1 \end{pmatrix} \equiv \frac{1}{4}\begin{pmatrix} 1 \\ 1 \\ 1 \\ 1 \end{pmatrix}(1\ 1\ 1\ 1).$$

Problem 52. (i) Given the three spin-$\frac{1}{2}$ matrices

$$S_1^{(1/2)} = \frac{1}{2}\begin{pmatrix} 0 & 1 \\ 1 & 0 \end{pmatrix}, \quad S_2^{(1/2)} = \frac{1}{2}\begin{pmatrix} 0 & -i \\ i & 0 \end{pmatrix}, \quad S_3^{(1/2)} = \frac{1}{2}\begin{pmatrix} 1 & 0 \\ 0 & -1 \end{pmatrix}$$

each with eigenvalues $-1/2$, $+1/2$. Find the eigenvalues and normalized eigenvectors of

$$S^{(1/2)} := \frac{1}{\sqrt{3}}(S_1^{(1/2)} + S_2^{(1/2)} + S_3^{(1/2)}).$$

Find the eigenvalues of $S^{(1/2)} \otimes S^{(1/2)}$.

(ii) Given the three spin-1 matrices

$$S_1^{(1)} = \frac{1}{\sqrt{2}}\begin{pmatrix} 0 & 1 & 0 \\ 1 & 0 & 1 \\ 0 & 1 & 0 \end{pmatrix}, \quad S_2^{(1)} = \frac{1}{\sqrt{2}}\begin{pmatrix} 0 & -i & 0 \\ i & 0 & -i \\ 0 & i & 0 \end{pmatrix}, \quad S_3^{(1)} = \begin{pmatrix} 1 & 0 & 0 \\ 0 & 0 & 0 \\ 0 & 0 & -1 \end{pmatrix}$$

with eigenvalues -1, 0, $+1$. Find the eigenvalues of

$$S^{(1)} := \frac{1}{\sqrt{3}}(S_1^{(1)} + S_2^{(1)} + S_3^{(1)}).$$

Find the eigenvalues of $S^{(1)} \otimes S^{(1)}$.

Solution 52. (i) The eigenvalues of $S^{(1/2)}$ are also $-1/2$ and $+1/2$. The normalized eigenvector for $-1/2$ is

$$\frac{1}{\sqrt{2(3 - \sqrt{3})}}\begin{pmatrix} 1 - \sqrt{3} \\ 1 + i \end{pmatrix}.$$

The normalized eigenvector for $+1/2$ is

$$\frac{1}{\sqrt{2(3+\sqrt{3})}}\begin{pmatrix}1+\sqrt{3}\\1+i\end{pmatrix}.$$

The eigenvalues of $S^{(1/2)}\otimes S^{(1/2)}$ are $-1/4$ (twice) and $+1/4$ (twice).
(ii) The eigenvalues of $S^{(1)}$ are -1, 0, $+1$. The eigenvalues of $S^{(1)}\otimes S^{(1)}$ are -1 (twice) 0 (5 times) $+1$ (twice).

Problem 53. Let $|0\rangle$, $|1\rangle$ be the standard basis of the Hilbert space $\mathbb{C}^2$. Find the 4×4 projection matrix Π_+ with

$$\Pi_+(|0\rangle\otimes|0\rangle)=|0\rangle\otimes|0\rangle,\quad \Pi_+(|1\rangle\otimes|1\rangle)=|1\rangle\otimes|1\rangle$$

$$\Pi_+(|0\rangle\otimes|1\rangle)=\frac{1}{2}(|0\rangle\otimes|1\rangle+|1\rangle\otimes|0\rangle),\quad \Pi_+(|1\rangle\otimes|0\rangle)=\frac{1}{2}(|1\rangle\otimes|0\rangle-|0\rangle\otimes|1\rangle).$$

Find the 4×4 projection matrix Π_- with

$$\Pi_-(|0\rangle\otimes|0\rangle)=\begin{pmatrix}0\\0\\0\\0\end{pmatrix},\quad \Pi_-(|1\rangle\otimes|1\rangle)=\begin{pmatrix}0\\0\\0\\0\end{pmatrix},$$

$$\Pi_-(|0\rangle\otimes|1\rangle)=\frac{1}{2}(|0\rangle\otimes|1\rangle-|1\rangle\otimes|0\rangle),\quad \Pi_-(|1\rangle\otimes|0\rangle)=\frac{1}{2}(|1\rangle\otimes|0\rangle-|0\rangle\otimes|1\rangle).$$

Calculate $\Pi_+ + \Pi_-$, $\Pi_+\Pi_-$ and

$$\Pi_+\begin{pmatrix}z_1\\z_2\\z_3\\z_4\end{pmatrix},\quad \Pi_-\begin{pmatrix}z_1\\z_2\\z_3\\z_4\end{pmatrix}.$$

Solution 53. From

$$\Pi_+\begin{pmatrix}1\\0\\0\\0\end{pmatrix}=\begin{pmatrix}1\\0\\0\\0\end{pmatrix},\quad \Pi_+\begin{pmatrix}0\\0\\0\\1\end{pmatrix}=\begin{pmatrix}0\\0\\0\\1\end{pmatrix},$$

$$\Pi_+\begin{pmatrix}0\\1\\0\\0\end{pmatrix}=\frac{1}{2}\begin{pmatrix}0\\1\\1\\0\end{pmatrix},\quad \Pi_+\begin{pmatrix}0\\0\\1\\0\end{pmatrix}=\frac{1}{2}\begin{pmatrix}0\\1\\1\\0\end{pmatrix}.$$

we obtain the projection matrix

$$\Pi_+ = \begin{pmatrix} 1 & 0 & 0 & 0 \\ 0 & 1/2 & 1/2 & 0 \\ 0 & 1/2 & 1/2 & 0 \\ 0 & 0 & 0 & 1 \end{pmatrix}.$$

From

$$\Pi_- \begin{pmatrix} 1 \\ 0 \\ 0 \\ 0 \end{pmatrix} = \begin{pmatrix} 0 \\ 0 \\ 0 \\ 0 \end{pmatrix}, \qquad \Pi_- \begin{pmatrix} 0 \\ 0 \\ 0 \\ 1 \end{pmatrix} = \begin{pmatrix} 0 \\ 0 \\ 0 \\ 0 \end{pmatrix},$$

$$\Pi_- \begin{pmatrix} 0 \\ 1 \\ 0 \\ 0 \end{pmatrix} = \frac{1}{2} \begin{pmatrix} 0 \\ 1 \\ -1 \\ 0 \end{pmatrix}, \qquad \Pi_- \begin{pmatrix} 0 \\ 0 \\ 1 \\ 0 \end{pmatrix} = \frac{1}{2} \begin{pmatrix} 0 \\ -1 \\ 1 \\ 0 \end{pmatrix}.$$

we obtain the projection matrix

$$\Pi_- = \begin{pmatrix} 0 & 0 & 0 & 0 \\ 0 & 1/2 & -1/2 & 0 \\ 0 & -1/2 & 1/2 & 0 \\ 0 & 0 & 0 & 0 \end{pmatrix}.$$

Then $\Pi_+ + \Pi_- = I_4$ and $\Pi_+\Pi_- = 0_4$. Now

$$\Pi_+ \begin{pmatrix} z_1 \\ z_2 \\ z_3 \\ z_4 \end{pmatrix} = \begin{pmatrix} z_1 \\ (z_2 + z_3)/2 \\ (z_2 + z_3)/2 \\ z_4 \end{pmatrix}, \qquad \Pi_- \begin{pmatrix} z_1 \\ z_2 \\ z_3 \\ z_4 \end{pmatrix} = \begin{pmatrix} 0 \\ (z_2 - z_3)/2 \\ (-z_2 + z_3)/2 \\ 0 \end{pmatrix}.$$

Hence Π_+ projects into a three-dimensional subspace and Π_- projects into a one-dimensional subspace.

Problem 54. Consider the Hilbert space $\mathbb{R}^2$ and the standard basis

$$e_1 = \begin{pmatrix} 1 \\ 0 \end{pmatrix}, \qquad e_2 = \begin{pmatrix} 0 \\ 1 \end{pmatrix}.$$

The 2×2 matrices

$$X = \begin{pmatrix} 0 & 1 \\ 0 & 0 \end{pmatrix}, \qquad Y = \begin{pmatrix} 0 & 0 \\ 1 & 0 \end{pmatrix}, \qquad H = \begin{pmatrix} 1 & 0 \\ 0 & -1 \end{pmatrix}$$

form a basis of the *Lie algebra* $s\ell(2, \mathbb{R})$. Find the vectors

$$Xe_1, \ Xe_2, \ Ye_1, \ Ye_2, \ He_1, \ He_2.$$

Find the vectors $X(Y e_1)$, $Y(X e_2)$.

Solution 54. We obtain

$$X e_1 = \begin{pmatrix} 0 \\ 0 \end{pmatrix}, \quad X e_2 = \begin{pmatrix} 1 \\ 0 \end{pmatrix} = e_1, \quad Y e_1 = \begin{pmatrix} 0 \\ 1 \end{pmatrix} = e_2, \quad Y e_2 = \begin{pmatrix} 0 \\ 0 \end{pmatrix}$$

and the eigenvalue equations

$$H e_1 = \begin{pmatrix} 1 \\ 0 \end{pmatrix} = e_1, \quad H e_2 = \begin{pmatrix} 0 \\ 1 \end{pmatrix} = -e_2.$$

We find $X(Y e_1) = e_1$, $Y(X e_2) = e_2$.

Problem 55. Consider the Hilbert spaces $\mathbb{C}^2$ and $\mathbb{C}^4$. Let

$$|e_0\rangle = \begin{pmatrix} 1 \\ 0 \end{pmatrix}, \quad |e_1\rangle = \begin{pmatrix} 0 \\ 1 \end{pmatrix}$$

be the standard basis in $\mathbb{C}^2$. Let S_2 be the *symmetric group*. Find

$$|\psi\rangle = \frac{1}{\sqrt{2!}} \sum_{\sigma \in S_2} |e_{\sigma(0)}\rangle \otimes |e_{\sigma(1)}\rangle$$

Solution 55. We have

$$|\psi\rangle = \frac{1}{\sqrt{2}}(|e_0\rangle \otimes |e_1\rangle + |e_1\rangle \otimes |e_0\rangle) = \frac{1}{\sqrt{2}}\left(\begin{pmatrix} 1 \\ 0 \end{pmatrix} \otimes \begin{pmatrix} 0 \\ 1 \end{pmatrix} + \begin{pmatrix} 0 \\ 1 \end{pmatrix} \otimes \begin{pmatrix} 1 \\ 0 \end{pmatrix} \right)$$

$$= \frac{1}{\sqrt{2}} \begin{pmatrix} 0 \\ 1 \\ 1 \\ 0 \end{pmatrix}.$$

Problem 56. Consider the Hilbert space $\mathbb{C}^n$. Let A be an $n \times n$ normal matrix (i.e. $AA^* = A^*A$) and

$$A = \sum_{j=1}^{n} \lambda_j \mathbf{v}_j \mathbf{v}_j^*$$

be the *spectral representation* of A, i.e. λ_j $(j = 1, \ldots, n)$ are the eigenvalues of A and $\mathbf{v}_1, \ldots, \mathbf{v}_n$ are the normalized eigenvectors which form an orthonormal basis in the Hilbert space $\mathbb{C}^n$. Find $\exp(i\alpha A)$.

Solution 56. Since $\mathbf{v}_j \mathbf{v}_k = 0$ for $j \neq k$ we have

$$\exp(i\alpha A) = \exp\left(i\alpha \sum_{j=1}^{n} \lambda_j \mathbf{v}_j \mathbf{v}_j^*\right) = \prod_{j=1}^{n} \exp(i\alpha \lambda_j \mathbf{v}_j \mathbf{v}_j^*).$$

Now $(\mathbf{v}_j \mathbf{v}_j^*)^2 = \mathbf{v}_j \mathbf{v}_j^*$. Then $\exp(i\alpha \lambda_j \mathbf{v}_j \mathbf{v}_j^*) = I_n + (e^{i\alpha \lambda_j} - 1)\mathbf{v}_j \mathbf{v}_j^*$. It follows that

$$\exp(i\alpha A) = \prod_{j=1}^{n} (I_n + (e^{i\alpha \lambda_j} - 1)\mathbf{v}_j \mathbf{v}_j^*) = \sum_{j=1}^{n} e^{i\alpha \lambda_j} \mathbf{v}_j \mathbf{v}_j^*.$$

Problem 57. Let $d \geq 2$ and $|\psi\rangle$, $|\phi\rangle$ be two normalized states in the Hilbert state $\mathbb{C}^d$ and $|0\rangle, |1\rangle, \ldots, |d-1\rangle$ be the standard basis in the Hilbert space $\mathbb{C}^d$. Consider the matrix (*swap gate*)

$$S := \sum_{j,k=0}^{d-1} ((|j\rangle\langle k|) \otimes (|k\rangle\langle j|)).$$

Find $S(|\psi\rangle \otimes |\phi\rangle)$.

Solution 57. We have

$$S = \sum_{j,k=0}^{d-1} ((|j\rangle\langle k|) \otimes (|k\rangle\langle j|)(|\psi\rangle \otimes |\phi\rangle) = \sum_{j,k=0}^{d-1} (|j\rangle\langle k|\psi\rangle) \otimes (|k\rangle\langle j|\phi\rangle)$$

$$= \sum_{j,k=1}^{d-1} \langle k|\psi\rangle\langle j|\phi\rangle(|j\rangle \otimes |k\rangle) = \sum_{j,k=0}^{d-1} \langle j|\psi\rangle\langle k|\phi\rangle(|k\rangle \otimes |j\rangle)$$

$$= |\phi\rangle \otimes |\psi\rangle.$$

Problem 58. Consider the column vector $\mathbf{v}$

$$\mathbf{v} = \begin{pmatrix} i & -i & 0 & 0 & -i & i \end{pmatrix}^T$$

in the Hilbert space $\mathbb{C}^6$.
(i) Can the vector $\mathbf{v}$ be written as

$$\mathbf{v} = \begin{pmatrix} a_1 \\ a_2 \\ a_3 \end{pmatrix} \otimes \begin{pmatrix} b_1 \\ b_2 \end{pmatrix} ?$$

(ii) Can the vector $\mathbf{v}$ be written as

$$\mathbf{v} = \begin{pmatrix} c_1 \\ c_2 \end{pmatrix} \otimes \begin{pmatrix} d_1 \\ d_2 \\ d_3 \end{pmatrix} ?$$

Solution 58. (i) We have

$$\mathbf{v} = \begin{pmatrix} 1 \\ 0 \\ -1 \end{pmatrix} \otimes \begin{pmatrix} i \\ -i \end{pmatrix}.$$

(ii) From the system of equation

$$\begin{pmatrix} c_1 d_1 \\ c_1 d_2 \\ c_1 d_3 \\ c_2 d_1 \\ c_2 d_2 \\ c_2 d_3 \end{pmatrix} = \begin{pmatrix} i \\ -i \\ 0 \\ 0 \\ -i \\ i \end{pmatrix}$$

we find a contradiction and hence $\mathbf{v}$ cannot be written in this form.

Problem 59. Consider the Hilbert spaces $\mathbb{C}^2$ and $\mathbb{C}^4$. Let

$$\mathbf{a} = \begin{pmatrix} a_1 \\ a_2 \end{pmatrix}, \quad \mathbf{b} = \begin{pmatrix} b_1 \\ b_2 \end{pmatrix}$$

be normalized vectors in $\mathbb{C}^2$ and

$$\mathbf{v} = \frac{1}{\sqrt{2}} \left(\begin{pmatrix} 1 \\ 0 \end{pmatrix} \otimes \begin{pmatrix} a_1 \\ a_2 \end{pmatrix} + \begin{pmatrix} 0 \\ 1 \end{pmatrix} \otimes \begin{pmatrix} b_1 \\ b_2 \end{pmatrix} \right)$$

$$\mathbf{w} = \frac{1}{\sqrt{2}} \left(\begin{pmatrix} 1 \\ 0 \end{pmatrix} \otimes I_2 - \begin{pmatrix} 1 \\ 0 \end{pmatrix} \otimes I_2 \right) \Rightarrow \mathbf{w}^* = \frac{1}{\sqrt{2}} ((1\ 0) \otimes I_2 - (0\ 1) \otimes I_2)$$

where I_2 is the 2×2 identity matrix. Find the *partial scalar product* $\mathbf{w}^*\mathbf{v}$ and $\|\mathbf{w}^*\mathbf{v}\|^2$.

Solution 59. With

$$(1\ 0) \otimes I_2 = \begin{pmatrix} 1\ 0\ 0\ 0 \\ 0\ 1\ 0\ 0 \end{pmatrix}, \quad (0\ 1) \otimes I_2 = \begin{pmatrix} 0\ 0\ 1\ 0 \\ 0\ 0\ 0\ 1 \end{pmatrix}$$

we obtain

$$\mathbf{w}^*\mathbf{v} = \frac{1}{2} \left(\begin{pmatrix} a_1 \\ a_2 \end{pmatrix} - \begin{pmatrix} b_1 \\ b_2 \end{pmatrix} \right) = \frac{1}{2} \begin{pmatrix} a_1 - b_1 \\ a_2 - b_2 \end{pmatrix}.$$

Then

$$\mathbf{v}^*\mathbf{w} = \frac{1}{2}\left(\bar{a}_1 - \bar{b}_1\ \bar{a}_2 - \bar{b}_2\right)$$

and with $\bar{a}_1 a_1 + \bar{a}_2 a_2 = 1$, $\bar{b}_1 b_1 + \bar{b}_2 b_2 = 1$ we obtain

$$(\mathbf{v}^*\mathbf{w})(\mathbf{w}^*\mathbf{v}) = \frac{1}{4}(2 - \bar{a}_1 b_1 - a_1 \bar{b}_1 - \bar{a}_2 b_2 - a_2 \bar{b}_2) = \frac{1}{2}(1 + \Re(\mathbf{a}^*\mathbf{b})).$$

Problem 60. Let $z \in \mathbb{C}$. Consider the normalized state (*spin-$\frac{1}{2}$ coherent state*)

$$|z\rangle = \frac{1}{(1+|z|^2)^{1/2}}\begin{pmatrix}1\\z\end{pmatrix}.$$

in the Hilbert space $\mathbb{C}^2$.

(i) Find $|z\rangle\langle z|$ and $\mathrm{tr}(|z\rangle\langle z|)$. Show that (*completeness relation*)

$$\frac{1}{\pi}\int_{\mathbb{C}} dz \frac{2}{(1+|z|^2)^2}|z\rangle\langle z| = I_2$$

where I_2 is the 2×2 unit matrix and

$$\frac{1}{\pi}\cdot\frac{2s+1}{(1+|z|^2)^2}$$

is the *weight factor* with $s = 1/2$. With $z = re^{i\phi}$ ($r \geq 0, \phi \in \mathbb{R}$) the integration over the complex plane $\mathbb{C}$ is given by

$$\int_{\mathbb{C}}(\cdots)dz = \int_{r=0}^{\infty}\int_{\phi=0}^{2\pi}(\cdots)r dr d\phi.$$

Solution 60. (i) We have

$$|z\rangle\langle z| = \frac{1}{1+|z|^2}\begin{pmatrix}1 & \bar{z}\\ z & z\bar{z}\end{pmatrix} = \frac{1}{1+r^2}\begin{pmatrix}1 & re^{-i\phi}\\ re^{i\phi} & r^2\end{pmatrix}$$

and $\mathrm{tr}(|z\rangle\langle z|) = 1$.

(ii) For the matrix element $(1,1)$ we have

$$\frac{2}{\pi}\int_{r=0}^{\infty}\int_0^{2\pi}\frac{1}{(1+r^2)^2}\cdot\frac{1}{1+r^2}r dr d\phi = 4\int_{r=0}^{\infty}\frac{r}{(1+r^2)^3}dr = 1.$$

For the matrix elements $(1,2)$ and $(2,1)$ we have

$$\int_{\phi=0}^{2\pi}e^{-i\phi}d\phi = 0, \qquad \int_{\phi=0}^{2\pi}e^{i\phi}d\phi = 0.$$

Finally for the matrix element $(2, 2)$ we have

$$\frac{2}{\pi} \int_{r=0}^{\infty} \int_{0}^{2\pi} \frac{1}{(1+r^2)^2} \cdot \frac{r^2}{1+r^2} r dr d\phi = 4 \int_{r=0}^{\infty} \frac{r^3}{(1+r^2)^3} dt = 1.$$

Hence we obtain the 2×2 unit matrix.

Problem 61. Consider the Hilbert space $\mathbb{C}^n$ and $\lambda_j \in \mathbb{C}$ with $j = 1, \ldots, n$. Let $|a_j\rangle$ $(j = 1, \ldots, n)$ be an orthonormal basis in the Hilbert space $\mathbb{C}^n$ and $|b_j\rangle$ be another orthonormal basis in $\mathbb{C}^n$. Consider the $n \times n$ matrices

$$A = \sum_{j=1}^{n} \lambda_j |a_j\rangle\langle a_j|, \qquad B = \sum_{j=1}^{n} \lambda_j |b_j\rangle\langle b_j|.$$

Find the commutator $[A, B]$. Then apply it to the special case $n = 2$ with $\lambda_1 = -1/2$, $\lambda_2 = 1/2$ and

$$|a_1\rangle = \begin{pmatrix} 1 \\ 0 \end{pmatrix}, \quad |a_2\rangle = \begin{pmatrix} 0 \\ 1 \end{pmatrix}, \qquad |b_1\rangle = \frac{1}{\sqrt{2}} \begin{pmatrix} 1 \\ 1 \end{pmatrix}, \quad |b_2\rangle = \frac{1}{\sqrt{2}} \begin{pmatrix} 1 \\ -1 \end{pmatrix}.$$

The two bases are *mutually unbiased*.

Solution 61. We obtain

$$[A, B] = \sum_{j,k=1}^{n} \lambda_j \lambda_k \left(\langle a_j | b_k \rangle |a_j\rangle\langle b_k| - \langle b_k | a_j \rangle |b_k\rangle\langle a_j| \right).$$

For the special case with $n = 2$ we have

$$\langle a_1 | b_1 \rangle = \langle b_1 | a_1 \rangle = \frac{1}{\sqrt{2}}, \quad \langle a_1 | b_2 \rangle = \langle b_2 | a_1 \rangle = \frac{1}{\sqrt{2}},$$

$$\langle a_2 | b_1 \rangle = \langle b_1 | a_2 \rangle = \frac{1}{\sqrt{2}}, \quad \langle a_2 | b_2 \rangle = \langle b_2 | a_2 \rangle = -\frac{1}{\sqrt{2}}.$$

Problem 62. Let $n \geq 2$. Consider the Hilbert spaces $\mathbb{C}^n$ and $\mathbb{C}^{n^2}$. Let P be the $n^2 \times n^2$ permutation matrix such that

$$\begin{pmatrix} x_1 \\ x_2 \\ \vdots \\ x_n \end{pmatrix} \otimes \begin{pmatrix} y_1 \\ y_2 \\ \vdots \\ y_n \end{pmatrix} = P \left(\begin{pmatrix} y_1 \\ y_2 \\ \vdots \\ y_n \end{pmatrix} \otimes \begin{pmatrix} x_1 \\ x_2 \\ \vdots \\ x_n \end{pmatrix} \right)$$

for $\mathbf{x}, \mathbf{y} \in \mathbb{C}^n$. Let A be an $n \times n$ matrix over $\mathbb{C}$. Find the commutator $[A \otimes A, P]$.

Solution 62. We obtain $P(A \otimes A)P^T = A \otimes A$ with $P^T = P^{-1}$. Hence

$$[A \otimes A, P] = O_{n^2 \times n^2}$$

i.e. the $n^2 \times n^2$ matrices P and $A \otimes A$ commute.

Problem 63. Consider a complex Hilbert space $\mathcal{H}$ and $|\phi_1\rangle, |\phi_2\rangle \in \mathcal{H}$. Let $c_1, c_2 \in \mathbb{C}$. An *antilinear operator* K in this Hilbert space $\mathcal{H}$ is characterized by

$$K(c_1|\phi_1\rangle + c_2|\phi_2\rangle) = c_1^* K|\phi_1\rangle + c_2^* K|\phi_2\rangle.$$

A *comb* is an antilinear operator K with zero expectation value for all states $|\psi\rangle$ of a certain complex Hilbert space $\mathcal{H}$. This means

$$\langle\psi|K|\psi\rangle = \langle\psi|LC|\psi\rangle = \langle\psi|L|\psi^*\rangle = 0$$

for all states $|\psi\rangle \in \mathcal{H}$, where L is a linear operator and C is the complex conjugation.
(i) Consider the two-dimensional Hilbert space $\mathcal{H} = \mathbb{C}^2$. Find a unitary 2×2 matrix U such that $\langle\psi|UC|\psi\rangle = 0$.
(ii) Consider the Pauli spin matrices σ_1, σ_2, σ_3 with $\sigma_0 = I_2$. Find

$$\sum_{\mu=0}^{3} \sum_{\nu=0}^{3} \langle\psi|\sigma_\mu C|\psi\rangle g^{\mu,\nu} \langle\psi|\sigma_\nu C|\psi\rangle$$

where $g^{\mu,\nu} = \mathrm{diag}(-1, 1, 0, 1)$.

Solution 63. (i) We find $U = \sigma_2$ since

$$\langle\psi|\sigma_2 C|\psi\rangle = \langle\psi|\sigma_2|\psi^*\rangle = \begin{pmatrix} \psi_1^* & \psi_2^* \end{pmatrix} \begin{pmatrix} 0 & -i \\ i & 0 \end{pmatrix} \begin{pmatrix} \psi_1^* \\ \psi_2^* \end{pmatrix} = \begin{pmatrix} \psi_1^* & \psi_2^* \end{pmatrix} \begin{pmatrix} -i\psi_2^* \\ i\psi_1^* \end{pmatrix} = 0.$$

(ii) We have

$$\sum_{\mu=0}^{3} \sum_{\nu=0}^{3} \langle\psi|\sigma_\mu C|\psi\rangle g^{\mu,\nu} \langle\psi|\sigma_\nu C|\psi\rangle = -\langle\psi|\sigma_0|\psi^*\rangle^2 + \langle\psi|\sigma_1|\psi^*\rangle + \langle\psi|\sigma_3|\psi^*\rangle = 0.$$

3.3 Programming Problems

Problem 1. Let $|0\rangle$, $|1\rangle$ be an orthonormal basis in the Hilbert space $\mathbb{C}^2$.
(i) Show that

$$\Pi_+ = \frac{1}{2} \sum_{j=0}^{1} \sum_{k=0}^{1} |j\rangle\langle k| \equiv \frac{1}{2}(|0\rangle\langle 0| + |0\rangle\langle 1| + |1\rangle\langle 0| + |1\rangle\langle 1|)$$

is a projection matrix.
(ii) Show that

$$\Pi_- = \frac{1}{2} \sum_{j=0}^{1} \sum_{k=0}^{1} (-1)^{j+k}|j\rangle\langle k| \equiv \frac{1}{2}(|0\rangle\langle 0| - |0\rangle\langle 1| - |1\rangle\langle 0| + |1\rangle\langle 1|)$$

is a projection matrix.
(iii) Find $\Pi_+\Pi_-$ and $\Pi_+ + \Pi_-$.
(iv) Find Π_+ and Π_- for the orthonormal basis

$$|0\rangle = \begin{pmatrix} \cos(\theta) \\ \sin(\theta) \end{pmatrix}, \quad |1\rangle = \begin{pmatrix} -\sin(\theta) \\ \cos(\theta) \end{pmatrix}.$$

Solution 1. (i) We have $\Pi_+ = \Pi_+^*$ (the matrix is hermitian) and with $\langle 0|1\rangle = \langle 1|0\rangle = 0$ we obtain $\Pi_+^2 = \Pi_+$. Hence Π_+ is a projection matrix.
(ii) We have $\Pi_- = \Pi_-^*$ (the matrix is hermitian) and with $\langle 0|1\rangle = \langle 1|0\rangle = 0$ we obtain $\Pi_-^2 = \Pi_-$. Hence Π_- is a projection matrix.
(iii) We have $\Pi_+\Pi_- = 0_2$ and $\Pi_+ + \Pi_- = I_2$.
(iv) Applying the Maxima program

```
/* projection1.mac */
v0k: matrix([cos(theta)],[sin(theta)]);
v0b: matrix([cos(theta),sin(theta)]);
v1k: matrix([-sin(theta)],[cos(theta)]);
v1b: matrix([-sin(theta),cos(theta)]);
Pip: (v0k . v0b + v0k . v1b + v1k . v0b + v1k . v1b)/2;
Pip: trigsimp(Pip);
Pip2: Pip . Pip; Pip2: trigsimp(Pip2);
Diffp21: Pip2-Pip; Diffp21: trigsimp(Diffp21);
Pim: (v0k . v0b - v0k . v1b - v1k . v0b + v1k . v1b)/2;
Pim: trigsimp(Pim);
Pim2: Pim . Pim; Pim2: trigsimp(Pim2);
Diffm21: Pim2 - Pim; Diffm21: trigsimp(Diffm21);
Prodpm: Pip . Pim; Prodpm: trigsimp(Prodpm);
Sumpm: Pip + Pim; Sumpm: trigsimp(Sumpm);
```

we find the projection matrices

$$\Pi_+ = \begin{pmatrix} (1 - 2\cos(\theta)\sin(\theta))/2 & (2\cos^2(\theta) - 1)/2 \\ (2\cos^2(\theta) - 1)/2 & (1 + \cos(\theta)\sin(\theta))/2 \end{pmatrix}$$

$$\Pi_- = \begin{pmatrix} (1 + 2\cos(\theta)\sin(\theta))/2 & (1 - 2\cos^2(\theta))/2 \\ (1 - 2\cos^2(\theta))/2 & (1 - \cos(\theta)\sin(\theta))/2 \end{pmatrix}.$$

Problem 2. Consider the *permutation matrix*

$$P = \begin{pmatrix} 0 & 0 & 0 & 1 \\ 0 & 0 & 1 & 0 \\ 1 & 0 & 0 & 0 \\ 0 & 1 & 0 & 0 \end{pmatrix}$$

which is a unitary matrix. Find the hermitian matrix H such that $P = \exp(iH)$. Apply the spectral decomposition of P.

Solution 2. The *spectral decomposition* provides that

$$P = \lambda_1 |v_1\rangle\langle v_1| + \lambda_2 |v_2\rangle\langle v_2| + \lambda_3 |v_3\rangle\langle v_3| + \lambda_4 |v_4\rangle\langle v_4|$$

where λ_1, λ_2, λ_3, λ_4 are the eigenvalues of P and $|v_1\rangle$, $|v_2\rangle$, $|v_3\rangle$, $|v_4\rangle$ are the corresponding normalized eigenvectors which form an orthonormal basis in the Hilbert space $\mathbb{C}^4$. The eigenvalues of P are

$$\lambda_1 = +1, \quad \lambda_2 = -1, \quad \lambda_3 = i, \quad \lambda_4 = -i$$

with the corresponding normalized eigenvectors

$$|v_1\rangle = \frac{1}{2}\begin{pmatrix} 1 \\ 1 \\ 1 \\ 1 \end{pmatrix}, \quad |v_2\rangle = \frac{1}{2}\begin{pmatrix} 1 \\ 1 \\ -1 \\ -1 \end{pmatrix}, \quad |v_3\rangle = \frac{1}{2}\begin{pmatrix} i \\ -i \\ 1 \\ -1 \end{pmatrix}, \quad |v_4\rangle = \frac{1}{2}\begin{pmatrix} 1 \\ -1 \\ i \\ -i \end{pmatrix}.$$

Hence the spectral decomposition of P is

$$P = 1 \cdot \frac{1}{4}\begin{pmatrix} 1 & 1 & 1 & 1 \\ 1 & 1 & 1 & 1 \\ 1 & 1 & 1 & 1 \\ 1 & 1 & 1 & 1 \end{pmatrix} - 1 \cdot \frac{1}{4}\begin{pmatrix} 1 & 1 & -1 & -1 \\ 1 & 1 & -1 & -1 \\ -1 & -1 & 1 & 1 \\ -1 & -1 & 1 & 1 \end{pmatrix}$$

$$+ i \cdot \frac{1}{4}\begin{pmatrix} 1 & -1 & i & -i \\ -1 & 1 & -i & i \\ -i & i & 1 & -1 \\ i & -i & -1 & 1 \end{pmatrix} - i \cdot \frac{1}{4}\begin{pmatrix} 1 & -1 & -i & i \\ -1 & 1 & i & -i \\ i & -i & 1 & -1 \\ -i & i & -1 & 1 \end{pmatrix}.$$

Utilizing that

$$\ln(1) = 0, \quad \ln(-1) = \ln(e^{i\pi}) = i\pi, \quad \ln(i) = \ln(e^{i\pi/2}) = i\pi/2,$$

$$\ln(-i) = \ln(e^{-i\pi/2}) = -i\pi/2$$

and the *Maxima* program

```
/* PermSpectral.mac */
v1: matrix([1],[1],[1],[1])/2;      v2: matrix([1],[1],[-1],[-1])/2;
v3: matrix([%i],[-%i],[1],[-1])/2;  v4: matrix([1],[-1],[%i],[-%i])/2;
v1T: transpose(v1); v1TC: conjugate(v1T);
v2T: transpose(v2); v2TC: conjugate(v2T);
v3T: transpose(v3); v3TC: conjugate(v3T);
v4T: transpose(v4); v4TC: conjugate(v4T);
/* checking I4 */
I4: v1 . v1TC + v2 . v2TC + v3 . v3TC + v4 . v4TC;
/* spectral representation */
P: 1*(v1 . v1TC)-1*(v2 . v2TC)+%i*(v3 . v3TC)-%i*(v4 . v4TC);
/* skewhermitian matrix K */
K: %i*%pi*(v2 . v2TC)-(%i*%pi)/2*(v3 . v3TC)-(%i*%pi)/2*(v4 . v4TC);
K: ratsimp(K);
/* Hamilton operator */
H: %i*K;
```

we obtain

$$H = \begin{pmatrix} 0 & -\pi/2 & \pi/4 & \pi/4 \\ -\pi/2 & 0 & \pi/4 & \pi/4 \\ \pi/4 & \pi/4 & 0 & -\pi/2 \\ \pi/4 & \pi/4 & -\pi/2 & 0 \end{pmatrix}.$$

Problem 3. Let e_1, e_2, e_3 be the standard basis (column vectors) in the Hilbert space $\mathbb{C}^3$. Find the 9×9 permutation matrix P acting in the Hilbert space $\mathbb{C}^9$

$$P = \sum_{j=1}^{3} \sum_{k=1}^{3} (e_j \otimes e_k)(e_k \otimes e_j)^T$$

where ^{T} denotes the transpose. Find the eigenvalues of P. Let A, B be 3×3 matrices. Show that $P(A \otimes B)P^T = B \otimes A$.

Solution 3. Applying the *Maxima* program

```
/* swap.mac */
n: 3;
e[1]: matrix([1],[0],[0]); e[2]: matrix([0],[1],[0]);
e[3]: matrix([0],[0],[1]);
P: matrix([0,0,0,0,0,0,0,0,0],[0,0,0,0,0,0,0,0,0],[0,0,0,0,0,0,0,0,0],
   [0,0,0,0,0,0,0,0,0],[0,0,0,0,0,0,0,0,0],[0,0,0,0,0,0,0,0,0],
   [0,0,0,0,0,0,0,0,0],[0,0,0,0,0,0,0,0,0],[0,0,0,0,0,0,0,0,0]);
for j:1 thru n do
 for k:1 thru n do
(
Rjk: kronecker_product(e[j],e[k]),
Rkj: kronecker_product(e[k],e[j]),
RkjT: transpose(Rkj),
P: P + Rjk . RkjT
)$
print("P=",P);
/* inverse of P wich is P^T */
PT: transpose(P); Check: P . PT;
/* eigenvalues of P */
Lambda: eigenvalues(P);
/* P A \otimes P^T = B \otimes A */
A: matrix([a11,a12,a13],[a21,a22,a23],[a31,a32,a33]);
B: matrix([b11,b12,b13],[b21,b22,b23],[b31,b32,b33]);
R1: P . kronecker_product(A,B) . PT;
R2: kronecker_product(B,A);
D: R1-R2;
```

we find the 9×9 permutation matrix

$$P = (1) \oplus \begin{pmatrix} 0\ 0\ 1\ 0\ 0\ 0\ 0 \\ 0\ 0\ 0\ 0\ 0\ 1\ 0 \\ 1\ 0\ 0\ 0\ 0\ 0\ 0 \\ 0\ 0\ 0\ 1\ 0\ 0\ 0 \\ 0\ 0\ 0\ 0\ 0\ 0\ 1 \\ 0\ 1\ 0\ 0\ 0\ 0\ 0 \\ 0\ 0\ 0\ 0\ 1\ 0\ 0 \end{pmatrix} \oplus (1)$$

where $\oplus$ denotes the direct sum. The eigenvalues are -1 ($3\times$) and $+1$ ($6\times$).

Problem 4. Let $n \geq 2$. Consider the unitary matrix (*Fourier matrix*) $F = (F_{jk})$

$$F_{jk} = \frac{1}{\sqrt{n}} \exp(2\pi ijk/n), \quad j,k = 0,1,\dots,n-1.$$

For $n = 2$ we have the *Hadamard matrix*

$$F_2 = \frac{1}{\sqrt{2}} \begin{pmatrix} 1 & 1 \\ 1 & -1 \end{pmatrix}.$$

Find F for $n = 4$, $n = 8$, $n = 16$. Let S_1, S_2, S_3 be the spin matrices for spin-$\frac{1}{2}$. Consider the Hamilton operator (8×8 hermitian matrix)

$$\hat{K} = \frac{\hat{H}}{\hbar\omega} = S_1 \otimes S_1 \otimes I_2 + I_2 \otimes S_1 \otimes S_1 + S_1 \otimes I_2 \otimes S_1$$
$$+ S_2 \otimes S_2 \otimes I_2 + I_2 \otimes S_2 \otimes S_2 + S_2 \otimes I_2 \otimes S_2$$
$$+ S_3 \otimes S_3 \otimes I_2 + I_2 \otimes S_3 \otimes S_3 + S_3 \otimes I_2 \otimes S_3.$$

Find the eigenvalues of $\hat{K}$. Find $R_8 = F_8 \hat{K} F_8^*$.

Solution 4. We apply the Maxima program

```
/* Fouriergenerate.mac */
n: 4;
F4: matrix([0,0,0,0],[0,0,0,0],[0,0,0,0],[0,0,0,0]);
for j:1 thru n do
for k:1 thru n do
(
F4[j][k]: exp(2*%pi*%i*(j-1)*(k-1)/n)/2
)$
F4; F4: demoivre(F4);

n: 8;
F8: matrix([0,0,0,0,0,0,0,0],[0,0,0,0,0,0,0,0],
[0,0,0,0,0,0,0,0],[0,0,0,0,0,0,0,0],
[0,0,0,0,0,0,0,0],[0,0,0,0,0,0,0,0],[0,0,0,0,0,0,0,0]);
for j:1 thru n do
for k:1 thru n do
(
F8[j][k]: exp(2*%pi*%i*(j-1)*(k-1)/n)/(2*sqrt(2))
)$
F8; F8: demoivre(F8);

n: 16;
F16: matrix(
[0,0,0,0,0,0,0,0,0,0,0,0,0,0,0,0],[0,0,0,0,0,0,0,0,0,0,0,0,0,0,0,0],
[0,0,0,0,0,0,0,0,0,0,0,0,0,0,0,0],[0,0,0,0,0,0,0,0,0,0,0,0,0,0,0,0],
[0,0,0,0,0,0,0,0,0,0,0,0,0,0,0,0],[0,0,0,0,0,0,0,0,0,0,0,0,0,0,0,0],
[0,0,0,0,0,0,0,0,0,0,0,0,0,0,0,0],[0,0,0,0,0,0,0,0,0,0,0,0,0,0,0,0],
[0,0,0,0,0,0,0,0,0,0,0,0,0,0,0,0],[0,0,0,0,0,0,0,0,0,0,0,0,0,0,0,0],
[0,0,0,0,0,0,0,0,0,0,0,0,0,0,0,0],[0,0,0,0,0,0,0,0,0,0,0,0,0,0,0,0],
[0,0,0,0,0,0,0,0,0,0,0,0,0,0,0,0],[0,0,0,0,0,0,0,0,0,0,0,0,0,0,0,0],
[0,0,0,0,0,0,0,0,0,0,0,0,0,0,0,0],[0,0,0,0,0,0,0,0,0,0,0,0,0,0,0,0]);
for j:1 thru n do
for k:1 thru n do
(
```

```
F16[j][k]: exp(2*%pi*%i*(j-1)*(k-1)/n)/4
)$
F16; F16: demoivre(F16); F16: trigsimp(F16);

I2: matrix([1,0],[0,1]);        S1: matrix([0,1],[1,0])/2;
S2: matrix([0,-%i],[%i,0])/2;  S3: matrix([1,0],[0,-1])/2;
T11: kronecker_product(S1,kronecker_product(S1,I2));
T12: kronecker_product(I2,kronecker_product(S1,S1));
T13: kronecker_product(S1,kronecker_product(I2,S1));
T21: kronecker_product(S2,kronecker_product(S2,I2));
T22: kronecker_product(I2,kronecker_product(S2,S2));
T23: kronecker_product(S2,kronecker_product(I2,S2));
T31: kronecker_product(S3,kronecker_product(S3,I2));
T32: kronecker_product(I2,kronecker_product(S3,S3));
T33: kronecker_product(S3,kronecker_product(I2,S3));
K: T11+T12+T13+T21+T22+T23+T31+T32+T33;
Eig: eigenvalues(K);
F8T: transpose(F8); F8TC: conjugate(F8T);
R: F8 . K . F8TC; R: expand(R);
D: determinant(K);
DR: determinant(R); DR: ratsimp(DR);
```

For F_4 we obtain

$$F_4 = \frac{1}{2}\begin{pmatrix} 1 & 1 & 1 & 1 \\ 1 & i & -1 & -i \\ 1 & -1 & 1 & -1 \\ 1 & -i & -1 & i \end{pmatrix}$$

and for F_8 we find

$$\begin{pmatrix} 1 & 1 & 1 & 1 & 1 & 1 & 1 & 1 \\ 1 & (1+i)/\sqrt{2} & i & (-1+i)/\sqrt{2} & -1 & (-1-i)/\sqrt{2} & -i & (1-i)/\sqrt{2} \\ 1 & i & -1 & -i & 1 & i & -1 & -i \\ 1 & (-1+i)/\sqrt{2} & -i & 1/\sqrt{2} & -1 & (1-i)/\sqrt{2} & i & (-1-i)/\sqrt{2} \\ 1 & -1 & 1 & -1 & 1 & -1 & 1 & -1 \\ 1 & (-1-i)/\sqrt{2} & i & (1-i)/\sqrt{2} & -1 & (1+i)/\sqrt{2} & -i & (-1+i)/\sqrt{2} \\ 1 & -i & -1 & i & 1 & -i & -1 & i \\ 1 & (1-i)/\sqrt{2} & -i & (-1-i)/\sqrt{2} & -1 & (-1+i)/\sqrt{2} & i & (1+i)/\sqrt{2} \end{pmatrix}$$

The 8×8 matrix $\hat{K}$ is given by

$$\hat{K} = (3/4) \oplus \frac{1}{4} \begin{pmatrix} -1 & 2 & 0 & 2 & 0 & 0 \\ 2 & -1 & 0 & 2 & 0 & 0 \\ 0 & 0 & -1 & 0 & 2 & 2 \\ 2 & 2 & 0 & -1 & 0 & 0 \\ 0 & 0 & 2 & 0 & -1 & 2 \\ 0 & 0 & 2 & 0 & 2 & -1 \end{pmatrix} \oplus (3/4)$$

where $\oplus$ denotes the direct sum. The eigenvalues are $3/4$ ($4\times$) and $-3/4$ ($4\times$).

Problem 5. Consider the Hilbert space $\mathbb{C}^4$ and the *spin singlet state* of two particles

$$|\psi\rangle = \frac{1}{\sqrt{2}} \begin{pmatrix} 0 \\ 1 \\ -1 \\ 0 \end{pmatrix}$$

which is fully entangled. Let σ_1, σ_2, σ_3 be the Pauli spin matrices and

$$S = \frac{1}{\sqrt{2}}(-\sigma_2 - \sigma_3), \qquad T = \frac{1}{\sqrt{2}}(-\sigma_2 + \sigma_3).$$

Show that

$$\langle\psi|(\sigma_3 \otimes S)|\psi\rangle + \langle\psi|(\sigma_2 \otimes S)|\psi\rangle + \langle\psi|(\sigma_2 \otimes T)|\psi\rangle - \langle\psi|(\sigma_3 \otimes T)|\psi\rangle > 2.$$

Solution 5. The Maxima program

```
/* Bell.mac */
sig3: matrix([1,0],[0,-1]); sig2: matrix([0,-%i],[%i,0]);
psi: matrix([0],[1],[-1],[0])/sqrt(2); psiT: transpose(psi);
S: (-sig2-sig3)/sqrt(2); T: (-sig2+sig3)/sqrt(2);
E: psiT . (kronecker_product(sig3,S)) . psi
+ psiT . (kronecker_product(sig2,S)) . psi
+ psiT . (kronecker_product(sig2,T)) . psi
- psiT . (kronecker_product(sig3,T)) .psi;
E: ratsimp(E);
```

provides $2\sqrt{2}$ which indicates that the *Bell inequality* is violated.

3.4 Supplementary Problems

Problem 1. Consider the Hilbert space $\mathbb{C}^2$. Show that the vectors

$$\mathbf{v}_1 = \frac{1}{\sqrt{1 + 2\sinh^2(\alpha)}} \begin{pmatrix} \cosh(\alpha) \\ \sinh(\alpha) \end{pmatrix}, \quad \mathbf{v}_2 = \frac{1}{\sqrt{1 + 2\sinh^2(\alpha)}} \begin{pmatrix} \sinh(\alpha) \\ -\cosh(\alpha) \end{pmatrix}$$

form an orthonormal basis in $\mathbb{C}^2$. Expand

$$\frac{1}{\sqrt{2}} \begin{pmatrix} 1 \\ -1 \end{pmatrix}$$

with respect to this orthonormal basis.

Problem 2. (i) Show that the vectors

$$\frac{1}{2}\begin{pmatrix} 1 \\ 1 \\ 1 \\ -1 \end{pmatrix}, \quad \frac{1}{2}\begin{pmatrix} 1 \\ 1 \\ -1 \\ 1 \end{pmatrix}, \quad \frac{1}{2}\begin{pmatrix} 1 \\ -1 \\ 1 \\ 1 \end{pmatrix}, \quad \frac{1}{2}\begin{pmatrix} -1 \\ 1 \\ 1 \\ 1 \end{pmatrix}$$

form an orthonormal basis in the Hilbert space $\mathbb{C}^4$.

(ii) Let $|0\rangle$, $|1\rangle$ be an orthonormal basis in $\mathbb{C}^2$. Show that

$$|0\rangle \otimes |0\rangle, \ |1\rangle \otimes |1\rangle, \ \frac{1}{\sqrt{2}}(|0\rangle \otimes |1\rangle + |1\rangle \otimes |0\rangle), \ \frac{1}{\sqrt{2}}(|0\rangle \otimes |1\rangle - |1\rangle \otimes |0\rangle)$$

is an orthonormal basis in the Hilbert space $\mathbb{C}^4$.

(iii) Let $\mathbf{v}_0$, $\mathbf{v}_1$, $\mathbf{v}_2$, $\mathbf{v}_3$ be an orthonormal basis in the Hilbert space $\mathbb{C}^4$. Show that the vectors

$$\mathbf{u}_0 = \frac{1}{2}(\mathbf{v}_0 + \mathbf{v}_1 + \mathbf{v}_2 + \mathbf{v}_3), \quad \mathbf{u}_1 = \frac{1}{2}(\mathbf{v}_0 - \mathbf{v}_1 + \mathbf{v}_2 - \mathbf{v}_3),$$

$$\mathbf{u}_2 = \frac{1}{2}(\mathbf{v}_0 + \mathbf{v}_1 - \mathbf{v}_2 - \mathbf{v}_3), \quad \mathbf{u}_3 = \frac{1}{2}(\mathbf{v}_0 - \mathbf{v}_1 - \mathbf{v}_2 + \mathbf{v}_3)$$

also form an orthonormal basis in $\mathbb{C}^4$.

Problem 3. (i) Show that the 2×2 matrices

$$A = \frac{1}{\sqrt{2}} \begin{pmatrix} 1 & 0 \\ 0 & 1 \end{pmatrix}, \quad B = \frac{1}{\sqrt{2}} \begin{pmatrix} 0 & 1 \\ 1 & 0 \end{pmatrix},$$

$$C = \frac{1}{\sqrt{2}} \begin{pmatrix} 0 & -i \\ i & 0 \end{pmatrix}, \quad D = \frac{1}{\sqrt{2}} \begin{pmatrix} 1 & 0 \\ 0 & -1 \end{pmatrix}$$

form an orthonormal basis in the Hilbert space $M(2,\mathbb{C})$. Show that the sixteen 4×4 matrices

$$A \otimes A, \quad A \otimes B, \quad A \otimes C, \quad A \otimes D,$$

$$B \otimes A, \quad B \otimes B, \quad B \otimes C, \quad B \otimes D,$$

$$C \otimes A, \quad C \otimes B, \quad C \otimes C, \quad C \otimes D,$$

$$D \otimes A, \quad D \otimes B, \quad D \otimes C, \quad D \otimes D$$

form an orthonormal basis in the Hilbert space $M(4,\mathbb{C})$.
(ii) Do the 2×2 matrices

$$\frac{1}{\sqrt{2}} \begin{pmatrix} -1 & 1 \\ 1 & 1 \end{pmatrix}, \quad \frac{1}{\sqrt{2}} \begin{pmatrix} 1 & -1 \\ 1 & 1 \end{pmatrix}, \quad \frac{1}{\sqrt{2}} \begin{pmatrix} 1 & 1 \\ -1 & 1 \end{pmatrix}, \quad \frac{1}{\sqrt{2}} \begin{pmatrix} 1 & 1 \\ 1 & -1 \end{pmatrix}$$

form a basis in the Hilbert space $M(2,\mathbb{R})$?

Problem 4. (i) Consider the Hilbert space $\mathbb{R}^2$ and $x \in \mathbb{R}$. Show that the vectors

$$\begin{pmatrix} 1 \\ x \end{pmatrix}, \quad \begin{pmatrix} 0 \\ 1 \end{pmatrix}$$

form a basis in this Hilbert space. Normalize the vectors.
(ii) Consider the Hilbert space $\mathbb{R}^3$ and $x \in \mathbb{R}$. Show that the vectors

$$\begin{pmatrix} 1 \\ x \\ x^2 \end{pmatrix}, \quad \begin{pmatrix} 0 \\ 1 \\ 2x \end{pmatrix}, \quad \begin{pmatrix} 0 \\ 0 \\ 2 \end{pmatrix}$$

form a basis in this Hilbert space. Normalized the vectors.
(iii) Consider the Hilbert space $\mathbb{R}^4$ and $x \in \mathbb{R}$. Show that the vectors

$$\begin{pmatrix} 1 \\ x \\ x^2 \\ x^3 \end{pmatrix}, \quad \begin{pmatrix} 0 \\ 1 \\ 2x \\ 3x^2 \end{pmatrix}, \quad \begin{pmatrix} 0 \\ 0 \\ 2 \\ 6x \end{pmatrix}, \quad \begin{pmatrix} 0 \\ 0 \\ 0 \\ 6 \end{pmatrix}$$

form a basis in this Hilbert space. Normalized the vectors. Apply the Gram-Schmidt technique to find an orthonormal basis.
(iv) Extend to $\mathbb{R}^n$.

Problem 5. (i) Consider the linear operator

$$A = \begin{pmatrix} 2 & 0 & 0 \\ 0 & 0 & 1 \\ 0 & 1 & 0 \end{pmatrix}$$

acting in the Hilbert space $\mathbb{R}^3$. Find $\|A\| := \sup_{\|\mathbf{x}\|=1} \|A\mathbf{x}\|$ using the method of the Lagrange multiplier.
(ii) Consider the Hilbert space $\mathbb{R}^3$. Find the spectrum (eigenvalues and normalized eigenvectors) of the 3×3 matrix

$$B = \begin{pmatrix} 1 & 2 & 3 \\ 1 & 2 & 3 \\ 1 & 2 & 3 \end{pmatrix}.$$

Find $\|B\| := \sup_{\|\mathbf{x}\|=1} \|B\mathbf{x}\|$, where $\|\cdot\|$ denotes the norm and $\mathbf{x} \in \mathbb{R}^3$.
(iii) Find the spectrum (eigenvalues and normalized eigenvectors) of the 3×3 matrix

$$C = \begin{pmatrix} 3 & 3 & 3 \\ 3 & 3 & 3 \\ 3 & 3 & 3 \end{pmatrix}.$$

Find

$$\|C\|_1 := \sup_{\|\mathbf{x}\|=1} \|C\mathbf{x}\|, \qquad \|C\|_2 := \sqrt{\operatorname{tr}(CC^*)}.$$

Compare the norms with the eigenvalues. Find $\exp(C)$.

Problem 6. Consider the Hilbert space $\mathbb{R}^3$. Let $\mathbf{x} \in \mathbb{R}^3$, where $\mathbf{x}$ is considered as a column vector and $\mathbf{x} \neq \mathbf{0}$. Find the matrix $\mathbf{x}\mathbf{x}^T$ and the real number $\mathbf{x}^T\mathbf{x}$. Show that the matrix $\mathbf{x}\mathbf{x}^T$ admits only one nonzero eigenvalue given by $\mathbf{x}^T\mathbf{x}$.

Problem 7. Show that the 2×2 matrices

$$A = \begin{pmatrix} 1 & 0 \\ 0 & 0 \end{pmatrix}, \quad B = \begin{pmatrix} 1 & 1 \\ 0 & 0 \end{pmatrix}, \quad C = \begin{pmatrix} 1 & 1 \\ 1 & 0 \end{pmatrix}, \quad D = \begin{pmatrix} 1 & 1 \\ 1 & 1 \end{pmatrix}$$

form a basis in the Hilbert space $M(2, \mathbb{R})$. Apply the *Gram-Schmidt technique* to obtain an orthonormal basis.

Problem 8. (i) An orthonormal basis in the Hilbert space $\mathbb{C}^2$ is given by

$$\mathbf{v}_1 = \frac{1}{\sqrt{2}} \begin{pmatrix} 1 \\ i \end{pmatrix}, \quad \mathbf{v}_2 = \frac{1}{\sqrt{2}} \begin{pmatrix} 1 \\ -i \end{pmatrix}.$$

Use this orthonormal basis and the Kronecker product to find an orthonormal basis in the Hilbert space $\mathbb{C}^4$.

(ii) An orthonormal basis in the Hilbert space $\mathbb{R}^3$ is given by

$$\mathbf{v}_1 = \frac{1}{\sqrt{2}} \begin{pmatrix} 1 \\ 0 \\ 1 \end{pmatrix}, \quad \mathbf{v}_2 = \begin{pmatrix} 0 \\ 1 \\ 0 \end{pmatrix}, \quad \mathbf{v}_3 = \frac{1}{\sqrt{2}} \begin{pmatrix} 1 \\ 0 \\ -1 \end{pmatrix}.$$

Use this orthonormal basis and the Kronecker product to find an orthonormal basis in the Hilbert space $\mathbb{R}^9$.

(iii) A basis in the Hilbert space $M(2, \mathbb{C})$ of 2×2 matrices is given by the *Pauli spin matrices* including the 2×2 identity matrix

$$\sigma_0 = \begin{pmatrix} 1 & 0 \\ 0 & 1 \end{pmatrix}, \quad \sigma_1 = \begin{pmatrix} 0 & 1 \\ 1 & 0 \end{pmatrix}, \quad \sigma_2 = \begin{pmatrix} 0 & -i \\ i & 0 \end{pmatrix}, \quad \sigma_3 = \begin{pmatrix} 1 & 0 \\ 0 & -1 \end{pmatrix}.$$

Use this orthonormal basis and the Kronecker product to find a basis in the Hilbert space $M(4, \mathbb{C})$.

Problem 9. Let A be an $n \times n$ matrix over $\mathbb{C}$. Consider the eigenvalue problem $A\mathbf{v} = \lambda\mathbf{v}$. Show that

$$\lambda = \frac{\mathbf{v}^* A \mathbf{v}}{\mathbf{v}^* \mathbf{v}}.$$

Problem 10. Consider the Hilbert space $\mathbb{C}^4$. Let $\phi \in \mathbb{R}$. Consider the 4×4 matrix

$$A(\phi) = \begin{pmatrix} 0 & 1 & 0 & 0 \\ 0 & 0 & 1 & 0 \\ 0 & 0 & 0 & 1 \\ e^{i\phi} & 0 & 0 & 0 \end{pmatrix}.$$

Show that the matrix is unitary. Find the eigenvalues and normalized eigenvectors. Then write down the spectral decomposition of $A(\phi)$. Of course first one has to check whether the spectral decomposition can be applied. Can the matrix be written as Kronecker product of two 2×2 matrices?

Problem 11. Let $d \geq 2$ and $|0\rangle, |1\rangle, \ldots, |d-1\rangle$ be an orthonormal basis in the Hilbert space $\mathbb{C}^d$. Consider the Bell state in $\mathbb{C}^d$

$$|\psi\rangle = \frac{1}{\sqrt{d}} \sum_{j=0}^{d-1} |j\rangle \otimes |j\rangle.$$

For any $d \times d$ matrix over $\mathbb{C}$ we define

$$|v(A)\rangle := (A \otimes I_d)|\psi\rangle.$$

Let B be another $d \times d$ matrix over $\mathbb{C}$. Show that

$$\langle v(A)|v(B)\rangle = \frac{1}{d}\langle A, B\rangle \equiv \frac{1}{d}\mathrm{tr}(A^*B).$$

Problem 12. Consider the Hilbert space $\mathbb{C}^3$ and the 3×3 symmetric matrices A and B over $\mathbb{R}$ and the commutator

$$A = \begin{pmatrix} 0 & 1 & 0 \\ 1 & 0 & 1 \\ 0 & 1 & 0 \end{pmatrix}, \quad B = \begin{pmatrix} 1 & 0 & 1 \\ 0 & 1 & 0 \\ 1 & 0 & 1 \end{pmatrix}, \quad [A, B] = \begin{pmatrix} 0 & 1 & 0 \\ -1 & 0 & -1 \\ 0 & 1 & 0 \end{pmatrix}.$$

Show that the three matrices admit the eigenvalue 0 with the normalized eigenvector

$$\frac{1}{\sqrt{2}}\begin{pmatrix} 1 \\ 0 \\ -1 \end{pmatrix}.$$

Show that the two other eigenvalues of A are $-\sqrt{2}$, $\sqrt{2}$, of B are 1, 2 and of $[A, B]$ are i, $-i$.

Problem 13. (i) Given two points in the Hilbert space $\mathbb{R}^2$ in *polar coordinates* (r_1, θ_1), (r_2, θ_2). Show that the square of the distance between these points is given by

$$D^2 = r_1^2 + r_2^2 - 2r_1r_2 \cos(\theta_2 - \theta_1).$$

(ii) Consider the Hilbert space (Euclidean space) $\mathbb{E}^3$. Calculate the shortest distance (Euclidean) between the surfaces

$$x_3 = \frac{1}{2}, \quad (x_1 - 2)^2 + (x_2 - 2)^2 + (x_3 - 2)^2 = 1.$$

Apply the Lagrange multiplier method.

Problem 14. Let $|0\rangle$, $|1\rangle$, $|2\rangle$, $|3\rangle$ be an orthonormal basis in $\mathbb{C}^4$.
(i) Is the 4×4 matrix $V = |0\rangle\langle 1| + |1\rangle\langle 2| + |2\rangle\langle 3| + |3\rangle\langle 0|$ unitary?
(ii) Is the 4×4 matrix $W = |0\rangle\langle 1| + e^{-i\pi/4}|1\rangle\langle 2| + e^{i\pi/2}|2\rangle\langle 3| + e^{-i\pi/4}|3\rangle\langle 0|$ unitary?
(iii) Is the 4×4 matrix $T = |0\rangle\langle 0| + e^{-i\pi/4}|1\rangle\langle 1| + e^{i\pi/2}|2\rangle\langle 2| + e^{-i\pi/4}|3\rangle\langle 3|$ unitary?

(iv) Is the 4×4 matrix

$$U = \sum_{j,k=0}^{3} (-1)^{jk} |j\rangle\langle k|$$

unitary?

(v) Is the 4×4 matrix $|0\rangle\langle 2| + |1\rangle\langle 3|$ hermitian?

(vi) Is the 4×4 matrix $|0\rangle\langle 3| + |1\rangle\langle 2|$ hermitian?

Problem 15. Consider the Hilbert space $\mathbb{R}^4$. Let $\alpha, \beta \in \mathbb{R}$. Find the minimum of the function $f : \mathbb{R}^2 \to \mathbb{R}$

$$f(\alpha, \beta) = \left\| \frac{1}{\sqrt{2}} \begin{pmatrix} 1 \\ 0 \\ 0 \\ 1 \end{pmatrix} - \begin{pmatrix} \cos(\alpha) \\ \sin(\alpha) \end{pmatrix} \otimes \begin{pmatrix} \cos(\beta) \\ \sin(\beta) \end{pmatrix} \right\|^2$$

where $\| \cdot \|$ denotes the Euclidean norm.

Problem 16. Consider an orthonormal basis in the Hilbert space $\mathbb{C}^3$, for example

$$\mathbf{e}_{-1} = \frac{1}{\sqrt{2}} \begin{pmatrix} 1 \\ 0 \\ 1 \end{pmatrix}, \quad \mathbf{e}_0 = \begin{pmatrix} 0 \\ 1 \\ 0 \end{pmatrix}, \quad \mathbf{e}_1 = \frac{1}{\sqrt{2}} \begin{pmatrix} 1 \\ 0 \\ -1 \end{pmatrix}.$$

Do the nine vectors in the Hilbert space $\mathbb{C}^9$

$$t_0^{(0)} = \frac{1}{\sqrt{3}} (\mathbf{e}_{-1} \otimes \mathbf{e}_{+1} - \mathbf{e}_0 \otimes \mathbf{e}_0 + \mathbf{e}_{+1} \otimes \mathbf{e}_{-1})$$

$$t_{+1}^{(1)} = \frac{1}{\sqrt{2}} (\mathbf{e}_0 \otimes \mathbf{e}_{+1} - \mathbf{e}_{+1} \otimes \mathbf{e}_0)$$

$$t_0^{(1)} = \frac{1}{\sqrt{2}} (\mathbf{e}_{-1} \otimes \mathbf{e}_{+1} - \mathbf{e}_{+1} \otimes \mathbf{e}_{-1})$$

$$t_{-1}^{(1)} = \frac{1}{\sqrt{2}} (-\mathbf{e}_0 \otimes \mathbf{e}_{-1} + \mathbf{e}_{-1} \otimes \mathbf{e}_0)$$

$$t_{+2}^{(2)} = \mathbf{e}_{+1} \otimes \mathbf{e}_{+1}$$

$$t_{+1}^{(2)} = \frac{1}{\sqrt{2}} (\mathbf{e}_0 \otimes \mathbf{e}_{+1} + \mathbf{e}_{+1} \otimes \mathbf{e}_0)$$

$$t_0^{(2)} = \frac{1}{\sqrt{6}} (\mathbf{e}_{+1} \otimes \mathbf{e}_{-1} + 2\mathbf{e}_0 \otimes \mathbf{e}_0 + \mathbf{e}_{-1} \otimes \mathbf{e}_{+1})$$

$$t_{-1}^{(2)} = \frac{1}{\sqrt{2}} (\mathbf{e}_{-1} \otimes \mathbf{e}_0 + \mathbf{e}_0 \otimes \mathbf{e}_{-1})$$

$$t_{-2}^{(2)} = \mathbf{e}_{-1} \otimes \mathbf{e}_{-1}$$

form an orthonormal basis in $\mathbb{C}^9$?

Problem 17. (i) Let $u_j : j = 1, \ldots, m$ be an orthonormal basis in $\mathbb{R}^m$ and $v_k : k = 1, \ldots, n$ be an orthonormal basis in $\mathbb{R}^n$. Find solutions of the equation

$$\sum_{j=1}^{m} \sum_{k=1}^{n} c_j d_k u_j \otimes v_k = \sum_{j=1}^{m} \sum_{k=1}^{n} t_{j,k} u_j \otimes v_k.$$

(ii) Consider the Hilbert space $\mathbb{C}^d$ with $d \geq 2$ and the orthonormal basis $|0\rangle, |1\rangle, \ldots, |d-1\rangle$. Study the equation

$$\sum_{j_1=0}^{d-1} \sum_{j_2=0}^{d-1} \sum_{j_3=0}^{d-1} t_{j_1,j_2,j_3} |j_1\rangle \otimes |j_2\rangle \otimes |j_3\rangle = \sum_{k_1=0}^{d-1} \sum_{k_2=0}^{d-1} \sum_{k_3=0}^{d-1} a_{k_1} b_{k_2} c_{k_3} |k_1\rangle \otimes |k_2\rangle \otimes |k_3\rangle.$$

Show that

$$t_{0,0,0} = a_0 b_0 c_0, \quad t_{0,0,1} = a_0 b_0 c_1, \quad t_{0,1,0} = a_0 b_1 c_0, \quad t_{0,1,1} = a_0 b_1 c_1,$$

$$t_{1,0,0} = a_1 b_0 c_0, \quad t_{1,0,1} = a_1 b_0 c_1, \quad t_{1,1,0} = a_1 b_1 c_0, \quad t_{1,1,1} = a_1 b_1 c_1.$$

Let $t_{0,0,0} = 1/\sqrt{2}$, $t_{1,1,1} = 1/\sqrt{2}$ and all other coefficients are 0. Can one find a_j, b_k, c_ℓ such that the eight conditions are satisfied?

Problem 18. Consider the normalized vectors in the Hilbert space $\mathbb{R}^2$

$$\mathbf{v}_1 = \begin{pmatrix} 1 \\ 0 \end{pmatrix}, \quad \mathbf{v}_2 = \frac{1}{2} \begin{pmatrix} 1 \\ \sqrt{3} \end{pmatrix}, \quad \mathbf{v}_3 = \frac{1}{2} \begin{pmatrix} -1 \\ \sqrt{3} \end{pmatrix}.$$

Calculate

$$A_{\mathbf{v}_j,\mathbf{v}_k} := \arccos((|\langle \mathbf{v}_j | \mathbf{v}_k \rangle|), \quad j, k = 1, 2, 3.$$

Find the eigenvalues and normalized eigenvectors of the matrix $A = (A_{\mathbf{v}_j,\mathbf{v}_k})$.

Problem 19. Consider the Hilbert space of the $n \times n$ matrices over $\mathbb{C}$. Let A be an $n \times n$ matrix over $\mathbb{C}$. Show that the following statements are equivalent:

(i) A is normal, i.e. $AA^* = A^*A$
(ii) $\|A\mathbf{v}\| = \|A^*\mathbf{v}\|$ for all $\mathbf{v} \in \mathbb{C}^n$, where $\| \cdot \|$ is the Euclidean vector norm
(iii) There exists a unitary $n \times n$ matrix U such that UAU^* is a diagonal matrix

(iv) The $n \times n$ matrices $B := \frac{1}{2}(A + A^*)$ and $C := \frac{1}{2i}(A - A^*)$ commute.

Problem 20. (i) Consider the Hilbert space $\mathbb{C}^n$ and the Banach space $\mathbb{C}^n$. Let $\mathbf{x}, \mathbf{y} \in \mathbb{C}^n$. Show that (*Hölder inequality*)

$$\sum_{j=1}^n |x_j y_j| \leq \|\mathbf{x}\|_p \|\mathbf{y}\|_q$$

where $1 \leq p \leq \infty$ and $1/p + 1/q = 1$.
(ii) Consider the Hilbert space $\mathbb{C}^n$ and the Banach space $\mathbb{C}^n$. Let $\mathbf{x}, \mathbf{y} \in \mathbb{C}^n$. Show that (*Minkowski inequality*)

$$\|\mathbf{x} + \mathbf{y}\|_p \leq \|\mathbf{x}\|_p + \|\mathbf{y}\|_p$$

where $1 \leq p \leq \infty$.

Problem 21. Let $|0\rangle$, $|1\rangle$ be an orthonormal basis in the Hilbert space $\mathbb{C}^2$. Show that the vectors

$$|0\rangle \otimes |0\rangle, \quad |0\rangle \otimes |1\rangle, \quad |1\rangle \otimes |0\rangle, \quad |1\rangle \otimes |1\rangle$$

form an orthonormal basis in the Hilbert space $\mathbb{C}^4$. Do the four vectors

$$\frac{1}{\sqrt{2}}(|0\rangle \otimes |0\rangle + |1\rangle \otimes |1\rangle), \quad \frac{1}{\sqrt{2}}(|0\rangle \otimes |0\rangle - |1\rangle \otimes |1\rangle),$$

$$\frac{1}{\sqrt{2}}(|0\rangle \otimes |1\rangle + |1\rangle \otimes |0\rangle), \quad \frac{1}{\sqrt{2}}(|0\rangle \otimes |1\rangle - |1\rangle \otimes |0\rangle)$$

form an orthonormal basis in the Hilbert space $\mathbb{C}^4$?

Problem 22. Let A be a hermitian $n \times n$ matrix and $\Im(z) \neq 0$. Show that $(I_n - zA)^{-1}$ exists. Let

$$A = \begin{pmatrix} 1 & 1 \\ 1 & 1 \end{pmatrix}.$$

Find $(I_2 - zA)^{-1}$.

Problem 23. Let $n \in \mathbb{N}$. In the Hilbert space $\mathcal{H} = \mathbb{C}^{n \times n}$ any bounded linear operator T is represented by an $n \times n$ matrix $T = (t_{j,k})$. The eigenvalues of the operator T are obtained as the roots of the algebraic equation

$$\det(\lambda \delta_{j,k} - t_{j,k}) = 0.$$

This equation is also called *secular equation* or *characteristic equation*. Find the eigenvalues and normalized eigenvectors of the permutation matrix

$$P_{3421} = \begin{pmatrix} 0 & 0 & 1 & 0 \\ 0 & 0 & 0 & 1 \\ 0 & 1 & 0 & 0 \\ 1 & 0 & 0 & 0 \end{pmatrix}.$$

Find the spectral decomposition of P_{3421}. Let $\alpha \in \mathbb{R}$. Find $\exp(i\alpha P_{3421})$. Is the matrix $\exp(i\alpha P_{3421})$ unitary? Note that $P_{3421}^4 = I_4$. Find $\langle P_{3421}, P_{3421} \rangle = \mathrm{tr}(P_{3421} P_{3421}^*)$, $\det(P_{3421})$ and $\det(P_{3421} P_{3421}^*)$.

Problem 24. (i) Consider the Hilbert space $\mathbb{C}^4$ and the five 4×4 matrices

$$B_0 = \begin{pmatrix} 1 & 0 & 0 & 0 \\ 0 & 1 & 0 & 0 \\ 0 & 0 & 1 & 0 \\ 0 & 0 & 0 & 1 \end{pmatrix},$$

$$B_1 = \frac{1}{2}\begin{pmatrix} 1 & 1 & 1 & 1 \\ 1 & 1 & -1 & -1 \\ 1 & -1 & -1 & 1 \\ 1 & -1 & 1 & -1 \end{pmatrix}, \quad B_2 = \begin{pmatrix} 1 & 1 & 1 & 1 \\ 1 & 1 & -1 & -1 \\ -i & i & i & -i \\ i & -i & i & -i \end{pmatrix},$$

$$B_3 = \begin{pmatrix} 1 & 1 & 1 & 1 \\ i & -i & i & -i \\ -1 & -1 & 1 & 1 \\ i & -i & -i & i \end{pmatrix}, \quad B_4 = \begin{pmatrix} 1 & 1 & 1 & 1 \\ i & -i & i & -i \\ i & i & -i & i \\ -1 & -1 & 1 & 1 \end{pmatrix}.$$

The first matrix B_0 is the 4×4 identity matrix and the column vectors provide the standard basis in the Hilbert space $\mathbb{C}^4$. Show that the column vectors of each of these matrices form a set of mutually unbiased bases.
(ii) Let $\omega := \exp(2\pi i/5)$. Consider the Hilbert space $\mathbb{C}^5$ and the six 5×5 matrices

$$B_0 = \begin{pmatrix} 1 & 0 & 0 & 0 & 0 \\ 0 & 1 & 0 & 0 & 0 \\ 0 & 0 & 1 & 0 & 0 \\ 0 & 0 & 0 & 1 & 0 \\ 0 & 0 & 0 & 0 & 1 \end{pmatrix}, \quad B_1 = \frac{1}{\sqrt{5}}\begin{pmatrix} 1 & 1 & 1 & 1 & 1 \\ 1 & \omega & \omega^2 & \omega^3 & \omega^4 \\ 1 & \omega^2 & \omega^4 & \omega & \omega^3 \\ 1 & \omega^3 & \omega & \omega^4 & \omega^2 \\ 1 & \omega^4 & \omega^3 & \omega^2 & \omega \end{pmatrix},$$

$$B_2 = \frac{1}{\sqrt{5}}\begin{pmatrix} 1 & 1 & 1 & 1 & 1 \\ \omega & \omega^2 & \omega^3 & \omega^4 & 1 \\ \omega^4 & \omega & \omega^3 & 1 & \omega^2 \\ \omega^4 & \omega^2 & 1 & \omega^3 & \omega \\ \omega & 1 & \omega^4 & \omega^3 & \omega^2 \end{pmatrix}, \quad B_3 = \frac{1}{\sqrt{5}}\begin{pmatrix} 1 & 1 & 1 & 1 & 1 \\ \omega^3 & \omega^4 & 1 & \omega & \omega^2 \\ \omega^2 & \omega^4 & \omega & \omega^3 & 1 \\ \omega^2 & 1 & \omega^3 & \omega & \omega^4 \\ \omega^3 & \omega^2 & \omega & 1 & \omega^4 \end{pmatrix},$$

$$B_4 = \frac{1}{\sqrt{5}} \begin{pmatrix} 1 & 1 & 1 & 1 & 1 \\ \omega^2 & \omega^3 & \omega^4 & 1 & \omega \\ \omega^3 & 1 & \omega^2 & \omega^4 & \omega \\ \omega^3 & \omega & \omega^2 & \omega^4 & 1 \\ \omega^2 & \omega & 1 & \omega^4 & \omega^3 \end{pmatrix}, \quad B_5 = \frac{1}{\sqrt{5}} \begin{pmatrix} 1 & 1 & 1 & 1 & 1 \\ \omega^4 & 1 & \omega & \omega^2 & \omega^3 \\ \omega & \omega^3 & 1 & \omega^2 & \omega^4 \\ \omega & \omega^4 & \omega^2 & 1 & \omega^3 \\ \omega^4 & \omega^3 & \omega^2 & \omega & 1 \end{pmatrix}.$$

The first matrix B_0 is the 5×5 identity matrix and the column vectors provide the standard basis in the Hilbert space $\mathbb{C}^5$. Show that the columns vectors of each of these matrices form a set of mutually unbiased bases.

Problem 25. Let $|0\rangle$, $|1\rangle$ be the standard basis in the Hilbert space $\mathbb{C}^2$. Consider the *Bell state* in the Hilbert space $\mathbb{C}^4$

$$|\psi\rangle = \frac{1}{\sqrt{2}}(|0\rangle_A \otimes |0\rangle_B + |1\rangle_A \otimes |1\rangle_B) \equiv \frac{1}{\sqrt{2}} \begin{pmatrix} 1 \\ 0 \\ 0 \\ 1 \end{pmatrix}$$

where A refers to *Alice* and B refers to *Bob*. Let

$$\Pi_0 = |0\rangle\langle 0| \equiv \begin{pmatrix} 1 & 0 \\ 0 & 0 \end{pmatrix}, \quad \Pi_1 = |1\rangle\langle 1| \equiv \begin{pmatrix} 0 & 0 \\ 0 & 1 \end{pmatrix}$$

be projection matrices with $\Pi_0 + \Pi_1 = I_2$. Show that *measurement* of the first qubit (Alice) provide

$$p_1(0) = \langle \psi|(\Pi_0 \otimes I_2)^*(\Pi_0 \otimes I_2)|\psi\rangle = \frac{1}{2}.$$

Show that the *post-measurement state* $|\phi\rangle$ is given by

$$|\phi\rangle = \frac{1}{\sqrt{p_1(0)}}(\Pi_0 \otimes I_2)|\psi\rangle = |0\rangle \otimes |0\rangle = \begin{pmatrix} 1 \\ 0 \\ 0 \\ 0 \end{pmatrix}.$$

Show that the measurement of qubit two (Bob) will then result with certainty in the same result, namely $\langle \phi|(I_2 \otimes \Pi_0)^*(I_2 \otimes \Pi_0)|\phi\rangle = 1$.

Problem 26. The *Lie algebra* $su(n)$ consists of all $n \times n$ matrices X with the conditions $X^* = -X$ and $\text{tr}(X) = 0$, i.e. the matrices are skew-hermitian and trace is equal to 0. Consider the case $n = 3$ and the three (hermitian) spin-1 matrices

$$S_1 = \frac{1}{\sqrt{2}} \begin{pmatrix} 0 & 1 & 0 \\ 1 & 0 & 1 \\ 0 & 1 & 0 \end{pmatrix}, \quad S_2 = \frac{1}{\sqrt{2}} \begin{pmatrix} 0 & -i & 0 \\ i & 0 & -i \\ 0 & i & 0 \end{pmatrix}, \quad S_3 = \begin{pmatrix} 1 & 0 & 0 \\ 0 & 0 & 0 \\ 0 & 0 & -1 \end{pmatrix}.$$

The trace of these matrices is equal to 0. The five (hermitian) *quadrupole matrices* are

$$U_1 = \begin{pmatrix} 0\ 0\ 1 \\ 0\ 0\ 0 \\ 1\ 0\ 0 \end{pmatrix}, \quad U_2 = \begin{pmatrix} 0\ 0\ -i \\ 0\ 0\ 0 \\ i\ 0\ 0 \end{pmatrix}, \quad Q_0 = \frac{1}{\sqrt{3}} \begin{pmatrix} 1\ 0\ 0 \\ 0\ -2\ 0 \\ 0\ 0\ 1 \end{pmatrix},$$

$$V_1 = \frac{1}{\sqrt{2}} \begin{pmatrix} 0\ 1\ 0 \\ 1\ 0\ -1 \\ 0\ -1\ 0 \end{pmatrix}, \quad V_2 = \frac{1}{\sqrt{2}} \begin{pmatrix} 0\ -i\ 0 \\ i\ 0\ i \\ 0\ -i\ 0 \end{pmatrix}.$$

Show that multiplying these eight hermitian matrices (with trace equal to 0) by $i = \sqrt{-1}$ we obtain eight skew-hermitian matrices with trace equal to 0 and thus a basis of the underlying vector space of the Lie algebra $su(3)$. Show that if we add the skew-hermitian matrix $i \cdot I_3$ to the eight matrices we obtain a basis for the Lie algebra $u(3)$.

(ii) The general *quadrupolar interaction* Hamilton operator of two spin-1 nuclei can be written as

$$\hat{H} = \hbar\omega_0(Q_0 \otimes Q_0) + \hbar\omega_1(V_1 \otimes V_1 + V_2 \otimes V_2) + \hbar\omega_3(U_1 \otimes U_1 + U_2 \otimes U_2).$$

Find the eigenvalues and normalized eigenvectors of the Hamilton operator. Are the eigenvectors entangled?

Problem 27. Consider the Hilbert space $M(3, \mathbb{R})$ with the scalar product $\langle A, B \rangle = \text{tr}(AB^*)$. Do the nine matrices

$$A_1 = \frac{1}{\sqrt{3}} \begin{pmatrix} 1\ 0\ 0 \\ 0\ 0\ 1 \\ 0\ -1\ 0 \end{pmatrix}, \quad A_2 = \frac{1}{\sqrt{3}} \begin{pmatrix} 0\ 1\ 0 \\ 1\ 0\ 0 \\ 0\ 0\ -1 \end{pmatrix}, \quad A_3 = \frac{1}{\sqrt{3}} \begin{pmatrix} 0\ 0\ 1 \\ 0\ 1\ 0 \\ -1\ 0\ 0 \end{pmatrix},$$

$$A_4 = \frac{1}{\sqrt{3}} \begin{pmatrix} 1\ 0\ 0 \\ 0\ 0\ -1 \\ 0\ 1\ 0 \end{pmatrix}, \quad A_5 = \frac{1}{\sqrt{3}} \begin{pmatrix} 0\ 1\ 0 \\ -1\ 0\ 0 \\ 0\ 0\ 1 \end{pmatrix}, \quad A_6 = \frac{1}{\sqrt{3}} \begin{pmatrix} 0\ 0\ 1 \\ 0\ -1\ 0 \\ 1\ 0\ 0 \end{pmatrix},$$

$$A_7 = \frac{1}{\sqrt{3}} \begin{pmatrix} -1\ 0\ 0 \\ 0\ 0\ 1 \\ 0\ 1\ 0 \end{pmatrix}, \quad A_8 = \frac{1}{\sqrt{3}} \begin{pmatrix} 0\ -1\ 0 \\ 1\ 0\ 0 \\ 0\ 0\ 1 \end{pmatrix}, \quad A_9 = \frac{1}{\sqrt{3}} \begin{pmatrix} 0\ 0\ -1 \\ 0\ 1\ 0 \\ 1\ 0\ 0 \end{pmatrix}$$

form a basis in this Hilbert space?

Problem 28. Let H be a 2×2 hermitian matrix. Assume that one of the eigenvalues λ_1, λ_2 of H is equal to 0, say $\lambda_1 = 0$. Can H be written as

$$H = U \begin{pmatrix} 0\ 0 \\ 0\ \lambda_2 \end{pmatrix} U^*$$

where U is a unitary matrix?

Problem 29. Let A be an $n \times n$ hermitian matrix over $\mathbb{C}$. Assume that the eigenvalues of A, λ_1, λ_2, ..., λ_n are nondegenerate and hence the normalized eigenvectors $\mathbf{v}_j$ ($j = 1, \ldots, n$) form an orthonormal basis in the Hilbert space $\mathbb{C}^n$. Let B be an $n \times n$ matrix over $\mathbb{C}$. Assume that $[A, B] = 0_n$. Show that

$$\mathbf{v}_k^* B \mathbf{v}_j = 0 \quad \text{for} \quad k \neq j.$$

Problem 30. Consider the vector space of $n \times n$ matrices over $\mathbb{C}$. Let A, B be $n \times n$ matrices with $[A, B]_+ \equiv AB + BA = 0_n$ i.e. the matrices anticommute. Let $\mathbf{v}$ be a normalized eigenvector of A and B. What can be said about the corresponding eigenvalue?

Problem 31. Consider the 4×4 unitary matrix

$$U(\phi) = \frac{1}{2} \begin{pmatrix} ie^{i\phi} & e^{i\phi} & e^{i\phi} & -ie^{i\phi} \\ e^{-i\phi} & ie^{-i\phi} & -ie^{-i\phi} & e^{-i\phi} \\ e^{-i\phi} & -ie^{-i\phi} & ie^{-i\phi} & e^{-i\phi} \\ -ie^{i\phi} & e^{i\phi} & e^{i\phi} & ie^{i\phi} \end{pmatrix}$$

acting in the Hilbert space $\mathbb{C}^4$. Show that for $\phi = \pi/4$ the unitary matrix $U(\phi/4)$ takes product states into maximally entangled states. First try the case

$$U(\pi/4) \left(\begin{pmatrix} 1 \\ 0 \end{pmatrix} \otimes \begin{pmatrix} 1 \\ 0 \end{pmatrix} \right).$$

Problem 32. Consider the Hilbert space $\mathbb{C}^3$ and e_{x_1}, e_{x_2}, e_{x_3} be an orthonormal basis in $\mathbb{C}^3$. Is ξ_1, ξ_0, ξ_{-1} given by

$$\begin{pmatrix} \xi_1 \\ \xi_0 \\ \xi_{-1} \end{pmatrix} = \frac{1}{\sqrt{2}} \begin{pmatrix} -1 & -i & 0 \\ 0 & 0 & \sqrt{2} \\ 1 & -i & 0 \end{pmatrix} \begin{pmatrix} e_{x_1} \\ e_{x_2} \\ e_{x_3} \end{pmatrix}$$

an orthonormal basis in the Hilbert space $\mathbb{C}^3$?

Problem 33. Let $\alpha, \beta, \theta \in \mathbb{R}$. Consider the 2×2 matrix

$$U(\alpha, \beta, \theta) = \begin{pmatrix} e^{i\alpha} \cos(\theta) & e^{-i\beta} \sin(\theta) \\ e^{i\beta} \sin(\theta) & -e^{-i\alpha} \cos(\theta) \end{pmatrix}.$$

(i) Show that the matrix is unitary.

(ii) Show that the 4×4 matrix $U(\alpha, \beta, \theta) \otimes U(\alpha', \beta', \theta')$ is unitary.

(iii) Show that the 4×4 matrix $U(\alpha, \beta, \theta) \oplus U(\alpha', \beta', \theta')$ is unitary, where $\oplus$ denotes the direct sum.

Problem 34. Let $d \geq 2$ and $|0\rangle, |1\rangle, \ldots, |d-1\rangle$ be an orthonormal basis in the Hilbert space $\mathbb{C}^d$.

(i) Consider the vectors in the Hilbert space $\mathbb{C}^{d^2}$

$$|v\rangle = \sum_{j=0}^{d-1} |j\rangle \otimes |j\rangle, \qquad |w\rangle = \sum_{k=0}^{d-1} (-1)^k |k\rangle \otimes |k\rangle$$

Find the scalar product $\langle v|w\rangle$.

(ii) Consider the $d \times d$ matrices

$$T_1 = \sum_{j,k=0}^{d-1} |j\rangle\langle k| \otimes |j\rangle\langle k| \otimes |k\rangle\langle j|, \quad T_2 = \sum_{j,k=0}^{d-1} |j\rangle\langle j| \otimes |k\rangle\langle k| \otimes |k\rangle\langle j|.$$

Find the commutator $[T_1, T_2]$.

Problem 35. Let $n \geq 2$ and $X_1, X_2, \ldots, X_n$ be an orthonormal basis in the Hilbert space of $n \times n$ matrices over $\mathbb{C}$ with scalar product $\langle A, B \rangle := \operatorname{tr}(AB^*)$. Let $Y \in SL(n, \mathbb{C})$, i.e. $\det(Y) = 1$. Then we can consider the map from $SL(n, \mathbb{C})$ to the vector space of $n^2 \times n^2$ matrices $\Lambda = (\Lambda_{\mu,\nu})$, $(\mu, \nu = 0, 1, \ldots, n^2 - 1)$

$$\Lambda_{\mu,\nu} = \operatorname{tr}(Y X_\nu Y^* X_\mu), \quad \mu, \nu = 0, 1, \ldots, n^2 - 1.$$

Consider the special case $n = 2$. Then the scaled *Pauli spin matrices*

$$\frac{1}{\sqrt{2}}\sigma_0, \quad \frac{1}{\sqrt{2}}\sigma_1, \quad \frac{1}{\sqrt{2}}\sigma_2, \quad \frac{1}{\sqrt{2}}\sigma_3$$

form an orthonormal basis in this Hilbert space with $\sigma_0 = I_2$. Then

$$\Lambda_{\mu,\nu} = \frac{1}{2}\operatorname{tr}(A\sigma_\nu A^*\sigma_\mu), \quad \mu, \nu = 0, 1, 2, 3.$$

Let $A_1 = \begin{pmatrix} 1 & 1 \\ 0 & 1 \end{pmatrix}$. Find Λ_1. Let $A_2 = \begin{pmatrix} 1 & 0 \\ 1 & 1 \end{pmatrix}$. Find Λ_2. Find the commutators $[A_1, A_2]$ and $[\Lambda_1, \Lambda_2]$. Discuss.

Problem 36. Consider the Hilbert space $\mathbb{C}^8$. Find the eigenvalues and normalized eigenvectors of the hermitian matrix

$$\hat{H} = \hbar\omega_1(\sigma_1 \otimes \sigma_1 \otimes \sigma_1) + \hbar\omega_2(\sigma_2 \otimes \sigma_2 \otimes \sigma_2) + \hbar\omega_3(\sigma_3 \otimes \sigma_3 \otimes \sigma_3)$$

with trace equal to 0.

Problem 37. The *Hadamard matrices* $H(n)$ of any dimension are generated by

$$H(n) = \begin{pmatrix} H(n-1) & H(n-1) \\ H(n-1) & -H(n-1) \end{pmatrix}, \quad n = 1, 2, \dots$$

with $H(0) = 1$. Hence

$$H(1) = \begin{pmatrix} 1 & 1 \\ 1 & -1 \end{pmatrix}.$$

Show that the inverse matrix is given by

$$H^{-1}(n) = \frac{1}{2^n} H(n), \quad n = 0, 1, 2, \dots$$

Show that the columns in a Hadamard matrix are pairwise orthogonal.

Problem 38. Consider the Hilbert space $\mathbb{C}^2$ and the normalized vector $|\mathbf{z}\rangle \in \mathbb{C}^2$, i.e.

$$|\mathbf{z}\rangle = \begin{pmatrix} z_0 \\ z_1 \end{pmatrix} \implies \langle \mathbf{z}| = (\, \bar{z}_0 \;\; \bar{z}_1 \,), \quad \langle \mathbf{z}|\mathbf{z}\rangle = z_0 \bar{z}_0 + z_1 \bar{z}_1 = 1.$$

Note that $|\mathbf{z}\rangle\langle \mathbf{z}|$ is a density matrix (pure state). Consider the skew-symmetric matrix over $\mathbb{R}$

$$\Gamma := \begin{pmatrix} 0 & -1 \\ 1 & 0 \end{pmatrix} \equiv -i\sigma_2$$

which is an element of the (compact) Lie group $SU(2)$. One defines the *anti-unitary map*

$$|\mathbf{z}\rangle \mapsto |\mathbf{z}] := \Gamma|\bar{\mathbf{z}}\rangle \equiv \begin{pmatrix} 0 & -1 \\ 1 & 0 \end{pmatrix} \begin{pmatrix} \bar{z}_0 \\ \bar{z}_1 \end{pmatrix} \equiv \begin{pmatrix} -\bar{z}_1 \\ \bar{z}_0 \end{pmatrix}.$$

Note that $|\mathbf{z}]$ is also normalized and the vectors $|\mathbf{z}\rangle$, $|\mathbf{z}]$ form an orthonormal basis in the Hilbert space $\mathbb{C}^2$. Now $[\mathbf{z}| = (-z_1 \;\; z_0)$ and $[\mathbf{z}|\mathbf{z}\rangle = 0$. Note that

$$|\mathbf{z}\rangle = \begin{pmatrix} e^{i\phi} \cos(\theta) \\ \sin(\theta) \end{pmatrix} \implies |\mathbf{z}] = \begin{pmatrix} -\sin(\theta) \\ e^{-i\phi} \cos(\theta) \end{pmatrix}.$$

Consider the two normalized vectors in $\mathbb{C}^2$

$$|\mathbf{z}_1\rangle = \begin{pmatrix} z_{1,0} \\ z_{1,1} \end{pmatrix}, \quad |\mathbf{z}_2\rangle = \begin{pmatrix} z_{2,0} \\ z_{2,1} \end{pmatrix}$$

and

$$|\mathbf{z}_1] = \begin{pmatrix} -\overline{z}_{1,1} \\ \overline{z}_{1,0} \end{pmatrix}, \quad |\mathbf{z}_2] = \begin{pmatrix} -\overline{z}_{2,1} \\ \overline{z}_{2,0} \end{pmatrix}.$$

Show that the matrix

$$U = |\mathbf{z}\rangle[\mathbf{z}_2] - |\mathbf{z}_1]\langle\mathbf{z}_2| = \begin{pmatrix} -z_{1,0}z_{2,1} + \overline{z}_{1,1}\overline{z}_{2,0} & z_{1,0}z_{2,0} + \overline{z}_{1,1}\overline{z}_{2,1} \\ -z_{1,1}z_{2,1} - \overline{z}_{1,0}\overline{z}_{2,0} & z_{1,1}z_{2,0} - \overline{z}_{1,0}\overline{z}_{2,1} \end{pmatrix}$$

is unitary with determinant equal to 1, i.e. the matrix is an element of the Lie group $SU(2)$.

Problem 39. (i) Consider the orthonormal basis

$$\mathbf{v}_1 = \frac{1}{\sqrt{2}}\begin{pmatrix} 1 \\ 1 \end{pmatrix}, \quad \mathbf{v}_2 = \frac{1}{\sqrt{2}}\begin{pmatrix} 1 \\ -1 \end{pmatrix}$$

in the Hilbert space $\mathbb{C}^2$ and the orthonormal basis

$$\mathbf{u}_1 = \frac{1}{\sqrt{2}}\begin{pmatrix} 1 \\ 0 \\ 1 \end{pmatrix}, \quad \mathbf{u}_2 = \begin{pmatrix} 0 \\ 1 \\ 0 \end{pmatrix}, \quad \mathbf{u}_3 = \frac{1}{\sqrt{2}}\begin{pmatrix} 1 \\ 0 \\ -1 \end{pmatrix}$$

in the Hilbert space $\mathbb{C}^3$. Find the orthonormal bases in the Hilbert space $\mathbb{C}^6$

$$\{\, \mathbf{v}_1 \otimes \mathbf{u}_1, \ \mathbf{v}_1 \otimes \mathbf{u}_2, \ \mathbf{v}_1 \otimes \mathbf{u}_3, \ \mathbf{v}_2 \otimes \mathbf{u}_1, \ \mathbf{v}_2 \otimes \mathbf{u}_2 \ \mathbf{v}_2 \otimes \mathbf{u}_3 \,\}$$

$$\{\, \mathbf{u}_1 \otimes \mathbf{v}_1, \ \mathbf{u}_1 \otimes \mathbf{v}_2, \ \mathbf{u}_2 \otimes \mathbf{v}_1, \ \mathbf{u}_2 \otimes \mathbf{v}_2, \ \mathbf{u}_3 \otimes \mathbf{v}_1, \ \mathbf{u}_3 \otimes \mathbf{v}_2 \,\}.$$

How are the two orthonormal bases are related?

(ii) Let $|0\rangle$, $|1\rangle$ be an orthonormal basis in the Hilbert space $\mathbb{C}^2$ and $|\widetilde{0}\rangle$, $|\widetilde{1}\rangle$, $|\widetilde{2}\rangle$ be an orthonormal basis in $\mathbb{C}^3$. Are the normalized states in the Hilbert space $\mathbb{C}^6$

$$\frac{1}{\sqrt{2}}(|0\rangle \otimes |\widetilde{0}\rangle + |1\rangle \otimes |\widetilde{1}\rangle), \quad \frac{1}{\sqrt{2}}(|0\rangle \otimes |\widetilde{0}\rangle + |2\rangle \otimes |\widetilde{2}\rangle), \quad \frac{1}{\sqrt{2}}(|1\rangle \otimes |\widetilde{1}\rangle + |2\rangle \otimes |\widetilde{2}\rangle)$$

entangled, i.e. can the vectors be written as Kronecker product of vectors in $\mathbb{C}^2$ and $\mathbb{C}^3$ or as a Kronecker product of vectors in $\mathbb{C}^3$ and $\mathbb{C}^2$.

(iii) Consider the vectors in the Hilbert spaces $\mathbb{R}^2$ and $\mathbb{R}^3$, respectively and

$$\mathbf{v} = \begin{pmatrix} v_1 \\ v_2 \end{pmatrix} \in \mathbb{R}^2, \quad \mathbf{u} = \begin{pmatrix} u_1 \\ u_2 \\ u_3 \end{pmatrix} \in \mathbb{R}^3.$$

Find the condition such that $\mathbf{u} \otimes \mathbf{v} = \mathbf{v} \otimes \mathbf{u}$. Find solutions to these conditions.

Problem 40. Let $n \geq 2$ and $GL(n, \mathbb{C})$ be the *general linear group*, i.e. the set of all $n \times n$ invertible matrices over $\mathbb{C}$. Consider the Hilbert spaces $\mathbb{C}^n$ and $\mathbb{C}^{n^k}$, where $k \geq 2$. Let $v_1, \ldots, v_k \in \mathbb{C}^n$. Let $g \in GL(n, \mathbb{C})$. One defines the map

$$\widetilde{g}(v_1 \otimes v_2 \otimes \cdots \otimes v_k) = (gv_1) \otimes (gv_2) \otimes \cdots \otimes (gv_k), \quad \widetilde{g} = g \otimes g \otimes \cdots \otimes g$$

and the map

$$\pi(v_1 \otimes v_2 \otimes \cdots \otimes v_k) = v_{\pi^{-1}(1)} \otimes v_{\pi^{-1}(2)} \otimes \cdots \otimes v_{\pi^{-1}(k)}$$

where π is an element of the *symmetric group* S_k. Show that the two maps commute (*Schur-Weyl duality*).

Problem 41. Consider the 2×2 symmetric matrix over $\mathbb{R}$

$$A_2 = \begin{pmatrix} 2 & -1 \\ -1 & 1 \end{pmatrix}.$$

Show that the eigenvalues are given by $\lambda_1 = \frac{1}{2}(3 - \sqrt{5})$, $\lambda_2 = \frac{1}{2}(3 + \sqrt{5})$ with the corresponding normalized eigenvectors

$$v_1 = \frac{1}{\sqrt{1 + (3 + \sqrt{5})/2}} \begin{pmatrix} 1 \\ (1 + \sqrt{5})/2 \end{pmatrix},$$

$$v_2 = \frac{1}{\sqrt{1 + (3 - \sqrt{5})/2}} \begin{pmatrix} 1 \\ (1 - \sqrt{5})/2 \end{pmatrix}.$$

Let $n \geq 2$. Consider the $n \times n$ symmetric tridiagonal matrix over $\mathbb{R}$

$$A_n = \begin{pmatrix} 2 & -1 & 0 & \cdots & \cdots & \cdots \\ -1 & 2 & -1 & \cdots & \cdots & \cdots \\ 0 & -1 & 2 & \cdots & \cdots & \cdots \\ \vdots & \vdots & \vdots & \ddots & \cdots & \cdots \\ 0 & 0 & \cdots & \cdots & 2 & -1 \\ 0 & 0 & \cdots & \cdots & -1 & 1 \end{pmatrix}.$$

Show that the vector $\mathbf{v} \in \mathbb{R}^n$ ($\mathbf{v} = (v_j)$) with

$$v_j = \sin\left(\frac{j\pi}{2n + 1}\right), \quad j = 1, 2, \ldots, n$$

is an eigenvector of A_n with eigenvalue

$$\lambda = 4\sin^2\left(\frac{\pi}{4n + 2}\right) > 0.$$

Note that for $n = 2$ we have $\lambda = 4\sin^2(\pi/10) = \frac{1}{2}(3 - \sqrt{5})$.

Problem 42. (i) Let S_1, S_2, S_3 be the spin-$\frac{1}{2}$ matrices, $\mathbf{n}$ be a normalized vector in $\mathbb{R}^3$ and $\mathbf{n} \cdot \mathbf{S} := n_1 S_1 + n_2 S_2 + n_3 S_3$. Show that

$$\exp(i\theta\mathbf{n} \cdot \mathbf{S}) = I_2 \cos(\theta/2) + 2i(\mathbf{n} \cdot \mathbf{S}) \sin(\theta/2).$$

(ii) Let S_1, S_2, S_3 be the spin-1 matrices with $S_3 = \mathrm{diag}(1, 0, -1)$, $\mathbf{n}$ be a normalized vector in $\mathbb{R}^3$ and $\mathbf{n} \cdot \mathbf{S} = n_1 S_1 + n_2 S_2 + n_3 S_3$. Show that

$$\exp(i\theta\mathbf{n} \cdot \mathbf{S}) = I_3 + i(\mathbf{n} \cdot \mathbf{S}) \sin(\theta) + (\mathbf{n} \cdot \mathbf{S})^2 (\cos(\theta) - 1).$$

Problem 43. Let $n \geq 2$. Consider the Hilbert space $\mathbb{C}^n$. Let A be an $n \times n$ hermitian matrix. Note that the eigenvalues of a hermitian matrix are real. Owing to the *Cayley-Hamilton theorem* one has

$$\prod_{k=1}^{n} (A - \lambda_k I_n) = 0_n$$

where $\lambda_1, \lambda_2, \ldots, \lambda_n$ are the eigenvalues of A, I_n the $n \times n$ identity matrix and 0_n the $n \times n$ zero matrix. Assume that the eigenvalues of A are pairwise distinct. One defines the $n \times n$ matrices

$$\Pi_\ell(A) := \prod_{k=1, k \neq \ell}^{n} \left(\frac{A - \lambda_k I_n}{\lambda_\ell - \lambda_k} \right), \quad \ell = 1, 2, \ldots, n.$$

Show that the matrices $\Pi_\ell(A)$ ($\ell = 1, 2, \ldots, n$) are *projection matrices*. Show that

$$\Pi_\ell(A)A = A\Pi_\ell(A) = \lambda_\ell \Pi_\ell(A), \quad \Pi_\ell(A)f(A) = f(A)\Pi_\ell(A) = f(\lambda_\ell)\Pi_\ell(A)$$

with $f : \mathbb{C} \to \mathbb{C}$ an analytic function. Show that

$$\Pi_m(A)\Pi_\ell(A) = \delta_{\ell,m}\Pi_\ell(A), \quad \sum_{\ell=1}^{n} \Pi_\ell(A) = I_n.$$

The projection operators $\Pi_\ell(A)$ ($\ell = 1, \ldots, n$) form a complete set of projection operators. Let $n = 2$. Consider the *spin matrices* for spin-$\frac{1}{2}$

$$S_1 = \frac{1}{2}\begin{pmatrix} 0 & 1 \\ 1 & 0 \end{pmatrix}, \quad S_2 = \frac{1}{2}\begin{pmatrix} 0 & -i \\ i & 0 \end{pmatrix}, \quad S_3 = \frac{1}{2}\begin{pmatrix} 1 & 0 \\ 0 & -1 \end{pmatrix}.$$

All three admit the eigenvalues $\lambda_1 = 1/2$ and $\lambda_2 = -1/2$. Show that the corresponding projection matrices are

$$\Pi_1^{(1)} = \frac{1}{2}\begin{pmatrix} 1 & 1 \\ 1 & 1 \end{pmatrix}, \quad \Pi_2^{(1)} = \frac{1}{2}\begin{pmatrix} 1 & -1 \\ -1 & 1 \end{pmatrix}$$

$$\Pi_1^{(2)} = \frac{1}{2}\begin{pmatrix} 1 & -i \\ i & 1 \end{pmatrix}, \quad \Pi_2^{(2)} = \frac{1}{2}\begin{pmatrix} 1 & i \\ -i & 1 \end{pmatrix}$$

$$\Pi_1^{(3)} = \begin{pmatrix} 1 & 0 \\ 0 & 0 \end{pmatrix}, \quad \Pi_2^{(3)} = \begin{pmatrix} 0 & 0 \\ 0 & 1 \end{pmatrix}.$$

Problem 44. Let $|0\rangle$, $|1\rangle$ be an orthonormal basis in the Hilbert space $\mathbb{C}^2$. Show that

$$\Pi_+ = \frac{1}{4}\sum_{j=0}^{1}\sum_{k=0}^{1}((|j\rangle \otimes |k\rangle)(\langle j| \otimes \langle k|) + (|j\rangle \otimes |k\rangle)(\langle k| \otimes \langle j|)$$
$$+ (|k\rangle \otimes |j\rangle)(\langle j| \otimes \langle k|) + (|k\rangle \otimes |j\rangle)(\langle k| \otimes \langle j|))$$

is a 4×4 projection matrix. Show that

$$\Pi_- = \frac{1}{4}\sum_{j=0}^{1}\sum_{k=0}^{1}((|j\rangle \otimes |k\rangle)(\langle j| \otimes \langle k|) - (|j\rangle \otimes |k\rangle)(\langle k| \otimes \langle j|)$$
$$- (|k\rangle \otimes |j\rangle)(\langle j| \otimes \langle k|) + (|k\rangle \otimes |j\rangle)(\langle k| \otimes \langle j|))$$

is a 4×4 projection matrix. Show that $\Pi_+\Pi_- = 0_4$.

Problem 45. Let $|0\rangle$, $|1\rangle$, $|2\rangle$, $|3\rangle$ be an orthonormal basis in the Hilbert space $\mathbb{C}^4$. Find the hermitian 4×4 matrix

$$\frac{\hat{H}}{\hbar\omega} = |0\rangle\langle 2| + |2\rangle\langle 0| + |0\rangle\langle 3| + |3\rangle\langle 0| + |1\rangle\langle 2| + |2\rangle\langle 1|$$
$$+ |1\rangle\langle 3| + |3\rangle\langle 1| + |2\rangle\langle 3| + |3\rangle\langle 2|$$

and their eigenvalues and normalized eigenvectors.

Problem 46. Consider the orthonormal basis $|u_1\rangle = \begin{pmatrix} 1 \\ 0 \end{pmatrix}$, $|u_2\rangle = \begin{pmatrix} 0 \\ 1 \end{pmatrix}$ in the Hilbert space $\mathbb{R}^2$, the orthonormal basis

$$|v_1\rangle = \begin{pmatrix} 1 \\ 0 \\ 0 \end{pmatrix}, \quad |v_2\rangle = \begin{pmatrix} 0 \\ 1 \\ 0 \end{pmatrix}, \quad |v_3\rangle = \begin{pmatrix} 0 \\ 0 \\ 1 \end{pmatrix}$$

in the Hilbert space $\mathbb{R}^3$ and the orthonormal basis $|w_1\rangle = \begin{pmatrix} 1 \\ 0 \end{pmatrix}$, $|w_2\rangle = \begin{pmatrix} 0 \\ 1 \end{pmatrix}$ in the Hilbert space $\mathbb{R}^2$. Find the vector

$$T = \sum_{j=1}^{2}\sum_{k=1}^{3}\sum_{\ell=1}^{2} t_{j,k,\ell}|u_j\rangle \otimes |v_k\rangle \otimes |w_\ell\rangle$$

in the Hilbert space $\mathbb{R}^{12}$ with

$$t_{j,k,\ell} = \frac{1}{j \cdot k \cdot \ell}.$$

Find the 12×12 matrix TT^* and the eigenvalues of the matrix.

Problem 47. Consider the Hilbert space $\mathbb{C}^8$ and the Hamilton operator

$$\hat{H} = \hbar\omega(\sigma_1 \otimes \sigma_3 \otimes \sigma_1)$$

with σ_1, σ_2, σ_3 be the Pauli spin matrices. Let

$$U(t) = \exp(i\hat{H}t/\hbar) = \exp(i\omega t\sigma_1 \otimes \sigma_3 \otimes \sigma_1)$$

which is a unitary 8×8 matrix. Show that

$$\exp(i\omega t\sigma_1 \otimes \sigma_3 \otimes \sigma_1) \equiv$$

$$e^{-i\alpha I_2\otimes I_2\otimes \sigma_2}e^{i\alpha\sigma_3\otimes I_2\otimes \sigma_3}e^{i\alpha\sigma_1\otimes I_2\otimes I_2}e^{i\omega t\sigma_3\otimes \sigma_3\otimes I_2}e^{-i\alpha\sigma_1\otimes I_2\otimes I_2}e^{-i\alpha\sigma_3\otimes I_2\otimes \sigma_3}$$

$$\cdot\, e^{i\alpha I_2\otimes I_2\otimes \sigma_2}$$

with $\alpha = \pi/4$.

Problem 48. *Trotter formula* [77]. Let A be the generator of a contractive C_0-semigroup $\exp(tA)_{t\geq 0}$ on a Banach space E, and let $B \in \mathcal{L}(E)$ be a linear dissipative operator, where $\mathcal{L}(E)$ denotes the vector space of all linear bounded maps $E \to E$. Then $A+B$ generates a C_0-semigroup which is given by Trotter's formula

$$\exp(t(A + B)) = \lim_{n\to\infty} \left(\exp\left(\frac{t}{n}A\right)\exp\left(\frac{t}{n}B\right)\right)^n$$

where the limit is taken in the strong operator topology. Thus the formula in particular applies if A and B are $n \times n$ matrices.

Let A, B be $n \times n$ matrices over $\mathbb{C}$ with $A^2 = B^2 = I_n$ and $[A, B]_+ = 0_n$. Let $z \in \mathbb{C}$. The *Lie-Trotter formula* is given by

$$\exp(z(A + B)) = \lim_{p\to\infty} \left(e^{zA/p}e^{zB/p}\right)^p.$$

Calculate $e^{z(A+B)}$ using the right-hand side. Note that

$$e^{zA/p} = I_n \cosh(z/p) + A\sinh(z/p), \quad e^{zB/p} = I_n \cosh(z/p) + B\sinh(z/p).$$

Chapter 4

Hilbert Space $L_2(\Omega)$

4.1 Introduction

$L_2(\Omega)$ is the space of Lebesgue square-integrable functions on Ω, where Ω is a Lebesgue measurable subset of $\mathbb{R}^n$ and $n \in \mathbb{N}$. If $f \in L_2(\Omega)$, then

$$\int_\Omega |f|^2 \, dm < \infty.$$

The integration is performed in the Lebesgue sense. The *scalar product* in $L_2(\Omega)$ is defined as

$$\langle f, g \rangle := \int_\Omega f(x)\bar{g}(x) \, dm$$

where $\bar{g}$ denotes the complex conjugate of g. It can be shown that this pre-Hilbert space is complete. Therefore $L_2(\Omega)$ is a Hilbert space. Instead of dm we also write dx in the following. If the Riemann integral exists then it is equal to the Lebesgue integral. However, the Lebesgue integral exists also in cases in which the Riemann integral does not exist.

The scalar product implies the norm

$$\|f\|^2 = \langle f, f \rangle = \int_\Omega f(x)f^*(x)dx \geq 0.$$

Consider the Hilbert space $L_2([a, b])$ and the set of polynomials

$$\{1, x, x^2, \ldots, x^n, \ldots\}.$$

Applying the *Gram-Schmidt technique* and the inner product

$$\langle f, g \rangle = \int_a^b f(x)g(x)\omega(x)dx, \qquad \omega(x) > 0$$

we obtain the first four orthogonal polynomials when

(i) $a = -1$, $b = 1$, $\omega(x) = 1$ (*Legendre polynomials*)
(ii) $a = -1$, $b = 1$, $\omega(x) = (1 - x^2)^{-1/2}$ (*Chebyshev polynomials*)
(iii) $a = 0$, $b = +\infty$, $\omega(x) = e^{-x}$ (*Laguerre polynomials*)
(iv) $a = -\infty$, $b = +\infty$, $\omega(x) = e^{-x^2}$ (*Hermite polynomials*).

The *Legendre functions*

$$\phi_n(x) = \frac{(2n+1)^{1/2}}{2^{n+1/2}n!} \frac{d^n}{dx^n}((x^2 - 1)^n), \quad n = 0, 1, 2, \ldots$$

form an orthonormal basis in the Hilbert space $L_2([-1, +1])$. The Legendre polynomials $P_j(x)$ can also be given by the *generating function*

$$K(z, \zeta) = \frac{1}{\sqrt{1 - 2z\zeta + \zeta^2}} = \sum_{j=0}^{\infty} P_j(z)\zeta^j.$$

Let $b > a$ and $n = 1, 2, \ldots$. Consider the Hilbert space $L_2([a, b])$. The functions

$$\phi_n(x) = \sqrt{\frac{2}{b-a}} \sin\left(\frac{n\pi(x-a)}{b-a}\right), \quad n \in \mathbb{N}$$

form an orthonormal basis in the Hilbert space $L_2([a, b])$.

For the Hilbert space $L_2([0, a])$ with $a > 0$ we have the orthonormal bases

$$B_1 = \left\{ \frac{1}{\sqrt{a}} \exp(2\pi i x n/a) \; : \; n \in \mathbb{Z} \right\}$$

$$B_2 = \left\{ \frac{1}{\sqrt{a}}, \; \sqrt{\frac{2}{a}} \cos(2\pi x n/a), \; \sqrt{\frac{2}{a}} \sin(2\pi x n/a) \; : \; n \in \mathbb{N} \right\}$$

$$B_3 = \left\{ \sqrt{\frac{2}{a}} \sin(\pi x n/a) \; : \; n \in \mathbb{N} \right\}$$

$$B_4 = \left\{ \frac{1}{\sqrt{a}}, \; \sqrt{\frac{2}{a}} \cos(\pi x n/a) \; : \; n \in \mathbb{N} \right\}.$$

For the Hilbert space $L_2(\prod_{j=1}^{n} (-\pi, \pi))$ an orthonormal basis is

$$B = \left\{ \frac{1}{(2\pi)^{n/2}} \exp(i\mathbf{k} \cdot \mathbf{x}) \; : \; \mathbf{k} \cdot \mathbf{x} := k_1 x_1 + \cdots + k_n x_n \right\}$$

where $|x_j| < \pi$ and $k_j \in \mathbb{Z}$.

Let $a > 0$, $b > 0$. Consider the Hilbert space $L_2([0,a] \times [0,b])$. An orthonormal basis is given by

$$\left\{ \frac{2}{\sqrt{ab}} \sin(m\pi x/a) \sin(n\pi x/b) \ : \ m, n \in \mathbb{N} \right\}.$$

A function f in this Hilbert space can be expanded as

$$f(x_1, x_2) = \sum_{n=1}^{\infty} \sum_{m=1}^{\infty} c_{n,m} \sin(n\pi x/a) \sin(m\pi x/b)$$

where the expansion coefficients $c_{n,m}$ are given by

$$c_{n,m} = \frac{-4ab}{\pi^2(a^2n^2 + b^2m^2)} \int_0^a \int_0^b f(x_1, x_2) dx_1 dx_2.$$

For the Hilbert space $L_2([0,a] \times [0,a] \times [0,a])$ an orthonormal basis is

$$B = \left\{ \frac{1}{a^{3/2}} e^{i2\pi \mathbf{n} \cdot \mathbf{x}/a} \ : \ \mathbf{n} \cdot \mathbf{x} := n_1 x_1 + n_2 x_2 + n_3 x_3 \right\}$$

where $a > 0$ and $n_j \in \mathbb{Z}$.

For the Hilbert space $L_2(\mathbb{S}^2)$ where

$$\mathbb{S}^2 := \{(x_1, x_2, x_3) \in \mathbb{R}^3 \ : \ x_1^2 + x_2^2 + x_3^2 = 1\}$$

a basis is given by

$$Y_{l,m}(\theta, \phi) := \frac{(-1)^{l+m}}{2^l l!} \sqrt{\frac{2l+1}{4\pi} \cdot \frac{(l-m)!}{(l+m)!}} \sin^m \theta \frac{d^{l+|m|}(\sin(\theta))^{2l}}{d(\cos(\theta))^{l+|m|}} e^{im\phi}$$

where $l = 0, 1, 2, 3, \ldots$, $m = -l, -l+1, \ldots, +l$ and $0 \le \phi < 2\pi$, $0 \le \theta < \pi$. The functions $Y_{l,m}$ are called *spherical harmonics*. The orthogonality relation is given by

$$\langle Y_{l,m}, Y_{l',m'} \rangle := \int_{\theta=0}^{\pi} \int_{\phi=0}^{2\pi} Y_{l,m}(\theta, \phi) \bar{Y}_{l',m'}(\theta, \phi) \overbrace{\sin(\theta) \, d\theta \, d\phi}^{d\Omega} = \delta_{l,l'} \delta_{m,m'}.$$

The first few spherical harmonics are given by

$$Y_{0,0}(\theta, \phi) = \frac{1}{\sqrt{4\pi}}, \quad Y_{1,0}(\theta, \phi) = \sqrt{\frac{3}{4\pi}} \cos(\theta),$$

$$Y_{1,1}(\theta, \phi) = -\sqrt{\frac{3}{8\pi}} \sin(\theta) e^{i\phi}, \quad Y_{1,-1}(\theta, \phi) = \sqrt{\frac{3}{8\pi}} \cos(\theta) e^{-i\phi}.$$

For the Hilbert space $L_2(\mathbb{R})$ an orthonormal basis is given by

$$B = \left\{ \frac{(-1)^k}{2^{\frac{k}{2}}\sqrt{k!}\sqrt[4]{\pi}} e^{x^2/2} \frac{d^k}{dx^k} e^{-x^2} \quad k = 0, 1, 2, \dots \right\}.$$

The functions

$$H_n(x) := (-1)^n e^{x^2} \frac{d^n}{dx^n} e^{-x^2}, \quad n = 0, 1, 2, \dots$$

are called the *Hermite polynomials*. For the first four Hermite polynomials we find $H_0(x) = 1$, $H_1(x) = 2x$, $H_2(x) = 4x^2 - 2$, $H_3(x) = 8x^3 - 12x$. Any function $f \in L_2(\mathbb{R})$ can be written as a combination of an even and an odd function.

For the Hilbert space $L_2([0, \infty))$ an orthonormal basis is given by

$$B = \left\{ e^{-x/2} L_n(x) \ : \ n = 0, 1, 2, \dots \right\}$$

where

$$L_n(x) := e^x \frac{d^n}{dx^n} (x^n e^{-x}).$$

The functions L_n are called *Laguerre polynomials*. For the first four Laguerre polynomials we find $L_0(x) = 1$, $L_1(x) = -x + 1$, $L_2(x) = x^2 - 4x + 2$, $L_3(x) = -x^3 + 9x^2 - 18x + 6$.

Consider the two Hilbert spaces $\mathcal{H}_1 = L_2((a, b))$ and $\mathcal{H}_2 = L_2((c, d))$. Then the tensor product Hilbert space $\mathcal{H}_1 \otimes \mathcal{H}_2$ is seen to be

$$L_2((a, b) \times (c, d))$$

the space of the functions $f(x_1, x_2)$ with $a < x_1 < b$, $c < x_2 < d$ and

$$\int_c^d \int_a^b |f(x_1, x_2)|^2 dx_1 dx_2 < \infty.$$

The inner product is defined by

$$\langle f, g \rangle := \int_c^d \int_a^b f(x_1, x_2) \bar{g}(x_1, x_2) dx_1 dx_2.$$

Let $b > a$ and $\{ \phi_n : n \in \mathbb{N} \}$ be an orthonormal basis in the Hilbert space $L_2([a, b])$. Let $d > c$ and $\{ \psi_n : n \in \mathbb{N} \}$ be an orthonormal basis in the Hilbert space $L_2([c, d])$. Then the set $\{ \phi_n \psi_m : n \in \mathbb{N}, m \in \mathbb{N} \}$ is an orthonormal basis in the Hilbert space $L_2([a \leq x_1 \leq b] \times [c \leq x_2 \leq d])$.

Consider the Hilbert space $L_2(\mathbb{T})$ of square integrable functions f on the unit circle $\mathbb{S}^1$ with the scalar product

$$\langle f, h \rangle := \int_0^{2\pi} f(e^{i\theta}) \overline{h}(e^{i\theta}) d\theta.$$

The action of a Lie group element $g \in SU(1,1)$ on a function $f \in L_2(\mathbb{T})$ is defined by

$$T(g)f(e^{i\theta}) = |\beta e^{i\theta} + \overline{\alpha}|^{2\ell} f\left(\frac{\alpha e^{i\theta} + \overline{\beta}}{\beta e^{i\theta} + \overline{\alpha}}\right).$$

Mercer's Theorem. [51] Let T be a positive, integral operator on the Hilbert space $L_2([a,b])$ with continuous *kernel* $K(s,t) = \overline{K}(t,s)$ on the domain $[a,b] \times [a,b]$ ($|a|, |b| < \infty$). Then the kernel $K(s,t)$ can be represented by the *bilinear series*

$$K(s,t) = \sum_{j=1}^{\infty} \lambda_j \phi_j(s) \overline{\phi}_j(t)$$

absolutely and uniformly convergent on $[a,b] \times [a,b]$, where $\lambda_j \geq 0$ ($j = 1, 2, \dots$) are the eigenvalues of the operator T and ϕ_j ($j = 1, 2, \dots$) are the corresponding orthonormal eigenfunctions.

The *Plancherel decomposition formula* for the Lie group $SL(2, \mathbb{C})$ tells us that the Hilbert space $L_2(SL(2, \mathbb{C}))$ functions with respect to the Haar measure on the Lie group $SL(2, \mathbb{C})$ uniquely decompose in term of matrix elements of the group element in the unitary irreducible representations of $SL(2, \mathbb{C})$ of the principal series. Such irreducible representations are labeled by the two numbers (n, ρ), where $n \in \mathbb{N}/2$ and $\rho \in \mathbb{R}$.

Stone theorem. [74] A C_0-semigroup of unitary operators (i.e. $T_\tau^* = T_\tau^{-1}$ for all $\tau \in \mathbb{R}^+$) is equal to $\{e^{i\tau A}\}$ for some self-adjoint operator A.

Consider the compact Lie group $SO(n, \mathbb{R})$. The action of $R \in SO(n, \mathbb{R})$ on $L_2(\mathbb{R}^n, \mu_G)$ is given by

$$(\Gamma_R f)(\mathbf{x}) := f(R^{-1}\mathbf{x}), \quad f \in L_2(\mathbb{R}^n, \mu_G).$$

4.2   Solved Problems

Problem 1. Consider the Hilbert space $L_2([0,1])$ and the polynomials

$$p_0 = 1, \ p_1 = x, \ p_2 = x^2.$$

Apply the *Gram-Schmidt technique* to these polynomials.

Solution 1. Let $m, n \in \mathbb{N}_0$. Then

$$\langle x^m, x^n \rangle = \int_0^1 x^m x^n dx = \int_0^1 x^{m+n} dx = \frac{x^{n+m+1}}{m+n+1}\Big|_0^1 = \frac{1}{n+m+1}$$

and therefore

$$\langle 1, 1 \rangle = 1, \ \langle x, x \rangle = \frac{1}{3}, \ \langle x^2, x^2 \rangle = \frac{1}{5}, \ \langle 1, x \rangle = 1/2, \ \langle 1, x^2 \rangle = 1/3.$$

The Gram-Schmidt technique provides

$$q_0 = p_0$$

$$q_1 = p_1 - \frac{\langle q_0, p_1 \rangle}{\langle q_0, q_0 \rangle} q_0$$

$$q_2 = p_2 - \frac{\langle q_1, p_1 \rangle}{\langle q_1, q_1 \rangle} q_1 - \frac{\langle q_0, p_2 \rangle}{\langle q_0, q_0 \rangle} q_0.$$

After normalizing q_0, q_1, q_2 we obtain

$$\tilde{q}_0 = 1, \ \tilde{q}_1 = 2\sqrt{3}(x - 1/2), \ \tilde{q}_2 = 6\sqrt{5}(x^2 - x + 1/6).$$

Problem 2. Let $a, b, c, d \in \mathbb{R}$. Consider the Hilbert space $L_2([0,1])$. Find a non-trivial polynomial p

$$p(x) = ax^3 + bx^2 + cx + d$$

such that $\langle p, 1 \rangle = 0$, $\langle p, x \rangle = 0$, $\langle p, x^2 \rangle = 0$.

Solution 2. Using the ansatz for a polynomial p we obtain

$$\langle p, 1 \rangle = \int_0^1 p(x) dx = \frac{a}{4} + \frac{b}{3} + \frac{c}{2} + d = 0$$

$$\langle p, x \rangle = \int_0^1 p(x) x \, dx = \frac{a}{5} + \frac{b}{4} + \frac{c}{3} + \frac{d}{2} = 0$$

$$\langle p, x^2 \rangle = \int_0^1 p(x) x^2 dx = \frac{a}{6} + \frac{b}{5} + \frac{c}{4} + \frac{d}{3} = 0.$$

This linear system of three equations can be written in matrix form

$$\begin{pmatrix} 1/4 & 1/3 & 1/2 \\ 1/5 & 1/4 & 1/3 \\ 1/6 & 1/5 & 1/4 \end{pmatrix} \begin{pmatrix} a \\ b \\ c \end{pmatrix} = \begin{pmatrix} -d \\ -d/2 \\ -d/3 \end{pmatrix}.$$

The inverse of the matrix on the left-hand side exists. If we assume that $d \neq 0$ we find a non-trivial polynomial.

Problem 3. Consider the Hilbert space $L_2([0,1])$. Find a non-trivial function

$$f(x) = ax^3 + bx^2 + cx + d$$

such that $\langle f(x), x \rangle = 0$, $\langle f(x), x^2 \rangle = 0$, $\langle f(x), x^3 \rangle = 0$.

Solution 3. We obtain the three linear equations

$$\frac{a}{5} + \frac{b}{4} + \frac{c}{3} + \frac{d}{2} = 0, \quad \frac{a}{6} + \frac{b}{5} + \frac{c}{4} + \frac{d}{3} = 0, \quad \frac{a}{7} + \frac{b}{6} + \frac{c}{5} + \frac{d}{4} = 0.$$

Solving for a, b and c in terms of d yields

$$a = -\frac{35}{4}d, \quad b = 15d, \quad c = -\frac{15}{2}d.$$

Selecting d (for example $d = 1$) gives a solution.

Problem 4. Consider the Hilbert space $L_2([0,1])$ and $f \in L_2([0,1])$. Assume that for all $n \in \mathbb{N}_0$

$$\langle x^n, f(x) \rangle \equiv \int_0^1 x^n f(x)dx = \frac{1}{n+2}.$$

Show that $f(x) = x$ almost everywhere on $[0,1]$.

Solution 4. A basis in the Hilbert space $L_2([0,1])$ is given by

$$\{ 1, x, x^2, x^3, \ldots \}.$$

Note that this basis is not an orthonormal basis. Since we have a basis we can write

$$f(x) = \sum_{j=0}^{\infty} c_j x^j, \quad c_j \in \mathbb{R}.$$

It follows that

$$\int_0^1 x^n f(x)dx = \int_0^1 x^n \left(\sum_{j=0}^{\infty} c_j x^j \right) dx = \sum_{j=0}^{\infty} c_j \int_0^1 x^{n+j} dx$$

$$= \sum_{j=0}^{\infty} c_j \frac{x^{n+j+1}}{n+j+1} \bigg|_0^1 = \sum_{j=0}^{\infty} c_j \frac{1}{n+j+1}.$$

Hence

$$\sum_{j=0}^{\infty} c_j \frac{1}{n+j+1} = \frac{1}{n+2}$$

with the solution $c_0 = 0$, $c_1 = 1$, $c_2 = c_3 = \cdots = 0$.

Problem 5. Consider the Hilbert space $L_2([0,1])$ and

$$\phi_n(x) = e^{2\pi i n x}, \quad n \in \mathbb{Z}.$$

Find the scalar product $\langle \phi_n(x), \phi_m(x) \rangle$ with $n, m \in \mathbb{Z}$.

Solution 5. For $n = m$ we have

$$\langle \phi_n, \phi_n \rangle = \int_0^1 e^{2\pi i x (n-n)} dx = \int_0^1 dx = 1.$$

For $n \neq m$ we have

$$\langle \phi_n, \phi_m \rangle = \int_0^1 e^{2\pi i x (n-m)} dx = \frac{1}{2\pi i (n-m)} (e^{2\pi i (n-m)} - 1)$$

$$= \frac{1}{2\pi i (n-m)} (\cos(2\pi(n-m)) - 1) = 0.$$

Problem 6. Consider the Hilbert space $L_2([-\pi, \pi])$.
(i) Obviously $\cos(x) \in L_2([-\pi, \pi])$. Find the norm $\| \cos(x) \|$. Find nontrivial functions $f, g \in L_2([-\pi, \pi])$ such that

$$\langle f(x), \cos(x) \rangle = 0, \quad \langle g(x), \cos(x) \rangle = 0, \quad \langle f(x), g(x) \rangle = 0.$$

(ii) Find

$$\| \cos(x) - \sin(x) \|^2 \equiv \langle \cos(x) - \sin(x), \cos(x) - \sin(x) \rangle.$$

Solution 6. (i) We have

$$\| \cos(x) \|^2 = \langle \cos(x), \cos(x) \rangle = \int_{-\pi}^{\pi} \cos(x) \overline{\cos(x)} dx$$

$$= \int_{-\pi}^{\pi} \cos^2(x) dx = \int_{-\pi}^{\pi} \left(\frac{1}{2} + \frac{1}{2} \cos(2x) \right) dx$$

$$= \left. \frac{x}{2} + \frac{1}{4} \sin(2x) \right|_{-\pi}^{\pi} = \pi.$$

Thus the norm of $\cos(x)$ is $\| \cos(x) \| = \sqrt{\pi}$. Consider $f(x) = x$ and $g(x) = 1$. Then since $\cos(x)$ is an even function, i.e. $\cos(-x) = \cos(x)$ and we integrate from $-\pi$ to π

$$\langle f(x), \cos(x) \rangle = 0, \quad \langle g(x), \cos(x) \rangle = 0, \quad \langle g(x), f(x) \rangle = 0.$$

(ii) We obtain

$$\langle \cos(x) - \sin(x), \cos(x) - \sin(x) \rangle = \langle \cos(x), \cos(x) \rangle - \langle \sin(x), \cos(x) \rangle$$

$$- \langle \cos(x), \sin(x) \rangle + \langle \sin(x), \sin(x) \rangle$$

$$= \langle \cos(x), \cos(x) \rangle + \langle \sin(x), \sin(x) \rangle$$

$$= \int_{-\pi}^{\pi} \cos^2(x) dx + \int_{-\pi}^{\pi} \sin^2(x) dx$$

$$= \int_{-\pi}^{\pi} dx = 2\pi.$$

Problem 7. Consider the Hilbert space $L_2([-1, 1])$. Normalize the functions $f(x) = 1$ and $g(x) = x$ in this Hilbert space.

Solution 7. Since

$$\langle f, f \rangle = \int_{-1}^{+1} dx = 2$$

the normalized function is $\widetilde{f}(x) = 1/\sqrt{2}$. Since

$$\langle g, g \rangle = \int_{-1}^{+1} x^2 dx = \left. \frac{x^3}{3} \right|_{-1}^{+1} = \frac{2}{3}$$

the normalized function is

$$\widetilde{g}(x) = \frac{\sqrt{3}}{\sqrt{2}} x.$$

Problem 8. Consider the Hilbert space $L_2([-1,1])$. The *Legendre polynomials* are defined as

$$P_0(x) := 1, \qquad P_n(x) := \frac{1}{2^n n!} \frac{d^n}{dx^n} (x^2 - 1)^n.$$

(i) The first first four elements are given by

$$P_0(x) = 1, \quad P_1(x) = x, \quad P_2(x) = \frac{1}{2}(3x^2 - 1), \quad P_3(x) = \frac{1}{2}(5x^3 - 3x).$$

Show that the four elements are pairwise orthogonal.
(ii) Find the scalar product $\langle P_j(x), P_k(x) \rangle$.

Solution 8. (i) We have

$$\langle P_0, P_1 \rangle = \langle 1, x \rangle = \int_{-1}^{1} x \, dx = 0$$

$$\langle P_0, P_2 \rangle = \langle 1, \frac{1}{2}(3x^2 - 1) \rangle = \int_{-1}^{1} \frac{1}{2}(3x^2 - 1) \, dx = 0$$

$$\langle P_0, P_3 \rangle = \langle 1, \frac{1}{2}(5x^3 - 3x) \rangle = \int_{-1}^{1} \frac{1}{2}(5x^3 - 3x) \, dx = 0$$

$$\langle P_1, P_2 \rangle = \langle x, \frac{1}{2}(3x^2 - 1) \rangle = \int_{-1}^{1} x \frac{1}{2}(3^2 - 1) \, dx = 0$$

$$\langle P_1, P_3 \rangle = \langle x, \frac{1}{2}(5x^3 - 3x) \rangle = \int_{-1}^{1} x \frac{1}{2}(5x^3 - 3x) \, dx = 0$$

$$\langle P_2, P_3 \rangle = \langle \frac{1}{2}(3x^2 - 1), \frac{1}{2}(5x^3 - 3x) \rangle = 0.$$

(ii) We obtain

$$\langle P_j(x), P_k(x) \rangle = \frac{2\delta_{k,j}}{2k + 1}$$

where $\delta_{k,j}$ is the Kronecker delta.

Problem 9. Consider the function $f \in L_2([0,1])$

$$f(x) = \begin{cases} x & \text{for } 0 \leq x \leq 1/2 \\ 1 - x & \text{for } 1/2 \leq x \leq 1 \end{cases}$$

A basis in the Hilbert space is given by

$$B := \left\{ 1, \sqrt{2} \cos(\pi n x) \ : \ n = 1, 2, \ldots \right\}.$$

Find the Fourier expansion of f with respect to this basis. From this expansion show that

$$\frac{\pi^2}{8} = \sum_{k=0}^{\infty} \frac{1}{(2k+1)^2}.$$

Solution 9. We have for ϕ_0

$$\langle f(x), \phi_0(x) \rangle = \int_0^{1/2} x\, dx + \int_{1/2}^1 (1-x)\, dx = \frac{x^2}{2}\Big|_0^{\frac{1}{2}} + \left(x - \frac{x^2}{2}\right)\Big|_{\frac{1}{2}}^1 = \frac{1}{4}.$$

For $n = 1, 2, \ldots$ we have

$$\langle f(x), \phi_n(x) \rangle = \int_0^{\frac{1}{2}} \sqrt{2}x \cos(\pi n x)\, dx + \int_{\frac{1}{2}}^1 (\sqrt{2}\cos(\pi n x) - \sqrt{2}x\cos(\pi n x))\, dx$$

$$= \frac{\sqrt{2}}{\pi n} x \sin(\pi n x)\Big|_0^{\frac{1}{2}} - \int_0^{\frac{1}{2}} \frac{\sqrt{2}}{\pi n} \sin(\pi n x)\, dx + \frac{\sqrt{2}}{\pi n} \sin(\pi n x)\Big|_{\frac{1}{2}}^1$$

$$- \frac{\sqrt{2}}{\pi n} x \sin(\pi n x)\Big|_{\frac{1}{2}}^1 + \int_{\frac{1}{2}}^1 \frac{\sqrt{2}}{\pi n} \sin(\pi n x)\, dx$$

$$= \frac{1}{\sqrt{2}\pi n} + \frac{\sqrt{2}}{\pi^2 n^2} \cos(\pi n x)\Big|_0^{\frac{1}{2}} + \frac{\sqrt{2}}{\pi n} \sin(\pi n) - \frac{\sqrt{2}}{\pi n} \sin(\pi n/2)$$

$$- \frac{\sqrt{2}}{\pi n} \sin(\pi n) + \frac{1}{\sqrt{2}\pi n} \sin(\pi n/2) - \frac{\sqrt{2}}{\pi^2 n^2} \cos(\pi n x)\Big|_{\frac{1}{2}}^1$$

$$= \frac{\sqrt{2}}{\pi^2 n^2} (2\cos(\pi n/2) - 1 - (-1)^n).$$

Thus for n odd $\langle f(x), \phi_n(x) \rangle = 0$. For $n = 2j$ even ($j \in \mathbb{N}$)

$$\langle f(x), \phi_{2j}(x) \rangle = \frac{\sqrt{2}}{4\pi^2 j^2} (2\cos(\pi j) - 2).$$

For j even $\langle f(x), \phi_{2j}(x) \rangle = 0$. For $j = 2k+1$ odd ($k \in \mathbb{N}_0$)

$$\langle f(x), \phi_{4k+2}(x) \rangle = -\frac{\sqrt{2}}{\pi^2 (2k+1)^2}.$$

Thus

$$f(x) = \frac{1}{4} - \sum_{k=0}^{\infty} \frac{2}{\pi^2 (2k+1)^2} \cos(2\pi(2k+1)x).$$

We have $f(0) = 0$,

$$f(0) = 0 = \frac{1}{4} - \sum_{k=0}^{\infty} \frac{2}{\pi^2 (2k+1)^2}$$

from which follows

$$\frac{\pi^2}{8} = \sum_{k=0}^{\infty} \frac{1}{(2k+1)^2}.$$

Problem 10. Consider the Hilbert space $L_2([0,1])$. Show that the linear operator $T : L_2([0,1]) \to L_2([0,1])$ defined by

$$Tf(x) = xf(x)$$

is a bounded self-adjoint linear operator without eigenvalues. Is the operator compact?

Solution 10. Self-adjointness of the operator T follows from the fact that x is real. Self-adjointness also follows from the integral representation on the scalar product

$$\langle Tf(x), g(x) \rangle = \int_0^1 xf(x)\bar{g}(x)dx, \qquad g \in L_2([0,1])$$

where the integral is in the Lebesgue sense. Now

$$R_\lambda(T)f(x) = (x - \lambda)^{-1}f(x)$$

shows that we have $\sigma(T) = [0,1]$, and for $\lambda \in [0,1]$ we find that

$$T_\lambda f(x) = (x - \lambda)f(x) = 0$$

implies $f(x) = 0$ for all $x \neq \lambda$, that is $f = 0$, i.e. the zero element in $L_2([0,1])$. Thus λ cannot be an eigenvalue of the linear operator T. The operator T has no eigenvectors and therefore (applying the spectral theorem) cannot be compact.

Problem 11. Consider the Hilbert space $L_2([0,1])$. The *shifted Legendre polynomials*, defined on the interval $[0,1]$, are obtained from the Legendre polynomial by the transformation $y = 2x - 1$. The shifted Legendre polynomials are given by the recurrence formula

$$P_j(x) = \frac{(2j+1)(2x-1)}{j+1}P_j(x) - \frac{j}{j+1}P_{j-1}(x) \qquad j = 1, 2, \dots$$

and $P_0(x) = 1$, $P_1(x) = 2x - 1$. They are elements of the Hilbert space $L_2([0,1])$. A function u in the Hilbert space $L_2([0,1])$ can be approximated in the form of a series with $n + 1$ terms

$$u(x) = \sum_{j=0}^{n} c_j P_j(x)$$

where the coefficients $c_j \in \mathbb{R}$, $j = 0, 1, \ldots, n$. Consider the *Volterra integral equation* of first kind

$$\lambda \int_0^x \frac{y(t)}{(x-t)^\alpha} dt = f(x), \qquad 0 \le t \le x \le 1$$

with $0 < \alpha < 1$ and $f \in L_2([0, 1])$. Consider the ansatz

$$y_n(x) = a_0 x^\alpha + \sum_{j=0}^n c_j P_j(x)$$

to find an approximate solution to the Volterra integral equation of first kind ($\alpha = 1/2$)

$$\lambda \int_0^x \frac{y(t)}{\sqrt{x-t}} dt = f(x)$$

where

$$f(x) = \frac{2}{105} \sqrt{x} (105 - 56x^2 + 48x^3).$$

Solution 11. Inserting the ansatz in the Volterra integral equation yields

$$\lambda a_0 \int_0^x \frac{t^\alpha}{(x-t)^\alpha} dt + \lambda \sum_{j=0}^n c_j \int_0^x \frac{P_j(t)}{(x-t)^\alpha} dt = f(x)$$

where $\alpha = 1/2$. Since

$$\int_0^x \frac{t^n}{(x-t)^\alpha} dt = \frac{\Gamma(n+1)\Gamma(1-\alpha)}{\Gamma(n+2-\alpha)} x^{n+1-\alpha}$$

and

$$\int_0^x \frac{t^\alpha}{(x-t)^\alpha} dt = \frac{\pi\alpha}{\sin(\pi\alpha)} x$$

we find

$$\int_0^x \frac{P_j(t)}{(x-t)^{(\alpha)}} dt = \sum_{k=0}^j a_{jk}^\alpha x^{k+1-\alpha}$$

where

$$a_{jk}^{(\alpha)} = (-1)^{j+k} \frac{(j+k)!\Gamma(1-\alpha)}{k!(j-k)!\Gamma(k+2-\alpha)}.$$

Thus we find

$$\lambda a_0 \frac{\pi\alpha}{\sin(\pi\alpha)} x + \lambda \sum_{j=0}^n \sum_{k=0}^j c_j a_{jk}^{(\alpha)} x^{k+1-\alpha} = f(x).$$

Since $\alpha = 1/2$ we have $\sin(\pi/2) = 1$, $\Gamma(1/2) = \sqrt{\pi}$ and the equation simplifies to

$$\frac{1}{2}\lambda a_0 \pi x + \lambda \sum_{j=0}^{n} \sum_{k=0}^{j} c_j a_{jk}^{(\alpha)} x^{k+1-\alpha} = f(x).$$

We consider this equation at the roots $y_j : (j = 1, 2, \ldots, n+1)$ of the shifted Legendre polynomials and $y_0 = 0$. Thus we find $n + 2$ equations for the coefficients c_j $(j = 0, 1, \ldots, n)$ and a_0. We have

$$P_2(x) = 6x^2 - 6x + 1, \qquad P_3(x) = 20x^3 - 30x^2 + 12x - 1.$$

We find $a_0 = 0$ and

$$c_0 = \frac{11}{12}, \quad c_1 = -\frac{1}{20}, \quad c_2 = \frac{1}{12}, \quad c_3 = \frac{1}{20}.$$

Problem 12. Consider the Hilbert space $L_2([a, b])$, where $a, b \in \mathbb{R}$ and $b > a$. Find the condition on a and b such that

$$\langle \cos(x), \sin(x) \rangle = 0$$

where $\langle \cdot, \cdot \rangle$ denotes the scalar product in $L_2([a, b])$.
Hint. Since $b > a$, we can write $b = x + \epsilon$, where $\epsilon > 0$.

Solution 12. We have

$$\langle \cos(x), \sin(x) \rangle = \int_{a}^{a+\epsilon} \cos(x) \sin(x) dx = \frac{1}{2} \sin^2(x) \Big|_{a}^{a+\epsilon}$$
$$= \frac{1}{2}(\sin^2(a + \epsilon) - \sin^2(a)).$$

Thus we have to solve $\sin^2(a + \epsilon) - \sin^2(a) = 0$. The solutions are $\epsilon = \pi, 2\pi, 3\pi, \ldots$.

Problem 13. Consider the Hilbert space $L_2([0, \pi])$. Let $\| \cdot \|$ be the norm induced by the scalar product of $L_2([0, \pi])$. Find the constants a, b such that

$$\| \sin(x) - (ax^2 + bx) \|$$

is a minimum.

Solution 13. We have to minimize the analytic function $f : \mathbb{R}^2 \to \mathbb{R}$

$$f(a,b) = \| \sin(x) - (ax^2 + bx) \|^2 = \int_0^\pi (\sin(x) - (ax^2 + bx))^2 dx.$$

Thus we have to calculate $\partial f/\partial a = 0$ and $\partial f/\partial b = 0$. We can either integrate first and then differentiate or we can differentiate first under the integration and then do the integration in the present case. Doing first the integration we obtain

$$f(a,b) = \frac{\pi}{2} - 2a(\pi^2 - 4) - 2b\pi + \frac{1}{5}a^2\pi^5 + \frac{1}{2}ab\pi^4 + \frac{1}{3}b^2\pi^3.$$

Then the differentiation yields

$$\frac{\partial f}{\partial a} = -2\pi^2 + 8 + \frac{2}{5}a\pi^5 + \frac{1}{2}b\pi^4 = 0$$

$$\frac{\partial f}{\partial b} = -2\pi + \frac{1}{2}a\pi^4 + \frac{2}{3}b\pi^3 = 0.$$

Solving these two equations we obtain

$$a = \frac{20}{\pi^3} - \frac{320}{\pi^5}, \qquad b = \frac{-12}{\pi^2} + \frac{240}{\pi^4}.$$

For the second derivatives we have

$$\frac{\partial^2 f}{\partial a \partial b} = \frac{1}{2}\pi^4, \quad \frac{\partial^2 f}{\partial a^2} = \frac{2}{5}\pi^5, \quad \frac{\partial^2 f}{\partial b^2} = \frac{2}{3}\pi^3.$$

The trace and determinant of the Hessian matrix are positive. Thus the Hessian matrix is positive definite and the solution is a minimum.

Problem 14. (i) Consider the functions

$$f(x) = \frac{1}{1 + x^2}, \qquad g(x) = \frac{x}{1 + x^2}.$$

Obviously $f, g \in L_2(\mathbb{R})$. Calculate the scalar product

$$\langle f, g \rangle = \int_{-\infty}^{\infty} f(x)g(x)dx.$$

(ii) Let $\omega > 0$. Consider the functions

$$f(t) = \frac{\sin(\omega t)}{\omega t}, \qquad g(t) = \frac{1 - \cos(\omega t)}{\omega t}$$

with $f(0) = 1$, $g(0) = 0$ and $f, g \in L_2(\mathbb{R})$. Calculate the scalar product

$$\langle f, g \rangle = \int_{-\infty}^{\infty} f(t)g(t)dt.$$

Solution 14. (i) Since f is even, i.e. $f(-x) = f(x)$ and g is odd, i.e.
$g(-x) = -g(x)$ we have $\langle f, g \rangle = 0$.
(ii) Since f is even, i.e. $f(-t) = f(t)$ and g is odd, i.e. $g(-t) = -g(t)$ we
have $\langle f, g \rangle = 0$.

Problem 15. Consider the Hilbert space $L_2([-\pi, \pi])$. Given the function

$$f(x) = \begin{cases} 1 & 0 < x \le \pi \\ 0 & x = 0 \\ -1 & -\pi \le x < 0 \end{cases}$$

Obviously $f \in L_2([-\pi, \pi])$. Find the Fourier expansion of f. An orthonormal basis B is given by

$$B := \left\{ \phi_k(x) = \frac{1}{\sqrt{2\pi}} \exp(ikx) \;:\; k \in \mathbb{Z} \right\}.$$

Find the approximation $a_0\phi_0(x) + a_1\phi_1(x) + a_{-1}\phi_{-1}(x)$, where a_0, a_1, a_{-1} are the Fourier coefficients.

Solution 15. The Fourier expansion is given by

$$f(x) = \sum_{k \in \mathbb{Z}} \langle f(x), \phi_k(x) \rangle \phi_k(x)$$

where $\langle \cdot, \cdot \rangle$ is the scalar product in $L_2([-\pi, \pi])$. We have $\langle f(x), \phi_0(x) \rangle = 0$.
For $k \ne 0$ we have

$$\langle f(x), \phi_k(x) \rangle = \lim_{b \to 0} \int_b^\pi \overline{\phi_k(x)} dx - \lim_{b \to 0} \int_{-\pi}^b \overline{\phi_k(x)} dx$$

$$= \left(\lim_{b \to 0} \frac{i}{k\sqrt{2\pi}} \exp(-ikx) \Big|_b^\pi - \frac{i}{k\sqrt{2\pi}} \exp(-ikx) \Big|_{-\pi}^b \right)$$

$$= \frac{i}{k\sqrt{2\pi}} \lim_{b \to 0} \left(2(-1)^k - 2\exp(ikb) \right)$$

$$= -\frac{2i}{k\sqrt{2\pi}} (1 - (-1)^k).$$

Thus

$$f(x) = -\sum_{k \in \mathbb{Z}} \frac{2i}{(2k+1)\pi} \exp(i(2k+1)x).$$

For the approximation we find

$$f(x) \approx a_0\phi_0(x) + a_1\phi_1(x) + a_{-1}\phi_{-1}(x) = -\frac{2i}{\pi}\exp(ix) + \frac{2i}{\pi}\exp(-ix)$$

$$= \frac{4}{\pi}\sin(x).$$

Problem 16. Consider the Hilbert space $L_2([0, 2\pi))$ and the functions

$$g(x) = \cos(x), \qquad f(x) = x.$$

Find the conditions on the coefficients of the polynomial

$$p(x) = a_3x^3 + a_2x^2 + a_1x + a_0$$

such that $\langle g(x), p(x) \rangle = 0$, $\langle f(x), p(x) \rangle = 0$. Solve the equations for a_3, a_2, a_1, a_0.

Solution 16. From the first condition we obtain

$$\int_0^{2\pi} p(x)\cos(x)dx = 12\pi^2 a_3 + 4\pi a_2 = 0.$$

Thus $a_2 = -3\pi a_3$. From the second condition we find

$$\langle f(x), p(x) \rangle = \frac{32}{5}\pi^5 a_3 + 4\pi^4 a_2 + \frac{8}{3}\pi^3 a_1 + 2\pi^2 a_0 = 0.$$

Hence

$$a_3 = \frac{5}{28}\left(\frac{8}{3\pi^2}a_1 + \frac{2}{\pi^3}a_0\right), \qquad a_2 = -\frac{15}{28}\left(\frac{8}{3\pi^2}a_1 + \frac{2}{\pi^3}a_0\right)$$

with a_1, a_0 arbitrary.

Problem 17. Consider the Hilbert space $L_2([-1/2, 1/2])$. The following sets

$$B_1 := \{\, \phi_k(x) = \exp(2\pi ikx),\ k \in \mathbb{Z} \,\}, \quad B_2 := \{\, \psi_k(x) = \sqrt{2}\sin(2\pi kx),\ k \in \mathbb{N} \,\}$$

each form an orthonormal basis in this Hilbert space. Expand the step function

$$f(x) = \begin{cases} -1 \text{ for } x \in [-1/2, 0] \\ \ \ 1 \ \text{ for } \ x \in [0, 1/2] \end{cases}$$

with respect to the basis B_1 and with respect to the basis B_2. Show that the two expansions are equivalent. Recall that

$$2\sin(x)\sin(y) \equiv \cos(x - y) - \cos(x + y).$$

Solution 17. We have for the basis B_1 $\langle f(x), \phi_0(x) \rangle = 0$ and for $k \neq 0$ we find

$$\langle f(x), \phi_k(x) \rangle = \frac{1}{i\pi k}(1 - (-1)^k).$$

For the basis B_2 we have

$$\langle f(x), \psi_k(x) \rangle = \frac{2}{\sqrt{2\pi}k}(1 - (-1)^k).$$

This leads us to the expansion for the first basis

$$f(x) = \sum_{k \in \mathbb{Z}, k \neq 0} \frac{1}{i\pi k}(1 - (-1)^k)e^{2\pi i k x}$$

$$= \sum_{k \in \mathbb{Z}, k \neq 0} \frac{1}{\pi i k}(1 - (-1)^k)(\cos(2\pi k x) + i\sin(2\pi k x))$$

$$= \frac{4}{\pi} \sum_{k=1, k \text{ odd}} \frac{1}{k} \sin(2\pi k x).$$

For the second basis we find the same result

$$f(x) = \sum_{k \in \mathbb{N}} \frac{2}{\sqrt{2\pi}k}(1 - (-1)^k)\sqrt{2}\sin(2\pi k x) = \frac{4}{\pi} \sum_{k=1, k \text{ odd}} \frac{1}{k} \sin(2\pi k x).$$

Problem 18. Consider the Hilbert space $L_2([-1,1])$. The *Chebyshev polynomials* are defined by

$$T_n(x) := \cos(n\cos^{-1}(x)), \quad n = 0, 1, 2, \ldots.$$

They are elements of the Hilbert space $L_2([-1,1])$. We have

$$T_0(x) = 1, \quad T_1(x) = x, \quad T_2(x) = 2x^2 - 1, \quad T_3(x) = 4x^3 - 3x.$$

Calculate the scalar products $\langle T_0, T_1 \rangle$, $\langle T_1, T_2 \rangle$, $\langle T_2, T_3 \rangle$. Calculate the integrals

$$\int_{-1}^{1} \frac{T_m(x)T_n(x)}{\sqrt{1-x^2}} dx$$

for $(m,n) = (0,1), (m,n) = (1,2), (m,n) = (2,3)$.

Solution 18. We obtain

$$\langle T_0, T_1 \rangle = \int_{-1}^{1} x\, dx = 0, \quad \langle T_1, T_2 \rangle = \int_{-1}^{1} x(2x^2 - 1)\, dx = 0,$$

$$\langle T_2, T_3 \rangle = \int_{-1}^{1} (2x^2 - 1)(4x^3 - 3x)dx = 0.$$

Obviously $\langle T_n, T_{n+1} \rangle = 0$.

Problem 19. The *Chebyshev polynomials* $T_n(x)$ of the 1-st kind are defined for $x \in [-1, 1]$ and given by

$$T_n(x) = \cos(n \arccos(x)), \qquad n = 0, 1, 2, \dots$$

The Chebyshev polynomials $U_n(x)$ of the 2-nd kind are defined for $x \in [-1, 1]$ and given by

$$U_n(x) = \frac{\sin((n + 1) \arccos(x))}{\sqrt{1 - x^2}}, \qquad n = 0, 1, 2, \dots.$$

Consider the Hilbert spaces

$$\mathcal{H}_1 = L_2 \left([-1, 1], \frac{dx}{\pi \sqrt{1 - x^2}} \right), \qquad \mathcal{H}_2 = L_2 \left([-1, 1], \frac{2\sqrt{1 - x^2}dx}{\pi} \right)$$

which bases are formed by the Chebyshev polynomials of the 1-st and 2-nd types

$$\Phi_n^{(1)}(x) = \sqrt{2}T_n(x), \quad n \geq 1, \quad \Phi_0^{(1)} = T_0(x) = 1$$

$$\Phi_n^{(2)}(x) = U_n(x), \qquad n \geq 0.$$

Find a recursion relation for $\Phi_n^{(1)}$ and $\Phi_n^{(2)}$.

Solution 19. Since T_n and U_n satisfy the recursion relations

$$T_{n+1}(x) = 2xT_n(x) - T_{n-1}(x)$$

$$U_{n+1}(x) = 2xU_n(x) - U_{n-1}(x)$$

we obtain $(k = 1, 2)$

$$x\Phi_n^{(k)}(x) = c_n\Phi_{n+1}^{(k)} + c_{n-1}\Phi_{n-1}^{(k)}(x), \qquad n \geq 0, \quad c_{-1} = 0$$

where

$$\Phi_0^{(k)}(x) = 1, \qquad c_n = \frac{1}{2}, \quad n \geq 1, \quad c_0 = 1/\sqrt{2}.$$

Problem 20. Consider the Hilbert space $L_2([-\pi, \pi])$. An orthonormal basis in this Hilbert space is given by

$$B = \left\{ \frac{1}{\sqrt{2\pi}} e^{ikx} : |x| \leq \pi, \; k \in \mathbb{Z} \right\}.$$

Consider the function $f(x) = e^{iax}$ in this Hilbert space, where the constant a is real but not an integer. Apply *Parseval's relation*

$$\|f\|^2 = \sum_{k \in \mathbb{Z}} |\langle f, \phi_k \rangle|^2, \quad \phi_k(x) = \frac{1}{\sqrt{2}} e^{ikx}$$

to show that

$$\sum_{k=-\infty}^{\infty} \frac{1}{(a-k)^2} = \frac{\pi^2}{\sin^2(ax)}.$$

Solution 20. We have

$$\|f\|^2 = \int_{-\pi}^{\pi} e^{iax} e^{-iax} dx = \int_{-\pi}^{\pi} dx = 2\pi.$$

Now we have for the scalar product

$$\langle f(x), \phi_k(x) \rangle = \int_{-\pi}^{\pi} f(x) \overline{\phi}_k(x) dx = \frac{1}{\sqrt{2\pi}} \int_{-\pi}^{\pi} e^{i(a-k)x} dx$$

$$= \frac{1}{\sqrt{2}} \left. \frac{e^{i(a-k)x}}{i(a-k)} \right|_{-\pi}^{\pi} = \frac{1}{\sqrt{2}} \frac{2\sin((a-k)\pi)}{a-k}.$$

Owing to $\sin(k\pi) = 0$ and $\cos(k\pi) = (-1)^k$ we have

$$\sin((a-k)\pi) \equiv (-1)^k \sin(a\pi).$$

Thus

$$\left| \frac{1}{\sqrt{2}} \frac{2(-1)^k \sin(a\pi)}{a-k} \right|^2 = \frac{2}{\pi} \frac{\sin^2(a\pi)}{(a-k)^2}.$$

Therefore

$$2\pi = \sum_{k=-\infty}^{\infty} \frac{2}{\pi} \frac{\sin^2(a\pi)}{(a-k)^2}$$

and the result follows.

Problem 21. Consider the Hilbert space $L_2(\mathbb{R})$. Let

$$f_n(x) = \frac{x}{1 + nx^2}, \qquad n = 1, 2, \ldots$$

with $f_n \in L_2(\mathbb{R})$ for all n. Does the sequence $f_n(x)$ converge uniformly on the real line?

Solution 21. It is easy to see that $f_n(x) \to 0$ for all $x \in \mathbb{R}$. To test its uniform convergence we have to find the sequence

$$g_n := \sup_{x \in \mathbb{R}} |f_n(x) - 0|.$$

This function f_n is differentiable. Thus sup = its absolute maximum. The absolute maximum will be at its critical points, since it tends to 0 as $x \to \pm\infty$. We have

$$\frac{df_n(x)}{dx} = \frac{1 - nx^2}{(1 + nx^2)^2} = 0.$$

At $x = 1/\sqrt{n}$ we have a critical point. Thus, the absolute maximum is

$$|f_n(x)| = 1/(2\sqrt{n}).$$

It follows that $g_n = 1/(2\sqrt{n})$. Since $\lim_{n \to \infty} g_n = 0$ the convergence is uniform.

Problem 22. Let $n = 1, 2, \dots$. We define the functions $f_n \in L_2([0, \infty))$ by

$$f_n(x) = \begin{cases} \sqrt{n} & \text{for } n \le x \le n + 1/n \\ 0 & \text{otherwise} \end{cases}$$

(i) Calculate the norm $\|f_n - f_m\|$ implied by the scalar product. Does the sequence $\{f_n\}$ converge in the $L_2([0, \infty))$ norm?
(ii) Show that $f_n(x)$ converges pointwise in the domain $[0, \infty)$ and find the limit. Does the sequence converge pointwise uniformly?
(iii) Show that $\{f_n\}$ $(n = 1, 2, \dots)$ is an orthonormal system. Is it a basis in the Hilbert space $L_2([0, \infty))$?

Solution 22. (i) We have

$$\|f_n(x) - f_m(x)\|^2 = \int_n^{n+1/n} n\,dx + \int_m^{m+1/m} m\,dx = 2.$$

This result implies that the distance between any two different elements in the sequence is always $\sqrt{2}$. Therefore the sequence f_n is not Cauchy, i.e. the sequence does not converge in the norm of the Hilbert space $L_2([0, \infty))$.
(ii) The sequence f_n converges pointwise to 0 since for any $x \in [0, \infty)$ and given $\epsilon > 0$, $|f_n(x) - 0| < \epsilon$ for all $n \ge [x]$, where $[x]$ is the upper rounding of x. The sequence does not converge uniformly. Consider the sequence $\{g_n\}$ where

$$g_n := \sup_{x \in [0, \infty)} (|f_n(x) - 0|) = \sqrt{n}.$$

This sequence diverges since $\lim_{n\to\infty} g_n = \infty$. Thus it does not converges uniformly.

(iii) It is an orthonormal system since $\|f_n\| = 1$ and $\langle f_n, f_m \rangle = \delta_{0,|n-m|}$. It is not an orthonormal basis since it does not span the whole of $L_2([0, \infty))$. For example if we take the *gate function* with $f(x) = 1$ on the unit interval $[0, 1]$ and 0 otherwise. Then we have for the inner product

$$\langle f_n(x), f(x) \rangle = 0 \quad \text{for} \quad n = 1, 2, \dots.$$

Problem 23. Let $a > 0$ with dimension length. Consider the Hilbert space $L_2([0, a])$ and the linear ordinary differential equation of fourth order with two initial conditions and two boundary conditions

$$\frac{d^4u}{dx^4} - \lambda u = 0, \quad u(0) = \frac{du(0)}{dx} = 0, \quad \frac{d^2u(a)}{dx^2} = \frac{d^3u(a)}{dx^3} = 0.$$

Show that the solution of this linear differential equation with the initial and boundary conditions provides an orthonormal basis in the Hilbert space $L_2([0, a])$.

Solution 23. Obviously the solution of the linear differential with constant coefficients is given by

$$u(x) = C_1 \cosh(kx) + C_2 \sinh(kx) + C_3 \cos(kx) + C_4 \sin(kx).$$

Then $d^4u/dx^4 = k^4 u$. It follows that $k^4 u = \lambda u$ or $k^4 = \lambda$. Imposing the condition $u(0) = 0$ we obtain $C_3 = -C_1$ and imposing the condition $du(0) = 0$ we obtain $C_4 = -C_2$. From $d^2u(a)/dx^2 = 0$ and $d^3u(a)/dx^3 = 0$ and inserting $C_3 = -C_1$, $C_4 = -C_2$ we arrive at the matrix equation

$$\begin{pmatrix} \cosh(ka) + \cos(ka) & \sinh(ka) + \sin(ka) \\ \sinh(ka) - \sin(ka) & \cosh(ka) + \cos(ka) \end{pmatrix} \begin{pmatrix} C_1 \\ C_2 \end{pmatrix} = \begin{pmatrix} 0 \\ 0 \end{pmatrix}.$$

To obtain a nontrivial solution the determinant of the 2×2 matrix must be equal to 0. This provides the *transcendental equation*

$$\cosh(ka)\cos(ka) + 1 = 0 \quad \text{or} \quad \cos(ka) = -1/\cosh(ka).$$

We set $s = ka$ and thus $\cosh(s)\cos(s) + 1 = 0$. Let s_n $(n = 0, 1, 2, \dots)$ be the positive roots of $\cosh(s)\cos(s) + 1 = 0$ we obtain the eigenfunctions

$$u_n(x) = (\sinh(s_n) + \sin(s_n))(\cosh(s_n x/a) - \cos(s_n x/a))$$
$$- (\cosh(s_n) + \cos(s_n))(\sinh(s_n x/a) - \sin(s_n x/a))$$

with $\lambda_n = (s_n/a)^4$ and $n = 0, 1, 2, \ldots$. The functions u_n $(n \in \mathbb{N}_0)$ form orthogonal system in the Hilbert space $L_2([0, a])$ since $(n \neq m)$

$$(\lambda_n - \lambda_m) \int_0^a u_n(x) u_m(x) dx = \int_0^a \left(\frac{d^4 u_n(x)}{dx^4} u_m(x) - u_n(x) \frac{d^4 u_m(x)}{dx^4} \right) dx$$

$$= \left(\frac{d^3 u_n(x)}{dx^3} u_m(x) - \frac{d^2 u_n(x)}{dx^2} \frac{du_m(x)}{dx} + \frac{du_n(x)}{dx} \frac{d^2 u_m(x)}{dx^2} - u_n(x) \frac{d^3 u_m(x)}{dx^3} \right) \Big|_0^a = 0$$

using integration by parts. The orthogonal system is complete and thus form a basis in the Hilbert space $L_2([0, a])$.

Problem 24. Consider the Hilbert space $L_2(\mathbb{T})$. Let $f \in L_2(\mathbb{T})$. Give an example of a bounded linear functional.

Solution 24. For each $n \in \mathbb{Z}$ the functional $\varphi_n : L_2(\mathbb{T}) \to \mathbb{C}$

$$\varphi_n(f) = \frac{1}{\sqrt{2\pi}} \int_{\mathbb{T}} f(x) e^{-inx} dx$$

that maps a function f to its nth Fourier coefficient is a bounded linear functional. We have $\|\varphi_n\| = 1$ for every $n \in \mathbb{Z}$.

Problem 25. Consider the Hilbert space $L_2([-\pi, \pi])$ and the vector space of continuous real-valued functions $C([-\pi, \pi])$ on the interval $[-\pi, \pi]$. Let $k > 0$ and

$$f_k(x) = \begin{cases} 0 & \text{if } -\pi \leq x \leq 0 \\ kx & \text{if } 0 \leq x \leq 1/k \\ 1 & \text{if } 1/k \leq x \leq \pi \end{cases}$$

The sequence of functions f_k belongs to the vector space $C([-\pi, \pi])$. Then $f_n \to \chi$ in the norm of the Hilbert space $L_2([-\pi, \pi])$, where

$$\chi(x) := \begin{cases} 0 & \text{if } -\pi \leq x \leq 0 \\ 1 & \text{if } 0 < x \leq \pi \end{cases}$$

so that the sequence $\{f_k\}$ is a *Cauchy sequence* in the Hilbert space $L_2([-\pi, \pi])$. Show that $\|\chi - g\| > 0$ for every $g \in C([-\pi, \pi])$. Conclude that $C([-\pi, \pi])$ is not a Hilbert space.

Solution 25. Let $g \in C([-\pi, \pi])$ and thus $g \in L_2([-\pi, \pi])$. We have

$$\langle \chi - g, \chi - g \rangle = \langle \chi, \chi \rangle - \langle \chi, g \rangle - \langle g, \chi \rangle + \langle g, g \rangle.$$

Thus

$$\|\chi - g\|_2^2 = \int_0^\pi (1 - 2g + g^2) dx + \int_{-\pi}^0 g^2 dx$$

where $(1 - 2g + g^2) \equiv (1 - g)^2$. Since both $(1 - g)^2$ and g^2 are nonnegative, the sum of the two integrals is nonnegative for all $x \in [-\pi, \pi]$. The only possibility for this integral to be 0 is that $g = \chi$, which is not possible since $\chi \notin C([-\pi, \pi])$. This implies that the vector space $C([-\pi, \pi])$ is not complete. Consequently it is not a Hilbert space.

Problem 26. (i) Consider the Hilbert space $L_2([0, 1])$ with the scalar product $\langle \cdot, \cdot \rangle$. Let $f : [0, 1] \to [0, 1]$

$$f(x) := \begin{cases} 2x & \text{if } x \in [0, 1/2) \\ 2(1 - x) & \text{if } x \in [1/2, 1] \end{cases}$$

Thus $f \in L_2([0, 1])$. Calculate the moments μ_k, $k = 0, 1, 2, \ldots$ defined by

$$\mu_k := \langle f(x), x^k \rangle \equiv \int_0^1 f(x) x^k dx.$$

(ii) Show that

$$\sum_{k=0}^{\infty} |\mu_k|^2 < \pi \int_0^1 |f(x)|^2 dx.$$

Solution 26. (i) For $k = 0$ we have

$$\mu_0 = \int_0^1 f(x) dx = \frac{1}{2}.$$

For $k \geq 1$ we obtain

$$\mu_k = \int_0^{1/2} 2x x^k dx + \int_{1/2}^1 (2 - 2x) x^k dx$$

$$= 2 \int_0^{1/2} x^{k+1} dx + 2 \int_{1/2}^1 x^k dx - 2 \int_{1/2}^1 x^{k+1} dx$$

$$= 2 \frac{(1/2)^{k+2}}{k+2} + 2 \frac{1}{k+1} - 2 \frac{(1/2)^{k+1}}{k+1} - 2 \frac{1}{k+2} + 2 \frac{(1/2)^{k+2}}{k+2}$$

$$= -\left(\frac{1}{2}\right)^{k+1} \frac{1}{(k+2)(k+1)} + \frac{2}{(k+1)(k+2)}.$$

(ii) For the right-hand side we find

$$\int_0^1 |f(x)|^2 dx = \int_0^{1/2} 4x^2 dx + \int_0^1 2(1-x)2(1-x) dx = \frac{1}{3}.$$

Problem 27. Let $a, b \in \mathbb{R}$ and $-\infty < a < b < +\infty$. Let f be a function in the class C^1 (i.e. the derivative df/dt exists and is continuous) on the interval $[a, b]$. Thus f is also an element of the Hilbert space $L_2([a, b])$. Show that

$$\lim_{\omega \to \infty} \int_a^b f(t) \sin(\omega t) dt = 0. \tag{1}$$

Solution 27. Setting $u = f(t)$, $du = (df(t)/dt)dt$, $dv = \sin(\omega t)dt$, $v = -\frac{1}{\omega}\cos(\omega t)$ and using integration by parts yields

$$\int_a^b f(t) \sin(\omega t) dt = -\frac{1}{\omega} f(t) \cos(\omega t) \Big|_a^b + \frac{1}{\omega} \int_a^b \frac{df(t)}{dt} \cos(\omega t) dt$$

$$= \frac{1}{\omega}(f(a) \cos(\omega a) - f(b) \cos(\omega b)) + \frac{1}{\omega} \int_a^b \frac{df(t)}{dt} \cos(\omega t) dt.$$

Now

$$\left| \frac{1}{\omega}(\cos(\omega a) - f(b) \cos(\omega b)) \right| \leq \frac{1}{\omega}(|f(a) \cos(\omega a)| + |f(b) \cos(\omega b)|)$$

$$\leq \frac{1}{\omega}(|f(a)| + |f(b)|) \to 0$$

as $\omega \to 0$. Since $df(t)/dt$ is continuous on $[a, b]$ it is bounded there. Thus there is a real number M such that $|df(t)/dt| \leq M$ for $t \in [a, b]$. Thus

$$\left| \frac{1}{\omega} \int_a^b \frac{df(t)}{dt} \cos(\omega t) dt \right| \leq \frac{1}{\omega} \int_a^b |\frac{df(t)}{dt}| |\cos(\omega t)| dt \leq \frac{1}{\omega} \int_a^b M dt$$

$$= \frac{M(b - a)}{\omega} \to 0$$

as $\omega \to \infty$. Thus (1) follows.

Problem 28. Let $b > a$ and $f'(x) = df(x)/dx$. Consider the vector space

$$H_1((a, b)) := \{ f(x) \in L_2(a, b) \, : \, f'(x) \in L_2(a, b) \}$$

with the norm $g \in H_1(a, b))$

$$\|g\|_1 := \sqrt{\|g\|_0^2 + \|\partial g/\partial x\|_0^2}.$$

Consider the Hilbert space $L_2((-\pi, \pi))$ and $f.(x) = \sin(x)$. Find the norm $\|f\|_1$.

Solution 28. We have

$$\|f\|_1^2 = \langle f, f \rangle_0 + \langle f', f' \rangle_0 = \int_{-\pi}^{\pi} (f(x)^2 + f'(x)^2) dx.$$

Now since $f'(x) = \cos(x)$ we obtain

$$\langle f, f \rangle_0 = \int_{-\pi}^{\pi} \sin^2(x) dx = \pi, \qquad \langle f', f' \rangle_0 = \int_{-\pi}^{\pi} \cos^2(x) dx = \pi.$$

It follows that $\|f\|_1^2 = \sqrt{2\pi}$.

Problem 29. Consider the Hilbert space $L_2([0, \infty))$. The *Laguerre polynomials* are defined by

$$L_n(x) = e^x \frac{d^n}{dx^n} (x^n e^{-x}), \qquad n = 0, 1, 2, \dots.$$

The first five Laguerre polynomials are given by

$$L_0(x) = 1, \quad L_1(x) = 1 - x, \quad L_2(x) = 2 - 4x + x^2,$$
$$L_3(x) = 6 - 18x + 9x^2 - x^3, \quad L_4(x) = 24 - 96x + 72x^2 - 16x^3 + x^4.$$

Show that the function

$$\phi_n(x) = \frac{1}{n!} e^{-x/2} L_n(x)$$

form an orthonormal system in the Hilbert space $L_2([0, \infty))$.

Solution 29. Since $L_m(x)$ is a polynomial of degree m in x we have for $m \neq n$

$$\int_0^\infty e^{-x} L_m(x) L_n(x) dx = 0.$$

For $m = n$ we have

$$\int_0^\infty e^{-x} (L_n(x))^2 dx = (-1)^n \int_0^\infty e^{-x} x^n L_n(x) dx = \int_0^\infty n! x^n e^{-x} dx = (n!)^2.$$

Thus

$$\int_0^\infty e^{-x} \phi_m(x) \phi_n(x) dx = \delta_{m,n}.$$

Problem 30. Consider the Hilbert space $L_2([0, 1])$. Let $\mathcal{P}^n$ be the $n + 1$-dimensional real linear space of all polynomial of maximal degree n in the variable x, i.e.

$$\mathcal{P}^n = \mathrm{span}\{\, 1, x, x^2, \dots, x^n \,\}.$$

The linear space $\mathcal{P}^n$ can be spanned by various systems of basis functions. An important basis is formed by the *Bernstein polynomials*

$$\{B_0^n(x), B_1^n(x), \ldots, B_n^n(x)\}$$

of degree n with

$$B_i^n(x) := x^i(1-x)^{n-i}, \qquad i = 0, 1, \ldots, n.$$

The Bernstein polynomials have a unique dual basis

$$\{D_0^n(x), D_1^n(x), \ldots, D_n^n(x)\}$$

which consists of the $n+1$ dual basis functions

$$D_i^n(x) = \sum_{j=0}^n c_{ij} B_j^n(x).$$

The dual basis functions satisfy $\langle D_i^n(x), B_j^n(x) \rangle = \delta_{ij}$.
(i) Calculate the scalar product $\langle B_i^m(x), B_j^n(x) \rangle$.
(ii) Find the coefficients c_{ij}.

Solution 30. (i) We obtain

$$\langle B_i^m(x), B_j^n(x) \rangle = \frac{\binom{m}{i}\binom{n}{j}}{(m+n+1)\binom{m+n}{i+j}}.$$

(ii) From the condition $\langle D_i^n(x), B_j^n(x) \rangle = \delta_{ij}$ we obtain

$$c_{p,q} = \frac{(-1)^{p+q}}{\binom{n}{p}\binom{n}{q}} \sum_{j=0}^{\min(p,q)} (2j+1)\binom{n+j+1}{n-p}\binom{n-j}{n-p}\binom{n+j+1}{n-q}\binom{n-j}{n-q}$$

where $p, q = 0, 1, \ldots, n$.

Problem 31. Consider the compact abelian Lie group $U(1)$

$$U(1) := \{ e^{2\pi i\theta} : 0 \le \theta < 1 \}.$$

The Hilbert space $L_2(U(1))$ is the space $L_2([0,1])$ consisting of all measurable functions $f(\theta)$ with period 1 such that

$$\int_0^1 |f(\theta)|^2 d\theta < \infty.$$

Now the set of functions $\{ e^{2\pi im\theta} : m \in \mathbb{Z} \}$ form an orthonormal basis for the Hilbert space $L_2([0,1])$. Thus every $f \in L_2([0,1])$ can be expressed uniquely as

$$f(\theta) = \sum_{m=-\infty}^{+\infty} c_m e^{2\pi im\theta}, \qquad c_m = \int_0^1 f(\theta) e^{-2\pi im\theta} d\theta.$$

Calculate

$$\int_0^1 |f(\theta)|^2 d\theta.$$

Solution 31. Since

$$\bar{f}(\theta) = \sum_{m=-\infty}^{\infty} c_m^* e^{-2\pi i m \theta}$$

we find

$$\int_0^1 |f(\theta)|^2 d\theta = \sum_{m=-\infty}^{+\infty} |c_m|^2.$$

Problem 32. The Hilbert space $L_2(\mathbb{R})$ is the vector space of measurable functions defined almost everywhere on $\mathbb{R}$ such that $|f|^2$ is integrable. $H_1(\mathbb{R})$ is the vector space of functions with first derivatives in $L_2(\mathbb{R})$. Give two examples of such a function.

Solution 32. An example is

$$f(x) = \exp(-x^2/2) \quad \Rightarrow \quad \frac{df}{dx} = -xe^{-x^2/2}.$$

These functions are even analytic. As a second example consider

$$g(x) = \begin{cases} 0 & \text{for } x \leq 0 \\ x^2 e^{-x} & \text{for } x \geq 0 \end{cases}$$

The first derivative exists and is an element of $L_2(\mathbb{R})$.

Problem 33. Consider the Hilbert space $L_2([-\pi, \pi])$. The set of functions

$$\left\{ \frac{1}{\sqrt{2\pi}} e^{-inx} \right\}_{n \in \mathbb{Z}}$$

is an orthonormal basis for $L_2([-\pi, \pi])$. Let

$$K(x, t) = \frac{1}{\sqrt{2\pi}} e^{itx}.$$

For t fixed find the Fourier expansion of this function.

Solution 33. We have

$$\frac{1}{\sqrt{2\pi}}e^{itx} = \frac{1}{2\pi}\sum_{n=-\infty}^{\infty}\langle e^{itx}, e^{inx}\rangle\frac{e^{inx}}{\sqrt{2\pi}} = \sum_{n=-\infty}^{\infty}\frac{\sin(\pi(t-n))}{\pi(t-n)}\frac{e^{inx}}{\sqrt{2\pi}}.$$

Problem 34. Let $a > 0$. Consider the Hilbert space $L_2(\mathbb{R})$ and the function $f(x) = \exp(-a|x|) \in L_2(\mathbb{R})$. Normalize the function.

Solution 34. We have

$$1 = \|f\| = \sqrt{\langle f, f\rangle}$$
$$= C^2\int_{-\infty}^{\infty}e^{-a|x|}e^{-a|x|}dx = C^2\left(\int_0^{\infty}e^{-2a|x|}dx + \int_{-\infty}^0 e^{2ax}dx\right)$$
$$= C^2\left(\left.\frac{e^{-2ax}}{-2a}\right|_0^{\infty} + \left.\frac{e^{2ax}}{2a}\right|_{-\infty}^0\right)$$
$$= C^2\frac{1}{a}.$$

It follows that $C = \sqrt{a}$ and the normalized function is $\tilde{f}(x) = e^{-a|x|}/\sqrt{a}$.

Problem 35. Consider the Hilbert space $L_2(\mathbb{R}^2)$ with the basis

$$\psi_{m,n}(x_1, x_2) = NH_m(x_1)H_n(x_2)e^{-(x_1^2+x_2^2)/2}$$

where $m, n = 0, 1, \ldots$ and N is the normalization factor. Consider the two-dimensional potential

$$V(x_1, x_2) = \frac{a}{4}(x_1^4 + x_2^4) + cx_1x_2.$$

(i) Find all linear transformation $T : \mathbb{R}^2 \to \mathbb{R}^2$ such that $V(T\mathbf{x}) = V(\mathbf{x})$.
(ii) Show that these 2×2 matrices form a group. Is the group abelian.
(iii) Find the conjugacy classes and the irreducible representations.
(iv) Consider the Hilbert space $L_2(\mathbb{R}^2)$ with the orthogonal basis

$$\psi_{m,n}(x_1, x_2) = H_m(x_1)e^{-x_1^2/2}H_n(x_2)e^{-x_2^2/2}$$

where $m, n = 0, 1, 2, \ldots$. Find the invariant subspaces from the projection operators of the irreducible representations.

Solution 35. (i) From $V(T\mathbf{x}) = V(\mathbf{x})$ we find the conditions

$$\frac{a}{4}((t_{11}x_1 + t_{12}x_2)^4 + (t_{21}x_1 + t_{12}x_2)^4 + c(t_{11}x_1 + t_{12}x_2)(t_{21}x_1 + t_{22}x_2) =$$

$$\frac{a}{4}(x_1^4 + x_2^4) + cx_1x_2.$$

This provides the 5 conditions

$$(x_1)^4 : t_{11}^4 + t_{21}^4 = 1$$
$$(x_2)^4 : t_{12}^4 + t_{22}^4 = 1$$
$$(x_1)^2 : t_{11}t_{21} = 0$$
$$(x_2)^2 : t_{12}t_{22} = 0$$
$$(x_1x_2) : t_{12}t_{21} + t_{11}t_{22} = 1.$$

To solve the system of equations we have to do a case study. If $t_{11} \neq 0$, then $t_{21} = 0$, $t_{22} \neq 0$ and $t_{12} = 0$. Thus we find the two matrices

$$I_2 = \begin{pmatrix} 1 & 0 \\ 0 & 1 \end{pmatrix}, \quad F = \begin{pmatrix} -1 & 0 \\ 0 & -1 \end{pmatrix}.$$

If $t_{11} = 0$, then we find $t_{22} = 0$, $t_{12} \neq 0$, $t_{21} \neq 0$ and finally

$$T = \begin{pmatrix} 0 & 1 \\ 1 & 0 \end{pmatrix}, \quad TF = \begin{pmatrix} 0 & -1 \\ -1 & 0 \end{pmatrix}.$$

(ii) These four matrices form an abelian group under matrix multiplication. The group table is

	I	F	T	FT
I	I	F	T	FT
F	F	I	FT	T
T	T	FT	I	F
FT	FT	T	F	I

Thus we have the *Klein four group*.

(iii) For an abelian group there can only be one element in each conjugacy class. Thus the conjugacy classes are $\{I\}$, $\{F\}$, $\{T\}$, $\{FT\}$. The following one-dimensional (and therefore irreducible) representations can be easily seen

$$R1 : \quad I, F, T, FT \mapsto 1$$

$$R2 : \quad I \to 1, \ F \to 1, \ T \to -1, \ FT \to -1$$

$$R3 : \quad I \to 1, \ F \to -1, \ T \to 1, \ FT \to -1$$

$$R4 : \quad I \to 1, \ F \to -1, \ T \to -1, \ FT \to 1.$$

Thus the character table is

	{ I }	{ F }	{ T }	{ FT }
R_1	1	1	1	1
R_2	1	1	-1	-1
R_3	1	-1	1	-1
R_4	1	-1	-1	1

(iv) From the character table we obtain the following four projection operators

$$\Pi_1 = \frac{1}{4}(O_I + O_F + O_T + O_{FT}), \quad \Pi_2 = \frac{1}{4}(O_I + O_F - O_T - O_{FT}),$$

$$\Pi_3 = \frac{1}{4}(O_I - O_F + O_T - O_{FT}), \quad \Pi_4 = \frac{1}{4}(O_I - O_F - O_T + O_{FT}).$$

A basis for the Hilbert space $L_2(\mathbb{R}^2)$ is given by

$$\psi_{m,n}(x_1, x_2) = N H_m(x_1) e^{-x_1^2/2} H_n(x_2) e^{-x_2^2/2}$$

where H_m, H_n are the Hermite polynomials and $m, n = 0, 1, \dots$. Note that $H_n(-x) = (-1)^n H(x)$. For the subspace given by Π_1 we find the basis

$$N e^{-(x_1^2 + x_2^2)/2}(H_m(x_1)H_n(x_2) + H_m(x_2)H_n(x_1))$$

for $m < n$ and m and n both even or odd and

$$N e^{-(x_1^2 + x_2^2)/2} H_m(x_1) H_m(x_2).$$

For the subspace with Π_2 we obtain

$$N e^{-(x^2 + y^2)/2}(H_m(x_1)H_n(x_2) - H_m(x_2)H_n(x_1))$$

where $m < n$ and m and n have the same parity. For the subspace with Π_3 we obtain

$$N e^{-(x_1^2 + x_2^2)/2}(H_m(x_1)H_n(x_2) + H_m(x_2)H_n(x_1))$$

where $m < n$ and m and n have the opposite parity. For the subspace with Π_4 we obtain

$$N e^{-(x_1^2 + x_2^2)/2}(H_m(x_1)H_n(x_2) - H_m(x_2)H_n(x_1))$$

where $m < n$ and m and n have the opposite parity.

Problem 36. Consider the Hilbert space $L_2([0, 2\pi])$. The linear operator $Lf(x) := df(x)/dx$ acts on a dense subset of $L_2([0, 2\pi])$. Show that this linear operator is not bounded.

Solution 36. Let $n \in \mathbb{N}$. Consider the differentiable function

$$f(x) = \sin(n\pi) \in L_2([0, 2\pi]).$$

We obtain for the square of the norm

$$\|Lf\|^2 = n^2 \int_0^{2\pi} \cos^2(nx)dx = n^2\pi.$$

Now the square of the norm of f is

$$\|f\|^2 = \int_0^{2\pi} \sin^2(nx)dx = \pi.$$

Therefore $\|Lf\| = n\|f\|$. Consequently there is no constant C which is bounded yet bigger than every integer n.

Problem 37. Let $b > a$. Consider the Hilbert space $L_2([a,b])$ and the second order ordinary differential equations

$$\frac{d^2u}{dx^2} + \lambda u = 0$$

with the boundary conditions $u(a) = u(b) = 0$. Solve the differential equation with this boundary condition. Discuss.

Solution 37. The general solution of the differential equations is

$$u(x) = C_1 \sin(\sqrt{\lambda}(x - a)) + C_2 \cos(\sqrt{\lambda}(x - a)).$$

Imposing the boundary condition $u(a) = 0$ provides $C_2 = 0$ and consequently $u(x) = C_1 \sin(\sqrt{\lambda}(x - a))$. From $u(b) = 0$ we obtain $C_1 \sin(\sqrt{\lambda}(b - a)) = 0$ with $C_1 \neq 0$ (to avoid the trivial solution). From $\sin(\sqrt{\lambda}(b-a)) = 0$ we obtain the eigenvalues

$$\lambda_n = \frac{n^2\pi^2}{(b-a)^2}, \quad n = 1, 2, \ldots$$

and the eigenfunctions

$$u_n(x) = C_n \sin\left(\frac{n\pi(x - a)}{b - a}\right).$$

The constants C_n we obtain from the normalization condition

$$\|u_n\|^2 = C_n^2 \int_a^b \sin^2\left(\frac{n\pi(x - a)}{b - a}\right) dx = 1.$$

Hence $C_n = \sqrt{2/(b - a)}$ and

$$u_n(x) = \sqrt{\frac{2}{b - a}} \sin\left(\frac{n\pi(x - a)}{b - a}\right).$$

These functions form an orthonormal basis in the Hilbert space $L_2([a, b])$. Study the case where the boundary conditions $u(a) = u(b) = 0$ are replaced by $u'(a) = u'(b) = 0$, where $'$ denotes differentiation.

Problem 38. Let $a, b > 0$. Consider the linear partial differential equation

$$\frac{\partial^4 u}{\partial x_1^4} + 2\frac{\partial^4 u}{\partial x_1^2 \partial x_2^2} + \frac{\partial^4 u}{\partial x_2^4} + c^2\frac{\partial^2 u}{\partial t^2} = 0.$$

For the space coordinates x_1, x_2 we have the domain $0 \leq x_1 \leq a$, $0 \leq x_2 \leq b$ and the boundary conditions that $u(0) = u(a) = u(b) = 0$. We consider the Hilbert space $L_2([0, a] \times [0, b])$ for the space coordinates. Find a solution of the partial differential equation.

Solution 38. Let $m, n = 1, 2, \ldots$. The function

$$\phi_{n,m}(x_1, x_2) = \sin(n\pi x_1/a)\sin(m\pi x_2/b)$$

satisfies

$$\frac{\partial^4 \phi}{\partial x_1^4} + 2\frac{\partial^4 \phi}{\partial x_1^2 \partial x_2^2} + \frac{\partial^4 \phi}{\partial x_2^4} = \lambda^2 \phi$$

with (eigenvalues)

$$\lambda_{n,m} = \frac{\pi^2(m^2 a^2 + n^2 b^2)}{a^2 b^2}.$$

The functions $\phi_{n,m}$ also form an orthonormal basis in the Hilbert space $L_2([0, a] \times [0, b])$. Now considering the ansatz

$$u(x_1, x_2, t) = \sum_{n=1}^{\infty}\sum_{m=1}^{\infty} \phi_{n,m}(x_1, x_2)\psi_{n,m}(t).$$

Inserting this ansatz into the partial differential equation we obtain

$$\sum_{n=1}^{\infty}\sum_{m=1}^{\infty} \phi_{n,m}(x_1, x_2)\left(c^2\frac{d^2\psi_{n,m}}{dt^2} + \lambda_{n,m}^2\psi_{n,m}\right) = 0.$$

Hence we obtain the ordinary second order differential equation

$$\frac{d^2\psi_{n,m}}{dt^2} + \frac{1}{c^2}\lambda_{n,m}^2\psi_{n,m} = 0$$

with the solution

$$\psi_{n,m}(t) = \gamma_{n,m}\sin(\lambda_{n,m}(t - \tau_{n,m})/c)$$

with arbitrary constants $\gamma_{n,m}$ and $\tau_{n,m}$. Thus we find the solution of the partial differential equation

$$u(x_1, x_2, t) = \sum_{n=1}^{\infty} \sum_{m=1}^{\infty} \gamma_{n,m} \sin(n\pi x/a) \sin(m\pi x/b) \sin(\lambda_{n,m}(t - \tau_{n,m})/c).$$

The constants $\gamma_{n,m}$, $\tau_{n,m}$ are utilized to satisfy the initial conditions

$$u(x_1, x_2, 0) = f(x_1, x_2), \qquad \partial u(x_1, x_2, 0)/\partial t = g(x_1, x_2).$$

Problem 39. The *Anosov map* is defined as follows: $\Omega = [0, 1)^2$,

$$\phi(x, y) = (x + y, x + 2y).$$

In matrix form we have

$$\begin{pmatrix} x \\ y \end{pmatrix} \mapsto \begin{pmatrix} 1 & 1 \\ 1 & 2 \end{pmatrix} \begin{pmatrix} x \\ y \end{pmatrix} \quad \text{mod } 1.$$

(i) Show that the map preserves Lebesgue measure.
(ii) Show that ϕ is invertible. Show that the entire sequence can be recovered from one term.
(iii) Show that ϕ is mixing utilizing the Hilbert space $L_2([0, 1] \times [0, 1])$.

Solution 39. (i) Since the *Jacobian determinant* of the matrix of the map is equal to one, i.e.

$$\det \begin{pmatrix} 1 & 1 \\ 1 & 2 \end{pmatrix} = 1$$

the map ϕ preserves the Lebesgue measure.
(ii) The inverse of the matrix which belongs to the Lie group $SL(2, \mathbb{R})$

$$\begin{pmatrix} 1 & 1 \\ 1 & 2 \end{pmatrix}$$

is given by

$$\begin{pmatrix} 2 & -1 \\ -1 & 1 \end{pmatrix}.$$

Since ϕ is invertible, the entire sequence can be recovered from one term. Thus ϕ maps Ω $1 - 1$ onto itself.
(iii) We observe that the n-th iterate is given by

$$\phi^{(n)}(x, y) = (a_{2n-2}x + a_{2n-1}y, a_{2n-1}x + a_{2n}y)$$

where the a_n are the *Fibonacci numbers* given by $a_0 = a_1 = 1$ and $a_{n+1} = a_n + a_{n-1}$ for $n \geq 1$. To check this notice that

$$a_{2n-2} + 2a_{2n-1} + a_{2n} = a_{2n} + a_{2n+1}.$$

Consider the Hilbert space $L_2([0,1) \times [0,1))$. Let f and g be two elements of this Hilbert space defined by

$$f(x,y) := \exp(2\pi i(px + qy)) \equiv \exp(2\pi i px)\exp(2\pi i qy)$$
$$g(x,y) := \exp(2\pi i(rx + sy)) \equiv \exp(2\pi i rx)\exp(2\pi i sy)$$

where $p, q, r, s \in \mathbb{Z}$. Since $(k \in \mathbb{Z})$

$$\int_0^1 \exp(2\pi i kx)dx = 0$$

unless $k = 0$. It follows that

$$\int_0^1 \int_0^1 f(x,y)g(\phi^{(n)}(x,y))dxdy = 0$$

unless $ra_{2n-2} + sa_{2n-1} + p = 0$, $ra_{2n-1} + sa_{2n} + q = 0$. Now the difference equation $b_{n+1} - b_n - b_{n-1} = 0$ has a two-parameter family of solutions given by

$$b_n = C_1\left(\frac{1+\sqrt{5}}{2}\right)^n + C_2\left(\frac{1-\sqrt{5}}{2}\right)^n$$

so the *Fibonacci numbers* are given by

$$a_n = \frac{1}{2}\left(\frac{1+\sqrt{5}}{2}\right)^n + \frac{1}{2}\left(\frac{1-\sqrt{5}}{2}\right)^n \quad \text{for } n \geq 0.$$

From the last formula we see that

$$\lim_{n\to\infty} a_{2n}/a_{2n-1} = \lim_{n\to\infty} a_{2n-1}/a_{2n-2} = \frac{1+\sqrt{5}}{2}.$$

The two equations for a_n cannot hold for infinitely many n unless $p = q = r = s = 0$. The last result implies that if

$$f(x,y) = \sum_{j=1}^k a_j \exp(2\pi i(p_j x + q_j y)), \quad g(x,y) = \sum_{j=1}^l b_j \exp(2\pi i(r_j x + s_j y)).$$

Then as $n \to \infty$,

$$\int_0^1 \int_0^1 f(x,y)g(\phi^{(n)}(x,y))dxdy \to 0.$$

Since the f and g for which the last equation holds are dense in

$$L_0^2([0,1]^2) := \left\{ f : \int_0^1 \int_0^1 f(x,y)dxdy = 0, \quad \int_0^1 \int_0^1 f^2(x,y)dxdy < \infty \right\}$$

it follows that the map ϕ is mixing.

Problem 40. Let Ψ be a complex-valued differentiable function of ϕ in the interval $[0, 2\pi]$ and $\Psi(0) = \Psi(2\pi)$, i.e. Ψ is an element of the Hilbert space $L_2([0, 2\pi])$. Assume that (normalization condition)

$$\int_0^{2\pi} \Psi^*(\phi)\Psi(\phi)d\phi = 1.$$

Calculate

$$\Im \left(\frac{\hbar}{i} \int_0^{2\pi} \Psi^*(\phi)\phi \frac{d}{d\phi}\Psi(\phi)d\phi \right)$$

where $\Im$ denotes the imaginary part.

Solution 40. Using integration by parts, $\Psi(2\pi) = \Psi(0)$ and the normalization condition we find

$$\int_0^{2\pi} \Psi^*(\phi)\phi \frac{d}{d\phi}\Psi(\phi)d\phi = \Psi^*(\phi)\phi\Psi(\phi)\big|_0^{2\pi} - \int_0^{2\pi} \Psi(\phi) \left(\frac{d}{d\phi}(\Psi^*(\phi)\phi) \right) d\phi$$

$$= 2\pi\Psi^*(0)\Psi(0) - 1 - \int_0^{2\pi} \Psi(\phi)\phi \frac{d}{d\phi}\Psi^*(\phi)d\phi.$$

Thus

$$\int_0^{2\pi} \Psi^*(\phi)\phi \frac{d}{d\phi}\Psi(\phi)d\phi + \int_0^{2\pi} \Psi(\phi)\phi \frac{d}{d\phi}\Psi^*(\phi)d\phi = 2\pi\Psi^*(0)\Psi(0) - 1$$

or

$$\int_0^{2\pi} \left(\Psi^*(\phi)\phi \frac{d}{d\phi}\Psi(\phi) + \Psi(\phi)\phi \frac{d}{d\phi}\Psi^*(\phi) \right) d\phi = 2\pi|\Psi(0)|^2 - 1.$$

It follows that

$$\Im \frac{\hbar}{i} \int_0^{2\pi} \Psi^*(\phi)\phi \frac{d}{d\phi}\Psi(\phi)d\phi = \frac{\hbar}{2}(1 - 2\pi|\Psi(0)|^2).$$

Problem 41. Consider the Hilbert space $L_2(\mathbb{R})$ and the *Schwartz space* $S(\mathbb{R})$. Note that $S(\mathbb{R}) \subset L_2(\mathbb{R})$. Show that the evolution equation of the *Airy operator* A defined by

$$(Af)(x) := \frac{d^2 f}{dx^2} + ixf(x)$$

acting in the Hilbert space $L_2(\mathbb{R})$ can be solved exactly. Take the domain of A to be Schwartz space $S(\mathbb{R})$.

Solution 41. Applying the Fourier transform F to A, i.e. $\hat{A} = FAF^{-1}$ we obtain

$$(\hat{A}g)(k) = -\frac{dg(k)}{dk} - k^2 g(k)$$

for all $g \in S(\mathbb{R})$. The evolution equation $dg_t/dt = \hat{A}g_t$ has the solution $g_t = \hat{T}_t g_0$ for all $t \geq 0$ with

$$(\hat{T}_t g)(k) = \exp(-k^2 t + kt^2 - t^3/3)g(k-t)$$

for all $g \in S(\mathbb{R})$. Then $\hat{T}_t$ is a bounded operator on $L_2(\mathbb{R})$ for all $t \geq 0$. The Airy operator has empty spectrum.

Problem 42. Consider the Hilbert space $L_2([-1,1])$ and the *Legendre polynomials*

$$P_j(x) = \frac{1}{2^j j!} \frac{d^j}{dx^j} (x^2 - 1)^j.$$

Show that these orthogonal polynomials are eigenfunctions of the Sturm-Liouville differential operator

$$(Lf)(x) := -\frac{d}{dx}\left((1-x^2)\frac{df}{dx} \right)$$

acting in $L_2([-1,1])$.

Solution 42. Note that

$$P_0(x) = 1, \quad P_1(x) = x, \quad P_2(x) = \frac{1}{2}(3x^2 - 1), \quad P_3(x) = \frac{1}{2}(5x^3 - 3x)$$

with

$$(LP_0)(x) = 0P_0, \quad (LP_1)(x) = 2P_1(x), \quad (LP_2)(x) = 6P_2(x), \quad (LP_3)(x) = 12P_3(x).$$

In general we have $(LP_n)(x) = n(n+1)P_n(x)$.

Problem 43. Let $a > 0$, $b > 0$ with dimension length. Consider the Hilbert space $L_2([0,a] \times [0,b])$. Find the minimum of

$$D[u] = \int_0^b \int_0^a \left(\left(\frac{\partial u}{\partial x_1}\right)^2 + \left(\frac{\partial u}{\partial x_2}\right)^2 \right) dx_1 dx_2$$

with the boundary conditions $u(0, x_2) = u(x_1, 0) = u(a, x_2) = u(x_1, b) = 0$
and the constraint

$$H[u] = \int_0^b \int_0^a u^2(x_1, x_2) dx_1 dx_2 = 1.$$

Note that

$$\left\{ \sin\left(\frac{m\pi x_1}{a}\right) \sin\left(\frac{n\pi x_2}{b}\right) : m, n \in \mathbb{N} \right\}$$

form an orthonormal basis in the Hilbert space $L_2([0, a] \times [0, b])$.

Solution 43. We can start with the ansatz

$$u(x_1, x_2) = \sum_{m=1}^{\infty} \sum_{n=1}^{\infty} c_{m,n} \sin\left(\frac{m\pi x_1}{a}\right) \sin\left(\frac{n\pi x_2}{b}\right)$$

with the real expansion coefficients $c_{m,n}$. Inserting this Fourier expansion
we obtain for D and H the expression

$$D = \pi^2 \frac{ab}{4} \sum_{m=1}^{\infty} \sum_{n=1}^{\infty} c_{m,n}^2 \left(\frac{m^2}{a^2} + \frac{n^2}{b^2}\right), \quad H = \frac{ab}{4} \sum_{m=1}^{\infty} \sum_{n=1}^{\infty} c_{m,n}^2 = 1.$$

Owing to the condition $H = 1$ we find that all expansion coefficients $c_{m,n}$
are 0 except $c_{1,1}$ which is

$$c_{1,1} = \sqrt{\frac{4}{ab}}.$$

It follows that

$$u(x_1, x_2) = \frac{2}{\sqrt{ab}} \sin\left(\frac{\pi x_1}{a}\right) \sin\left(\frac{\pi x_2}{b}\right)$$

with the minimum value for D as

$$d = \pi^2 \left(\frac{1}{a^2} + \frac{1}{b^2}\right).$$

Problem 44. Consider the Hilbert space $L_2([0, 1])$ and let $f, g : \mathbb{R} \to \mathbb{R}$ be
differentiable functions. Show that the differential operator $Df := df/dx$
does not have an adjoint in $L_2([0, 1])$.

Solution 44. We have

$$\langle Df, g \rangle = \int_0^1 \frac{df(x)}{dx} g(x) dx = f(x)g(x)\big|_0^1 - \int_0^1 f(x) \frac{dg(x)}{dx} dx$$
$$= f(1)g(1) - f(0)g(0) + \langle f, -dg/dx \rangle.$$

For all choices of the differentiable functions f, g one cannot define an adjoint except for the cases $f(1)g(1) - f(0)g(0) = 0$.

Problem 45. Consider the Hilbert space $L_2(\mathbb{S}^2)$ with

$$\mathbb{S}^2 := \{ (x_1, x_2, x_3) \in \mathbb{R}^3 \ : \ x_1^2 + x_2^2 + x_3^2 = 1 \}$$

and the operators

$$\hat{L}_3 = -i\hbar \frac{\partial}{\partial \phi}$$

$$\hat{L}_+ = \hbar e^{i\phi} \left(\frac{\partial}{\partial \theta} + i \cot(\theta) \frac{\partial}{\partial \phi} \right)$$

$$\hat{L}_- = \hbar e^{-i\phi} \left(-\frac{\partial}{\partial \theta} + i \cot(\theta) \frac{\partial}{\partial \phi} \right)$$

with the commutation relation

$$[\hat{L}_+, \hat{L}_-] = 2\hbar \hat{L}_3, \quad [\hat{L}_+, \hat{L}_3] = -\hbar \hat{L}_+, \quad [\hat{L}_-, \hat{L}_3] = \hbar \hat{L}_-.$$

An orthonormal basis in the Hilbert space $L_2(\mathbb{S}^2)$ is given by the *spherical harmonics* $Y_{\ell,m}(\theta, \phi)$ with $\ell = 0, 1, 2, \ldots$ and $m = -\ell, -\ell+1, \ldots, \ell-1, \ell$. For $\ell = 1$ we have $m = -1, 0, 1$ and the spherical harmonics are given by

$$Y_{1,-1}(\theta, \phi) = \sqrt{\frac{3}{8\pi}} \sin(\theta) e^{-i\phi} \quad Y_{1,1}(\theta, \phi) = -\sqrt{\frac{3}{8\pi}} \sin(\theta) e^{i\phi}$$

$$Y_{1,0}(\theta, \phi) = \sqrt{\frac{3}{4\pi}} \cos(\theta).$$

Solution 45. We have

$$\hat{L}_+ Y_{1,-1} = \sqrt{2}\hbar Y_{1,0}, \quad \hat{L}_+ Y_{1,0} = \sqrt{2}\hbar Y_{1,1}, \quad \hat{L}_+ Y_{1,1} = 0 Y_{1,1}$$

$$\hat{L}_- Y_{1,-1} = 0 Y_{1,-1}, \quad \hat{L}_- Y_{1,0} = \sqrt{2}\hbar Y_{1,-1}, \quad \hat{L}_- Y_{1,1} = \sqrt{2}\hbar Y_{1,0}$$

$$\hat{L}_3 Y_{1,-1} = -\hbar Y_{1,-1} \quad \hat{L}_3 Y_{1,0} = 0 Y_{1,0}, \quad \hat{L}_3 Y_{1,1} = \hbar Y_{1,1}.$$

Since

$$\langle Y_{1,-1}, Y_{1,-1} \rangle = 1, \quad \langle Y_{1,-1}, Y_{1,0} \rangle = 0, \quad \langle Y_{1,-1}, Y_{1,1} \rangle = 0$$

$$\langle Y_{1,0}, Y_{1,-1} \rangle = 0, \quad \langle Y_{1,0}, Y_{1,0} \rangle = 1, \quad \langle Y_{1,0}, Y_{1,1} \rangle = 0$$

$$\langle Y_{1,1}, Y_{1,-1} \rangle = 0, \quad \langle Y_{1,1}, Y_{1,0} \rangle = 0, \quad \langle Y_{1,1}, Y_{1,1} \rangle = 1$$

we find for the matrix representations of $\hat{L}_+$ and $\hat{L}_-$

$$M_+ = \hbar \begin{pmatrix} 0 & 0 & 0 \\ \sqrt{2} & 0 & 0 \\ 0 & \sqrt{2} & 0 \end{pmatrix}, \quad M_- = \hbar \begin{pmatrix} 0 & \sqrt{2} & 0 \\ 0 & 0 & \sqrt{2} \\ 0 & 0 & 0 \end{pmatrix}.$$

For the matrix representation of $\hat{L}_3$ we find

$$M_3 = \hbar \begin{pmatrix} -1 & 0 & 0 \\ 0 & 0 & 0 \\ 0 & 0 & 1 \end{pmatrix}.$$

The commutation relations are

$$[M_+, M_-] = 2\hbar M_3, \quad [M_+, M_3] = -\hbar M_+, \quad [M_-, M_3] = \hbar M_-.$$

We have a *irreducible representation*.

Let $\ell = 2$ with $m = -2, -1, 0, 1, 2$ and the five spherical harmonics $Y_{2,-2}$, $Y_{2,-1}$, $Y_{2,0}$, $Y_{2,1}$, $Y_{2,2}$. Find a representation of $\hat{L}_+$, $\hat{L}_-$, $\hat{L}_3$. Find a representation by 5×5 matrices.

Problem 46. Consider the Hilbert space $L_2([0,1])$. Let $n \in \mathbb{N}$ and $f_n \in L_2([0,1])$ given by

$$f_n(x) = \frac{1}{\sqrt{x + 1/n}}$$

and f be the pointwise limit of f_n. Show that f is integrable.

Solution 46. For all $n \in \mathbb{N}$ one has

$$f_{n+1}(x) = \frac{1}{\sqrt{x + 1/(n+1)}} \geq \frac{1}{\sqrt{x + 1/n}} = f_n(x)$$

and

$$\lim_{n \to \infty} f_n(x) = \frac{1}{\sqrt{x}}$$

almost everywhere. Applying the monotone convergence theorem we find that $1/\sqrt{x}$ is integrable and

$$\lim_{n \to \infty} \int_0^1 \frac{dx}{\sqrt{x + 1/n}} = \int_0^1 \frac{dx}{\sqrt{x}} = 2.$$

Problem 47. Consider the Hilbert space $L_2([0, \pi])$. Solve the *integral equation*

$$f(x) = \lambda \int_0^\pi \frac{\sin(x - x')}{\pi} f(x') dx' + x.$$

Note that

$$\sin(x - x') \equiv \sin(x) \cos(x') - \cos(x) \sin(x').$$

Solution 47. We write $f(x) = \lambda K f(x) + g(x)$ with $g(x) = x$ and the kernel

$$K f(x) := \int_0^\pi \frac{\sin(x - x')}{\pi} f(x') dx' = \sum_{j=1}^N \phi_j \langle \psi_j, f \rangle$$

and $\langle \cdot, \cdot \rangle$ the scalar product in the Hilbert space $L_2([0, \pi])$. Taking the scalar product with ψ_k provides

$$\langle \psi_k, f \rangle = \lambda \sum_{j=1}^N \langle \psi_k, \phi_j \rangle \langle \psi_j, f \rangle + \langle \psi_k, g \rangle, \quad k = 1, \dots, N.$$

We set

$$a_k := \langle \psi_k, f \rangle, \quad \alpha_{kj} := \langle \psi_k, \phi_j \rangle, \quad b_k := \langle \psi_k, g \rangle.$$

Hence

$$a_k - \lambda \sum_{j=1}^N \alpha_{kj} a_j = b_k \quad \Rightarrow \quad (I_N - \lambda A)\mathbf{a} = \mathbf{b}$$

with $A = (\alpha_{jk})$ $(k, j = 1, \dots, N)$. Owing to (2) we have $N + 2$ and

$$\phi_1(x) = \frac{\sin(x)}{\sqrt{\pi}}, \qquad \phi_2(x) = \frac{\cos(x)}{\sqrt{\pi}},$$

$$\psi_1(x) = \frac{\cos(x)}{\sqrt{\pi}}, \qquad \psi_2(x) = \frac{\sin(x)}{\sqrt{\pi}}$$

with

$$\alpha_{11} = \langle \psi_1, \phi_1 \rangle = \int_0^\pi \frac{\cos(x) \sin(x)}{\pi} dx = 0, \quad \alpha_{12} = \langle \psi_1, \phi_2 \rangle = \int_0^\pi \frac{\cos(x) \cos(x)}{\pi} dx = \frac{1}{2}$$

$$\alpha_{21} = \langle \psi_2, \phi_1 \rangle = \int_0^\pi \frac{\sin(x) \sin(x)}{\pi} dx = \frac{1}{2}, \quad \alpha_{22} = \langle \psi_2, \phi_2 \rangle = \int_0^\pi \frac{\cos(x) \sin(x)}{\pi} dx = 0$$

and

$$b_1 = \langle \psi_1, g(x) \rangle = \int_0^\pi \frac{x \cos(x)}{\sqrt{\pi}} dx = -\frac{2}{\sqrt{\pi}}, \quad b_2 = \langle \psi_2, g(x) \rangle = \int_0^\pi \frac{x \sin(x)}{\sqrt{\pi}} dx = \sqrt{\pi}.$$

Then we obtain the linear equation

$$\left(\begin{pmatrix} 1 & 0 \\ 0 & 1 \end{pmatrix} - \begin{pmatrix} \lambda/2 & 0 \\ 0 & \lambda/2 \end{pmatrix} \right) \begin{pmatrix} a_1 \\ a_2 \end{pmatrix} = \begin{pmatrix} -2/\sqrt{\pi} \\ \sqrt{\pi} \end{pmatrix}$$

with the solution

$$a_1 = \frac{2}{\sqrt{\pi}} \cdot \frac{\pi\lambda - 4}{4 - \lambda^2}, \qquad a_2 = \frac{4}{\sqrt{\pi}} \cdot \frac{\pi - \lambda}{4 - \lambda^2}.$$

From

$$f(x) = \lambda K f + x = \lambda \sum_{j=1}^2 \phi_j \langle \psi_j, f \rangle + x = \lambda a_1 \phi_1(x) + \lambda a_2 \phi_2(x)$$

we finally obtain

$$f(x) = \frac{\lambda}{\pi} \frac{\pi\lambda - 4}{4 - \lambda^2} \sin(x) + \frac{4\lambda}{\pi} \frac{\pi - \lambda}{4 - \lambda^2} \cos(x) + x.$$

Problem 48. Consider the Hilbert space $L_2([0, \infty))$. Let $J_0(x)$ be the *Bessel function* of first kind of order $n = 0$ defined by

$$J_0(x) = \sum_{k=0}^{\infty} \frac{(-1)^k (x/2)^{2k}}{k! \Gamma(k+1)}$$

with $J_0(0) = 1$. Let $a > 0$ and $b > 0$. Find the scalar product

$$\langle e^{-ax}, J_0(x) \rangle = \int_0^\infty e^{-ax} J_0(bx) dx.$$

Solution 48. We obtain

$$\int_0^\infty e^{-ax} J_0(bx) dx = \frac{1}{\sqrt{a^2 + b^2}}.$$

Problem 49. Consider the Hilbert space $L_2([0, \infty))$ and the differential operators

$$T_1 = -\frac{1}{2\gamma_1} x \frac{d^2}{dx^2} + \frac{\gamma_2}{2x} - \frac{\gamma_1}{2} x, \quad T_2 = -ix \frac{d}{dx}, \quad T_3 = -\frac{1}{2\gamma_1} x \frac{d^2}{dx^2} + \frac{\gamma_2}{2x} + \frac{\gamma_1}{2} x$$

with $\gamma_1 >, \gamma_2 > 0$. Find the commutators $[T_1, T_2]$, $[T_2, T_3]$, $[T_3, T_1]$. Discuss. One defines (*ladder operators*)

$$T_+ := T_1 + iT_2, \qquad T_- := T_1 - iT_2.$$

Find the commutators $[T_+, T_-]$, $[T_+, T_3]$, $[T_-, T_3]$. Find the *Casimir operator*

$$T^2 = T_3^2 - T_1^2 - T_2^2.$$

Let $L_\nu(2\gamma_1, x)$ be the generalized *Laguerre polynomials* and $\tau(\tau+1) = \gamma_1\gamma_2$. Consider the basis

$$|\tau, \nu + \tau + 1\rangle = \left(\frac{\Gamma(\nu+1)}{\Gamma(\nu + 2\tau + 2)}\right)^{1/2} (2\gamma_1 x)^{\tau+1} e^{-\gamma_1 x} L_\nu^{(2\tau+1)}(2\gamma_1 x)$$

in the Hilbert space $L_2([0, \infty))$ with $\nu \in \mathbb{N}_0$. Find the states

$$T_3|\tau, \nu + \tau + 1\rangle, \quad T_+|\tau, \nu + \tau + 1\rangle, \quad T_-|\tau, \nu + \tau + 1\rangle.$$

Solution 49. We have

$$[T_1, T_2] = -iT_3, \quad [T_2, T_3] = iT_1, \quad [T_3, T_1] = iT_2$$

and

$$[T_+, T] = -2T_3 \quad [T_+, T_3] = -T_+, \quad [T_-, T_3] = T_-.$$

Hence T_1, T_2, T_3 form a basis of the simple Lie algebra $so(2,1)$. The *Casimir operator* with $[T^2, T_1] = [T^2, T_2] = [T^2, T_3] = 0$ is given by

$$T^2 = T_3^2 - T_1^2 - T_2^2 = \gamma_1\gamma_2.$$

Now

$$T_3|\tau, \nu + \tau + 1\rangle = (\nu + \tau + 1)|\tau + 1\rangle \quad \text{eigenvalue equation}$$

$$T_+|\tau, \nu + \tau + 1\rangle = (\nu + 1)^{1/2}(\nu + 2t + 2)^{1/2}|\tau, \nu + \tau + 2\rangle$$

$$T_-|\tau, \nu + \tau + 1\rangle = (\nu + 2\tau + 1)^{1/2}\nu^{1/2}|\tau, \nu + \tau\rangle.$$

Problem 50. Let $F: \mathbb{R} \to \mathbb{R}$ a polynomial vector field with

$$F(x) := \sum_{j=1}^{l} a_j x^j = a_0 + a_1 x + a_2 x^2 + \cdots + a_l x^l.$$

Furthermore, we consider the differential equation $dx/dt = F(x)$ and assume that the range of its solution $x(t)$ is $\Omega := \text{range}(x)$ is $\mathbb{R}$, $\mathbb{R}^{+(-)}$ or $[a, b] \subset \mathbb{R}$. Now let $L_2(\Omega)$ the Hilbert space with the (weighted) inner product

$$(f, g)_\Omega := \int_\Omega f(x)g(x)\omega_\Omega(x)dx$$

with the weighting function $\omega(x)$ and an orthogonal basis

$$\Omega = \mathbb{R} : \text{Hermite polynomials}\{p_n\} \text{ and } \omega_\Omega(x) = e^{-x^2}$$
$$\Omega = \mathbb{R}^{+(-)} : \text{Laguerre polynomials}\{p_n\} \text{ and } \omega_\Omega(x) = e^{-|x|}$$
$$\Omega = [a, b] : \text{Jacobi polynomials}\{p_n\} \text{ and } \omega_\Omega(x) = 1$$

Reformulate the (nonlinear) differential $dx/dt = F(x)$ as infinite system of linear differential equations (*Carleman linearization* in Hilbert space)

$$\frac{d\mathbf{p}_\Omega}{dt} = \mathbf{A}\mathbf{p}_\Omega$$

where $\mathbf{p}_\Omega = (p_0, p_1, \ldots, p_n, \ldots)^T$ with $\mathbf{p}_\Omega(x(t)) \in \mathbb{R}^\infty$ and an infinite matrix $\mathbf{A}_\Omega \in \mathbb{R}^{\infty \times \infty}$. Determine the coefficients of matrix $\mathbf{A}_\Omega = (\alpha_{n,m})$.

Solution 50. Let $P \colon \mathbb{R} \to \mathbb{R}$ a polynomial function where $P \in L_2(\Omega)$ then

$$P(x) = \sum_{n=0}^\infty \beta_n p_n(x)$$

where $\{p_n\}$ is the corresponding orthogonal basis in $L_2(\Omega)$ and $\beta_n = (P, p_n)_\Omega$. Furthermore, we have

$$\frac{dp_n(x(t))}{dt} = \frac{dp_n}{dx}\frac{dx(t)}{dt} = \frac{dp_n}{dx}F(x(t))$$

where the r.h.s. is obviously a polynomial in x of degree $l + (n-1)$. Therefore, we have with the original differential equation

$$\frac{dp_n(x)}{dt} = \frac{dp_n}{dx}F(x) = \sum_{m=0}^{l+n-1} \alpha_{n,m} p_m(x)$$

for $n = 0, 1, 2, \ldots$ where

$$\alpha_{n,m} := \left(\frac{dp_n}{dx}F, p_m\right)_\Omega$$

for $m = 0, 1, \ldots, l + n - 1$.

4.3 Programming Problems

Problem 1. Consider the Hilbert space $L_2(\mathbb{R})$ and $a > 0$. Then

$$f(x) = e^{-a|x|} \in L_2(\mathbb{R}), \quad g(x) = xe^{-a|x|} \in L_2(\mathbb{R}).$$

Normalize the functions. Find the scalar product $\langle f, g \rangle$.

Solution 1. Applying the Maxima program

```
/* hilbert2.mac */
assume(a > 0);
f1: exp(-a*x);
f12: f1*f1;
r1: integrate(f12,x,0,inf);
f2: exp(a*x);
f22: f2*f2;
r2: integrate(f22,x,-inf,0);
r: r1 + r2;
g1: x*exp(-a*x);
g12: g1*g1;
s1: integrate(g12,x,0,inf);
g2: x*exp(a*x);
g22: g2*g2;
s2: integrate(g22,x,-inf,0);
s: s1 + s2;
```

we have

$$\langle f, f \rangle = \|f\|^2 = \int_{\mathbb{R}} e^{-2a|x|} dx = \frac{1}{a}, \quad \langle g, g \rangle = \|g\|^2 = \int_{\mathbb{R}} x^2 e^{-2a|x|} dx = \frac{1}{2a^3}.$$

Since $f(x) = f(-x)$ and $g(x) = -g(x)$ we have $\langle f, g \rangle = 0$.

Problem 2. Let V be a finite dimensional vector space over the real numbers with a positive definite scalar product $\langle \cdot, \cdot \rangle$. We assume that $\dim(V) = n$. Suppose that $w_1, w_2, \ldots, w_n$ are a basis of V. We define

$$v_1 = w_1$$

$$v_2 = w_2 - \frac{\langle v_1, w_2 \rangle}{\langle v_1, v_1 \rangle} v_1$$

$$v_3 = w_3 - \frac{\langle v_2, w_3 \rangle}{\langle v_2, v_2 \rangle} v_2 - \frac{\langle v_1, w_3 \rangle}{\langle v_1, v_1 \rangle} v_1$$

$$\cdots = \cdots$$

$$v_n = w_n - \frac{\langle v_{n-1}, w_n \rangle}{\langle v_{n-1}, v_{n-1} \rangle} v_{n-1} - \cdots - \frac{\langle v_1, w_n \rangle}{\langle v_1, v_1 \rangle} v_1.$$

Then the unit vectors

$$u_j := \frac{v_j}{\|v_j\|}$$

where

$$\|v_j\| := \sqrt{\langle v_j, v_j \rangle}$$

are mutually orthogonal and are an orthonormal basis of V. This is called the *Gram-Schmidt technique*.

Consider the Hilbert space $L_2([-1,1])$. The scalar product is given by

$$\langle f, g \rangle := \int_{-1}^{1} f(x)g^*(x)dx.$$

A basis in the Hilbert space $L_2([-1,1])$ is given by

$$\{1,\ x,\ x^2,\ x^3,\ \dots\}.$$

However, this basis is not orthogonal and not normalized. We select the subset

$$\{\,w_0 = 1,\ w_1 = x,\ w_2 = x^2,\ w_3 = x^3\,\}.$$

Evaluate the corresponding v_0, v_1, v_2, v_3 to obtain the normalized Legendre polynomials. The normalized Legendre polynomials form an orthonormal basis in the Hilbert space $L_2([-1,1])$.

Solution 2. We apply the following Maxima program

```
/* Schmidt.mac */
n: 3;

sp(u,v) :=
block(subst(x=1,integrate(u*v,x))-subst(x=-1,integrate(u*v,x)))$

sch(k) :=
block(if k=0 then return (1/sqrt(2))
      else
      s: x^k-sum(sp(x^k,sch(j))*sch(j)/sp(sch(j),sch(j)),j,0,k-1),
      return (s/sqrt(sp(s,s)))
)$

for k:0 thru n do
print(sch(k));
```

The output is the normalized Legendre polynomials

$$p_0(x) = \frac{1}{\sqrt{2}}, \qquad p_1(x) = \sqrt{\frac{3}{2}}x,$$

$$p_2(x) = \frac{3\sqrt{5}}{2\sqrt{2}}(x^2 - 1/3), \qquad p_3(x) = \frac{5\sqrt{7}}{2\sqrt{2}}(x^3 - 3x/5).$$

Obviously we could also run the program for higher n than $n = 3$.

Problem 3. Consider the (real) Hilbert space $L_2([-1,1])$ with the scalar product

$$\langle f(x), g(x) \rangle = \int_{-1}^{+1} f(x)g(x)dx.$$

Consider the linear operator L defined by

$$Lu(x) := \int_{-1}^{+1} K(x,y)u(y)dy$$

with the *kernel*

$$K(x,y) = x + 3xy^2 + 3x^2y + y.$$

Let $u(x) = c_0 + c_1x + c_2x^2$ with $c_0, c_1, c_2 \in \mathbb{R}$.

Solution 3. Applying the Maxima program

```
/* inthilbert2.mac */
K: x + 3*x*y*y + 3*x*x*y + y;
u: c0 + c1*y + c2*y*y;
T: K*u;
R: integrate(T,y,-1,+1); R: expand(R); R: ratsimp(R);
```

we obtain

$$\frac{1}{15}(30c_1x^2 + (28c_2 + 60c_0)x + 10c_1).$$

Problem 4. Let $a > 0$. Consider the Hilbert space $L_2(T)$, where T is a $30°$-$60°$-$90°$ triangle in the plane $\mathbb{R}^2$. The boundary of the triangle is given by

$$x = 0, \quad y = 0, \quad \sqrt{3}x + y = \sqrt{3}a.$$

Show that the wave function

$$u(x,y) = \cos(2\pi(5x + \sqrt{3}y)/(3a)) - \cos(2\pi(5x - \sqrt{3}y)/(3a))$$
$$+ \cos(2\pi(-x - 3\sqrt{3}y)/(3a)) - \cos(2\pi(-x + 3\sqrt{3}y)/(3a))$$
$$+ \cos(2\pi(-4x + 2\sqrt{3}y)/(3a)) - \cos(2\pi(-4x - 2\sqrt{3}y)/(3a))$$

vanishes at the boundary.

Solution 4. We apply the Maxima program

```
/* triangle.mac */
assume(a > 0);
u: cos(2*%pi*(5*x+sqrt(3)*y)/(3*a))-cos(2*%pi*(5*x-sqrt(3)*y)/(3*a))
+ cos(2*%pi*(-x-3*sqrt(3)*y)/(3*a))-cos(2*%pi*(-x+3*sqrt(3)*y)/(3*a))
+ cos(2*%pi*(-4*x+2*sqrt(3)*y)/(3*a))-cos(2*%pi*(-4*x-2*sqrt(3)*y)/(3*a));
tx: subst(0,x,u); tx: trigsimp(tx);
ty: subst(0,y,u); ty: trigsimp(ty);
tl: subst(sqrt(3)*(a-x),y,u); tl: trigsimp(tl); tl: expand(tl);
```

Problem 5. Consider the Hilbert space $L_2([0,1])$ and the second order differential equation with boundary conditions

$$\frac{d^2u}{dx^2} = 2\cos(\pi x/2), \quad u(0) = \frac{du(0)}{dx} = 0.$$

Show that $u(x) = (2/\pi)^2 \sin^2(\pi x/2) \in L_2([0,1])$ is a solution.

Solution 5. Applying the Maxima program

```
/* diffso.mac */
u: 4/(%pi*%pi)*sin(%pi*x/2)*sin(%pi*x/2);
u0: subst(0,x,u);
ud: diff(u,x);
ud0: subst(0,x,ud);
udd: diff(ud,x);
udd: subst((1+cos(%pi*x))/2,cos(%pi*x/2)*cos(%pi*x/2),udd);
udd: subst((1-cos(%pi*x))/2,sin(%pi*x/2)*sin(%pi*x/2),udd);
udd: trigsimp(udd);
R: udd-2*cos(%pi*x);
```

we find that $u(x)$ is a solution.

Problem 6. An orthonormal basis in the Hilbert space $L_2([0,1])$ is given by

$$B := \left\{ \phi_n(x) = e^{2\pi i x n} : n \in \mathbb{Z} \right\}.$$

Let

$$f(x) = \begin{cases} 2x & 0 \le x < 1/2 \\ 2(1-x) & 1/2 \le x < 1 \end{cases}$$

Is $f \in L_2([0,1])$? Find the first two expansion coefficients of the Fourier expansion of f with respect to the basis given above.

Solution 6. Applying the Maxima program

```
/* L2Fexp.mac */
declare(n,integer);
t1: integrate(2*x*exp(2*%pi*%i*n*x),x,0,1/2);
t2: integrate(2*(1-x)*exp(2*%pi*%i*n*x),x,1/2,1);
t: t1 + t2; t: demoivre(t); t: expand(t);
```

we obtain

$$c_n = \frac{(-1)^n - 1}{\pi^2 n^2}, \quad n \in \mathbb{Z} \setminus \{0\}$$

and $c_0 = 1/2$.

4.4   Supplementary Problems

Problem 1. (i) Consider the Hilbert space $L_2([0,1])$. Let $n \geq 2$ and consider the function $f_n : [0,1] \to \mathbb{R}$

$$f_n(x) = \begin{cases} n^2 x & \text{for } 0 \leq x \leq 1/n \\ -n^2(x - 2/n) & \text{for } 1/n \leq x \leq 2/n \\ 0 & \text{for } 2/n \leq x \leq 1. \end{cases}$$

Show that $\int_0^1 f_n(x)dx = 1$. Study $\lim_{n\to\infty} f_n$.

(ii) Consider the Hilbert space $L_2([0,1])$ and the function $(n \in \mathbb{N})$

$$f_n(x) = \begin{cases} 4n^2 x & \text{if } 0 \leq x \leq 1/(2n) \\ -4n^2 x + 4n & \text{if } 1/(2n) < x < 1/n \\ 0 & \text{if } 1/n \leq x \leq 1 \end{cases}$$

So the maximum of the function f_n is at $x = 1/(2n)$ with $f(1/(2n)) = 2n$. Show that

$$\lim_{n\to\infty} \int_0^1 f_n(x)dx \neq \int_0^1 \lim_{n\to\infty} f_n(x)dx.$$

(iii) Consider the Hilbert space $L_2([-1,1])$ and the function $f_n \in L_2([-1,1])$ $(n \in \mathbb{N})$

$$f_n(x) \begin{cases} 1 & \text{for } -1 \leq x \leq 0 \\ \sqrt{1 - nx} & \text{for } 0 \leq x \leq 1/n \\ 0 & \text{for } 1/n \leq x \leq 1. \end{cases}$$

Show that

$$\|f_n(x) - f_m(x)\| \leq \frac{1}{n} + \frac{1}{m}.$$

Note that if $n \to \infty$, $m \to \infty$, then $\|f_n(x) - f_m(x)\|^2 \to 0$.

(iv) Consider the Hilbert space $L_2([-1,1])$. Consider the sequence

$$f_n(x) = \begin{cases} -1 & \text{if } -1 \leq x \leq -1/n \\ nx & \text{if } -1/n \leq x \leq 1/n \\ +1 & \text{if } 1/n \leq x \leq 1 \end{cases}$$

where $n = 1, 2, \ldots$. Show that $\{f_n(x)\}$ is a sequence in $L_2([-1,1])$ that is a *Cauchy sequence* in the norm of $L_2([-1,1])$. Show that $f_n(x)$ converges in the norm of $L_2([-1,1])$ to

$$\text{sgn}(x) = \begin{cases} -1 & \text{if } -1 \leq x < 0 \\ +1 & \text{if } 0 < x \leq 1 \end{cases}.$$

Use this sequence to show that the space $C([-1,1])$ is a subspace of $L_2([-1,1])$ that is not closed.

Problem 2. Consider the Hilbert space $L_2([0,\infty))$ and the function $f \in L_2([0,\infty))$

$$f(x) = \exp(-x^{1/4})\sin(x^{1/4}).$$

Show that

$$\langle f, x^n \rangle = \int_0^\infty f(x)x^n dx = 0 \quad \text{for } n = 0,1,2,\ldots.$$

Problem 3. Consider the Hilbert space $L_2([-1,1])$. Let $n_1, n_2, m_1, m_2 \in \mathbb{N}$ with $n_1 \neq n_2$, $m_1 \neq m_2$ and

$$f_1(x) = ax^{n_1} + bx^{n_2}, \quad f_2(x) = cx^{m_1} + dx^{m_2}.$$

Find the conditions on $a, b, c, d \in \mathbb{R}$ such that $\langle f_1, f_2 \rangle = 0$. Note that integration yields

$$\langle f_1, f_2 \rangle = \int_{-1}^1 (acx^{n_1+m_1} + adx^{n_1+m_2} + bcx^{n_2+m_1} + bdx^{n_2+m_2})dx$$

$$= \frac{ac}{n_1 + m_1 + 1}(1 - (-1)^{n_1+m_1}) + \frac{ad}{n_1 + m_2 + 1}(1 - (-1)^{n_1+m_2})$$

$$+ \frac{bc}{n_2 + m_1 + 1}(1 - (-1)^{n_2+m_1}) + \frac{bd}{n_2 + m_2 + 1}(1 - (-1)^{n_2+m_2}).$$

Problem 4. Consider the Hilbert space $L_2([-\pi, \pi])$. Find the series

$$f(\theta) = \sum_{n=-\infty}^{\infty} c_n e^{in\theta}$$

where $\theta \in [-\pi, \pi]$ and

$$c_0 = 1, \quad c_1 = c_{-1} = 1, \quad c_2 = c_{-2} = \frac{1}{2}, \ldots, c_n = c_{-n} = \frac{1}{n}.$$

Problem 5. (i) Let $0 \leq r < 1$. Consider the Hilbert space $L_2([0, 2\pi])$ and $f(\theta) \in L_2([0, 2\pi])$. Show that

$$\frac{1}{2\pi} \int_0^{2\pi} f(\theta)d\theta + \frac{1}{\pi} \int_0^{2\pi} \sum_{j=1}^{\infty} r^j f(\theta) \cos(j(\phi - \theta))d\theta$$

$$= \frac{1}{2\pi} \int_0^{2\pi} f(\theta) \frac{1 - r^2}{1 - 2r \cos(\phi - \theta) + r^2} d\theta.$$

(ii) Consider the Hilbert space $L_2([0, 2\pi])$, the curve in the plane expressed in polar coordinates $r(\theta) = 1 + r \cos(2\theta)$ and

$$r(\theta) = \sum_{n=-\infty}^{+\infty} C_n \exp(in\theta), \quad C_n = \frac{1}{2\pi} \int_0^{2\pi} r(\theta) \exp(-in\theta) d\theta.$$

Find the coefficients C_n.

Problem 6. Consider the real axis $\mathbb{R}$ and a cover

$$\mathbb{R} = \cup_{j=-\infty}^{j=\infty} I_j, \quad I_j = [\epsilon_j, \epsilon_{j+1}), \quad \epsilon_j < \epsilon_{j+1}$$

with $d_j := \epsilon_{j+1} - \epsilon_j = |I_j|$. Let f_j be a window function supported in the interval $[\epsilon_j - d_{j-1}/2, \epsilon_{j+1} + d_{j+1}/2)$ such that

$$\sum_{j=-\infty}^{\infty} f_j^2(x) = 1$$

and $f_j^2(x) = 1 - f_j^2(2\epsilon_{j+1} - x)$ for x near ϵ_{j+1}. Show that the functions

$$g_{j,k}(x) = \frac{2}{\sqrt{2d_j}} f_j(x) \sin\left(\frac{\pi}{2d_j}(2k + 1)(x - \epsilon_j) \right), \quad k = 0, 1, 2, \ldots$$

form an orthonormal basis of the Hilbert space $L_2(\mathbb{R})$ subordinate to the partition f_j.

Problem 7. Let $\ell_1 > 0$, $\ell_2 > 0$, $\ell_3 > 0$ with dimension length. Consider the Hilbert space $L_2([0, \ell_1] \times [0, \ell_2] \times [0, \ell_3])$. Let $n_1, n_2, n_3 \in \mathbb{N}_0$.
(i) Do the functions

$$u(x_1, x_2, x_3) = \cos(k_1 x_1) \sin(k_2 x_2) \sin(k_3 x_3)$$
$$v(x_1, x_2, x_3) = \sin(k_1 x_1) \cos(k_2 x_2) \sin(k_3 x_3)$$
$$w(x_1, x_2, x_3) = \sin(k_1 x_1) \sin(k_2 x_2) \cos(k_3 x_3)$$

in this Hilbert space form an orthonormal basis, where

$$k_1 = \frac{n_1 \pi}{\ell_1}, \quad k_2 = \frac{n_2 \pi}{\ell_2}, \quad k_3 = \frac{n_3 \pi}{\ell_3}.$$

(ii) Show that $u(x_1, x_2, x_3)$, $v(x_1, x_2, x_3)$, $w(x_1, x_2, x_3)$ are 0 at the boundaries of the box (rectangular cavity) $[0, \ell_1] \times [0, \ell_2] \times [0, \ell_3]$.

(iii) Show that the functions (A, B, C are constants)

$$\widetilde{u}(x_1, x_2, x_3, t) = Au(x_1, x_2, x_3)e^{i\omega t}$$
$$\widetilde{v}(x_1, x_2, x_3, t) = Bv(x_1, x_2, x_3)e^{i\omega t}$$
$$\widetilde{w}(x_1, x_2, x_3, t) = Cw(x_1, x_2, x_3)e^{i\omega t}$$

satisfy the wave equation

$$\frac{1}{c^2}\frac{\partial^2 \psi}{\partial t^2} = \frac{\partial^2 \psi}{\partial x_1^2} + \frac{\partial^2 \psi}{\partial x_2^2} + \frac{\partial^2 \psi}{\partial x_3^2}$$

with (*dispersion relation*) $\omega_{\mathbf{k}} = c(k_1^2 + k_2^2 + k_3^2)^{1/2}$.

Problem 8. Let $\ell = 0, 1, 2, \ldots$. Show that

$$\sum_{m=-\ell}^{+\ell} Y_{\ell,m}(\theta, \phi) Y_{\ell,m}^*(\theta, \phi) = \frac{2\ell + 1}{4\pi}.$$

Problem 9. Let $Ai(x)$ be the *Airy function*, $dAi(x)/dx$ be the derivative and a_n ($n = 1, 2, \ldots$) be the zeros of the Airy functions. The Airy function can be defined as

$$Ai(x) = \frac{1}{\pi}\int_0^\infty \cos\left(\frac{1}{3}t^3 + xt\right) dt.$$

Show that the functions

$$\frac{Ai(x + a_n)}{dAi(x = a_n)/dx}, \quad n = 1, 2, \ldots$$

form an orthonormal basis in the Hilbert space $L_2([0, \infty))$.

Problem 10. Let $b > a$. Consider the Hilbert space $L_2([a, b])$. Do the functions (*Chebyshev polynomial*)

$$T_n(x) = \frac{(b-a)^n}{2^{2n-1}}\cos\left(n \arccos\left(\frac{2x}{b-a} - \frac{b+a}{b-a}\right)\right)$$

form an orthonormal basis in $L_2([a, b])$?

Problem 11. Starting with the set of polynomials $\{1, x, x^2, \ldots, x^n, \ldots\}$ use the Gram-Schmidt procedure the scalar product (inner product)

$$\langle f, g \rangle = \int_{-1}^1 (f(x)g(x) + f'(x)g'(x))dx$$

to find the first five orthogonal polynomials, where f' denotes derivative.

Problem 12. Let $I = [0, 1)$, μ the Lebesgue measure, $f_0(x) = 1$ and for $n \geq 1$

$$f_n(x) = \begin{cases} +1 \text{ for } 2^{n-1}x = y \in [0, 1/2) \text{ (mod 1)} \\ -1 \text{ for } 2^{n-1}x = y \in [1/2, 1) \text{ (mod 1)} \end{cases}$$

Show that the functions $f_n : [0, 1) \to \mathbb{R}$ form an orthonormal sequence in the Hilbert space $L_2([0, 1))$. The functions are called *Rademacher functions* [58].

Problem 13. Consider the Hilbert space $L_2([-\pi, \pi])$ and the functions

$$f(x) = |\sin(x)|, \qquad g(x) = |\cos(x)|.$$

Find the distance $\|f(x) - g(x)\|$ in this Hilbert space.

Problem 14. The n-th *Rademacher function* $f_n : [0, 1] \to \mathbb{R}$ is defined by

$$f_n(x) := \mathrm{sgn}(\sin(2^n \pi x))$$

where $n = 0, 1, 2, \ldots$ and sgn denotes the sign function. The sgn function is defined as

$$\mathrm{sgn}(x) := \begin{cases} 1 \text{ for } x > 1 \\ 0 \text{ for } x = 0 \\ -1 \text{ for } x < 0 \end{cases}.$$

Show that $\{ f_n : n = 0, 1, 2, \ldots \}$ is an orthonormal sequence in the Hilbert space $L_2([0, 1])$. Is this orthonormal sequence an orthonormal basis in $L_2([0, 1])$?

Problem 15. Consider the Hilbert space $L_2(\mathbb{R})$. Is

$$\phi_n(x) = \frac{1}{\sqrt{\pi(1 + x^2)}} e^{2in \arctan(x)}, \quad n \in \mathbb{Z}$$

an orthonormal basis in $L_2(\mathbb{R})$?

Problem 16. Consider the function $f : \mathbb{R} \to \mathbb{R}$

$$f(x) = \frac{x}{\sinh(x)}$$

with $f(0) = 1$. Show that $f \in L_2(\mathbb{R})$. Show that

$$\frac{x}{\sinh(x)} = 1 + 2 \sum_{j=1}^{\infty} (-1)^j \frac{x^2}{x^2 + (j\pi)^2}.$$

Problem 17. Consider the Hilbert space $L_2(\mathbb{R})$. Let $j, k = 1, 2, \ldots$. Consider the functions

$$f_j(x) = x^j e^{-|x|}, \qquad f_k(x) = x^k e^{-|x|}.$$

Find the scalar product

$$\langle f_j(x), f_k(x) \rangle := \int_{-\infty}^{\infty} f_j(x) \bar{f}_k(x) dx = \int_{-\infty}^{\infty} f_j(x) f_k(x) dx.$$

With $a > 0$ and $n \in \mathbb{N}$ one has

$$\int_0^{\infty} x^n e^{-ax} dx = \frac{\Gamma(n+1)}{a^{n+1}}.$$

Problem 18. Consider the function

$$f(x) = \sum_{j=0}^{\infty} \frac{1}{2^j} \cos(jx).$$

Is f an element of the Hilbert space $L_2([-\pi, \pi])$? Note that

$$\sum_{j=0}^{\infty} \frac{1}{2^j} = 2.$$

Problem 19. Let $m, n \in \mathbb{N}$. Consider the Hilbert space $L_2([-1, 1])$ and the functions

$$f_n(x) = \frac{1}{1 + nx^2}, \qquad n = 1, 2, \ldots$$

which are elements in this Hilbert space. Find $\| f_n(x) - f_m(x) \|$.

Problem 20. Let n be a positive integer. Consider the Hilbert space $L_2([0, n])$ and the function $f(x) = e^{-x}$. Find $a, b \in \mathbb{R}$ such that

$$\| f(x) - (ax^2 + bx) \|$$

is a minimum. The norm in the Hilbert space $L_2([0, n])$ is induced by the scalar product.

Problem 21. Consider the Hilbert space $L_2([0, \pi])$. The functions

$$\phi_j(x) = \sqrt{\frac{2}{\pi}} \sin(jx), \qquad j \in \mathbb{N}$$

form an orthonormal basis in this Hilbert space. Consider the *integral kernel*

$$K(x, y) = \sum_{j=1}^{\infty} \frac{1}{j^2} \phi_j(x) \phi_j(y).$$

Show that this series converges uniformly. Show that K is a continuous function on $[0, \pi] \times [0, \pi]$ which vanishes on the boundary of the square.

Problem 22. Consider the Hilbert space $L_2(\mathbb{R})$. Let $k \in \mathbb{Z}$. For $k = 0$ we define $s_0 = 0$, for $k \geq 1$ we define

$$s_k := 1 + \frac{1}{2} + \frac{1}{3} + \cdots + \frac{1}{k}$$

and for $k < 0$ we define $s_k = -s_{-k}$. Let $\epsilon > 0$. Define the indicator functions W_k as

$$W_k(x) := \begin{cases} 1 & \text{for} \quad s_k < x/\epsilon \leq s_{k+1} \\ 0 & \text{otherwise} \end{cases}$$

Let $u \in L_2(\mathbb{R})$. Define the linear operator $\hat{O}$ as

$$(\hat{O}u)(x) := g(x)u(x)$$

where

$$g(x) = -\frac{x}{\epsilon} + \sum_{k \in \mathbb{Z}} \left(\frac{s_k + s_{k+1}}{2} \right) W_k(x).$$

(i) Show that $\hat{O}$ is a bounded self-adjoint operator for any $\epsilon > 0$.
(ii) Show that the norm of $\hat{O}$

$$\|\hat{O}\| = \sup_{\|u\|=1} \|\hat{O}u\|$$

is given by $1/2$.

Problem 23. (i) Consider the *momentum operator*

$$\hat{p} = -i \frac{d}{dx}$$

defined on $C_0^{\infty}([0, \infty))$. Show that the operator has no self-adjoint extensions on the Hilbert space $L_2([0, \infty))$.
(ii) Consider the Hilbert space $L_2(\mathbb{R})$ and the linear operator

$$\hat{p} = -i \frac{d}{dx} \quad \text{on} \quad D = \{f : f, \frac{df}{dx} \in L_2(\mathbb{R})\}.$$

Show that the spectrum is the whole real axis.

(iii) The one-dimensional *momentum operator* $\hat{p}$ can be defined on

$$\hat{p} := -\frac{i}{\hbar}\frac{d}{dx} \; : \; S(\mathbb{R}) \subset L_2(\mathbb{R}) \to L_2(\mathbb{R})$$

where $S(\mathbb{R})$ is the Schwartzian space. Let $\sigma > 0$ and

$$f(x) = \exp(-x^2/\sigma) \in S(\mathbb{R}).$$

Find $\hat{p}f$.

(iv) Consider the Hilbert space $L_2(\mathbb{R})$. Show that the spectrum of the *position operator* $\hat{x}$ is the real line denoted by $\mathbb{R}$.

(v) Consider the Hilbert space $L_2(\mathbb{R})$. Let $f \in L_2(\mathbb{R})$ and $\theta \in \mathbb{R}$. We define the operator $U(\theta)$ as

$$U(\theta)f(x) := e^{i\theta/2}f(xe^{i\theta}).$$

Is the operator $U(\theta)$ unitary?

Problem 24. Let R be a bounded region in n-dimensional space. Consider the eigenvalue problem

$$-\Delta u = \lambda u, \qquad u(q \in \partial R) = 0$$

where ∂R denotes the boundary of R.

(i) Show that all eigenvalues are real and positive.

(ii) Show that the eigenfunctions which belong to different eigenvalues are orthogonal.

Problem 25. Let $S = L_{2,\text{real}}(\mathbb{R}, dx/(1 + |x|^3))$ and $q \in S$. Consider the linear operator

$$L(q) := -\frac{d^2}{dx^2} + q.$$

Show that the operator $L(q)$ defines a selfadjoint operator in the Hilbert space $L_2(\mathbb{R}, dx)$.

Problem 26. Consider the Hilbert space $L_2(\mathbb{R})$ and $f \in L_2(\mathbb{R})$. One can define a function $g^{(\epsilon)}(\phi)$ on the unit circle $\mathbb{S}^1$ by

$$g^{(\epsilon)}(\phi) = \sum_{n \in \mathbb{Z}} e^{-i2\pi\epsilon n} f(\phi + 2\pi n), \quad \phi \in [0, 2\pi), \; \epsilon \in [0, 1].$$

(i) Show that

$$g^{(\epsilon)}(\phi + 2\pi) = e^{i2\pi\epsilon} g^{(\epsilon)}(\phi).$$

(ii) Show that

$$\int_0^{2\pi} d\epsilon \int_0^{2\pi} d\phi |g^{(\epsilon)}(\phi)|^2 = \int_0^{2\pi} d\phi \sum_{n \in \mathbb{Z}} |f(\phi + 2\pi n)|^2 = \int_{-\infty}^{+\infty} |f(x)|^2 dx.$$

(iii) Show that

$$f(\phi + 2\pi m) = \int_0^{2\pi} d\epsilon\, e^{in2\pi\epsilon} g^{(\epsilon)}(\phi) = \int_0^{2\pi} e^{in2\pi\epsilon} \sum_{m \in \mathbb{Z}} d\epsilon\, e^{-im2\pi\epsilon} f(\phi + 2\pi m).$$

Problem 27. Let $n \in \mathbb{N}$. Is $f_n : \mathbb{R}^+ \to \mathbb{R}$

$$f_n(x) = \frac{2n^2 x}{1 + n^2 x^2}, \qquad x \geq 0$$

an element of the Hilbert space $L_2([0, \infty))$?

Problem 28. Consider the Hilbert space $L_2([0, \infty))$. Find the minimum of

$$\left\| e^{-x} - \frac{1 + ax}{1 + bx + cx^2} \right\|$$

with respect to a, b, c.

Problem 29. (i) Consider the Hilbert space $L_2([0, \pi])$. Let $r \in (0, 1)$. The *Poisson kernel* is given by

$$P_r(\theta) = 1 + 2 \sum_{j=1}^{\infty} r^j \cos(j\theta) \equiv \frac{1 - r^2}{1 - 2r \cos(\theta) + r^2}.$$

Show that $r^2 + \cos(\theta)(1 - 2r) \geq 0$ if $0 \leq \theta \leq \pi$ and $1/2 \leq r \leq 1$. Show that

$$\theta^2 P_r(\theta) \leq \frac{(1 - r^2)\theta^2}{1 - \cos(\theta)}.$$

Show that

$$\lim_{r \to 1} \int_0^{\pi} \theta^2 P_r(\theta) d\theta = 0.$$

(ii) Consider the Hilbert space $L_2([-\pi, \pi])$ and $u(e^{i\alpha}) = e^{i\alpha}$. Find

$$\frac{1}{2\pi} \int_{-\pi}^{\pi} P_r(\theta - \alpha) u(e^{i\alpha}) d\alpha$$

where $r > 0$ and

$$P_r(\delta) := \frac{1 - r^2}{1 - 2r \cos(\delta) + r^2}$$

is the *Poisson kernel.*

Problem 30. Consider the function $f : \mathbb{R} \to \mathbb{R}$, $f(x) = \frac{1}{\pi} \arctan(x)$. Show that df/dx is an element of the Hilbert space $L_2(\mathbb{R})$.

Problem 31. (i) Consider the Hilbert space $L_2(\mathbb{R})$ and the linear operator

$$Tf(x) := xf(x)$$

with the domain

$$D(T) = \{\, f(x) : f \in L_2(\mathbb{R}), \ xf \in L_2(\mathbb{R}) \,\}.$$

Show that every real number λ is in the continuous spectrum $C_\sigma(T)$. Note that $(\lambda I - T)f(x) = 0$ implies $(\lambda - x)f(x) = 0$ almost everywhere and so $f(x) = 0$ almost everywhere.

(ii) Consider the Hilbert space $L_2(\mathbb{R})$ and the *multiplication operator* T

$$Tf(x) = xf(x)$$

with $f \in L_2(\mathbb{R})$. Show that the spectral resolution is

$$T = \int_{\mathbb{R}} \lambda \, dE(\lambda)$$

where

$$E(\lambda)f(x) = \begin{cases} f(x) & \text{for } x \le \lambda \\ 0 & \text{for } x > \lambda \end{cases}.$$

Show that

$$\int_{\mathbb{R}} \lambda^2 d\|E(\lambda)f\|^2 = \|Tf\|^2, \quad \int_{\mathbb{R}} \lambda d\langle E(\lambda)f, g \rangle = \langle Tf, g \rangle.$$

Problem 32. Consider the Hilbert space $L_2([0, 1])$. Show that the boundary value problem for the second order ordinary differential equation

$$\frac{d^2u}{dx^2} + \lambda u = 0$$

with $u(0) = u(1) = 0$ is equivalent to the integral equation

$$u(x) = -\lambda \int_0^1 K(x, y)u(y)dy$$

where the *kernel* is given

$$K(x, y) = \begin{cases} y(x - 1) & 0 \le y < x \le 1 \\ x(y - 1) & 0 \le x < y \le 1 \end{cases}$$

Thus we have a symmetric, continuous kernel. Show that the eigenfunctions of this operator are given by

$$\{\sin(n\pi x) : n \in \mathbb{N}\}.$$

These functions form an orthonormal basis in the Hilbert space $L_2([0,1])$. Show that the range of K is the set of continuous function $u(x)$ with the conditions $u(0) = u(1) = 0$.

Problem 33. Consider the function $f \in C_0^\infty(\mathbb{R}^n)$ and $f \in L_2(\mathbb{R}^n)$

$$f(\mathbf{x}) = \begin{cases} \exp((1/(\mathbf{x}^2 - 1)) & \text{for } \mathbf{x} \equiv \sum_{j=1}^n x_j^2 < 1 \\ 0 & \text{for} \quad \mathbf{x}^2 \geq 1. \end{cases}$$

Find

$$\int_{\mathbb{R}^n} f(\mathbf{x}) dx_1 \cdots dx_n.$$

Is the function f analytic?

Problem 34. Consider the Hilbert space $L_2([-1,1])$. The *Legendre functions* are given by

$$\phi_j(x) := \frac{(2j+1)^{1/2}}{2^{j+1/2} j!} \frac{d^j}{dx^j}((x^2 - 1)^j), \quad j = 0, 1, 2, \ldots.$$

They form an orthonormal basis in the Hilbert space $L_2([-1,1])$.
(i) Find the scalar product $\langle f, \phi_j \rangle$ for each j with $f(x) = x\chi_{[0,1]}(x)$, where χ is the *indicator function*.
(ii) Find the function $g \in L_2([-1,1])$ which minimizes

$$\int_{-1}^1 |\exp(-x) - g(x)|^2 dx.$$

(iii) Show that if $j \geq 1$ and $c_0, c_1, \ldots, c_{j-1}$ are real constants, then

$$\int_{-1}^1 (x^n + c_{j-1}x^{j-1} + c_{j-2}x^{j-2} + \cdots + c_0)^2 dx \geq \frac{2^{2j+1}(j!)^4}{(2j)!(2j+1)!}.$$

Problem 35. Let R be a bounded region in the n-dimensional Euclidean space. Consider the eigenvalue problem in the Hilbert space $L_2(R)$

$$-\delta u(\mathbf{q}) = \lambda u(\mathbf{q}), \quad u(\mathbf{q} \in \partial R) = 0$$

where ∂R denotes the smooth boundary of R. Show that all eigenvalues are real and positive. Show that the eigenfunctions which belong to different eigenvalues are orthogonal.

Problem 36. Consider the Hilbert space $L_2([0, \infty))$ and the *shift operator*

$$Sf(x) := f(x+1)$$

with $f \in L_2([0, \infty))$. Find the spectrum of S and of its adjoint operator.

Problem 37. Consider the linear operators acting on a subset of $L_2(\mathbb{R})$

$$S_1 = S \cosh(\zeta) - \frac{B}{2} \sinh^2(\zeta) - \sinh(\zeta) \frac{d}{d\zeta}$$

$$S_2 = i \left(-S \sinh(\zeta) + \frac{B}{2} \sinh(\zeta) \cosh(\zeta) + \cosh(\zeta) \frac{d}{d\zeta} \right)$$

$$S_3 = \frac{B}{2} \sinh(\zeta) + \frac{d}{d\zeta}.$$

Show that three operators satisfy $[S_1, S_2] = iS_3$, $[S_2, S_3] = iS_1$, $[S_3, S_1] = iS_2$. Consider the Hamilton operator

$$\hat{H} = -S_3^2 - BS_1.$$

Find the commutators $[S_1, \hat{H}]$, $[S_2, \hat{H}]$, $[S_3, \hat{H}]$. Show that

$$S_1^2 + S_2^2 + S_3^2 = S(S+1)\mathbf{1}.$$

Problem 38. Consider the Hilbert space $L_2([-\pi, \pi])$. Show that the *convolution kernel* defined by

$$K(t, x) := \frac{1}{2\pi} \sum_{n \in \mathbb{Z}} \exp(-n^2 t + inx)$$

is periodic, smooth and positive in x.

Problem 39. Consider the Hilbert space $L_2([0, \pi])$ and the linear bounded operator T acting on $L_2([0, \pi])$ defined by

$$(Tf)(x) := \int_0^\pi K(x, y) f(y) dy$$

with

$$K(x, y) := \begin{cases} x(\pi - y)/\pi & \text{for } x \leq y \\ (\pi - x)y/\pi & \text{for } x \geq y \end{cases}$$

Let $f(y) = \sin(y)$. Find

$$\int_0^\pi K(x,y)f(y)dy.$$

Show that

$$\int_0^\pi K(x,x)dx = \frac{\pi^2}{6}.$$

Problem 40. Let $L_2(\mathbb{R}, \mu)$ be the Hilbert space of complex functions which are square integrable over $\mathbb{R}$ with respect to the Lebesgue measure μ. Let $f \in L_2(\mathbb{R}, \mu)$. The *Hilbert transform* Hf is defined by

$$(Hf)(x) := P\frac{1}{\pi} \int_\mathbb{R} \frac{f(y)}{x-y} d\mu(y)$$

at every point $x \in \mathbb{R}$ for which the integral on the right hand side exists in the principal value sense. Show that Hf exists almost everywhere on $\mathbb{R}$ and is an element of $L_2(\mathbb{R}, \mu)$. Show that the Hilbert transform is an isometric map and satisfies $H^2 = -id$.

Problem 41. Consider the Hilbert space $L_2(\mathbb{R})$. Find the *autocorrelation function*

$$f(\delta) = \int_\mathbb{R} u(x+\delta)u^*(x)dx$$

for $u(x) = \exp(-x^2/2)$.

Problem 42. Consider the Hilbert space $L_2([-1,1])$ and the integral operator K acting in $L_2([-1,1])$

$$K\phi(x) := \int_{-1}^{+1} \frac{\phi(x')}{|x-x'|^\gamma}dx', \quad |x| \le 1$$

for $0 < \gamma < 1$. Show that the *integral kernel* can be written as

$$\frac{1}{|x-x'|^\gamma} = \frac{1}{\Gamma(\gamma)} \int_0^\infty t^{\gamma-1}e^{-|x-x'|t}dt$$

with $\Gamma(\gamma)$ the gamma function. Show that the integral equation

$$\phi(x) = f(x) + \lambda \int_{-1}^{+1} \frac{\phi(x')}{|x-x'|^\gamma}dx'$$

can be written as an integro-differential equation. Show that integral operator K is compact and positive.

Problem 43. Consider the *integral equation*

$$\lambda u(x) = \int_{\mathbb{R}} K(x,y)u(y)dy$$

with the *kernel*

$$K(x,y) = \exp(-(x^2 - 2\cos(\theta)xy + y^2))$$

where $0 < \theta < \pi/2$. Show that the solution is $(n = 0, 1, 2, \dots)$

$$u_n(x) = \left(\frac{2|\sin(\theta)|}{\pi}\right)^{1/4} \frac{1}{(2^n n!)^{1/2}} \exp(-|\sin(\theta)|x^2) H_n(x(2|\sin(\theta)|)^{1/2})$$

with the eigenvalues

$$\lambda_n = \sqrt{\frac{\pi}{1 + |\sin(\theta)|}} \left(\frac{\cos(\theta)}{1 + |\sin(\theta)|}\right)^n.$$

Problem 44. Consider the (real) Hilbert space $L_2([-1,1])$ with the scalar product

$$\langle f, g \rangle = \int_{-1}^{+1} f(x)g(x)dx$$

and the linear operators

$$Lu(x) := \int_{-1}^{+1} K(x,y)u(y)dy, \quad L^*u(x) := \int_{-1}^{+1} K(y,x)u(y)dy$$

with the kernel

$$K(x,y) = x + 3x^2 y + xy^2.$$

Note that 1, x, x^2 are linearly independent in $L_2([-1,1])$. Consider

$$p(x) = a_0 + a_1 x + a_2 x^2.$$

Find $Lp(x)$, $L^*p(x)$ and the scalar product $\langle Lp(x), L^*p(x) \rangle$.

Problem 45. Consider the Hilbert space $L_2(\mathbb{R})$.
(i) Can one construct an orthonormal basis from

$$e^{-x^4/4}, \quad xe^{-x^4/4}, \quad x^2 e^{-x^4/4}, \quad x^3 e^{-x^4/4}, \quad \dots$$

(ii) Can one construct an orthonormal basis from

$$e^{x^2/2 - x^4/2}, \quad xe^{x^2/2 - x^4/4}, \quad x^2 e^{x^2 - x^4/4}, \quad x^3 e^{x^2 - x^4/4}, \quad \dots$$

(iii) Can one construct an orthonormal basis in the Hilbert space $L_2(\mathbb{R})$ starting from $(\sigma > 0)$

$$e^{-|x|/\sigma}, \quad xe^{-|x|/\sigma}, \quad x^2 e^{-|x|/\sigma}, \quad x^3 e^{-|x|/\sigma}, \ldots?$$

Problem 46. We define $\mathrm{sinc}(x) := \sin(\pi x)/(\pi x)$. Let $c > 0$. Consider the Fredholm homogeneous integral equation

$$\lambda_j \phi_j(x) = \int_{-c/2}^{c/2} \phi_j(y) \mathrm{sinc}^2(x-y) dy.$$

Show that a denumerably infinite set of eigenfunctions $\phi_j(x)$ exists with real eigenvalues λ_j. Show that both the eigenvalues λ_j and eigenfunctions $\phi_j(x)$ depend on c.

Problem 47. Consider the Hilbert space $L_2([-1,1])$. The *Chebyshev polynomials* are defined as

$$T_n(x) := \cos(n \arccos(x)), \quad |x| \le 1$$

with $n = 0, 1, 2, \ldots$. Show that

$$\langle T_m(x), T_n(x) \rangle = \int_{-1}^{+1} T_m(x) T_n(x) \frac{dx}{\sqrt{1-x^2}} = \delta_{n,m}$$

with $n, m \in \mathbb{N}_0$. Show that (*semigroup property*) $T_n(T_m(x)) = T_{n \cdot m}(x)$.

Problem 48. Consider the Hilbert space $L_2([0,\infty))$ and $f \in L_2([0,\infty))$ and the *integral equation*

$$\phi(x) = \lambda \int_0^\infty K(x,y)\phi(y) dy + f(x).$$

Show that $\lambda = \sqrt{2/\pi}$ is an eigenvalue of the kernel $\cos(xy)$ with $0 < x, y < \infty$. Show that the corresponding eigenfunction is

$$\phi(x) = f(x) + \sqrt{\frac{2}{\pi}} \int_0^\infty \cos(xy) f(y) dy$$

where $f \in L_2([0,\infty))$.

Problem 49. Consider the Hilbert space $L_2([0,1])$. Show that the Green's function of the boundary value problem

$$\frac{d^2 u}{dx^2} + f(x) = 0, \quad u(0) = u(1) = 0$$

is given by

$$G(x, y) = \begin{cases} x(1 - y) & 0 \leq x \leq y \\ (1 - x)y & y \leq x \leq 1 \end{cases}$$

Problem 50. A partial isometry is an operator whose restriction to the orthogonal complement of its null space is an isometry. Consider the Hilbert space $L_2([0, 1))$. The β-adic Renyi map f on the unit interval $[0, 1)$ is given by

$$f : [0, 1) \to [0, 1) \; : \; x \mapsto f(x) = \beta x \pmod 1$$

with an integer $\beta \geq 2$. The probability densities over the unit interval evolve according to the Frobenius-Perron operator U

$$U\rho(x) := \sum_{y, f(y)=x} \frac{1}{|f'(y)|} \rho(y) = \frac{1}{\beta} \sum_{k=0}^{\beta-1} \rho\left(\frac{x + k}{\beta}\right).$$

Show that the Frobenius-Perron operator is a partial isometry on the Hilbert space $L_2([0, 1))$ of the probability densities over the unit interval $[0, 1)$. Note that the Frobenius-Perron operator is the adjoint of the isometric Koopman operator $U^\dagger$, i.e. $U^\dagger \rho(x) = \rho(f(x))$.

Problem 51. Let $n \geq 2$. Consider the Hilbert space $L_2(\mathbb{R}^n)$ and the compact Lie group $O(n)$ which consists of all $n \times n$ real orthogonal matrices O with $O^T O = I_n$, i.e. $O^T = O^{-1}$. For $f \in L_2(\mathbb{R}^n)$ one defines

$$U_O f(\mathbf{x}) := f(O^{-1}\mathbf{x}).$$

(i) Show that $U_O f$ is well defined, i.e. show that if $f = g$ a.e. then $U_O f = U_O g$ a.e..

(ii) Show that $U_O : L_2(\mathbb{R}^n) \to L_2(\mathbb{R}^n)$ is unitary and satisfies $U_{O_1} U_{O_2} = U_{O_1 O_2}$.

(iii) Let $n = 2$, $f(x_1, x_2) = (x_1^2 + x_2^2)e^{-(x_1^2+x_2^2)} \in L_2(\mathbb{R}^2)$ and

$$O(\alpha) = \begin{pmatrix} \cos(\alpha) & \sin(\alpha) \\ \sin(\alpha) & -\cos(\alpha) \end{pmatrix}$$

with $\det(O(\alpha)) = -1$. Find $U_O f(\mathbf{x})$.

Problem 52. Consider the function $f : \mathbb{R} \to \mathbb{R}$

$$f(x) = \frac{1 - \cos(2\pi x)}{x}.$$

Using L'Hospital rules we have $f(0) = 0$. Is $f \in L_2(\mathbb{R})$? Since $f(x) = -f(-x)$ we only have to prove that

$$\int_0^\infty f(x)f(x)dx < \infty.$$

Note that $f(0)f(0) = 0$.

Problem 53. Let $f \in H_1(a, b)$. Then for $a \leq x < y \leq b$ we have

$$f(y) = f(x) + \int_x^y f'(s)ds.$$

(i) Show that $f \in C([a, b])$.
(ii) Show that $|f(y) - f(x)| \leq \|f\|_1 \sqrt{|y - x|}$.

Problem 54. Let s be a nonnegative integer, $x \in \mathbb{R}$ and h_n $(n = 0, 1, 2, \dots)$ be

$$h_n(x) = \frac{(-1)^n}{2^{n/2}\sqrt{n!}\sqrt[4]{\pi}} \exp(x^2/2)\frac{d^n e^{-x^2}}{dx^n}.$$

The analytic functions h_n form an orthonormal basis in the Hilbert space $L_2(\mathbb{R})$. Consider the sequence

$$f_s(x) = \frac{1}{\sqrt{s+1}} \sum_{n=0}^{s} e^{in\theta} h_n(x)$$

where $s = 0, 1, 2, \dots$. Show that the sequence converges weakly but not strongly to 0.

Problem 55. Consider the Hilbert space $L_2([0, 2\pi))$ with the scalar product

$$\langle f_1, f_2 \rangle = \frac{1}{2\pi} \int_0^{2\pi} f_1(e^{i\theta})\overline{f_2(e^{i\theta})}d\theta.$$

(i) Let $f_1(z) = z$ and $f_2(z) = z^2$. Find $\langle f_1, f_2 \rangle$.
(ii) Let $f_1(z) = z^2$ and $f_2(z) = \sin(z)$. Find $\langle f_1, f_2 \rangle$.

Problem 56. Let $H_n(x)$ $(n = 0, 1, \dots)$ be the *Hermite polynomials*. Show that (*Mehler's formula*)

$$\exp(-(u^2 + v^2 - 2uvz)/(1 - z^2)) = (1 - z^2)^{1/2} \exp(-(u^2 + v^2)) \sum_{n=0}^{\infty} \frac{z^n}{n!} H_n(u)H_n(v).$$

Problem 57. Consider the Hilbert space $L_2([0, \infty))$ and

$$f(x) = \sqrt{\frac{2}{\pi}} \int_0^\infty g(y) \cos(yx) dy, \qquad g(x) = \sqrt{\frac{2}{\pi}} \int_0^\infty f(y) \cos(yx) dy.$$

Let $g : \mathbb{R}^+ \to \mathbb{R}$, $g(y) = e^{-y}$. Find $f(x)$.

Problem 58. Consider the Hilbert space $L_2([0, \infty))$. Can one find $a, b, c \in \mathbb{R}$ and $d > 0$ such that

$$f(x) = (ax^2 + bx + c)e^{-d|x|}$$

satisfies

$$\int_0^\infty f(x) dx = 1, \qquad \int_0^\infty x f(x) dx = 1.$$

Problem 59. Consider the Hilbert space $L_2(\mathbb{R}^3, d\mathbf{x})$ and let

$$\mathbb{S}^2 := \{ (x_1, x_2, x_3) \in \mathbb{R}^3 : x_1^2 + x_2^2 + x_3^2 = 1 \}.$$

In spherical coordinates this Hilbert space has the decomposition

$$L_2(\mathbb{R}^3, d\mathbf{x}) = L_2(\mathbb{R}^+, r^2 dr) \otimes L_2(\mathbb{S}^2, \sin(\theta) d\theta d\phi).$$

Let $\hat{I}$ be the identity operator in the Hilbert space $L_2(\mathbb{S}^2, \sin(\theta) d\theta d\phi)$. Then the *radial momentum operator*

$$\hat{P}_r := -i\hbar \frac{1}{r} \left(\frac{\partial}{\partial r} \right) r$$

is identified with the closure of the operator $\hat{P}_r \otimes \hat{I}$ defined on

$$D(\hat{P}_r) \otimes L_2(\mathbb{S}^2, \sin(\theta) d\theta d\phi)$$

where

$$D(\hat{P}_r) =$$

$$\left\{ f \in L_2(\mathbb{R}^+, r^2 dr) : f \in AC(\mathbb{R}^+), \frac{1}{r} \frac{d}{dr} rf(r) \in L_2(\mathbb{R}^+, r^2 dr) \lim_{r \to 0} r|f(r)| = 0 \right\}$$

and for each $f \in D(\hat{P}_r)$

$$\hat{P}_r f(r) = -i\hbar \frac{1}{r} \frac{d}{dr} (rf(r))$$

where $\hat{P}_r$ is maximal symmetric in $L_2(\mathbb{R}^+, r^2 dr)$. Show that $\hat{P}_r$ is not self-adjoint.

Chapter 5

Hilbert Space $\ell_2(\mathbb{I})$

5.1 Introduction

The Hilbert space $\ell_2(\mathbb{N})$ is defined as

$$\ell_2(\mathbb{N}) = \{ (x_1, x_2, \dots) : x_j \in \mathbb{C} \quad \sum_{j=1}^{\infty} x_j x_j^* < \infty \}.$$

We also consider $\ell_2(\mathbb{N}_0)$, i.e. we start counting from 0. One can also consider other countable sets, for example $\ell_2(\mathbb{Z})$, $\ell_2(\mathbb{N} \times \mathbb{N})$ etc. In general we write $\ell_2(\mathbb{I})$, where $\mathbb{I}$ is a countable index set. Hence $\ell_2(\mathbb{I})$ includes the Hilbert spaces $\mathbb{R}^n$ and $\mathbb{C}^n$.

The standard basis for the Hilbert space $\ell_2(\mathbb{N}_0)$ is given by

$$(1,0,0,0,\dots)^T, \quad (0,1,0,0,\dots)^T, \quad (0,0,1,0,\dots)^T, \dots .$$

The Hilbert space $\ell_2(\mathbb{Z})$ is defined as

$$\ell_2(\mathbb{Z}) := \{ \{ a_j \}_{j \in \mathbb{Z}} \; a_j \in \mathbb{C} : \sum_{j \in \mathbb{Z}} |a_j|^2 < \infty \}.$$

The counting could be 0, 1, −1, 2, −2, ... An orthonormal basis in the Hilbert space $\ell_2(\mathbb{Z})$ is given by

$$(e_n)_{n \in \mathbb{Z}} = (\dots, \delta_{n,-2}, \delta_{n,-1}, \delta_{n,0}, \delta_{n,+1}, \delta_{n,+2}, \dots) : n \in \mathbb{Z}.$$

Let $\mathbb{Z}$ be the set of all integers. The vector space $\ell_\infty(\mathbb{Z})$ consists of all sequences $\mathbf{u}$ with

$$\sup_{n \in \mathbb{Z}} |u_n| < \infty.$$

The vector space $\ell_1(\mathbb{Z})$ consists of all sequences with

$$\|\mathbf{u}\| := \sum_{n \in \mathbb{Z}} |x_n| < \infty.$$

We have
$$\ell_1(\mathbb{Z}) \subset \ell_2(\mathbb{Z}) \subset \cdots \subset \ell_p(\mathbb{Z}) \subset \cdots \subset \ell_\infty(\mathbb{Z}).$$
The linear space $\ell_\infty(\mathbb{N}_0)$ is a Banach space $\mathcal{B}$ defined by
$$\ell_\infty(\mathbb{N}_0) := \{\, x \in \ell(\mathbb{N}_0) \,:\, \|z\| := \sup_{n \in \mathbb{N}_0} |z_n| < \infty \,\}$$
where $\ell(\mathbb{N}_0)$ is the linear space of infinite sequences $z = (z_0, z_1, z_2, \ldots)$, where $z_j \in \mathbb{C}$.

The Hilbert space $L_2(\Omega)$ can be mapped into $\ell_2(\mathbb{I})$. Every separable Hilbert space $\mathcal{H}$ is isometrically isomorphic to $\ell_2(\mathbb{N}_0)$.

Let $\mathcal{H}$ be a separable Hilbert space and $\mathbb{I}$ be a countable index set. A bounded linear operator T over $\mathcal{H}$ is called to be of trace class if for some orthonormal basis $\{e_k\}_{k \in \mathbb{I}}$ of $\mathcal{H}$, the sum of the positive terms
$$\|T\|_1 = \sum_{k \in \mathbb{I}} \langle (T^*T)^{1/2} e_k, e_k \rangle < \infty$$
is finite. A *trace class operator* is a compact operator.

Let $\mathbb{I}$ be a countable index set. A *Hilbert Schmidt operator* is a bounded operator T on a Hilbert space $\mathcal{H}$ with finite Hilbert Schmidt norm
$$\|T\|_{HS}^2 = \mathrm{tr}(T^*T) = \sum_{j \in \mathbb{I}} \|Te_j\|^2 < \infty.$$
Let $\mathcal{H}$ be a Hilbert space and $K(\mathcal{H})$ be the set of bounded operators. An operator $T \in L(\mathcal{H})$ is called to be a compact operator if the image of each bounded set under the map T is relatively compact. For every compact self-adjoint operator on a complex Hilbert space, there exists an orthonormal basis of the Hilbert space given by the eigenvectors of the operator T.

The eigenvalues of a non-zero compact self-adjoint operator on an infinite dimensional separable Hilbert space are real, bounded and countably many with the only possible accumulation point 0. The multiplicity of each nonzero is finite. The normalized eigenvectors form an orthonormal basis in the infinite dimensional separable Hilbert space. Note that the eigenvalues appear in positive and negative pairs.

Let $\mathbb{I}$ be a countable index set. A family $(\phi_j)_{j \in \mathbb{I}}$ in a Hilbert space $\mathcal{H}$ is called a *Bessel family* if the map
$$C : f \mapsto (\langle f, \phi_j \rangle)_{j \in \mathbb{I}}$$

is bounded from $\mathcal{H}$ into $\ell_2(\mathbb{I})$, i.e. if and only if there exists some positive constant $K > 0$ such that

$$\|Cf\|_{\ell_2(\mathbb{I})}^2 = \sum_{j\in\mathbb{I}} |\langle f, \phi_j\rangle|^2 \leq K\|f\|_{\mathcal{H}}^2$$

for all $f \in \mathcal{H}$.

Let $\mathbb{I}$ be a countable index set. A family $(\phi_j)_{j\in\mathbb{I}}$ in a Hilbert space $\mathcal{H}$ is called a *frame* if there exist constants $C_1 > 0$, $C_2 > 0$ such that for all $f \in \mathcal{H}$ we have the inequality

$$C_1\|f\|^2 \leq \sum_{j\in\mathbb{I}} |\langle f, \phi_j\rangle|^2 \leq C_2\|f\|^2.$$

Let $\mathcal{H}$ be a Hilbert space. Then

$$\mathrm{spec}(T) = \mathrm{spec}(T^*)$$

for every bounded operator $T : \mathcal{H} \to \mathcal{H}$.

Let T be a compact self-adjoint operator acting in the separable Hilbert space $\ell_2(\mathbb{N})$. Then there exists an orthonormal basis $\{e_j\}_{j=1}^{\infty}$ in $\ell_2(\mathbb{N})$ and real numbers λ_j which converge to 0 as $j \to \infty$ such that

$$Te_j = \lambda_j e_j$$

for all $j = 1, 2, \ldots$. Each non-zero eigenvalue of T has finite multiplicity.

5.2   Solved Problems

Problem 1. (i) Is

$$\mathbf{x} = (1, 1/2, 1/3, ..., 1/n, ...)$$

an element of $\ell_2(\mathbb{N})$?
(ii) Is

$$\mathbf{x} = (1, 1/\sqrt{2}, 1/\sqrt{3}, \ldots, 1/\sqrt{n}, \ldots)$$

an element of $\ell_2(\mathbb{N})$?
(iii) Let $n \in \mathbb{N}$. Consider the Hilbert space $\ell_2(\mathbb{Z})$ and

$$v = (\ldots, -1/n, \ldots, -1/3, -1/2, 0, 1/2, 1/3, \ldots, 1/n, \ldots)$$

Show that v is an element of $\ell_2(\mathbb{Z})$.

Solution 1. (i) Yes. We have

$$\sum_{j=1}^{\infty} x_j x_j = \sum_{j=1}^{\infty} \frac{1}{j^2} = \frac{\pi^2}{6} < \infty.$$

(ii) No. We have

$$\sum_{j=1}^{\infty} x_j x_j = \sum_{j=1}^{\infty} \frac{1}{j} = \infty.$$

(iii) Since

$$2 \sum_{j \in \mathbb{Z} \setminus \{0\}} \frac{1}{j^2} < \infty$$

v is an element of $\ell_2(\mathbb{Z})$.

Problem 2. Let $n \in \mathbb{N}_0$. Show that

$$(1/\sqrt{3}, \sqrt{2}/3, 2/(3\sqrt{3}), 2\sqrt{2}/9, \cdots \sqrt{2^n}/\sqrt{3^{n+1}} \cdots)^T$$

is an element of $\ell_2(\mathbb{N}_0)$.

Solution 2. We have

$$S = \frac{1}{3} + \frac{2}{9} + \frac{4}{27} + \frac{8}{81} + \cdots + \frac{2^n}{3^{n+1}} + \cdots = 1 < \infty.$$

Problem 3. Let $n, m \in \mathbb{N}$. Consider the vectors in $\ell_2(\mathbb{N})$

$$v_1 = (1, 1/2, \ldots, 1/m, 0, 0, \ldots), \quad v_2 = (1, 1/2, \ldots, 1/n, 0, 0, \ldots).$$

Find the square of the distance

$$\|v_1 - v_2\|^2 = \langle v_1 - v_2, v_1 - v_2 \rangle.$$

Solution 3. If $m = n$, then obviously $\|v_1 - v_2\|^2 = 0$. Assume now $m < n$. Then

$$\langle v_1 - v_2, v_1 - v_2 \rangle = \langle v_1, v_1 \rangle - \langle v_1, v_2 \rangle - \langle v_2, v_1 \rangle + \langle v_2, v_2 \rangle$$

$$= \sum_{j=1}^{m} \frac{1}{j^2} - \sum_{j=1}^{m} \frac{1}{j^2} - \sum_{j=1}^{m} \frac{1}{j^2} + \sum_{j=1}^{n} \frac{1}{j^2}$$

$$= \sum_{j=1}^{n} \frac{1}{j^2} - \sum_{j=1}^{m} \frac{1}{j^2} = \sum_{j=m+1}^{n} \frac{1}{j^2}.$$

Problem 4. Consider the Hilbert space $L_2([0, 1])$, the Hilbert space $\ell_2(\mathbb{N})$ and the function $f \in L_2([0, 1])$

$$f(x) = \begin{cases} x & 0 \leq x \leq 1/2 \\ 2(1 - x) & 1/2 \leq x \leq 1 \end{cases}.$$

Show that $(c_0, c_1, c_2, \ldots)$ with

$$c_n := \langle x^n, f(x) \rangle, \quad n \in \mathbb{N}$$

is an element of $\ell_2(\mathbb{N})$. Note that $1, x, x^2, x^3, \ldots$ is basis in the Hilbert space $L_2([0, 1])$.

Solution 4. For $n = 0$ we have

$$c_0 = \int_0^1 f(x)dx = \int_0^{1/2} x\, dx + \int_{1/2}^1 2(1 - x)dx = \frac{1}{2}.$$

For $n \geq 1$ we find

$$c_n = x^n f(x)dx = \int_0^{1/2} x^{n+1}dx + \int_{1/2}^1 2x^n(1 - x)dx = \frac{2^{n+3} - n - 5}{(4n^2 + 12n + 8)2^n}$$

utilizing the interactive Maxima program

```
/* xpowern.mac */
declare(n,integer);
t1: integrate(x^(n+1),x,0,1/2);
t2: integrate(2*(x^n)*(1-x),x,1/2,1);
S: t1 + t2; S: ratsimp(S);
```

For large n we have

$$c_n \approx \frac{2}{n^2}$$

and hence $(c_0, c_1, c_2, \dots)$ is an element of $\ell_2(\mathbb{N})$.

Problem 5. Let $\ell_2(\mathbb{N}_0)$ be the collection of the sequences $x = \{x(j) : j = 0, 1, 2, \dots\}$ of complex numbers with the constraint

$$\sum_{j=0}^{\infty} |x(j)|^2 < \infty.$$

For two such sequences x and y the scalar product is

$$\langle x, y \rangle := \sum_{j=0}^{\infty} x(j)\overline{y(j)}.$$

(i) Show that $\ell_2(\mathbb{N}_0)$ is a vector space and the sum defining the scalar product in $\ell_2(\mathbb{N}_0)$ is absolutely convergent.

(ii) Show that $\ell_2(\mathbb{N}_0)$ is complete.

Solution 5. (i) Obviously $\ell_2(\mathbb{N}_0)$ is closed under multiplication with a complex number. Let $x, y \in \ell_2(\mathbb{N}_0)$. Then the *Cauchy-Schwarz inequality*

$$\left| \sum_{j=0}^{N} x(j) \cdot \overline{y(j)} \right| \le \sum_{j=0}^{N} (|x(j)| \cdot |y(j)|) \le \left| \sum_{j=0}^{N} |x(j)|^2 \right|^{1/2} \cdot \left| \sum_{j=0}^{N} |y(j)|^2 \right|^{1/2}$$

$$\le \|x\|_{\ell_2(\mathbb{N}_0)} \cdot \|y\|_{\ell_2(\mathbb{N}_0)}$$

providing the absolute convergence of the infinite sum for the scalar product $\langle x, y \rangle$ and

$$\sum_{j=0}^{N} |x(j) + y(j)|^2 \le \sum_{j=0}^{N} (|x(j)|^2 + 2|x(j)| \cdot |y(j)| + |y(j)|^2) < \infty.$$

(ii) Let $\{x_n\}$ be a Cauchy sequence of elements in $\ell_2(\mathbb{N}_0)$. Then for every $j \in \mathbb{N}_0$ we have

$$|x_m(j) - x_n(j)|^2 \le \sum_{j=0}^{\infty} |x_m(j) - x_n(j)|^2 = \|x_m - x_n\|_{\ell_2(\mathbb{N}_0)}^2$$

so that $x(j) = \lim_n x_n(j)$ exists for every $j \in \mathbb{N}_0$. To show that this limit is in $\ell_2(\mathbb{N}_0)$ we apply *Fatou's lemma*

$$\sum_{j=0}^{\infty} |x(j)|^2 = \sum_{j=0}^{\infty} |\lim_n (x_n(j))| = \sum_{j=0}^{\infty} |\liminf_n (x_n(j))| \leq \liminf_n \left(\sum_{j=0}^{\infty} |x_n(j)| \right)$$

$$= \liminf_n (\|x_n\|^2_{\ell_2(\mathbb{N}_0)}).$$

Since $\{x_n\}$ is a Cauchy sequence $\lim_n (\|x_n\|^2_{\ell_2(\mathbb{N}_0)})$ exists.

Problem 6. Let $\ell_2(\mathbb{N}_0)$ be the collection of sequences $f = \{f_j : j \in \mathbb{N}_0\}$ of complex numbers with the constraint

$$\sum_{j=0}^{\infty} |f_j|^2 < \infty.$$

One defines for two such sequences f_j, g_j the inner product (scalar product)

$$\langle f, g \rangle := \sum_{j=0}^{\infty} f_j \bar{g}_j.$$

Show that $\ell_2(\mathbb{N}_0)$ is a vector space. Show that the sum defining the inner product on $\ell_2(\mathbb{N})$ is absolutely convergent. Show that $\ell_2(\mathbb{N}_0)$ is complete.

Solution 6. Obviously $\ell_2(\mathbb{N}_0)$ is closed under scalar multiplication with a complex number $c \in \mathbb{C}$. For $f, g \in \ell_2(\mathbb{N}_0)$ the *Cauchy-Schwarz inequality* provides

$$\left| \sum_{n \leq N} f_j \bar{g}_j \right| \leq \sum_{n \leq N} |f_j| \cdot |g_j| \leq \left| \sum_{n \leq N} |f_j|^2 \right|^{1/2} \cdot \left| \sum_{n \leq N} |g_j|^2 \right|^{1/2} \leq \|f\|_{\ell_2(\mathbb{N}_0)} \cdot \|g\|_{\ell_2(\mathbb{N}_0)}$$

providing the absolute convergence of the infinite sum for $\langle f, g \rangle$. Then

$$\sum_{n \leq N} |f_j + g_j|^2 \leq \sum_{n \leq N} (|f_j|^2 + 2|f_j| \cdot |g_j| + |g_j|^2) < \infty.$$

Let $\{f^n\}$ be a Cauchy sequence of elements in $\ell_2(\mathbb{N}_0)$. Then for every $j \in \mathbb{N}_0$ one has

$$|f_j^{(m)} - f_j^{(n)}|^2 \leq \sum_{j \geq 0} |f_j^{(m)} - f_j^{(n)}|^2 = \|f^{(m)} - f^{(n)}\|^2_{\ell_2(\mathbb{N}_0)}.$$

Hence $f_j = \lim_{n \to \infty} f_j^{(n)}$ exists for every j. Now we show that the limit is an element of $\ell_2(\mathbb{N}_0)$ applying the Fatou lemma. One has

$$\sum_{j=0}^{\infty} |f_j|^2 = \sum_{j=0}^{\infty} |\lim_n f_j^{(n)}| = \sum_{j=0}^{\infty} |\liminf_n f_j^{(n)}| \leq \liminf_n \sum_{j=0}^{\infty} |f_j^{(n)}|$$

$$= \liminf_n \|f^{(n)}\|^2_{\ell_2(\mathbb{N}_0)}.$$

It follows that $\{f^{(n)}\}$ is a Cauchy sequence $\lim_n |f^{(n)}|^2_{\ell_2(\mathbb{N}_0)}$ exists.

Problem 7. Consider the Hilbert space $\ell_2(\mathbb{N}_0)$ and the orthonormal basis (*standard basis*)

$$e_0 = \begin{pmatrix} 1\ 0\ 0\ \cdots \end{pmatrix}, \quad e_1 = \begin{pmatrix} 0\ 1\ 0\ \cdots \end{pmatrix}, \quad e_2 = \begin{pmatrix} 0\ 0\ 1\ \cdots \end{pmatrix}, \quad \dots .$$

Show that $\|e_j - e_k\|^2 = 2$ if $j \neq k$ with $j, k \in \mathbb{N}_0$.

Solution 7. With $\langle e_j, e_k \rangle = 0$ for $j \neq k$ and $\langle e_j, e_j \rangle = 1$ we have

$$\begin{aligned}
\|e_j - e_k\|^2 &= \langle e_j - e_k, e_j - e_k \rangle \\
&= \langle e_j, e_j \rangle - \langle e_j, e_k \rangle - \langle e_k, e_j \rangle + \langle e_k, e_k \rangle \\
&= \langle e_j, e_j \rangle + \langle e_k, e_k \rangle \\
&= 2.
\end{aligned}$$

Problem 8. Consider the Hilbert space $\ell_2(\mathbb{N})$. Let $\mathbf{x} = (x_1, x_2, \dots)^T$ be an element of $\ell_2(\mathbb{N})$. We define the linear operator A in $\ell_2(\mathbb{N})$ as

$$A\mathbf{x} = (x_2, x_3, \dots)^T$$

i.e. x_1 is omitted and the $n+1$st coordinate replaces the nth for $n = 1, 2, \dots$. Then for the domain we have $\mathcal{D}(A) = \ell_2(\mathbb{N})$. Find $A^*\mathbf{y}$ and the domain of A^*, where $\mathbf{y} = (y_1, y, \dots)$. Is A unitary?

Solution 8. We have

$$A^*\mathbf{y} = (0, y_1, y_2, \dots)^T.$$

Although $\mathcal{D}(A^*) = \ell_2(\mathbb{N})$ and

$$\|A^*\mathbf{y}\| = \|\mathbf{y}\|$$

for all $\mathbf{y}$ in $\ell_2(\mathbb{N})$ we find that A^* is not unitary. The inverse is not defined for all elements on the Hilbert space. We note that $\|A\mathbf{x}\| < \|\mathbf{x}\|$ in general.

Problem 9. Consider the Hilbert space $\ell_2(\mathbb{N})$ and $\mathbf{x} = (x_1, x_2, \dots)^T \in \ell_2(\mathbb{N})$. The linear bounded operator V is defined by

$$V(x_1, x_2, x_3, \dots, x_{2n}, x_{2n+1}, \dots)^T =$$

$$(x_2, x_4, x_1, x_6, x_3, x_8, x_5, \dots, x_{2n+2}, x_{2n-1}, \dots)^T.$$

Show that the operator V is unitary.

Solution 9. We have

$$V^*(x_2, x_4, x_1, x_6, x_8, x_5, \ldots, x_{2n+2}, x_{2n-1}, \ldots)^T =$$

$$(x_1, x_2, x_3, x_4, x_5, x_6, \ldots, x_{2n}, x_{2n+1}, \ldots)^T$$

and $VV^* = V^*V = I_{\ell_2(\mathbb{N})}$. The point spectrum of V is empty and the continuous spectrum is the entire unit circle in the λ-plane. V can be considered as an infinite dimensional permutation matrix.

Problem 10. Consider the Hilbert space $\ell_2(\mathbb{Z})$ and let

$$\{ |n\rangle : n \in \mathbb{Z} \}$$

be an orthonormal basis in $\ell_2(\mathbb{Z})$. Hence $\langle m|n \rangle = \delta_{m,n}$. Consider the linear operator

$$T|n\rangle = -|n+1\rangle - |n-1\rangle, \quad n \in \mathbb{Z}.$$

Let $R = T + 2I$. Find $R|n\rangle$ and $\langle m|R|n\rangle$.

Solution 10. We have

$$R|n\rangle = (T + 2I)|n\rangle = T|n\rangle + 2I|n\rangle = -|n+1\rangle + 2|n\rangle - |n-1\rangle.$$

and

$$\langle m|R|n\rangle = -\delta_{m,n+1} + 2\delta_{m,n} - \delta_{m,n-1}.$$

Problem 11. Let $|n\rangle$ $(n \in \mathbb{Z})$ be an orthonormal basis in the Hilbert space $\ell_2(\mathbb{Z})$. Consider the self-adjoint operator

$$\hat{T} = \sum_{n=-\infty}^{+\infty} (|n\rangle\langle n+1| - 2|n\rangle\langle n| + |n+1\rangle\langle n|).$$

Find the matrix representation $\langle k|\hat{T}|m\rangle$. Truncate to a 3×3 matrix with $(-1, 0, +1)$ and find the eigenvalues. Truncate to a 5×5 matrix with $(-2, -1, 0, +1, +2)$ and find the eigenvalues.

Solution 11. With $\langle n|m\rangle = \delta_{n,m}$ we obtain

$$\langle k|\hat{T}|m\rangle = \sum_{n=-\infty}^{+\infty} (\delta_{k,n}\delta_{n+1,m} - 2\delta_{k,n}\delta_{n,m} + \delta_{k,n+1}\delta_{n,m})$$

which provides the matrix

$$
\begin{pmatrix}
& \vdots & \vdots & \vdots & \vdots & \vdots & & & & \\
\ldots & 0 & 1 & -2 & 1 & 0 & 0 & 0 & 0 & \ldots \\
\ldots & 0 & 0 & 1 & -2 & 1 & 0 & 0 & 0 & \ldots \\
\ldots & 0 & 0 & 0 & 1 & -2 & 1 & 0 & 0 & \ldots \\
\ldots & 0 & 0 & 0 & 0 & 1 & -2 & 1 & 0 & \ldots \\
\ldots & 0 & 0 & 0 & 0 & 0 & 1 & -2 & 1 & \ldots \\
& \vdots & \vdots & \vdots & \vdots & \vdots & \vdots & \vdots & &
\end{pmatrix} .
$$

Truncating at the 3×3 level we obtain the matrix

$$
\begin{pmatrix}
-2 & 1 & 0 \\
1 & -2 & 1 \\
0 & 1 & -2
\end{pmatrix}
$$

with the eigenvalues $-\sqrt{2} - 2$, -2, $\sqrt{2} - 2$. Truncating at the 5×5 level we obtain the matrix

$$
\begin{pmatrix}
-2 & 1 & 0 & 0 & 0 \\
1 & -2 & 1 & 0 & 0 \\
0 & 1 & -2 & 1 & 0 \\
0 & 0 & 1 & -2 & 1 \\
0 & 0 & 0 & 1 & -2
\end{pmatrix} ,
$$

with the eigenvalues $-\sqrt{3} - 2$, -3, -2, -1, $\sqrt{3} - 2$.

Problem 12. Consider the Hilbert space $\ell_2(\mathbb{N})$. Suppose that S and T are the right and left shift linear operators on this sequence space, defined by

$$
S(x_1, x_2, \ldots) = (0, x_1, x_2, \ldots), \qquad T(x_1, x_2, x_3, \ldots) = (x_2, x_3, x_4, \ldots).
$$

Show that $T = S^*$.

Solution 12. Let $x, y \in \ell_2(\mathbb{N})$. We have

$$
\langle x, Sy \rangle = \bar{x}_2 y_1 + \bar{x}_3 y_2 + \bar{x}_4 y_3 + \cdots = \langle Tx, y \rangle.
$$

Problem 13. Let Ω be the unit disk. A Hilbert space of analytic functions can be defined by

$$
\mathcal{H} := \left\{ f(z) \text{ analytic } |z| < 1 : \sup_{a < 1} \int_{|z| = a} |f(z)|^2 ds < \infty \right\}
$$

and the scalar product

$$\langle f, g \rangle := \lim_{a \to 1} \int_{|z|=a} \overline{f(z)} g(z) ds.$$

Let c_n $(n = 0, 1, 2, \ldots)$ be the coefficients of the power-series expansion of the analytic function f. Find the norm of f.

Solution 13. The norm is

$$\|f\| = \sqrt{\sum_{n=0}^{\infty} |c_n|^2}.$$

Problem 14. Let A be a square finite-dimensional matrix over $\mathbb{R}$ such that

$$AA^T = I_n. \tag{1}$$

Show that

$$A^T A = I_n. \tag{2}$$

Does (2) also hold for infinite dimensional matrices?

Solution 14. (i) Since $\det(A) = \det(A^T)$ and $\det(I_n) = 1$ we obtain from (1) that

$$(\det(A))^2 = 1.$$

Therefore the inverse of A exists and we have $A^T = A^{-1}$ with $A^{-1}A = AA^{-1} = I_n$.

(ii) The answer is no. Let

$$A = \begin{pmatrix} 0 & 1 & 0 & 0 & 0 & \ldots \\ 0 & 0 & 1 & 0 & 0 & \ldots \\ 0 & 0 & 0 & 1 & 0 & \ldots \\ \vdots & \vdots & \vdots & \vdots & \ddots & \ldots \end{pmatrix}.$$

Then the transpose matrix A^T of A is given by

$$A^T = \begin{pmatrix} 0 & 0 & 0 & 0 & 0 & \ldots \\ 1 & 0 & 0 & 0 & 0 & \ldots \\ 0 & 1 & 0 & 0 & 0 & \ldots \\ 0 & 0 & 1 & 0 & 0 & \ldots \\ \vdots & \vdots & \vdots & \vdots & \vdots & \ddots \end{pmatrix}.$$

It follows that $AA^T = \text{diag}(1,1,1,\ldots) \equiv I$, $A^T A = \text{diag}(0,1,1,\ldots)$, where I is the infinite-dimensional unit matrix. Hence $A^T A \neq AA^T$.

Problem 15. Show that an *isometric operator* need not be a unitary operator.

Solution 15. An isometric operator need not be a unitary operator since it may fail to be surjective. For example, let $\mathcal{H} = \ell_2(\mathbb{N})$. Consider the right-shift operator $T : \ell_2(\mathbb{N}) \to \ell_2(\mathbb{N})$ given by

$$(x_1, x_2, x_3, \ldots) \mapsto (0, x_1, x_2, x_3, \ldots)$$

where $\mathbf{x} = (x_j) \in \ell_2(\mathbb{N})$.

Problem 16. Consider the Hilbert space $\mathcal{H} = L_2(\mathbb{T})$. This is the vector space of 2π-periodic functions. Then $u(x) = 1/\sqrt{2}$ is a constant function which is normalized, i.e. $\|u\| = 1$. Show that the projection operator Π_u defined by

$$\Pi_u f := \langle u, f \rangle u$$

maps a function f to its mean. This means

$$\Pi_u f = \langle f \rangle, \qquad \langle f \rangle = \int_0^{2\pi} f(x)dx.$$

Solution 16. We have

$$\Pi_u f(x) = \langle u, f \rangle u = \int_0^{2\pi} u(x)\bar{f}(x)u(x)dx = \frac{1}{2\pi} \int_0^{2\pi} \bar{f}(x)dx = \langle f \rangle.$$

The corresponding orthogonal decomposition

$$f(x) = \langle f \rangle + \tilde{f}(x)$$

decomposes a function f into a constant mean part $\langle f \rangle$ and a fluctuating part $\tilde{f}$ with zero mean.

Problem 17. Consider the Hilbert space $\ell_2(\mathbb{N})$ and the symmetric infinite dimensional matrix (bounded self-adjoint operator)

$$T = \begin{pmatrix} 1 & 1/2 & 1/3 & \ldots & 1/n & \ldots \\ 1/2 & 0 & 0 & \ldots & 0 & \ldots \\ 1/3 & 0 & 0 & \ldots & 0 & \ldots \\ \vdots & \vdots & \vdots & & 0 & \ldots \\ 1/n & 0 & 0 & \ldots & 0 & \ldots \\ \vdots & \vdots & \vdots & \ldots & & \vdots \end{pmatrix}.$$

Find the spectrum of T. Consider first the finite dimensional $n \times n$ hermitian matrix with $n \geq 3$

$$T_n = \begin{pmatrix} 1 & 1/2 & 1/3 & \ldots & 1/n \\ 1/2 & 0 & 0 & \ldots & 0 \\ 1/3 & 0 & 0 & \ldots & 0 \\ \vdots & \vdots & \vdots & & \\ 1/n & 0 & 0 & \ldots & 0 \end{pmatrix}.$$

Solution 17. The rank of the matrix T_n is 2. For all $n \geq 0$ the matrix T_n has only two nonzero real eigenvalues λ_1 and λ_2 with $\lambda_1 + \lambda_2 = 1$. Hence $n - 2$ eigenvalues are 0. For $n \to \infty$ still only two real eigenvalues are nonzero. Now

$$\mathrm{tr}(T^2) = \sum_{n=1}^{\infty} \frac{1}{n^2} + \sum_{n=2}^{\infty} \frac{1}{n^2} = \frac{\pi^2}{3} - 1.$$

Hence $\lambda_1^2 + \lambda_2^2 = 1$. The set of equations

$$\lambda_1 + \lambda_2 = 1, \quad \lambda_1^2 + \lambda_2^2 = \frac{\pi^2}{3} - 1$$

admits the solution

$$\lambda_1 = 1 + \frac{\pi}{\sqrt{6}}, \quad \lambda_2 = -\frac{\pi}{\sqrt{6}}.$$

Problem 18. Consider the Hilbert spaces $\mathbb{C}^n$ and $\ell_2(\mathbb{N})$. Find the eigenvalues and normalized eigenvectors of the matrices

$$T_2 = \begin{pmatrix} 1 & 1 \\ 1 & 1 \end{pmatrix}, \quad T_3 = \begin{pmatrix} 1 & 0 & 1 \\ 0 & 0 & 0 \\ 1 & 0 & 1 \end{pmatrix},$$

$$T_4 = \begin{pmatrix} 1 & 0 & 0 & 1 \\ 0 & 0 & 0 & 0 \\ 0 & 0 & 0 & 0 \\ 1 & 0 & 0 & 1 \end{pmatrix}, \quad T_5 = \begin{pmatrix} 1 & 0 & 0 & 0 & 1 \\ 0 & 0 & 0 & 0 & 0 \\ 0 & 0 & 0 & 0 & 0 \\ 0 & 0 & 0 & 0 & 0 \\ 1 & 0 & 0 & 0 & 1 \end{pmatrix}.$$

Extend to arbitrary n. Then consider the case $n \to \infty$. Find the unitary matrix $\exp(i\alpha T_n)$ with $\alpha \in \mathbb{R}$. Then consider $n \to \infty$.

Solution 18. For T_2 we have

$$\lambda_1 = 0, \quad \lambda_2 = 2, \quad \mathbf{v}_1 = \frac{1}{\sqrt{2}} \begin{pmatrix} 1 \\ -1 \end{pmatrix}, \quad \mathbf{v}_2 = \frac{1}{\sqrt{2}} \begin{pmatrix} 1 \\ 1 \end{pmatrix}.$$

It follows that

$$\exp(i\alpha T_2) = I_2 + (e^{2i\alpha} - 1)\frac{1}{2}T_2.$$

For T_3 we have $\lambda_{1,2} = 0$, $\lambda_3 = 2$,

$$\mathbf{v}_1 = \frac{1}{\sqrt{2}} \begin{pmatrix} 1 \\ 0 \\ -1 \end{pmatrix}, \quad \mathbf{v}_2 = \begin{pmatrix} 0 \\ 1 \\ 0 \end{pmatrix}, \quad \mathbf{v}_3 = \frac{1}{\sqrt{2}} \begin{pmatrix} 1 \\ 0 \\ 1 \end{pmatrix}.$$

It follows that

$$\exp(i\alpha T_3) = I_3 + (e^{2i\alpha} - 1)\frac{1}{2}T_3.$$

For T_4 we have $\lambda_{1,2,3} = 0$, $\lambda_4 = 2$,

$$\mathbf{v}_1 = \frac{1}{\sqrt{2}} \begin{pmatrix} 1 \\ 0 \\ 0 \\ -1 \end{pmatrix}, \quad \mathbf{v}_2 = \begin{pmatrix} 0 \\ 1 \\ 0 \\ 0 \end{pmatrix}, \quad \mathbf{v}_3 = \begin{pmatrix} 0 \\ 0 \\ 1 \\ 0 \end{pmatrix}, \quad \mathbf{v}_4 = \frac{1}{\sqrt{2}} \begin{pmatrix} 1 \\ 0 \\ 0 \\ 1 \end{pmatrix}.$$

It follows that

$$\exp(i\alpha T_4) = I_4 + (e^{2i\alpha} - 1)\frac{1}{2}T_4$$

For T_5 we have $\lambda_{1,2,3,4} = 0$, $\lambda_5 = 2$,

$$\mathbf{v}_1 = \frac{1}{\sqrt{2}} \begin{pmatrix} 1 \\ 0 \\ 0 \\ 0 \\ -1 \end{pmatrix}, \quad \mathbf{v}_2 = \begin{pmatrix} 0 \\ 1 \\ 0 \\ 0 \\ 0 \end{pmatrix}, \quad \mathbf{v}_3 = \begin{pmatrix} 0 \\ 0 \\ 1 \\ 0 \\ 0 \end{pmatrix}, \quad \mathbf{v}_4 = \begin{pmatrix} 0 \\ 0 \\ 0 \\ 1 \\ 0 \end{pmatrix}, \quad \mathbf{v}_5 = \frac{1}{\sqrt{2}} \begin{pmatrix} 1 \\ 0 \\ 0 \\ 0 \\ 1 \end{pmatrix}.$$

It follows that

$$\exp(i\alpha T_5) = I_5 + (e^{2i\alpha} - 1)\frac{1}{2}T_5.$$

For general n there are $n - 1$ eigenvalues are equal to 0 and one eigenvalue is equal to 2. Then we find

$$\exp(i\alpha T_n) = I_n + (e^{2i\alpha} - 1)\frac{1}{2}T_n.$$

Problem 19. Consider the linear bounded self-adjoint operator in a Hilbert space $\ell_2(\mathbb{N})$

$$A = \begin{pmatrix} 0 & 1 & 0 & 0 & \cdots \\ 1 & 0 & 1 & 0 & \cdots \\ 0 & 1 & 0 & 1 & \cdots \\ & \ddots & & \ddots & \\ & & \ddots & & \ddots \\ & & & \ddots & & \ddots \end{pmatrix}$$

In other words

$$A_{ij} = \begin{cases} 1 & \text{if } i = j+1 \\ 1 & \text{if } i = j-1 \\ 0 & \text{otherwise} \end{cases}$$

with $i, j \in \mathbb{N}$. Find the spectrum.

Solution 19. We study first the finite dimensional $n \times n$ matrices A_n. The matrices arise when we truncate the infinite dimensional matrix at a finite level. Since the truncated matrices are symmetric the eigenvalues are real. Since $\text{tr}(A_n) = 0$ we have $\lambda_1 + \lambda_2 + \cdots + \lambda_n = 0$, where $\lambda_1, \dots, \lambda_n$ are the eigenvalues. Since $\det(A_n) = 0$ if n is odd and $\det(A_n) = \lambda_1 \lambda_2 \dots \lambda_n$ at least one of the eigenvalues has to be zero. Since $\det(A_n) = \pm 1$ if n is even, all eigenvalues must be nonzero. Let $n = 1$. Then we have $A_1 = 0$ and the eigenvalue is given by $\lambda = 0$. Let $n = 2$. Then we have

$$A_2 = \begin{pmatrix} 0 & 1 \\ 1 & 0 \end{pmatrix}.$$

The eigenvalues are given by $\{1, -1\}$. Let $n = 3$. Thus

$$A_3 = \begin{pmatrix} 0 & 1 & 0 \\ 1 & 0 & 1 \\ 0 & 1 & 0 \end{pmatrix}.$$

The eigenvalues are given by $\{\sqrt{2}, 0, -\sqrt{2}\}$. For $n = 4$ we have

$$A_4 = \begin{pmatrix} 0 & 1 & 0 & 0 \\ 1 & 0 & 1 & 0 \\ 0 & 1 & 0 & 1 \\ 0 & 0 & 1 & 0 \end{pmatrix}$$

with the eigenvalues $\{\frac{1}{2}(\sqrt{5}+1), \frac{1}{2}(\sqrt{5}-1), -\frac{1}{2}(\sqrt{5}-1), -\frac{1}{2}(\sqrt{5}+1)\}$. For $n = 5$ we have

$$
A_5 = \begin{pmatrix}
0 & 1 & 0 & 0 & 0 \\
1 & 0 & 1 & 0 & 0 \\
0 & 1 & 0 & 1 & 0 \\
0 & 0 & 1 & 0 & 1 \\
0 & 0 & 0 & 1 & 0
\end{pmatrix}.
$$

The eigenvalues are $\{\sqrt{3}, 1, 0, -1, -\sqrt{3}\}$.

Now let us investigate the matrix A. Let A_n be the $n \times n$ truncated matrix of A. Then the eigenvalue problem for A_n is given by $A_n \mathbf{u} = \lambda \mathbf{u}$ with

$$
A_n = \begin{pmatrix}
0 & 1 & 0 & 0 & \cdots & 0 & 0 & 0 \\
1 & 0 & 1 & 0 & \cdots & 0 & 0 & 0 \\
0 & 1 & 0 & 1 & \cdots & 0 & 0 & 0 \\
& & \ddots & & \ddots & & & \\
\vdots & & & \ddots & & \ddots & & \\
& & & & \ddots & & \ddots & \\
0 & 0 & 0 & 0 & \cdots & 1 & 0 & 1 \\
0 & 0 & 0 & 0 & \cdots & 0 & 1 & 0
\end{pmatrix}.
$$

First we calculate the eigenvalues of A_n. Then we study $n \to \infty$. The eigenvalue problem leads to $D_n(\lambda) = 0$ where

$$
D_n(\lambda) = \det \begin{pmatrix}
-\lambda & 1 & 0 & 0 & \cdots & 0 & 0 & 0 \\
1 & -\lambda & 1 & 0 & \cdots & 0 & 0 & 0 \\
0 & 1 & -\lambda & 1 & \cdots & 0 & 0 & 0 \\
& \vdots & & & & & & \\
0 & 0 & 0 & 0 & \cdots & 1 & -\lambda & 1 \\
0 & 0 & 0 & 0 & \cdots & 0 & 1 & -\lambda
\end{pmatrix}.
$$

We find a difference equation for $D_n(\lambda)$, where $n = 1, 2, \ldots$. We obtain

$$D_n(\lambda) = -\lambda \det \begin{pmatrix} -\lambda & 1 & 0 & 0 & \cdots & 0 & 0 & 0 \\ 1 & -\lambda & 1 & 0 & \cdots & 0 & 0 & 0 \\ 0 & 1 & -\lambda & 1 & \cdots & 0 & 0 & 0 \\ 0 & 0 & 1 & -\lambda & \cdots & 0 & 0 & 0 \\ & & \vdots & & & & & \\ 0 & 0 & 0 & 0 & \cdots & 1 & -\lambda & 1 \\ 0 & 0 & 0 & 0 & \cdots & 0 & 1 & -\lambda \end{pmatrix}$$

$$- \det \begin{pmatrix} 1 & 1 & 0 & 0 & \cdots & 0 & 0 & 0 \\ 0 & -\lambda & 1 & 0 & \cdots & 0 & 0 & 0 \\ 0 & 1 & -\lambda & 1 & \cdots & 0 & 0 & 0 \\ & & \vdots & & & & & \\ 0 & 0 & 0 & 0 & \cdots & 1 & -\lambda & 1 \\ 0 & 0 & 0 & 0 & \cdots & 0 & 1 & -\lambda \end{pmatrix}.$$

The first determinant on the right-hand side is equal to $D_{n-1}(\lambda)$. For the second determinant we find (expansion of the first row)

$$\det \begin{pmatrix} 1 & 1 & 0 & 0 & \cdots & 0 & 0 & 0 \\ 0 & -\lambda & 1 & 0 & \cdots & 0 & 0 & 0 \\ 0 & 1 & -\lambda & 1 & \cdots & 0 & 0 & 0 \\ & & \vdots & & & & & \\ 0 & 0 & 0 & 0 & \cdots & 1 & -\lambda & 1 \\ 0 & 0 & 0 & 0 & \cdots & 0 & 1 & -\lambda \end{pmatrix}.$$

We obtain a second order linear difference equation with constant coefficients

$$D_n(\lambda) = -\lambda D_{n-1}(\lambda) - D_{n-2}(\lambda)$$

with the "initial condition"

$$D_1(\lambda) = -\lambda, \qquad D_2(\lambda) = \det \begin{pmatrix} -\lambda & 1 \\ 1 & -\lambda \end{pmatrix} = \lambda^2 - 1.$$

To solve this linear difference equation we make the ansatz $D_n(\lambda) = e^{in\theta}$, where $n = 1, 2, \ldots$. Inserting this ansatz the into the difference equation yields

$$e^{in\theta} = -\lambda e^{i(n-1)\theta} - e^{i(n-2)\theta}.$$

It follows that $e^{i\theta} = -\lambda - e^{-i\theta}$. Consequently $\lambda = -2\cos(\theta)$. Thus the general solution to the difference equation is given by

$$D_n(\lambda) = C_1 \cos(n\theta) + C_2 \sin(n\theta)$$

where C_1, C_2 are constants and $\lambda = -2\cos(\theta)$. Imposing the initial condition $D_1(\lambda) = -\lambda$ and $D_2(\lambda) = \lambda^2 - 1$, it follows that

$$D_n(\lambda) = \frac{\sin((n+1)\theta)}{\sin(\theta)}.$$

Since $D_n(\lambda) = 0$ we find

$$\frac{\sin((n+1)\theta)}{\sin(\theta)} = 0.$$

The solutions to this equation are given by $\theta = \frac{k\pi}{n+1}$ with $k = 1, 2, \ldots, n$. Since $\lambda = -2\cos(\theta)$, we find the eigenvalues

$$\lambda_k = -2\cos\left(\frac{k\pi}{n+1}\right)$$

with $k = 1, 2, \ldots, n$. Hence

$$\text{(i)} \quad |\lambda_k| < 2$$

and $\lambda_k \neq \lambda_{k'}$ if $k \neq k'$. If $n \to \infty$, then

$$\text{(ii)} \quad \text{infinitely many } \lambda_k \text{ with } |\lambda_k| \leq 2$$

and

$$\text{(iii)} \quad \lambda_k - \lambda_{k+1} \to 0 \text{ for } n \to \infty.$$

Therefore $\text{spec}(A) = [-2, 2]$, i.e. we have a continuous spectrum.

Another approach to find the spectrum is as follows. Let $A = B + B^*$, where

$$B := \begin{pmatrix} 0 & 0 & 0 & 0 & \cdots \\ 1 & 0 & 0 & 0 & \cdots \\ 0 & 1 & 0 & 0 & \cdots \\ 0 & 0 & 1 & 0 & \cdots \\ & & \vdots & & \end{pmatrix} \quad \Rightarrow \quad B^* = \begin{pmatrix} 0 & 1 & 0 & 0 & \cdots \\ 0 & 0 & 1 & 0 & \cdots \\ 0 & 0 & 0 & 1 & \cdots \\ & & \vdots & & \end{pmatrix}.$$

Then $B^*B = I$, where I is the infinite unit matrix. Notice that $BB^* \neq I$. We use the following notation

$$Cf = \lambda f \quad \text{means} \quad \|Cf_n - \lambda f_n\| \to 0$$

as $n \to \infty$. Now

$$Bf = \lambda f \quad \Rightarrow \quad B^*Bf = B^*\lambda f \quad \Rightarrow \quad f = \lambda B^* f \quad \Rightarrow \quad B^* f = \frac{1}{\lambda} f.$$

From $Bf = \lambda f$ it also follows that

$$\|Bf\|^2 = \langle Bf, Bf \rangle = \langle f, B^*Bf \rangle = \langle f, f \rangle = \|f\|^2.$$

On the other hand

$$\langle Bf, Bf \rangle = \bar{\lambda}\lambda \langle f, f \rangle = |\lambda|^2 \|f\|^2.$$

Since $\|f\| > 0$ we find that $|\lambda|^2 = 1$. Therefore

$$Af = (B + B^*)f = (\lambda + \frac{1}{\lambda})f = (\lambda + \bar{\lambda})f = 2(\cos(\gamma))f.$$

This means

$$\bigwedge_{\gamma \in \mathbb{R}} \|(A - 2(\cos(\gamma))I)f_n\| \to 0$$

or

$$A \begin{pmatrix} \sin(\gamma) \\ \sin(2\gamma) \\ \sin(3\gamma) \\ \vdots \end{pmatrix} = 2\cos(\gamma) \begin{pmatrix} \sin(\gamma) \\ \sin(2\gamma) \\ \sin(3\gamma) \\ \vdots \end{pmatrix} \in \ell_\infty(\mathbb{N}).$$

The linear space $\ell_\infty(\mathbb{N})$ is a *Banach space* $\mathcal{B}$ defined as

$$\ell_\infty(\mathbb{N}) := \{ x \in \ell(\mathbb{N}) : \|x\| := \sup_{n \in \mathbb{N}} |x_n| < \infty \}$$

where $\ell(\mathbb{N})$ is the linear space of infinite sequences $x = (x_1, x_2, \ldots)$, where $x_j \in \mathbb{C}$. However, the eigenvector is not an element of the Hilbert space $\ell_2(\mathbb{N})$. For the first two rows we have the identities

$$\sin(2\gamma) \equiv 2\sin(\gamma)\cos(\gamma), \quad \sin(\gamma) + \sin(3\gamma) \equiv 2\cos(\gamma)\sin(2\gamma).$$

The norm is given by

$$\|A\| = \sup |\alpha|_{\alpha \in \text{Spec } A} = \max |\alpha|_{\alpha \in \text{Spec } A} = 2.$$

Problem 20. Consider the Hilbert space $L_2(U(1))$ and the Hilbert space $\ell_2(\mathbb{Z})$. The circle group

$$U(1) := \{ e^{i\theta} : \theta \in \mathbb{R} \}$$

is compact and abelian. The Hilbert space $L_2((U(1))$ is the Hilbert space $L_2([0, 2\pi))$ consisting of all functions f with period 2π and

$$\int_0^{2\pi} |f(\theta)|^2 d\theta < \infty.$$

The functions $\{\, e^{im\theta} \; : \; m \in \mathbb{Z} \,\}$ form an *orthonormal basis* in the Hilbert space $L_2([0, 2\pi))$ satisfying $e^{i(m\theta+2\pi)} = e^{im\theta}$. Then

$$f(\theta) = \sum_{m\in\mathbb{Z}} c_m e^{im\theta}, \qquad c_m = \frac{1}{2\pi} \int_0^{2\pi} f(\theta) e^{-im\theta} d\theta.$$

Let

$$f(\theta) = \sin(\theta) \equiv \frac{e^{i\theta} - e^{-i\theta}}{2i}.$$

Find c_m ($m \in \mathbb{Z}$) and show that $(\ldots, c_{-2}, c_{-1}, c_0, c_1, c_2, \ldots)$ is an element of the Hilbert space $\ell_2(\mathbb{Z})$.

Solution 20. With $e^{-im\theta} \equiv \cos(m\theta) - i\sin(m\theta)$ we have

$$
\begin{aligned}
c_m &= \frac{1}{2\pi} \int_0^{2\pi} \sin(\theta) e^{-im\theta} d\theta \\
&= \frac{1}{2\pi} \int_0^{2\pi} (\sin((1-m)\theta)d\theta + \sin((1+m)\theta)d\theta \\
&\quad - \frac{i}{4\pi} \int_0^{2\pi} (\cos((1-m)\theta)d\theta - \cos((1+m)\theta)d\theta).
\end{aligned}
$$

Hence

$$
c_m = \begin{cases}
-i/2 & \text{for} \quad m = 1 \\
i/2 & \text{for} \quad m = -1 \\
0 & \text{otherwise}
\end{cases}
$$

Thus $c_0 = 0$ and $(\ldots, 0, 0, i/2, 0, -i/2, 0, 0, \ldots)$. It follows that

$$\sum_{m\in\mathbb{Z}} |c_m|^2 = \frac{1}{2} < \infty$$

The *Parseval equality* is

$$\frac{1}{2\pi} \int_0^{2\pi} |f(\theta)|^2 d\theta = \sum_{m\in\mathbb{Z}} |c_m|^2$$

with

$$\frac{1}{2\pi} \int_0^{2\pi} \sin^2(\theta) d\theta = \frac{1}{2}.$$

Problem 21. Let $a > 0$. Consider the Hilbert spaces $L_2([0, a])$ and the Hilbert space $\ell_2(\mathbb{Z})$. An orthonormal basis in $L_2([0, a])$ is given by

$$B := \left\{ \phi_n(x) = \frac{1}{\sqrt{a}} \exp(2\pi i n x/a) \; : \; n \in \mathbb{Z} \right\}.$$

Then $f(x) = x$ is an element of $L_2([0, a])$. Let

$$c_n := \langle f(x), \phi_n(x) \rangle \equiv \langle x, \phi_n(x) \rangle = \int_0^a x \frac{1}{\sqrt{a}} e^{-2\pi i n/a} dx$$

where $n \in \mathbb{Z}$. Show that $(\ldots, c_2, c_1, c_0, c_1, c_2, \ldots)$ is an element of $\ell_2(\mathbb{Z})$.

Solution 21. For $n = 0$ we have

$$c_0 = \frac{1}{\sqrt{a}} \int_0^a x dx = \frac{1}{2} a \sqrt{a}.$$

For $n \neq 0$ we have

$$c_n = \frac{1}{\sqrt{a}} \int_0^a x(\cos(2\pi i n x/a) - i\sin(2\pi i n x/a)) dx = \frac{a\sqrt{a}(2\pi i n - 1)}{4\pi^2 n^2}.$$

Hence

$$\sum_{n \in \mathbb{Z}} |c_n|^2 < \infty.$$

Problem 22. Consider the Hilbert spaces $\ell_2(\mathbb{Z})$ and $L_2([0, 1])$. The Hilbert space $L_2([0, 1])$ admits the orthonormal basis

$$\{ \phi_n(x) = e^{2\pi i n x} : n \in \mathbb{Z} \}.$$

Consider the element in $\ell_2(\mathbb{Z})$

$$(c_n)_{n \in \mathbb{Z}} = (\ldots, 1/4, 1/3, 1/2, 1, 1/2, 1/3, 1/4, \ldots)$$

where 1 is at the position 0. Find the function

$$f(x) = \sum_{n \in \mathbb{Z}} c_n \phi_n(x)$$

in $L_2([0, 1])$.

Solution 22. The sum is given by

$$\sum_{n \in \mathbb{Z}} \frac{1}{|n| + 1} e^{2\pi i n x} = \sum_{n \in \mathbb{Z}} \frac{1}{|n| + 1} (\cos(2\pi i n x) + i\sin(2\pi i n x))$$

$$= 1 + 2 \sum_{n=1}^{\infty} \frac{1}{n + 1} \cos(2\pi i n x).$$

Problem 23. (i) Consider the 3×3 matrix T_3, the 4×4 matrix T_4 and the 5×5 matrix T_5

$$T_3 = \begin{pmatrix} 1 & 1 & 0 \\ 0 & 0 & 1 \\ 1 & 0 & 0 \end{pmatrix}, \quad T_4 = \begin{pmatrix} 1 & 1 & 0 & 0 \\ 0 & 0 & 1 & 0 \\ 0 & 0 & 0 & 1 \\ 1 & 0 & 0 & 0 \end{pmatrix}, \quad T_5 = \begin{pmatrix} 1 & 1 & 0 & 0 & 0 \\ 0 & 0 & 1 & 0 & 0 \\ 0 & 0 & 0 & 1 & 0 \\ 0 & 0 & 0 & 0 & 1 \\ 1 & 0 & 0 & 0 & 0 \end{pmatrix}.$$

Note that the matrices are nonnormal and invertible. Find the trace $\text{tr}(T_j)$, $\text{tr}(T_j^2)$, $(j = 3, 4, 5)$, determinant, characteristic equation and the eigenvalues λ_j.
(ii) Extend to arbitrary n. Then study the case $n \to \infty$.

Solution 23. (i) Consider T_3. Then

$$\text{tr}(T_3) = 1 = \sum_{j=1}^{3} \lambda_j, \qquad \text{tr}(T_3^2) = 1 = \sum_{j=1}^{3} \lambda_j^2.$$

The determinant is equal to $+1$. The characteristic equation takes the form

$$\lambda^3 - \lambda^2 - 1 = 0.$$

One eigenvalue is real. The eigenvalues are $\lambda_1 = 1.46557$, $\lambda_2 = -0.23279 + 0.79255i$, $\lambda_3 = -023279 - 0.79255i$.
Consider T_4. Then

$$\text{tr}(T_4) = 1 = \sum_{j=1}^{4} \lambda_j, \qquad \text{tr}(T_4^2) = 1 = \sum_{j=1}^{4} \lambda_j^2.$$

The determinant is equal to -1. The characteristic equation takes the form

$$\lambda^4 - \lambda^3 - 1 = 0.$$

Two eigenvalues are real. The eigenvalues are $\lambda_1 = 1.38028$, $\lambda_2 = -0.81917$, $\lambda_3 = 0.21945 + 0.91447i$, $\lambda_4 = 0.21945 - 0.91447i$.
Consider T_5. Then

$$\text{tr}(T_5) = 1 = \sum_{j=1}^{5} \lambda_j, \qquad \text{tr}(T_5^2) = 1 = \sum_{j=1}^{5} \lambda_j^2.$$

The determinant is equal to $+1$. The characteristic equation takes the form

$$\lambda^5 - \lambda^4 - 1 = 0.$$

One eigenvalue is real. The eigenvalues are $\lambda_1 = 1.32472$, $\lambda_2 = 1/2 + 0.86603i$, $\lambda_3 = 1/2 - 0.86603i$, $\lambda_4 = -0.66236 + 0.56228i$, $\lambda_5 = -0.66236 - 0.56228i$.

(ii) For arbitrary n we have

$$\mathrm{tr}(T_n) = 1 = \sum_{j=1}^n \lambda_j, \quad \mathrm{tr}(T_n^2) = 1 = \sum_{j=1}^n \lambda_j^2.$$

The determinant is $+1$ if n is odd and -1 if n is even. The characteristic equation is

$$\lambda^n - \lambda^{n-1} - 1 = 0.$$

For $n \to \infty$ we have a bounded operator (but not self-adjoint) in the Hilbert space $\ell_2(\mathbb{N})$.

Problem 24. Consider the Hilbert space $\ell_2(\mathbb{N}_0)$ and the self-adjoint bounded operator acting on $\ell_2(\mathbb{N}_0)$.

$$T = \begin{pmatrix} 0 & 1 & 0 & 0 & 0 & 0 & 0 & \dots \\ 1 & 0 & 1/2 & 0 & 0 & 0 & 0 & \dots \\ 0 & 1/2 & 0 & 1/3 & 0 & 0 & 0 & \dots \\ 0 & 0 & 1/3 & 0 & 1/4 & 0 & 0 & \dots \\ 0 & 0 & 0 & 1/4 & 0 & 1/5 & 0 & \dots \\ \vdots & \vdots & \vdots & \vdots & \vdots & \ddots & \vdots & \dots \end{pmatrix}.$$

Show that the operator is a Hilbert Schmidt operator. Note that $\mathrm{tr}(T) = 0$.

Solution 24. We obtain

$$T^*T = T^2 = \begin{pmatrix} 1 & * & * & * & * & \dots \\ * & 1+1/4 & * & * & * & * & \dots \\ * & * & 1/4+1/9 & * & * & \dots \\ * & * & * & 1/9+1/16 & * & \dots \\ \vdots & \vdots & \vdots & \vdots & \ddots & \vdots \end{pmatrix}.$$

Hence

$$\mathrm{tr}(T^2) = 2 + 2\left(\frac{1}{2^2} + \frac{1}{3^2} + \frac{1}{4^2} + \cdots\right) = 2 + 2\left(\frac{\pi^2}{6} - 1\right) = \frac{\pi^2}{3}.$$

Problem 25. Let $n \geq 2$. Given the real symmetric matrix A with the diagonal elements are all the same, say $a_{11} \in \mathbb{R}$ and the non-diagonal elements are all the same, say $a_{12} \in \mathbb{R}$. Find the spectrum for all n.

Solution 25. We can write

$$A = (a_{11} - a_{12})I_n + a_{12}J_n$$

where I_n is the $n \times n$ identity matrix and J_n is the rank 1 $n \times n$ matrix with all entries 1's. Note that $[I_n, J_n] = 0_n$. The eigenvalues are

$$a_{11} - a_{12} + na_{12} \quad (\text{1 times}), \quad a_{11} - a_{12} \quad (n-1)\text{-times}.$$

Hence we have a two-level system for all n.

Problem 26. Consider the Schwartz space $S(\mathbb{R})$, the Hilbert space $L_2([0, 2\pi])$ and the Hilbert space $\ell_2(\mathbb{Z})$. The Hilbert space $L_2([0, 2\pi])$ admits the orthonormal basis

$$B = \left\{ \frac{1}{\sqrt{2\pi}} e^{-ikx} : k \in \mathbb{Z} \right\}.$$

Let $\phi \in S(\mathbb{R})$ and $\hat{\phi}$ be the Fourier transform of ϕ, i.e.

$$\hat{\phi}(k) = \int_{-\infty}^{+\infty} e^{ikx} \phi(x) dx.$$

Then $\hat{\phi} \in S(\mathbb{R})$. Show that (*Poisson summation formula*)

$$\sum_{n \in \mathbb{Z}} \phi(2\pi n) \equiv \sum_{k \in \mathbb{Z}} \hat{\phi}(k).$$

Solution 26. We define

$$f(x) = \sum_{n \in \mathbb{Z}} \phi(x + 2\pi n).$$

Then $f \in C^\infty(\mathbb{R})$ and the series is absolutely convergent and $f(x + 2\pi) = f(x)$. The Fourier coefficients c_k ($k \in \mathbb{Z}$) of the function f with respect to the orthonormal basis B defined above are given by

$$c_k = \frac{1}{\sqrt{2\pi}} \int_0^{2\pi} f(x) e^{-ikx} dx = \sum_{n \in \mathbb{Z}} \frac{1}{\sqrt{2\pi}} \int_0^{2\pi} \phi(x + 2\pi n) e^{-ikx} dx$$

$$= \sum_{n \in \mathbb{Z}} \frac{1}{\sqrt{2\pi}} \int_{2\pi n}^{2\pi(n+1)} \phi(x) e^{-ikx} dx = \hat{\phi}(k).$$

Note that $(..., c_{-2}, c_{-1}, c_0, c_1, c_2, ...) \in \ell_2(\mathbb{C})$. With $f \in L_2([0, 2\pi))$ we have

$$f(x) = \sum_{n \in \mathbb{Z}} \phi(x + 2\pi n) = \lim_{s \to \infty} \sum_{k=-s}^{s} \hat{\phi}(k) e^{ikx}.$$

With $\hat{\phi} \in S(\mathbb{R})$ the series converges absolutely. It follows that

$$\sum_{n \in \mathbb{Z}} \phi(x + 2\pi n) \equiv \sum_{k \in \mathbb{Z}} \hat{\phi}(k) e^{ikx}.$$

Setting $x = 0$ we obtain Poisson's summation formula.

For example let $\tau > 0$. Then

$$\sum_{n \in \mathbb{Z}} \exp(-\tau (2\pi)^2 n^2) \equiv \sum_{k \in \mathbb{Z}} \frac{1}{\sqrt{2\tau}} \exp(-k^2/(4\tau)).$$

Problem 27. Consider the 3×3, 5×5 and 7×7 matrix

$$A_3 = \begin{pmatrix} 0 & 1 & 0 \\ 1 & 0 & 1/2 \\ 0 & 1/2 & 0 \end{pmatrix}, \quad A_5 = \begin{pmatrix} 0 & 1 & 0 & 0 & 0 \\ 1 & 0 & 1/2 & 0 & 0 \\ 0 & 1/2 & 0 & 1/3 & 0 \\ 0 & 0 & 1/3 & 0 & 1/4 \\ 0 & 0 & 0 & 1/4 & 0 \end{pmatrix},$$

$$A_7 = \begin{pmatrix} 0 & 1 & 0 & 0 & 0 & 0 & 0 \\ 1 & 0 & 1/2 & 0 & 0 & 0 & 0 \\ 0 & 1/2 & 0 & 1/3 & 0 & 0 & 0 \\ 0 & 0 & 1/3 & 0 & 1/4 & 0 & 0 \\ 0 & 0 & 0 & 1/4 & 0 & 1/5 & 0 \\ 0 & 0 & 0 & 0 & 1/5 & 0 & 1/6 \\ 0 & 0 & 0 & 0 & 0 & 1/6 & 0 \end{pmatrix}.$$

Obviously the matrices are hermitian. Show that A_3 admits the eigenvalue $\lambda = 0$. Find $\text{tr}(A_3)$, $\text{tr}(A_3^2)$. Show that A_5 admits the eigenvalue $\lambda = 0$. Find $\text{tr}(A_5)$, $\text{tr}(A_5^2)$. Find $\text{tr}(A_7)$, $\text{tr}(A_7^3)$. Extend to A_{2n+1} ($n \in \mathbb{N}$) and then $n \to \infty$.

Solution 27. From

$$A_3 \begin{pmatrix} x_1 \\ x_2 \\ x_3 \end{pmatrix} = \begin{pmatrix} 0 \\ 0 \\ 0 \end{pmatrix}$$

we obtain $x_1 + x_3/2 = 0$, $x_2 = 0$. With $x_1 = 1/2$ we obtain the eigenvector for A_3 of the eigenvalue 0

$$\begin{pmatrix} 1/2 \\ 0 \\ -1 \end{pmatrix}.$$

Now $\mathrm{tr}(A_3) = 0$ and

$$\mathrm{tr}(A_3^2) = 2(1 + 1/4) = \lambda_1^2 + \lambda_2^2 + \lambda_3^2 = \lambda_1^2 + \lambda_3^2$$

with $\lambda_2 = 0$. From

$$A_5 \begin{pmatrix} x_1 \\ x_2 \\ x_3 \\ x_4 \\ x_5 \end{pmatrix} = \begin{pmatrix} 0 \\ 0 \\ 0 \\ 0 \\ 0 \end{pmatrix}$$

we obtain the eigenvector with $x_1 = 1/2$

$$x_1 = 1/2, \quad x_2 = 0, \quad x_3 = -1, \quad x_4 = 0, \quad x_5 = 4/3$$

for $\lambda = 0$. Now $\mathrm{tr}(A_5) = 0$ and

$$\mathrm{tr}(A_5^2) = 2(1 + 1/4 + 1/9 + 1/16) = \lambda_1^2 + \lambda_2^2 + \lambda_3^2 + \lambda_4^2 + \lambda_5 = \lambda_1^2 + \lambda_2^2 + \lambda_4^2 + \lambda_5^2$$

with $\lambda_3 = 0$. From

$$A_7 \begin{pmatrix} x_1 \\ x_2 \\ x_3 \\ x_4 \\ x_5 \\ x_6 \\ x_7 \end{pmatrix} = \begin{pmatrix} 0 \\ 0 \\ 0 \\ 0 \\ 0 \\ 0 \\ 0 \end{pmatrix}$$

we obtain $x_2 = x_4 = x_6 = 0$ and with $x_1/2$ we find $x_3 = -1$, $x_5 = 4/3$, $x_7 = 8/5$. Hence 0 is an eigenvalue. Now $\mathrm{tr}(A_7) = 0$ and

$$\mathrm{tr}(A_7^2) = 2(1 + 1/2^2 + 1/3^2 + 1/4^2 + 1/5^2 + 1/6^2) = \sum_{j=1}^{2} \lambda_j^2$$

with $\lambda_4 = 0$. For the linear equation $A_{2n+1}\mathbf{x} = \mathbf{0}$ we find that $x_2 = x_4 = \cdots = x_{2n} = 0$ and $x_1 \neq 0$, $x_3 \neq 0$, ..., $x_{2n+1} \neq 0$ and hence 0 is an eigenvalue. For A_∞ we have

$$\mathrm{tr}(A_\infty^2) = 2(1 + 1/2^2 + 1/3^2 + 1/4^2 + 1/5^2 + \cdots) < \infty.$$

Problem 28. Consider the 3×3, 5×5 and 7×7 symmetric matrices over $\mathbb{R}$ with trace equal to 0

$$A_3 = \begin{pmatrix} 0 & 1 & 0 \\ 1 & 0 & 1 \\ 0 & 1 & 0 \end{pmatrix}, \quad A_5 = \begin{pmatrix} 0 & 0 & 0 & 1 & 0 \\ 0 & 0 & 1 & 0 & 1 \\ 0 & 1 & 0 & 1 & 0 \\ 1 & 0 & 1 & 0 & 0 \\ 0 & 1 & 0 & 0 & 0 \end{pmatrix},$$

$$A_7 = \begin{pmatrix} 0 & 0 & 0 & 0 & 0 & 1 & 0 \\ 0 & 0 & 0 & 0 & 1 & 0 & 1 \\ 0 & 0 & 0 & 1 & 0 & 1 & 0 \\ 0 & 0 & 1 & 0 & 1 & 0 & 0 \\ 0 & 1 & 0 & 1 & 0 & 0 & 0 \\ 1 & 0 & 1 & 0 & 0 & 0 & 0 \\ 0 & 1 & 0 & 0 & 0 & 0 & 0 \end{pmatrix}.$$

Show that one eigenvalue of these matrices is 0. Find the corresponding eigenvectors. Extend to A_{2n+1} with $n \in \mathbb{N}$ and $n \to \infty$.

Solution 28. We have $\det(A_3) = 0$. Hence at least one eigenvalue must be 0. Solving the eigenvalue problem $A_3 \mathbf{v}_3 = \mathbf{0}_3$ provides one eigenvector

$$\mathbf{v}_3 = \begin{pmatrix} 1 & 0 & -1 \end{pmatrix}^T.$$

We have $\det(A_5) = 0$. Hence at least one eigenvalue must be 0. Solving the eigenvalue problem $A_5 \mathbf{v}_5 = \mathbf{0}_5$ provides one eigenvector

$$\mathbf{v}_5 = \begin{pmatrix} 1 & 0 & -1 & 0 & 1 \end{pmatrix}^T.$$

We have $\det(A_7) = 0$. Hence at least one eigenvalue must be 0. Solving the eigenvalue problem $A_7 \mathbf{v}_7 = \mathbf{0}_7$ provides one eigenvector

$$\mathbf{v}_7 = \begin{pmatrix} 1 & 0 & -1 & 0 & 1 & 0 & -1 \end{pmatrix}.$$

We have $\det(A_{2n+1}) = 0$ for all $n \in \mathbb{N}$. One eigenvalue is equal to 0 with the eigenvector

$$\mathbf{v}_{2n+1} = \begin{pmatrix} 1 & 0 & -1 & 0 & 1 & \ldots & 0 & \pm 1 \end{pmatrix}$$

with $+1$ if n is odd and -1 if n is even. The operator A_∞ is bounded and self-adjoint and admits the eigenvalue 0.

Problem 29. Let $a > 0$. Consider the Hilbert spaces $L_2([0,a])$ and $\ell_2(\mathbb{Z})$. An orthonormal basis in $L_2([0,a])$ is given by

$$B := \left\{ \phi_n(x) = \frac{1}{\sqrt{a}} \exp(2\pi i x n/a) \; : \; n \in \mathbb{Z} \right\}.$$

Find

$$c_n := \langle x^2, \phi_n(x) \rangle, \quad n \in \mathbb{Z}.$$

Is $(\ldots, c_{-2}, c_{-1}, c_0, c_1, c_2, \ldots)$ an element of $\ell_2(\mathbb{Z})$?

Solution 29. For $n = 0$ we have

$$c_0 = \langle x^2, \phi_0(x) \rangle = \int_0^a x^2 \frac{1}{\sqrt{a}} dx = \frac{1}{\sqrt{a}} \frac{a^3}{3}.$$

For $n \neq 0$ we utilize that

$$\int x^2 e^{kx} dx = \frac{e^{kx}}{k} \left(x^2 - \frac{2x}{k} + \frac{2}{k^2} \right).$$

Then

$$c_n = \langle x^2, \phi_n(x) \rangle = -\frac{a^3}{2\pi i n \sqrt{a}} \left(1 + \frac{1}{\pi i n} \right).$$

Since $\sum_{n \in \mathbb{Z}, n \neq 0} 1/n^2 < \infty$ we find that $(\ldots, c_{-2}, c_{-1}, c_0, c_1, c_2, \ldots)$ is an element of the Hilbert space $\ell_2(\mathbb{Z})$.

Problem 30. Consider the Hilbert space $\mathbb{C}^2$, $\ell_2(\mathbb{Z})$ and $\ell_2(\mathbb{Z}) \otimes \mathbb{C}^2$. Let U_H be the *Hadamard gate* (*coin operator*)

$$U_H = \frac{1}{\sqrt{2}} \begin{pmatrix} 1 & 1 \\ 1 & -1 \end{pmatrix}$$

and I be the identity operator acting in $\ell_2(\mathbb{Z})$. Let $| \uparrow \rangle$, $| \downarrow \rangle$ be the standard basis in $\mathbb{C}^2$

$$| \uparrow \rangle = \begin{pmatrix} 1 \\ 0 \end{pmatrix}, \quad | \downarrow \rangle = \begin{pmatrix} 0 \\ 1 \end{pmatrix}.$$

The *shift operator* is defined as

$$S = \left(\sum_{j \in \mathbb{Z}} |j+1\rangle\langle j| \right) \otimes (| \uparrow \rangle\langle \uparrow |) + \left(\sum_{j \in \mathbb{Z}} |j-1\rangle\langle j| \right) \otimes (| \downarrow \rangle\langle \downarrow |)$$

acting in the Hilbert space $\ell_2(\mathbb{Z}) \otimes \mathbb{C}^2$. Then

$$T = S(I \otimes U_H)$$

is a unitary operator. Consider the initial state

$$|\psi_0\rangle = |0\rangle \otimes |\uparrow\rangle.$$

Find $T|\psi_0\rangle$ and $\langle\psi_0|T|\psi_0\rangle$.

Solution 30. We have

$$T|\psi_0\rangle = (S(I \otimes U_H))(|0\rangle \otimes |\uparrow\rangle) = S(I|0\rangle \otimes U_H|\uparrow\rangle)$$

$$= S\left(|0\rangle \otimes \frac{1}{\sqrt{2}}\begin{pmatrix}1\\1\end{pmatrix}\right)$$

$$= |1\rangle \otimes \frac{1}{\sqrt{2}}\begin{pmatrix}1\\0\end{pmatrix} + |-1\rangle \otimes \frac{1}{\sqrt{2}}\begin{pmatrix}0\\1\end{pmatrix}.$$

Then with $\langle\psi_0| = \langle0| \otimes (1\ 0)$ we obtain $\langle\psi_0|T|\psi_0\rangle = 0$.

Problem 31. Consider the *Banach-Gelfand triple*

$$\ell_1(\mathbb{N}_0) \subset \ell_2(\mathbb{N}_0) \subset \ell_\infty(\mathbb{N}_0)$$

and the infinite dimensional linear equation

$$\begin{pmatrix} 0 & 1/2 & 0 & 0 & 0 & 0 & 0 & 0 & \cdots \\ 1/2 & 0 & 2/3 & 0 & 0 & 0 & 0 & 0 & \cdots \\ 0 & 2/3 & 0 & 3/4 & 0 & 0 & 0 & 0 & \cdots \\ 0 & 0 & 3/4 & 0 & 4/5 & 0 & 0 & 0 & \cdots \\ 0 & 0 & 0 & 4/5 & 0 & 5/6 & 0 & 0 & \cdots \\ 0 & 0 & 0 & 0 & 5/6 & 0 & 6/7 & 0 & \cdots \\ 0 & 0 & 0 & 0 & 0 & 6/7 & 0 & 7/8 & \cdots \\ \vdots & \vdots & \vdots & \vdots & \vdots & \vdots & \vdots & \vdots & \ddots \end{pmatrix}\begin{pmatrix}t_0\\t_1\\t_2\\t_3\\t_4\\t_5\\t_6\\\vdots\end{pmatrix} = \begin{pmatrix}0\\0\\0\\0\\0\\0\\0\\0\\\vdots\end{pmatrix}.$$

The infinite dimensional matrix represents a linear bounded self-adjoint operator. Find all solutions of the linear equation with $t_0 \neq 0$. Is the vector **t** an element of $\ell_1(\mathbb{N}_0)$? Is the vector **t** an element of $\ell_2(\mathbb{N}_0)$? Is the vector **t** an element of $\ell_\infty(\mathbb{N}_0)$?

Solution 31. Obviously $t_1 = t_3 = t_5 = \cdots = 0$ and

$$\frac{1}{2}t_0 + \frac{2}{3}t_2 = 0, \quad \frac{3}{4}t_2 + \frac{4}{5}t_4 = 0, \quad \frac{5}{6}t_4 + \frac{6}{7}t_6 = 0, \quad \frac{n+1}{n+2}t_n + \frac{n+2}{n+3}t_{n+2} = 0$$

for n even ($n = 0, 2, 4, \dots$). Note that

$$\frac{(n+3)(n+1)}{(n+2)^2} < 1.$$

So the vector **t** is an element of the Banach space $\ell_\infty(\mathbb{N}_0)$, but not of the spaces $\ell_1(\mathbb{N}_0)$ and $\ell_2(\mathbb{N}_0)$. Furthermore 0 is an eigenvalue of the operator.

Problem 32. Consider the Banach-Gelfand triple

$$\ell_1(\mathbb{N}_0) \subset \ell_2(\mathbb{N}_0) \subset \ell_\infty(\mathbb{N}_0)$$

and the infinite dimensional matrices

$$A = \begin{pmatrix} 0 & 1 & 0 & 0 & 0 & 0 & 0 & \ldots \\ 1 & 0 & 1 & 0 & 0 & 0 & 0 & \ldots \\ 0 & 1 & 0 & 1 & 0 & 0 & 0 & \ldots \\ 0 & 0 & 1 & 0 & 1 & 0 & 0 & \ldots \\ 0 & 0 & 0 & 1 & 0 & 1 & 0 & \ldots \\ 0 & 0 & 0 & 0 & 1 & 0 & 1 & \ldots \\ 0 & 0 & 0 & 0 & 0 & 1 & 0 & \ldots \\ \vdots & \vdots & \vdots & \vdots & \vdots & \vdots & \vdots & \ddots \end{pmatrix}$$

$$B = \begin{pmatrix} 0 & 1 & 0 & 0 & 0 & 0 & 0 & \ldots \\ 1 & 0 & 1/\sqrt{2} & 0 & 0 & 0 & 0 & \ldots \\ 0 & 1/\sqrt{2} & 0 & 1/\sqrt{3} & 0 & 0 & 0 & \ldots \\ 0 & 0 & 1/\sqrt{3} & 0 & 1/\sqrt{4} & 0 & 0 & \ldots \\ 0 & 0 & 0 & 1/\sqrt{4} & 0 & 1/\sqrt{5} & 0 & \ldots \\ 0 & 0 & 0 & 0 & 1/\sqrt{5} & 0 & 1/\sqrt{6} & \ldots \\ 0 & 0 & 0 & 0 & 0 & 1/\sqrt{6} & 0 & \ldots \\ \vdots & \vdots & \vdots & \vdots & \vdots & \vdots & \vdots & \ddots \end{pmatrix}$$

$$C = \begin{pmatrix} 0 & 1 & 0 & 0 & 0 & 0 & 0 & \ldots \\ 1 & 0 & 1/2 & 0 & 0 & 0 & 0 & \ldots \\ 0 & 1/2 & 0 & 1/3 & 0 & 0 & 0 & \ldots \\ 0 & 0 & 1/3 & 0 & 1/4 & 0 & 0 & \ldots \\ 0 & 0 & 0 & 1/4 & 0 & 1/5 & 0 & \ldots \\ 0 & 0 & 0 & 0 & 1/5 & 0 & 1/6 & \ldots \\ 0 & 0 & 0 & 0 & 0 & 1/6 & 0 & \ldots \\ \vdots & \vdots & \vdots & \vdots & \vdots & \vdots & \vdots & \ddots \end{pmatrix}$$

$$D = \begin{pmatrix} 0 & 1 & 0 & 0 & 0 & 0 & 0 & \cdots \\ 1 & 0 & 1/2! & 0 & 0 & 0 & 0 & \cdots \\ 0 & 1/2! & 0 & 1/3! & 0 & 0 & 0 & \cdots \\ 0 & 0 & 1/3! & 0 & 1/4! & 0 & 0 & \cdots \\ 0 & 0 & 0 & 1/4! & 0 & 1/5! & 0 & \cdots \\ 0 & 0 & 0 & 0 & 1/5! & 0 & 1/6! & \cdots \\ 0 & 0 & 0 & 0 & 0 & 1/6! & 0 & \cdots \\ \vdots & \vdots & \vdots & \vdots & \vdots & \vdots & \vdots & \ddots \end{pmatrix}.$$

These matrices are bounded self-adjoint operators with trace equal to 0. Find

$$\text{tr}(A^2), \quad \text{tr}(B^2), \quad \text{tr}(C^2), \quad \text{tr}(D^2).$$

Solution 32. We obtain

$$\text{tr}(A^2) = 1 + 2\sum_{j=2}^{\infty} 1 = \infty = \sum_{j=0}^{\infty} \lambda_j^2$$

$$\text{tr}(B^2) = 2 + 2\sum_{j=2}^{\infty} \frac{1}{j} = \infty = \sum_{j=0}^{\infty} \lambda_j^2$$

$$\text{tr}(C^2) = 2 + 2\sum_{j=2}^{\infty} \frac{1}{j^2} < \infty = \sum_{j=0}^{\infty} \lambda_j^2$$

$$\text{tr}(D^2) = 2 + 2\sum_{j=2}^{\infty} \left(\frac{1}{j!} \cdot \frac{1}{j!} \right) < \infty = \sum_{j=0}^{\infty} \lambda_j^2.$$

Problem 33. Consider the Hilbert spaces $\ell_2(\mathbb{Z})$, $L_2([-\pi, \pi])$ and the linear bounded self-adjoint operator T defined by

$$Tf := xf(x), \qquad f \in L_2([-\pi, \pi])$$

in the Hilbert space $L_2([-\pi, \pi])$. In the Hilbert space $L_2([-\pi, \pi])$ an orthonormal basis $\mathcal{B}$ is given by

$$\mathcal{B} := \left\{ \phi_k(x) := \frac{1}{\sqrt{2\pi}} \exp(ikx), \quad k \in \mathbb{Z} \right\}.$$

This means that any $f \in L_2([-\pi, \pi])$ can be expanded with respect to this basis. Note that if $f \in L_2([-\pi, \pi])$, then xf is also an element of

$L_2([-\pi, \pi])$. Find the matrix representation of the operator T.

Solution 33. Let $k, l \in \mathbb{Z}$. For $k = l$ we have

$$T_{k,k} = \langle \phi_k, x\phi_k \rangle = \frac{1}{2\pi} \int_{-\pi}^{\pi} e^{ikx} x e^{-ikx} dx = \frac{1}{2\pi} \int_{-\pi}^{\pi} x \, dx = 0$$

where $\langle \, , \, \rangle$ denotes the scalar product in the Hilbert space $L_2([-\pi, \pi])$. Thus the diagonal elements are 0. For $k \neq l$ we find that

$$T_{k,l} = \frac{1}{2\pi} \int_{-\pi}^{\pi} e^{ikx} x e^{-ilx} dx = \frac{1}{2\pi} \int_{-\pi}^{\pi} x e^{i(k-l)x} dx = \frac{i}{(l-k)\cos((k+l)\pi)}.$$

Introducing the ordering $0, 1, -1, 2, -2, \ldots$ we find that the self-adjoint infinite dimensional matrix T has the form

$$T = \begin{pmatrix} 0 & -i & i & i/2 & \cdots \\ i & 0 & -i/2 & i/2 & \cdots \\ -i & i/2 & 0 & -i/3 & \cdots \\ -i/2 & -i/2 & i/3 & & \ddots \\ & & & \ddots & \end{pmatrix}.$$

The spectrum of the linear operator T is continuous. If we truncate the infinite dimensional matrix T at finite level (say $n \times n$ matrix), we find that the trace is this finite dimensional matrix is 0. Therefore

$$\lambda_1 + \lambda_2 + \cdots + \lambda_n = 0$$

where $\lambda_1, \ldots, \lambda_n$ denote the real eigenvalues of this finite-dimensional hermitian matrix.

5.3 Programming Problems

Problem 1. Let $\mathcal{H}$ be a separable Hilbert space and $\{e_k : k \in \mathbb{I}\}$ be an orthonormal basis of $\mathcal{H}$, where $\mathbb{I}$ is a countable index set. Let $v \in \mathcal{H}$ one defines (*Parseval isomorphism*) the expansion coefficients

$$v_k = \langle v, e_k \rangle, \quad k \in \mathbb{I}.$$

The map $v \mapsto \hat{v}$ is an isomorphism of the Hilbert spaces $\mathcal{H} \mapsto \ell_2(\mathbb{I})$. Consider the Hilbert space $L_2([0,1])$ and the $v \in L_2([0,1])$ given by

$$v(x) = \begin{cases} 2x & \text{for } 0 \le x \le 1/2 \\ 2(1-x) & \text{for } 1/2 \le x \le 1 \end{cases}.$$

An orthonormal basis in the Hilbert space $L_2([0,1])$ is given by

$$B := \{\, e_k = \exp(2\pi i k x) \, : \, k \in \mathbb{Z} \,\}.$$

Hence $\mathbb{I} = \mathbb{Z}$. Find the expansion coefficients $v_k = \langle v, e_k \rangle$. The function v can be reconstructed from

$$v = \sum_{k \in \mathbb{Z}} v_k e_k.$$

Solution 1. Applying the Maxima program

```
/* hilbert1.mac */
f1: 2*x; f2: 2*(1-x);
/* for k=0 */
a0: integrate(f1,x,0,1/2) + integrate(f2,x,1/2,1);
/* for k \ne 0 */
r1: integrate(f1*exp(-2*%i*%pi*k*x),x,0,1/2);
r2: integrate(f2*exp(-2*%i*%pi*k*x),x,1/2,1);
ak: r1 + r2; ak: trigsimp(ak);
```

and $\exp(2ik) = \exp(-2i\pi k) = 1$ we have $(k \in \mathbb{Z} \setminus \{0\})$

$$\langle v, e_k \rangle = \int_0^1 v(x) \exp(-2\pi i k x) dx$$

$$= \int_0^{1/2} 2x \exp(-2\pi i k x) dx + \int_{1/2}^1 (2(1-x) \exp(-2\pi i k x) dx$$

$$= \frac{e^{i\pi k} - 1}{\pi^2 k^2} = \frac{\cos(\pi k) - 1}{\pi^2 k^2}$$

and for $k = 0$ we have $\langle v, e_0 \rangle = 1/2$. For example $\langle v, e_1 \rangle = v_1 = -1/\pi^2$, $\langle v, e_{-1} \rangle = -1/\pi^2$.

Problem 2. Let $b > a$. Consider the Hilbert space $L_2([a,b])$ and the functions

$$\phi_n(x) := \sqrt{\frac{2}{b-a}} \sin\left(n\pi \frac{x-a}{b-a}\right), \quad n = 1, 2, \dots$$

which form an orthonormal basis in $L_2([a,b])$. Find

$$\langle \phi_m(x), x\phi_n(x) \rangle \equiv \int_a^b \phi_m(x) x \phi_n(x) dx, \quad m, n = 1, 2, \dots$$

i.e. find the matrix representation of x acting in the Hilbert space $\ell_2(\mathbb{N})$.

Solution 2. We have to do a case study with $m = n$ and $m \neq n$. We set $\gamma = \pi/(b-a)$ and $d = b - a$. For $m = n$ we have

$$\langle \phi_m(x), x\phi_m(x) \rangle = \frac{d + 2a}{2} = \frac{b+a}{2}.$$

For $m \neq n$ and setting $d = b - a$, $n = r + m$ we obtain

$$\langle \phi_m(x)), x\phi_n(x) \rangle = \frac{4d((m^2 + mr)(-1)^r - m^2 - mr)}{\pi^2 r^2 (r^2 + 4m + 4mr)}$$

$$= \frac{4d(m \cdot n((-1)^{n-m} - 1))}{\pi^2(n-m)^2(n+m)^2}.$$

with $m^2 + mr = m \cdot n$. The operator $\hat{x}$ is bounded and self-adjoint.

```
/* matrixrep.mac */
declare(m,integer); declare(n,integer); declare(r,integer);
I1: integrate(sqrt(2/d)*sin(m*g*(x-a))*x*sqrt(2/d)*sin(m*g*(x-a))),
x,a,a+d);
I1: subst(%pi/d,g,I1);
I1: ratsimp(I1); I1: trigsimp(I1);
Sm: sqrt(2/d)*(exp(%i*m*g*x)*exp(-%i*m*g*a)
-exp(-%i*m*g*x)*exp(%i*m*g*a))/(2*%i);
Sn: sqrt(2/d)*(exp(%i*(m+r)*g*x)*exp(-%i*(m+r)*g*a)
-exp(-%i*(m+r)*g*x)*exp(%i*(m+r)*g*a))/(2*%i);
I2: integrate(Sm*x*Sn,x,a,a+d);
I2: subst(%pi/d,g,I2); I2: trigsimp(I2);
I2: subst(1,exp(2*%i*%pi*m),I2);
I2: subst(1,exp(4*%i*%pi*m),I2);
I2: subst(1,exp(2*%i*%pi*r),I2);
I2: subst(1,exp(4*%i*%pi*r),I2);
I2: expand(I2); I2: ratsimp(I2);
I2: subst(cos(%pi*r),exp(%i*%pi*r),I2);
I2: subst(cos(%pi*r),exp(%i*%pi*r),I2);
I2: ratsimp(I2);
```

5.4 Supplementary Problems

Problem 1. Let p_j $(j = 1, 2, \dots)$ be the prime numbers

$$p_1 = 2, \quad p_2 = 3, \quad p_3 = 5,, \ p_4 = 7, \quad p_5 = 11, \dots .$$

Is

$$(1/\sqrt{2}, 1/\sqrt{3}, 1/\sqrt{5}, 1/\sqrt{7}, 1/\sqrt{11}, \dots)$$

an element of $\ell_2(\mathbb{N})$?

Problem 2. Let M be any $n \times n$ matrix. Let $\mathbf{x} = (x_1, x_2, \dots)^T$. The linear operator A is defined by

$$A\mathbf{x} = (w_1, w_2, \dots)^T$$

where

$$w_j = \sum_{k=1}^n M_{jk} x_k, \qquad j = 1, 2, \dots, n$$

$$w_j = x_j, \qquad j > n$$

and $\mathcal{D}(A) = \ell_2(\mathbb{N})$. Show that A is self-adjoint if the $n \times n$ matrix M is hermitian. Show that A is unitary if M is unitary.

Problem 3. Consider the Hilbert space $\ell_2(\mathbb{N}_0)$ of square summable sequences $u = (u_j)_{j=0}^\infty$ and the action of a difference expression

$$(Du)_j := a_{j-1} u_{j-1} + b_j u_j + a_j u_{j+1}, \quad a_j > 0, \quad b_j \in \mathbb{R}$$

where $j = 0, 1, 2, \dots$ and $a_{-1} = 0$. Show that if the (real) coefficients a_j and b_j are bounded, then D generates in the Hilbert space $\ell_2(\mathbb{N}_0)$ a bounded self-adjoint operator $\hat{D}$

$$(\hat{D}u)_j = (Du)_j, \quad u \in \ell_2(\mathbb{N}_0), \quad j = 0, 1, 2, \dots, \quad u_{-1} = 0.$$

These infinite dimensional matrices are called *Jacobi matrices*.

Problem 4. Consider the Hilbert space $\ell_2(\mathbb{N})$ and the linear operator T defined by

$$T : (x_1, x_2, x_3, \dots) \mapsto (0, 0, x_3, x_4, \dots).$$

Is T bounded? Is T self-adjoint? If so is T positive?

Problem 5. Let $\mathbb{Z}$ be the set of integers and $n, m \in \mathbb{Z}$. We consider the Hilbert space $\ell_2(\mathbb{Z})$ and denotes the standard basis by $|n\rangle$ $(n \in \mathbb{Z})$ and the dual one by $\langle m|$ $(m \in \mathbb{Z})$ with $\langle m|n\rangle = \delta_{m,n}$. Let $\mathbb{C}^2$ be the two-dimensional Hilbert space with the standard basis

$$|+\rangle = \begin{pmatrix} 1 \\ 0 \end{pmatrix}, \quad |-\rangle = \begin{pmatrix} 0 \\ 1 \end{pmatrix}, \quad \langle +| = (1\ 0), \quad \langle -| = (0\ 1).$$

We consider the product Hilbert space $\ell_2(\mathbb{Z}) \otimes \mathbb{C}^2$ and define the linear operator $\hat{S}$ acting on $\ell_2(\mathbb{Z}) \otimes \mathbb{C}^2$ as

$$\hat{S}(|n\rangle \otimes |\pm\rangle) = |n \pm 1\rangle \otimes |\pm\rangle$$

and the unitary operator $U(\theta_n)$ $(\theta_n \in \mathbb{R}), n \in \mathbb{Z}$ defined by

$$\hat{U}(\theta_n) = \sum_{n \in \mathbb{Z}} ((|n\rangle\langle n|) \otimes ((\cos(\theta_n)|+\rangle\langle +|) + (\sin(\theta_n)|+\rangle\langle -|)$$
$$+ (\sin(\theta_n)|-\rangle\langle +|) - (\cos(\theta_n)|-\rangle\langle -|)).$$

The underlying matrix is

$$\begin{pmatrix} \cos(\theta_n) & \sin(\theta_n) \\ \sin(\theta_n) & -\cos(\theta_n) \end{pmatrix}$$

with determinant -1. Setting $\theta_n = \pi/4$ for all $n \in \mathbb{Z}$ we obtain the Hadamard matrix. We can consider now the linear operator $\hat{S}\hat{U}(\theta_n)$. Assume that θ_n does not depend on n and we set $\theta_n = \theta$. Consider the normalized state $|0\rangle \otimes |+\rangle$ in the Hilbert space $\ell_2(\mathbb{Z}) \otimes \mathbb{C}^2$. Find the state $(\hat{S}\hat{U}(\theta))(|0\rangle \otimes |+\rangle)$ and

$$(\langle 0| \otimes \langle +|)(\hat{S}\hat{U}(\theta))(|0\rangle \otimes |+\rangle)), \quad (\langle 0| \otimes \langle -|)(\hat{S}\hat{U}(\theta))(|0\rangle \otimes |+\rangle)).$$

Problem 6. Let $\mathbf{u}, \mathbf{v} \in \ell_2(\mathbb{Z})$. Is the sequence $\mathbf{w}$ defined by (*convolution*)

$$w_n = \mathbf{u} \star \mathbf{v} = \sum_{k \in \mathbb{Z}} u_k v_{n-k}, \quad n \in \mathbb{Z}$$

an element of $\ell_2(\mathbb{Z})$?

Problem 7. If T is a self-adjoint, unbounded operator, then for every $\tau \in \mathbb{R}$ one has a well-defined bounded operator $\exp(i\tau T)$ with

$$\exp(i\tau_1 T)\exp(i\tau_2 T) = \exp(i(\tau_1 + \tau_2)T)$$

and $(\exp(i\tau T))^* = \exp(-i\tau T)$.

Problem 8. Given $u_0 \in \mathbb{C}$, $z \in \mathbb{C}$ and $a \in \mathbb{C} \setminus \{0\}$. Find the solution of the initial value problem of the difference equation

$$u_{\tau+1} = au_\tau + \delta_{-1,\tau} z, \quad \tau \in \mathbb{Z}$$

where $\delta_{-1,\tau}$ is the Kronecker delta. Note that

$$u_1 = au_0, \quad u_0 = au_{-1} + z \quad \Rightarrow \quad u_{-1} = \frac{1}{a}(u_0 - z).$$

What are the conditions on u_0, z, a such that (u_τ) $(\tau \in \mathbb{Z})$ is an element of the Hilbert space $\ell_2(\mathbb{Z})$?

Problem 9. Let $\alpha \in \mathbb{R}$. Consider the infinite dimensional vector (counting from 0)

$$(\cos(\alpha), \sin(\alpha), \cos(\alpha), \sin(\alpha), \cos(\alpha), \dots)$$

Is the vector an element of $\ell_1(\mathbb{N}_0)$? Is the vector an element of $\ell_2(\mathbb{N}_0)$? Is the vector an element of $\ell_\infty(\mathbb{N}_0)$?

Problem 10. Consider the Hilbert space $\ell_2(\mathbb{N})$ and the linear operator S defined by

$$S(x_1, x_2, x_3, \dots)^T := (0, x_1, x_2, x_3, \dots)^T.$$

Show that 0 is in the residual spectrum of S. Find the spectrum of the operator (3×3) in the Hilbert space $\mathbb{R}^3$

$$S_3 = \begin{pmatrix} 0 & 0 & 0 \\ 1 & 0 & 0 \\ 0 & 1 & 0 \end{pmatrix}.$$

Problem 11. Consider the sequence $\mathbf{x} = \{x_n\}_{n=0}^\infty$ and the Banach space

$$\ell_\infty(\mathbb{N}_0) := \{ \mathbf{x} : x_n \in \mathbb{R} \; \forall n \in \mathbb{N}_0 \; \|\mathbf{x}\|_\infty = \sup_{n \in \mathbb{N}_0} |x_n| < \infty \}.$$

Define the map $T : \ell_\infty(\mathbb{N}_0) \mapsto \ell_\infty(\mathbb{N}_0)$

$$(T\mathbf{x})_n := x_n - x_{n-1} \quad \text{for } n > 0.$$

Let $\mathbf{y} = (1, 1, \dots, 1, \dots) \in \ell_\infty(\mathbb{N}_0)$. Show that the equation $T\mathbf{x} = \mathbf{y}$ has no solution $\mathbf{x}$ in $\ell_\infty(\mathbb{N}_0)$.

Problem 12. Consider the Hilbert space $\ell_2(\mathbb{N}_0)$ and the linear bounded operator $T : \ell_2(\mathbb{N}_0) \mapsto \ell_2(\mathbb{N}_0)$

$$T(x_0, x_1, x_2, \dots)^T = (x_1, x_2, x_3, \dots)^T.$$

Show that

$$T^*(x_0, x_1, x_2, \dots)^T = (0, x_0, x_1, \dots)^T.$$

Problem 13. Let $n \in \mathbb{N}$. Consider the sequence

$$x_n = \frac{2^n n!}{n^n}.$$

We know that

$$\lim_{n \to \infty} \frac{2^n n!}{n^n} = 0.$$

Is $(x_1, x_2, x_3, \dots)$ an element of $\ell_2(\mathbb{N})$?

Problem 14. Consider a separable Hilbert space $\mathcal{H}$ with the orthonormal basis $\{e_j\}_{j=1}^\infty$. Show that the unilateral shift operator defined by

$$T : \ Te_j = e_{j+1}$$

is an *isometry* from $\mathcal{H}$ onto $\mathcal{H}\theta\{e_1\}$. Show that its adjoint

$$T^* : \ T^*e_j = e_{j-1} \text{ for } j > 1 \text{ and } T^*e_1 = 0$$

is a *partial isometry* from $\mathcal{H}\theta\{e_1\}$ onto $\mathcal{H}$. Show that the spectrum of T is the closed unit disk in the complex plane and it is purely continuous except for the points

$$\{ z \ : \ |z| < 1 \}$$

which belong to the residual spectrum. Show that every point z in the interior of the unit disk is a proper value of the operator T^* with the corresponding proper element

$$f(z) = (1 - |z|^2)^{1/2} \sum_{j=1}^\infty z^{j-1} e_j.$$

Show that the points on the circumference of the unit disk belong to the continuous spectrum of T^* and the residual spectrum is empty.

Problem 15. (i) Consider the Hilbert space $\ell_2(\mathbb{N})$ and the set

$$D := \left\{ (u_1, u_2, u_3, \dots)^T \ : \ (u_1, u_2, u_3, \dots)^T \text{ and } (u_1, 2u_2, 3u_3, 4u_4, \dots)^T \in \ell_2(\mathbb{N}) \right\}.$$

Is D dense in $\ell_2(\mathbb{N})$?

(ii) Show that the set of all $\mathbf{u} = (u_1, u_2, u_3, \dots)^T$ in the Hilbert space $\ell_2(\mathbb{N})$ with only finitely many nonzero components u_j is dense in $\ell_2(\mathbb{N})$.

Problem 16. Let $a > 0$. Consider the Hilbert space $L_2([0, a])$ and the Hilbert space $\ell_2(\mathbb{Z})$. An orthonormal basis B in $L_2([0, a])$ is given by

$$B = \{\, \phi_n(x) = \frac{1}{\sqrt{a}} \exp(2\pi ixn/a) \,:\, n \in \mathbb{Z} \,\}.$$

Let $f(x) = x$. Then $f \in L_2([0, a])$. Show that

$$(\cdots, c_{-2}, c_{-1}, c_0, c_1, c_2, \cdots)$$

with $c_j = \langle f(x), \phi_j(x) \rangle$ is an element of $\ell_2(\mathbb{Z})$.

Problem 17. (i) Show that $\ell_1(\mathbb{N}_0)$ is dense in $\ell_2(\mathbb{N}_0)$.
(ii) Show that $\ell_2(\mathbb{N}_0)$ is dense in $\ell_\infty(\mathbb{N}_0)$ in the sense of weak-convergence.

Problem 18. Let $m_0 \in \mathbb{N}$. Consider the Hilbert space $\ell_2(\mathbb{Z})$ and the sets

$$V_1 := \left\{ v = (v_m)_{m \in \mathbb{Z}} \in \ell_2(\mathbb{Z}) : v_m = \begin{cases} u_m & \text{for } |m| \leq m_0 \\ 0 & \text{for } |m| > m_0 \end{cases} u = (u_m)_{m \in \mathbb{Z}} \in \ell_2(\mathbb{Z}) \right\}$$

$$V_2 := \left\{ w = (w_m)_{m \in \mathbb{Z}} \in \ell_2(\mathbb{Z}) : w_m = \begin{cases} 0 & \text{for } |m| \leq m_0 \\ u_m & \text{for } |m| > m_0 \end{cases} u = (u_m)_{m \in \mathbb{Z}} \in \ell_2(\mathbb{Z}) \right\}$$

Show that V_1 and V_2 are two subspaces of the Hilbert space $\ell_2(\mathbb{Z})$ and $\dim(V_1) = 2m_0 + 1 < \infty$. Show that

$$V_1 \perp V_2 \quad \text{and} \quad \ell_2(\mathbb{Z}) = V_1 \oplus V_2.$$

Show that for any $u \in \ell_2(\mathbb{Z})$ there exist unique $v \in V_1$, $w \in V_2$ such that

$$u = v + w, \qquad u_m = v_m + w_m, \quad m \in \mathbb{Z}.$$

Show that one can define the linear bounded operator $\Pi : \ell_2(\mathbb{Z}) \to V_1$ by

$$\Pi u = v, \quad (I - \Pi)u = w \quad \text{for all } u \in \ell_2(\mathbb{Z})$$

with I the identity operator acting in the Hilbert space $\ell_2(\mathbb{Z})$.

Problem 19. Let $n \in \mathbb{N}$.
(i) Solve the equation

$$\frac{1}{n(n+1)} = \frac{x_0}{n} + \frac{x_1}{n+1}.$$

(ii) Solve the equation

$$\frac{1}{n(n+1)(n+2)} = \frac{x_0}{n} + \frac{x_1}{n+1} + \frac{x_2}{n+2}.$$

(iii) Extend to arbitrary n and $n \to \infty$. Is the sequence $(x_0, x_1, x_2, \dots)$ an element of $\ell_1(\mathbb{N}_0)$? Is the sequence $(x_0, x_1, x_2, \dots)$ an element of $\ell_2(\mathbb{N}_0)$? Is the sequence $(x_0, x_1, x_2, \dots)$ an element of $\ell_\infty(\mathbb{N}_0)$?

Problem 20. Let $n \geq 2$ and v be a normalized (column) vector in the Hilbert space $\mathbb{C}^n$. What can be said about the $n \times n$ matrix

$$T_n := I_n - 2vv^*$$

with respect to the eigenvalues, eigenvectors and other properties of the matrix. Then study the case $n \to \infty$. Is the operator V_∞ bounded? Is the operator V_∞ of the trace class?

Problem 21. Consider the infinite dimensional separable Hilbert space $\ell_2(\mathbb{N})$. Let T be the infinite dimensional (hermitian) *Jacobi matrix* acting in $\ell_2(\mathbb{N})$. The (hermitian) Jacobi matrix is defined by

$$(T)_{k,k+1} = (T)_{k+1,k} = t_k(0) \quad k \in \mathbb{N}$$

$$(T)_{j,k} = 0 \quad \text{(otherwise)}$$

with $\{\, t_k(0) > 0 : k \in \mathbb{N} \,\}$ satisfying $t_k(0) \to 0$ for $k \to \infty$. Show that T is a compact (=completely continuous) self-adjoint operator acting on the separable Hilbert space $\ell_2(\mathbb{N})$.

Problem 22. Let $n \geq 3$. Consider the hermitian matrices with trace equal to 0

$$A_3 = \begin{pmatrix} 0 & 1 & 1/2 \\ 1 & 0 & 0 \\ 1/2 & 0 & 0 \end{pmatrix}, \quad A_4 = \begin{pmatrix} 0 & 1 & 1/2 & 1/3 \\ 1 & 0 & 0 & 0 \\ 1/2 & 0 & 0 & 0 \\ 1/3 & 0 & 0 & 0 \end{pmatrix},$$

$$A_5 = \begin{pmatrix} 0 & 1 & 1/2 & 1/3 & 1/4 \\ 1 & 0 & 0 & 0 & 0 \\ 1/2 & 0 & 0 & 0 & 0 \\ 1/3 & 0 & 0 & 0 & 0 \\ 1/4 & 0 & 0 & 0 & 0 \end{pmatrix}, \quad A_6 = \begin{pmatrix} 0 & 1 & 1/2 & 1/3 & 1/4 & 1/5 \\ 1 & 0 & 0 & 0 & 0 & 0 \\ 1/2 & 0 & 0 & 0 & 0 & 0 \\ 1/3 & 0 & 0 & 0 & 0 & 0 \\ 1/4 & 0 & 0 & 0 & 0 & 0 \\ 1/5 & 0 & 0 & 0 & 0 & 0 \end{pmatrix}.$$

Find the eigenvalues for these matrices. Extend to arbitrary n. Show that $n-2$ eigenvalues are equal to 0. So show that we have a three level system. Find $\mathrm{tr}(A_n^2)$ and show that

$$\lim_{n\to\infty} \mathrm{tr}(A_n^2) < \infty$$

and thus that A_n $(n \to \infty)$ is a trace class operator.

Let $\tau > 0$. Suppose the linear operator T_τ is of trace class for all $\tau > 0$. Then there exists a complete orthonormal set $\{ e_j \}_{j=1}^\infty$ in the Hilbert space $\mathcal{H}$ and a non-decreasing sequence $\{ \lambda_j \}_{j=1}^\infty$ of real numbers such that

$$T_\tau e_j = e^{-\lambda_j \tau} e_j, \quad \sum_{j=1}^\infty e^{-\lambda_j \tau} < \infty$$

for all $j \in \mathbb{N}$ and all $\tau > 0$. The linear operator T_τ has a square integrable kernel K_τ given by

$$K_\tau(x_1, x_2) := \sum_{j=1}^\infty e^{-\lambda_j \tau} e_j(x_1) \bar{e}_j(x_2).$$

Problem 23. Let $0 < a < 1$. Find the spectrum of the 2×2, 3×3, 4×4 matrices

$$A_2 = \begin{pmatrix} a & a \\ 1 & 0 \end{pmatrix}, \quad A_3 = \begin{pmatrix} a & a & a \\ 1 & 0 & 0 \\ 0 & 1 & 0 \end{pmatrix}, \quad A_4 = \begin{pmatrix} a & a & a & a \\ 1 & 0 & 0 & 0 \\ 0 & 1 & 0 & 0 \\ 0 & 0 & 1 & 0 \end{pmatrix}.$$

Find $\mathrm{tr}(A_2^2)$, $\mathrm{tr}(A_3^2)$, $\mathrm{tr}(A_4^2)$. Find the spectrum of the infinite dimensional matrix

$$A_\infty = \begin{pmatrix} a & a & a & a & a & \ldots \\ 1 & 0 & 0 & 0 & 0 & \ldots \\ 0 & 1 & 0 & 0 & 0 & \ldots \\ 0 & 0 & 1 & 0 & 0 & \ldots \\ 0 & 0 & 0 & 1 & 0 & \ldots \\ \vdots & \vdots & \vdots & \vdots & \vdots & \ddots \end{pmatrix}.$$

Is $\mathrm{tr}(A_\infty^2) < \infty$?

Problem 24. Consider the symmetric 3×3, 5×5, 7×7 matrices over $\mathbb{R}$

$$A_3 = \begin{pmatrix} 1 & 0 & 1 \\ 0 & 1 & 0 \\ 1 & 0 & 1 \end{pmatrix}, \quad A_5 = \begin{pmatrix} 1 & 0 & 0 & 0 & 1 \\ 0 & 1 & 0 & 0 & 0 \\ 0 & 0 & 1 & 0 & 0 \\ 0 & 0 & 0 & 1 & 0 \\ 1 & 0 & 0 & 0 & 1 \end{pmatrix},$$

$$A_7 = \begin{pmatrix} 1 & 0 & 0 & 0 & 0 & 0 & 1 \\ 0 & 1 & 0 & 0 & 0 & 0 & 0 \\ 0 & 0 & 1 & 0 & 0 & 0 & 0 \\ 0 & 0 & 0 & 1 & 0 & 0 & 0 \\ 0 & 0 & 0 & 0 & 1 & 0 & 0 \\ 0 & 0 & 0 & 0 & 0 & 1 & 0 \\ 1 & 0 & 0 & 0 & 0 & 0 & 1 \end{pmatrix}.$$

Find the spectrum. Then study the general case n (n odd) and $n \to \infty$. Show that the operator is bounded.

Problem 25. Let $\{\phi_j\}_{j=0}^{\infty}$ be an arbitrary orthonormal basis in the Hilbert space $L_2(\Omega)$ and let the complex numbers $c_0,\ c_1,\ c_2,\ \ldots$ be such that

$$\sum_{j=0}^{\infty} |c_j|^2 < \infty$$

i.e. $(c_0, c_1, c_2, \ldots) \in \ell_2(\mathbb{N}_0)$. Then there exists a function $f \in L_2(\Omega)$ such that

$$c_j = \langle f, \phi_j \rangle \quad j = 0, 1, 2, \ldots$$

and $\|f\| = \sum_{j=0}^{\infty} |c_j|^2$. Consider the Hilbert space $L_2(\mathbb{R})$ with the orthonormal basis

$$\phi_j(x) = \frac{(-1)^j}{2^{j/2}\sqrt{j!}\pi^{1/4}} \exp(x^2/2) \frac{d^j}{dx^j} \exp(-x^2), \quad j = 0, 1, 2, \ldots$$

Let $c_0 = 1$, $c_1 = 1/2$, $c_2 = 1/3$, $\ldots$, $c_j = 1/(j+1)$, $\ldots$. Find the function f.

Problem 26. Consider the Hilbert space $\ell_2(\mathbb{Z})$ and the linear bounded operator $T = (T_{j,k})$ ($j, k \in \mathbb{Z}$) acting on $\ell_2(\mathbb{Z})$

$$T_{j,k} = \begin{cases} 1 & \text{for} & |j - k| = 1 \\ i & \text{for} & j = k > 0 \\ -i & \text{for} & j = k < 0 \\ 0 & \text{otherwise} \end{cases}$$

Show that 0 is an isolated eigenvalue of T. Show that the corresponding eigenvector is concentrated around the origin.

Problem 27. (i) Consider the Hilbert space $\mathbb{C}^3$ and the 3×3 matrix

$$A_3 = \begin{pmatrix} -i & 1 & 0 \\ 1 & 0 & 1 \\ 0 & 1 & i \end{pmatrix}$$

with trace equal to 0. We count the entries of the matrix with

$$(j, k) = (-1, -1), (-1, 0), (-1, +1), \ldots, (+1, +1).$$

Show that the eigenvalues of A_3 are $-1, 0, +1$. Find the normalized eigenvectors. Show that the normalized eigenvector for the eigenvalue $\lambda = 0$ is given by

$$\frac{1}{\sqrt{3}} \begin{pmatrix} 1 \\ i \\ -1 \end{pmatrix}.$$

(ii) Consider the Hilbert space $\mathbb{C}^5$ and the 5×5 matrix

$$A_5 = \begin{pmatrix} -i & 1 & 0 & 0 & 0 \\ 1 & -i & 1 & 0 & 0 \\ 0 & 1 & 0 & 1 & 0 \\ 0 & 0 & 1 & i & 1 \\ 0 & 0 & 0 & 1 & i \end{pmatrix}$$

with trace equal to zero. We count the matrix entries with

$$(j, k) = (-2, -2), (-2, -1), \ldots, (0, 0), \ldots, (+2, +2).$$

Show that the eigenvalues of A_5 are given by

$$-(3^{1/2}i + 1)^{1/2}, \quad +(3^{1/2}i + 1)^{1/2}, \quad -(1 - 3^{1/2}i)^{1/2}, \quad +(1 - 3^{1/2}i)^{1/2}, \quad 0.$$

Find the normalized eigenvectors of A_5. Show that the normalized eigenvector for the eigenvalue 0 is given by

$$\frac{1}{\sqrt{8}} \begin{pmatrix} 1 \\ i \\ -2 \\ -i \\ 1 \end{pmatrix}.$$

(iii) Extend to A_{2n+1} $(n \in \mathbb{N})$ and then $n \to \infty$.

Problem 28. The spectrum $\sigma(\hat{H})$ of a linear operator $\hat{H}$ is defined as the set of all λ for which the *resolvent*

$$R(\lambda) = (\lambda I - \hat{H})^{-1}$$

does not exist. If the linear operator $\hat{H}$ is self-adjoint, the spectrum is a subset of the real axis. The Lebesgue decomposition theorem states that

$$\sigma = \sigma_{pp} \cup \sigma_{ac} \cup \sigma_{sing}$$

where σ_{pp} is the countable union of points (the pure point spectrum), σ_{ac} is absolutely continuous with respect to Lebesgue measure and σ_{sing} is singular with respect to Lebesgue measure, i.e. it is supported on a set of measure zero. Consider the Hilbert space $\ell_2(\mathbb{Z})$ and the linear operator

$$\hat{H} = \cdots \otimes I_2 \otimes I_2 \otimes \sigma_3 \otimes \sigma_1 \otimes \sigma_3 \otimes I_2 \otimes I_2 \otimes \cdots$$

where the Pauli spin matrix σ_1 is at position 0. Find the spectrum of this linear operator.

Problem 29. Consider the linear operator T in the Hilbert space $L_2([0,1])$ defined by $Tf(x) := xf(x)$. Find the matrix elements $\langle p_i, Tp_j \rangle$ for $i, j = 0, 1, 2, 3$, where p_i are the (normalized) Legendre polynomials. Is the matrix T_{ij} symmetric?

Problem 30. (i) Consider the Hilbert space $L_2(\mathbb{R})$. Let T be the linear operator of pointwise multiplication on $L_2(\mathbb{R})$ given by

$$(Tf)x = xf(x) \quad \text{for} \quad f \in L_2(\mathbb{R}).$$

Find the matrix representation of operator T utilizing the basis given by

$$B = \left\{ \frac{(-1)^k}{2^{\frac{k}{2}}\sqrt{k!}\sqrt[4]{\pi}} e^{x^2/2} \frac{d^k}{dx^k} e^{-x^2} \ : \ k = 0, 1, 2, \ldots \right\}.$$

Find the spectrum of T.
(ii) Consider the Hilbert space $L_2(\mathbb{R})$. Find the matrix representation of $|x|$. Utilize as orthonormal basis

$$B = \left\{ \frac{(-1)^k}{2^{\frac{k}{2}}\sqrt{k!}\sqrt[4]{\pi}} e^{x^2/2} \frac{d^k}{dx^k} e^{-x^2} \ : \ k = 0, 1, 2, \ldots \right\}$$

in the Hilbert space $L_2(\mathbb{R})$.

Problem 31. Let $a > 0$ with dimension length. An orthonormal basis in the Hilbert space $L_2([-a, a])$ is given by

$$\{\, \phi_k(x) := \frac{1}{\sqrt{2a}} \exp(2\pi i k x/a) \,:\, k \in \mathbb{Z} \,\}.$$

(i) Find the expansion coefficients a_k $(k \in \mathbb{Z})$

$$a_k = \langle f, \phi \rangle = \int_{-a}^{a} f(x)\overline{\phi}_k(x)dx = \int_{-a}^{a} x \exp(-i2\pi k x)dx$$

with $f(x) = x$, i.e. $f \in L_2([-a, a])$.
(ii) Is a_k $(k \in \mathbb{Z})$ an element of $\ell_2(\mathbb{Z})$?
(iii) Let $g(x) = |x|$, i.e. $g \in L_2([-a, a])$. Find the expansion coefficients b_k $(k \in \mathbb{Z})$

$$b_k = \langle g, \phi_k \rangle = \int_{-a}^{a} g(x)\overline{\phi}_k(x)dx = \frac{1}{\sqrt{2a}} \int_{-a}^{a} |x| \exp(-i2\pi k x)dx.$$

Is (b_k) $(k \in \mathbb{Z})$ an element of $\ell_2(\mathbb{Z})$?
(iv) Find the matrix $C = (c_{j,k})$ $(j, k \in \mathbb{Z})$

$$c_{jk} = \langle \phi_j(x), g(x)\phi_k(x) \rangle, \quad j, k \in \mathbb{Z}.$$

Truncate to the 3×3 matrix with $j, k \in \{-1, 0, 1\}$ and find the eigenvalues and normalized eigenvalues. Truncate to the 5×5 matrix with $j, k \in \{-2, -1, 0, +1, +2\}$ and find the eigenvalues and normalized eigenvectors.

Problem 32. Let $\mathbb{N}_0$ be the natural numbers including 0. Consider the separable Hilbert space

$$\ell_2(\mathbb{N}_0) := \{\, \mathbf{x} = (x_0, x_1, \dots)^T \,:\, \sum_{j=0}^{\infty} |x_j|^2 < \infty \,\}$$

of the square summable complex sequences with the inner product $\langle \mathbf{x}, \mathbf{y} \rangle = \sum_{j=0}^{\infty} \overline{x}_j y_j$. An infinite tridiagonal matrix of the form

$$A = \begin{pmatrix} a_0 & b_0 & 0 & 0 & \cdots \\ b_0 & a_1 & b_1 & 0 & \cdots \\ 0 & b_1 & a_2 & b_2 & \cdots \\ 0 & 0 & b_2 & a_3 & \cdots \\ \vdots & \vdots & \vdots & \vdots & \ddots \end{pmatrix}$$

with a_j real, $b_j > 0$ and $j = 0, 1, 2, \dots$ is called a *Jacobi matrix*. Applied to a vector $\mathbf{x}$ in $\ell_2(\mathbb{N}_0)$ provides

$$(A\mathbf{x})_j = b_{j-1}x_{j-1} + a_j x_j + b_j x_{j+1}$$

with $b_{-1} = 0$, $x_{-1} = 0$, $j = 0, 1, 2, \ldots$. The domain of A is

$$\mathcal{D}(A) = \{\mathbf{x} : x_j = 0 \text{ for all } j \text{ sufficiently large}\}.$$

Let

$$\mathbf{e}_0 = (1, 0, 0, \ldots)^T, \qquad \mathbf{e}_1 = (0, 1, 0, \ldots)^T.$$

Find $\|A\mathbf{e}_0\|^2$ and $\|A\mathbf{e}_1\|^2$.

Problem 33. (i) Let T be a compact self-adjoint operator acting in the separable Hilbert space $\ell_2(\mathbb{N})$. Then there exists an orthonormal basis $\{e_j\}_{j=1}^\infty$ in $\ell_2(\mathbb{N})$ and real numbers λ_j which converge to 0 as $j \to \infty$ such that

$$Te_j = \lambda_j e_j$$

for all $j = 1, 2, \ldots$. Each non-zero eigenvalue of T has finite multiplicity. Apply the theorem to the operator

$$T = \begin{pmatrix} 0 & 1 & 0 & 0 & \cdots \\ 1 & 0 & 1/2 & 0 & \cdots \\ 0 & 1/2 & 0 & 1/3 & \cdots \\ 0 & 0 & 1/3 & 0 & \cdots \\ \vdots & \vdots & \vdots & \vdots & \ddots \end{pmatrix}.$$

Consider first the matrices

$$T_3 = \begin{pmatrix} 0 & 1 & 0 \\ 1 & 0 & 1/2 \\ 0 & 1/2 & 0 \end{pmatrix}, \quad T_4 = \begin{pmatrix} 0 & 1 & 0 & 0 \\ 1 & 0 & 1/2 & 0 \\ 0 & 1/2 & 0 & 1/3 \\ 0 & 0 & 1/3 & 0 \end{pmatrix},$$

$$T_5 = \begin{pmatrix} 0 & 1 & 0 & 0 & 0 \\ 1 & 0 & 1/2 & 0 & 0 \\ 0 & 1/2 & 0 & 1/3 & 0 \\ 0 & 0 & 1/3 & 0 & 1/4 \\ 0 & 0 & 0 & 1/4 & 0 \end{pmatrix}, \quad T_6 = \begin{pmatrix} 0 & 1 & 0 & 0 & 0 & 0 \\ 1 & 0 & 1/2 & 0 & 0 & 0 \\ 0 & 1/2 & 0 & 1/3 & 0 & 0 \\ 0 & 0 & 1/3 & 0 & 1/4 & 0 \\ 0 & 0 & 0 & 1/4 & 0 & 1/5 \\ 0 & 0 & 0 & 0 & 1/5 & 0 \end{pmatrix}.$$

Note that $\operatorname{tr}(T^2) < \infty$.

(ii) Consider the Banach-Gelfand triple

$$\ell_1(\mathbb{N}_0) \subset \ell_2(\mathbb{N}_0) \subset \ell_\infty(\mathbb{N}_0)$$

and the infinite dimensional linear equation

$$
\begin{pmatrix}
0 & 1 & 0 & 0 & 0 & 0 & 0 & 0 & \cdots \\
1 & 0 & 1/2 & 0 & 0 & 0 & 0 & 0 & \cdots \\
0 & 1/2 & 0 & 1/3 & 0 & 0 & 0 & 0 & \cdots \\
0 & 0 & 1/3 & 0 & 1/4 & 0 & 0 & 0 & \cdots \\
0 & 0 & 0 & 1/4 & 0 & 1/5 & 0 & 0 & \cdots \\
0 & 0 & 0 & 0 & 1/5 & 0 & 1/6 & 0 & \cdots \\
0 & 0 & 0 & 0 & 0 & 1/6 & 0 & 1/7 & \cdots \\
\vdots & \vdots & \vdots & \vdots & \vdots & \vdots & \vdots & \vdots & \ddots
\end{pmatrix}
\begin{pmatrix}
t_0 \\ t_1 \\ t_2 \\ t_3 \\ t_4 \\ t_5 \\ t_6 \\ \vdots
\end{pmatrix}
=
\begin{pmatrix}
0 \\ 0 \\ 0 \\ 0 \\ 0 \\ 0 \\ 0 \\ \vdots
\end{pmatrix}.
$$

The infinite dimensional matrix represents a linear bounded self-adjoint operator. Find all solutions of the linear equation with $t_0 \neq 0$. Is the vector $\mathbf{t}$ an element of $\ell_1(\mathbb{N}_0)$? Is the vector $\mathbf{t}$ an element of $\ell_2(\mathbb{N}_0)$? Is the vector $\mathbf{t}$ an element of $\ell_\infty(\mathbb{N}_0)$?

Problem 34. Let $\mathbb{Z}$ be the set of integers. Consider the Hilbert space $\ell_2(\mathbb{Z}^2)$. Let $(m_1, m_2) \in \mathbb{Z}^2$. Let $f(m_1, m_2)$ be an element of $\ell_2(\mathbb{Z}^2)$. Consider the unitary operators

$$U f(m_1, m_2) := e^{-2\pi i \alpha m_2} f(m_1 + 1, m_2),$$

$$V f(m_1, m_2) := e^{-2\pi i \beta m_1} f(m_1, m_2 + 1).$$

They are the so-called magnetic translation operators with phase α and β, respectively. Find the spectrum of U and V. Find the commutator $[U, V]$. The so-called *Harper operator* which is self-adjoint is defined by

$$\hat{H} := U + U^* + V + V^*.$$

Find the spectrum of $\hat{H}$. Consider the case α, β irrational and α, β rational. In the theory of noncommutative geometry, the magnetic translation operators U and V give rise to a noncommutative torus A_θ with $\theta = \alpha - \beta$.

Problem 35. Let P_j $(j = 0, 1, 2, \ldots)$ be the *Legendre polynomials*

$$P_0(x) = 1, \quad P_1(x) = x, \quad P_2(x) = \frac{1}{2}(3x^2 - 1), \ldots.$$

Calculate the infinite dimensional matrix $A = (A_{jk})$

$$A_{jk} = \int_{-1}^{+1} P_j(x) \frac{dP_k(x)}{dx} dx.$$

where $j, k = 0, 1, \ldots$. Consider the matrix A as a linear operator in the Hilbert space $\ell_2(\mathbb{N}_0)$. Show that

$$A = \begin{pmatrix} 0 & 0 & 0 & 0 & 0 & 0 & 0 & \ldots \\ 2 & 0 & 0 & 0 & 0 & 0 & 0 & \ldots \\ 0 & 2 & 0 & 0 & 0 & 0 & 0 & \ldots \\ 2 & 0 & 2 & 0 & 0 & 0 & 0 & \ldots \\ 0 & 2 & 0 & 2 & 0 & 0 & 0 & \ldots \\ 2 & 0 & 2 & 0 & 2 & 0 & 0 & \ldots \\ 0 & 2 & 0 & 2 & 0 & 2 & 0 & \ldots \\ \vdots & \vdots & \vdots & \vdots & \vdots & \vdots & \vdots & \ddots \end{pmatrix}.$$

Is $\|A\| < \infty$?

Chapter 6

Fourier Transform

6.1 Introduction

We consider the discrete and the continuous Fourier transform [45],[76]. In chapter 9 we consider the Fourier transform in the sense of generalized functions.

Consider the Hilbert space $\mathbb{C}^n$. Let $n \geq 2$ and $\omega := \exp(-2\pi i/n)$. The discrete Fourier transform matrices are given by

$$F_n := \frac{1}{\sqrt{n}} \begin{pmatrix} 1 & 1 & 1 & \cdots & 1 \\ 1 & \omega & \omega^2 & \cdots & \omega^{n-1} \\ 1 & \omega^2 & \omega^4 & \cdots & \omega^{2n-2} \\ \vdots & \vdots & \vdots & \ddots & \vdots \\ 1 & \omega^{n-1} & \omega^{2n-2} & \cdots & \omega^{(n-1)^2} \end{pmatrix}.$$

They are unitary matrices. The inverse of F_n is $F_n^{-1} = F_n^*$. The $n \times n$ Fourier matrix can also be written as $F_n = (F_{j\ell})_n$ with

$$F_{j\ell} = \frac{1}{\sqrt{n}} \exp(2\pi i j\ell/n), \quad j, \ell = 0, 1, \ldots, n-1.$$

For $n = 2$ and $n = 3$ we have the matrices

$$F_2 = \frac{1}{\sqrt{2}} \begin{pmatrix} 1 & 1 \\ 1 & -1 \end{pmatrix}, \quad F_3 = \frac{1}{\sqrt{3}} \begin{pmatrix} 1 & 1 & 1 \\ 1 & e^{-2\pi i/3} & e^{-4\pi i/3} \\ 1 & e^{-4\pi i/3} & e^{-2\pi i/3} \end{pmatrix}.$$

with

$$e^{-2\pi i/3} \equiv \cos(2\pi/3) - i\sin(2\pi/3) \equiv -\frac{1}{2} + \frac{i}{2}\sqrt{3}$$

$$e^{-4\pi i/3} \equiv \cos(4\pi/3) - i\sin(4\pi/3) \equiv -\frac{1}{2} - \frac{i}{2}\sqrt{3}.$$

For $n = 4$ we have

$$F_4 = \frac{1}{2} \begin{pmatrix} 1 & 1 & 1 & 1 \\ 1 & -i & -1 & i \\ 1 & -1 & 1 & -1 \\ 1 & i & -1 & -i \end{pmatrix}$$

and for $n = 5$

$$F_5 = \frac{1}{\sqrt{5}} \begin{pmatrix} 1 & 1 & 1 & 1 & 1 \\ 1 & \omega & \omega^2 & \omega^3 & \omega^4 \\ 1 & \omega^2 & \omega^4 & \omega^6 & \omega^8 \\ 1 & \omega^3 & \omega^6 & \omega^9 & \omega^{12} \\ 1 & \omega^4 & \omega^8 & \omega^{12} & \omega^{16} \end{pmatrix}$$

with $\omega := \exp(-2\pi i/5)$. Consider the Hilbert space $L_2(\mathbb{R}^n)$ and the Banach space $L_1(\mathbb{R}^n)$. Given $f \in L_2(\mathbb{R}^n)$ and $f \in L_1(\mathbb{R}^n)$. The Fourier transform of f is defined as

$$\hat{f}(\mathbf{k}) = \int_{\mathbb{R}^n} \exp(i\mathbf{k} \cdot \mathbf{x}) f(\mathbf{x}) d\mathbf{x}$$

where $\mathbf{k} \cdot \mathbf{x} := k_1 x_1 + \cdots + k_n x_n$ and $d\mathbf{x} := dx_1 dx_2 \ldots dx_n$. The Fourier transform is a contraction from $L_1(\mathbb{R}^n)$ to $L_\infty(\mathbb{R}^n)$. The transformation is unitary into the Hilbert space $L_2(\mathbb{R}^n)$. The inverse Fourier transformation is

$$f(\mathbf{x}) = \frac{1}{(2\pi)^n} \int_{\mathbb{R}^n} \hat{f}(\mathbf{k}) \exp(-i\mathbf{k} \cdot \mathbf{x}) d\mathbf{k}$$

where $d\mathbf{k} := dk_1 dk_2 \ldots dk_n$. Let $f, g \in L_1(\mathbb{R}) \cap L_2(\mathbb{R})$ and $\hat{f}$, $\hat{g}$ be the Fourier transforms. Then

$$2\pi \langle g, f \rangle = \langle \hat{g}, \hat{f} \rangle.$$

Let $f \in L_1(\mathbb{R}^n)$. Then

$$\left| \int_{\mathbb{R}^n} e^{-i\mathbf{x} \cdot \mathbf{y}} f(\mathbf{x}) d\mathbf{x} \right| \leq \int_{\mathbb{R}^n} |f(\mathbf{x})| d\mathbf{x}$$

for all $\mathbf{y} \in \mathbb{R}^n$. Consider the Hilbert space $L_2(\mathbb{R})$ and $f, g \in L_2(\mathbb{R})$. The *polarization identity*

$$\langle f, g \rangle \equiv \frac{1}{4} (|f + g|^2 - |f - g|^2 + i|f + ig|^2 - i|f - ig|^2)$$

implies that every *isometry* preserves the inner product. Since the Fourier transform is an isometry on $L_2(\mathbb{R})$ we have

$$2\pi \langle f, g \rangle = \langle \hat{f}, \hat{g} \rangle.$$

The Fourier transform of an integrable function is a continuous function.

Let $Q = [-1/2, 1/2]^d$ a d-dimensional cube. Then (*Whittaker-Kotelnikov-Shannon Sampling Theorem*) for any $f \in L_2(\mathbb{R}^d)$ with supp$(f) \subseteq Q$ one has

$$f(t) = \sum_{n \in \mathbb{Z}^d} f(n) T_n \text{sinc}(t)$$

with absolute and uniform convergence of the series and norm convergence in $L_2(\mathbb{R}^d)$.

Theorem. (Paley-Wiener) Let a and c be positive constants and f be an entire function with $|f(z)| \leq c \exp(2\pi a |z|)$ for all z and $\|f\|_{L_2(\mathbb{R})} < \infty$. Then there exists $F \in L_2(-a, a)$ so that

$$f(z) = \int_{-a}^{a} F(t) \exp(2\pi i t z) dt$$

for all z.

Theorem. Let $f \in L_1(\mathbb{R})$. Then

(a) $\mathcal{F}\{\bar{f}(x)\} = \overline{\mathcal{F}\{f(-x)\}}$

(b) $\mathcal{F}\{f(x - y)\} = \mathcal{F}\{f(x)\} e^{-iky}$

(c) $\mathcal{F}\{f(\alpha x)\} = (1/\alpha)\mathcal{F}\{f(x/\alpha)\}, \qquad \alpha > 0.$

Theorem. If f is a continuous piecewise differentiable function, $f, f' \in L_1(\mathbb{R})$, and $\lim_{|x| \to \infty} f(x) = 0$, then

$$\mathcal{F}\{f'\} = -ik\mathcal{F}\{f\}.$$

To prove this theorem we apply integration by parts.

Let $S(\mathbb{R}^n)$ be the Schwartzian space and $S'(\mathbb{R}^n)$ be the dual space. Then

$$\mathcal{F} : S(\mathbb{R}^n) \to S(\mathbb{R}^n) \quad \text{one-one, onto}$$

$$\mathcal{F} : L_2(\mathbb{R}^n) \to L_2(\mathbb{R}^n) \quad \text{one-one, onto, unitary}$$

$$\mathcal{F} : L_1(\mathbb{R}^n) \to C_0(\mathbb{R}^n) \quad \text{one-one, not onto}$$

$$\mathcal{F} : L_p(\mathbb{R}^n) \to L_q(\mathbb{R}^n) \quad \text{if } 1 \leq p \leq 2 \text{ and } 1/p + 1/q = 1$$

$$\mathcal{F} : S'(\mathbb{R}^n) \to S'(\mathbb{R}^n) \quad \text{one-one, onto}$$

6.2 Solved Problems

Problem 1. Consider the Hilbert space $\mathbb{C}^2$, the *standard basis*

$$\mathbf{e}_1 = \begin{pmatrix} 1 \\ 0 \end{pmatrix}, \qquad \mathbf{e}_2 = \begin{pmatrix} 0 \\ 1 \end{pmatrix}$$

and the 2×2 Fourier matrix (*Hadamard matrix*)

$$F_2 = \frac{1}{\sqrt{2}} \begin{pmatrix} 1 & 1 \\ 1 & -1 \end{pmatrix} \;\Rightarrow\; F_2^* = \frac{1}{\sqrt{2}} \begin{pmatrix} 1 & 1 \\ 1 & -1 \end{pmatrix} = F_2^{-1} = F_2.$$

Find the orthonormal basis $F_2\mathbf{e}_1$, $F_2\mathbf{e}_2$. Express the hermitian matrix

$$A = \begin{pmatrix} 0 & 1 \\ 1 & 0 \end{pmatrix}$$

in this basis.

Solution 1. We obtain the orthonormal basis

$$\frac{1}{\sqrt{2}} \begin{pmatrix} 1 \\ 1 \end{pmatrix}, \qquad \frac{1}{\sqrt{2}} \begin{pmatrix} 1 \\ -1 \end{pmatrix}.$$

The matrix A expressed in this basis is

$$\tilde{A} = F_2 A F_2^* = \begin{pmatrix} 1 & 0 \\ 0 & -1 \end{pmatrix}$$

i.e. the matrix is diagonal with eigenvalues $+1$ and -1.

Problem 2. Consider the Hilbert spaces $\mathbb{C}^2$, $\mathbb{C}^3$, $\mathbb{C}^4$ and the 2×2, 3×3, 4×4 Fourier matrices. Find the normalized vectors

$$\frac{1}{\sqrt{2}} \begin{pmatrix} 1 & 1 \\ 1 & -1 \end{pmatrix} \begin{pmatrix} 1/\sqrt{2} \\ 1/\sqrt{2} \end{pmatrix}, \quad \frac{1}{\sqrt{3}} \begin{pmatrix} 1 & 1 & 1 \\ 1 & e^{-2\pi i/3} & e^{-4\pi i/3} \\ 1 & e^{-4\pi i/3} & e^{-2\pi i/3} \end{pmatrix} \begin{pmatrix} 1/\sqrt{3} \\ 1/\sqrt{3} \\ 1/\sqrt{3} \end{pmatrix},$$

$$\frac{1}{2} \begin{pmatrix} 1 & 1 & 1 & 1 \\ 1 & i & -1 & -i \\ 1 & -1 & 1 & -1 \\ 1 & -i & -1 & i \end{pmatrix} \begin{pmatrix} 1/2 \\ 1/2 \\ 1/2 \\ 1/2 \end{pmatrix}.$$

Solution 2. We obtain

$$\frac{1}{\sqrt{2}} \begin{pmatrix} 1 & 1 \\ 1 & -1 \end{pmatrix} \begin{pmatrix} 1/\sqrt{2} \\ 1/\sqrt{2} \end{pmatrix} = \begin{pmatrix} 1 \\ 0 \end{pmatrix}$$

$$\frac{1}{\sqrt{3}} \begin{pmatrix} 1 & 1 & 1 \\ 1 & e^{-2\pi i/3} & e^{-4\pi i/3} \\ 1 & e^{-4\pi i/3} & e^{-2\pi i/3} \end{pmatrix} \begin{pmatrix} 1/\sqrt{3} \\ 1/\sqrt{3} \\ 1/\sqrt{3} \end{pmatrix} = \begin{pmatrix} 1 \\ 0 \\ 0 \end{pmatrix}$$

$$\frac{1}{2} \begin{pmatrix} 1 & 1 & 1 & 1 \\ 1 & i & -1 & -i \\ 1 & -1 & 1 & -1 \\ 1 & -i & -1 & i \end{pmatrix} \begin{pmatrix} 1/2 \\ 1/2 \\ 1/2 \\ 1/2 \end{pmatrix} = \begin{pmatrix} 1 \\ 0 \\ 0 \\ 0 \end{pmatrix}.$$

Problem 3. Consider the standard basis in the Hilbert space $\mathbb{C}^4$

$$e_1 = \begin{pmatrix} 1 \\ 0 \\ 0 \\ 0 \end{pmatrix}, \quad e_2 = \begin{pmatrix} 0 \\ 1 \\ 0 \\ 0 \end{pmatrix}, \quad e_3 = \begin{pmatrix} 0 \\ 0 \\ 1 \\ 0 \end{pmatrix}, \quad e_4 = \begin{pmatrix} 0 \\ 0 \\ 0 \\ 1 \end{pmatrix}$$

and the 4×4 discrete Fourier transform matrix

$$F_4 = \frac{1}{2} \begin{pmatrix} 1 & 1 & 1 & 1 \\ 1 & -i & -1 & i \\ 1 & -1 & 1 & -1 \\ 1 & i & -1 & -i \end{pmatrix}.$$

Find $v_j = F_4 e_j$ $(j = 1, 2, 3, 4)$.

Solution 3. We apply the Maxima program

```
/* Fourier0.mac */
e1: matrix([1],[0],[0],[0]); e2: matrix([0],[1],[0],[0]);
e3: matrix([0],[0],[1],[0]); e4: matrix([0],[0],[0],[1]);
F: (1/2)*matrix([1,1,1,1],[1,-%i,-1,%i],[1,-1,1,-1],[1,%i,-1,-%i]);
v1: F . e1; v2: F . e2; v3: F . e3; v4: F . e4;
```

The output is the orthonormal basis in $\mathbb{C}^4$

$$v_1 = \frac{1}{2} \begin{pmatrix} 1 \\ 1 \\ 1 \\ 1 \end{pmatrix}, \quad v_2 = \frac{1}{2} \begin{pmatrix} 1 \\ -i \\ -1 \\ i \end{pmatrix}, \quad v_3 = \frac{1}{2} \begin{pmatrix} 1 \\ -1 \\ 1 \\ -1 \end{pmatrix}, \quad v_4 = \frac{1}{2} \begin{pmatrix} 1 \\ i \\ -1 \\ -i \end{pmatrix}.$$

Apply the Fourier transform F_4 to the *Bell basis* which also forms an orthonormal basis in the Hilbert space $\mathbb{C}^4$

$$\frac{1}{\sqrt{2}}\begin{pmatrix}1\\0\\0\\1\end{pmatrix},\quad \frac{1}{\sqrt{2}}\begin{pmatrix}0\\1\\1\\0\end{pmatrix},\quad \frac{1}{\sqrt{2}}\begin{pmatrix}0\\1\\-1\\0\end{pmatrix},\quad \frac{1}{\sqrt{2}}\begin{pmatrix}1\\0\\0\\-1\end{pmatrix}.$$

Problem 4. Consider the three spin matrices S_1, S_2, S_3 for spin $\frac{1}{2}$

$$S_1^{(1/2)}=\frac{1}{2}\begin{pmatrix}0&1\\1&0\end{pmatrix},\quad S_2^{(1/2)}=\frac{1}{2}\begin{pmatrix}0&-i\\i&0\end{pmatrix},\quad S_3^{(1/2)}=\frac{1}{2}\begin{pmatrix}1&0\\0&-1\end{pmatrix}$$

and the 2×2 Fourier matrix (*Hadamard matrix*)

$$F_2=\frac{1}{\sqrt{2}}\begin{pmatrix}1&1\\1&-1\end{pmatrix}.$$

The spin matrices obey the commutation relation

$$[S_1^{(1/2)},S_2^{(1/2)}]=iS_3^{(1/2)},\ [S_2^{(1/2)},S_3^{(1/2)}]=iS_1^{(1/2)},\ [S_3^{(1/2)},S_1^{(1/2)}]=iS_2^{(1/2)}.$$

Find $F_2 S_1^{(1/2)} F_2^{-1}$, $F_2 S_2^{(1/2)} F_2^{-1}$, $F_2 S_3^{(1/2)} F_2^{-1}$.

Solution 4. With

$$F_2=\frac{1}{\sqrt{2}}\begin{pmatrix}1&1\\1&-1\end{pmatrix}=F_2^*=F_2^{-1}$$

we obtain

$$F_2 S_1^{(1/2)} F_2^{-1}=S_3^{(1/2)},\quad F_2 S_2^{(1/2)} F_2^{-1}=-S_2^{(1/2)},\quad F_2 S_3^{(1/2)} F_3^{-1}=S_1^{(1/2)}.$$

Hence

$$S_1^{(1/2)}\mapsto S_3^{(1/2)},\quad S_3^{(1/2)}\mapsto S_1^{(1/2)},\ S_2^{(1/2)}\mapsto -S_2^{(1/2)}.$$

Extend to the three spin matrices of spin-1.

Problem 5. The *circulant matrices* are diagonalized by the discrete Fourier transform matrices. For example

$$F_2\begin{pmatrix}0&1\\1&0\end{pmatrix}F_2^*=\frac{1}{\sqrt{2}}\begin{pmatrix}1&1\\1&-1\end{pmatrix}\begin{pmatrix}0&1\\1&0\end{pmatrix}\frac{1}{\sqrt{2}}\begin{pmatrix}1&1\\1&-1\end{pmatrix}=\begin{pmatrix}1&0\\0&-1\end{pmatrix}.$$

Let $n=3$. Apply the Fourier transform F_3 to the *circulant matrix*

$$P=\begin{pmatrix}0&1&0\\0&0&1\\1&0&0\end{pmatrix}$$

which is also a permutation matrix.

Solution 5. We apply the Maxima program

```
/* Fourier1.mac */
P: matrix([0,1,0],[0,0,1],[1,0,0]);
F: matrix([1,1,1],[1,-1/2-%i*sqrt(3)/2,-1/2+%i*sqrt(3)/2],
          [1,-1/2+%i*sqrt(3)/2,-1/2-%i*sqrt(3)/2])/sqrt(3);
FT: transpose(F); FTC: conjugate(FT);
D: F . P . FTC; D: ratsimp(D);
```

The output is the unitary 3×3 diagonal matrix

$$(1) \oplus \begin{pmatrix} -1/2 + \sqrt{3}i/2 & 0 \\ 0 & -1/2 + \sqrt{3}i/2 \end{pmatrix}$$

with the diagonal elements are the eigenvalues of P and $\oplus$ denotes the direct sum.

Problem 6. Given the six 3×3 permutation matrices

$$P_{123} = \begin{pmatrix} 1 & 0 & 0 \\ 0 & 1 & 0 \\ 0 & 0 & 1 \end{pmatrix}, \quad P_{132} = \begin{pmatrix} 1 & 0 & 0 \\ 0 & 0 & 1 \\ 0 & 1 & 0 \end{pmatrix}, \quad P_{213} = \begin{pmatrix} 0 & 1 & 0 \\ 1 & 0 & 0 \\ 0 & 0 & 1 \end{pmatrix},$$

$$P_{231} = \begin{pmatrix} 0 & 1 & 0 \\ 0 & 0 & 1 \\ 1 & 0 & 0 \end{pmatrix}, \quad P_{312} = \begin{pmatrix} 0 & 0 & 1 \\ 1 & 0 & 0 \\ 0 & 1 & 0 \end{pmatrix}, \quad P_{321} = \begin{pmatrix} 0 & 0 & 1 \\ 0 & 1 & 0 \\ 1 & 0 & 0 \end{pmatrix}$$

which form a non-commutative group under matrix multiplication. Find the Fourier transform $F_3 P F_3^*$. Do these six matrices form a group under matrix multiplication?

Solution 6. We apply the Maxima program

```
/* Fourier2.mac */
p[1]: matrix([1,0,0],[0,1,0],[0,0,1]); p[2]:
matrix([1,0,0],[0,0,1],[0,1,0]);
p[3]: matrix([0,1,0],[1,0,0],[0,0,1]); p[4]:
matrix([0,1,0],[0,0,1],[1,0,0]);
p[5]: matrix([0,0,1],[1,0,0],[0,1,0]); p[6]:
matrix([0,0,1],[0,1,0],[1,0,0]);
w: exp((%i*2*%pi)/3);
F: matrix([1,1,1],[1,w,w^2],[1,w^2,w])/sqrt(3);
for j thru 6 do print(expand(F . p[j] . ctranspose(F)));
```

The output is

$$\begin{pmatrix} 1 & 0 & 0 \\ 0 & 1 & 0 \\ 0 & 0 & 1 \end{pmatrix}, \quad \begin{pmatrix} 1 & 0 & 0 \\ 0 & 0 & 1 \\ 0 & 1 & 0 \end{pmatrix}, \quad \begin{pmatrix} 1 & 0 & 0 \\ 0 & 0 & (-1+\sqrt{3}i)/2 \\ 0 & (-1-\sqrt{3}i)/2 & 0 \end{pmatrix},$$

$$\begin{pmatrix} 1 & 0 & 0 \\ 0 & (-1-\sqrt{3}i)/2 & 0 \\ 0 & 0 & (-1+\sqrt{3}i)/2 \end{pmatrix},$$

$$\begin{pmatrix} 1 & 0 & 0 \\ 0 & (-1+\sqrt{3}i)/2 & 0 \\ 0 & 0 & (-1-\sqrt{3}i)/2 \end{pmatrix},$$

$$\begin{pmatrix} 1 & 0 & 0 \\ 0 & 0 & (-1-\sqrt{3}i)/2 \\ 0 & (-1+\sqrt{3}i)/2 & 0 \end{pmatrix}.$$

All these matrices can be written as $(1) \oplus M$, where M is a 2×2 matrices and $\oplus$ denotes the *direct sum*. These six 2×2 matrices form a non-commutative *group* under matrix multiplication.

Problem 7. Let $L > 0$ with dimension length. Consider the function
$$f(x) = \begin{cases} 1 & |x| < L \\ 0 & \text{otherwise} \end{cases}.$$
Then $f \in L_1(\mathbb{R})$ and $f \in L_2(\mathbb{R})$ and $\langle f, f \rangle = \|f\|^2 = 2L$.
(i) Find the Fourier transform
$$\hat{f}(k) = \int_{\mathbb{R}} f(x) e^{ikx} dx.$$
(ii) Find the inverse transform
$$f(x) = \frac{1}{2\pi} \int_{\mathbb{R}} \hat{f}(k) e^{-ikx} dk.$$
(iii) Show that $2\pi \langle f, f \rangle = \langle \hat{f}, \hat{f} \rangle$.

Solution 7. (i) We have
$$\hat{f}(k) = \int_{\mathbb{R}} f(x) e^{ikx} dx = \int_{-L}^{L} e^{ikx} dx = \left. \frac{e^{ikx}}{ik} \right|_{-L}^{L} = \frac{e^{ikL}}{ik} - \frac{e^{-ikL}}{ik}$$
$$= \frac{2}{k} \sin(Lk)$$
with $\hat{f}(0) = 2L$.
(ii) For the inverse transform we find
$$f(x) = \frac{1}{2\pi} \int_{\mathbb{R}} \hat{f}(k) e^{-ikx} dk = \frac{1}{\pi} \int_{\mathbb{R}} \frac{\sin(Lk)}{k} e^{-ikx} dk$$
$$= \frac{1}{\pi} \int_{\mathbb{R}} \frac{\sin(Lk) \cos(Lk)}{k} dk = \begin{cases} 1 & |x| < L \\ 0 & \text{otherwise} \end{cases}$$

(iii) We have

$$\langle \hat{f}(k), \hat{f}(k) \rangle = \int_{\mathbb{R}} \frac{4}{k^2} \sin^2(Lk) dk = 4L \int_{\mathbb{R}} \frac{\sin^2(s)}{s^2} ds = 4\pi L = 2\pi \langle f, f \rangle.$$

Problem 8. Let $k_c > 0$ with dimension meter^{-1}.
(i) Show that $f_{k_c} : \mathbb{R} \to \mathbb{R}$ given by

$$f_{k_c}(x) = \frac{\sin(k_c x)}{k_c x}$$

is an element of $L_2(\mathbb{R})$. Note that $f_{k_c}(0) = 1$ and

$$\int_0^\infty \frac{\sin^2(x)}{x^2} dx = \frac{\pi}{2}.$$

(ii) Find the Fourier transform of f_{k_c}, i.e.

$$\hat{f}_{k_c}(k) = \int_{\mathbb{R}} e^{ikx} f_{k_c}(x) dx.$$

Solution 8. (i) We have

$$\int_{\mathbb{R}} \frac{\sin^2(k_c x)}{(k_c x)^2} dx = \frac{1}{k_c} \int_{\mathbb{R}} \frac{\sin^2(s)}{s^2} ds = \frac{\pi}{k_c}.$$

(ii) We find

$$\hat{f}_{k_c}(k) = \int_{\mathbb{R}} e^{ikx} f_{k_c}(x) dx = \frac{1}{k_c} \int_{\mathbb{R}} \frac{\sin(k_c x) \cos(k_c x)}{x} dx$$

$$= \frac{2}{k_c} \int_0^\infty \frac{\sin(k_c x) \cos(k_c x)}{x} dx$$

$$= \begin{cases} 0 & k_c > k > 0 \\ \pi/k_c & 0 < k_c < k \\ \pi/(2k_c) & k = k_c > 0 \end{cases}$$

Problem 9. Consider the Hilbert space $L_2(\mathbb{R})$ and the Banach space $L_1(\mathbb{R})$. Let

$$f(x) = \begin{cases} -1 & \text{for} & -3/2 \le x < -1/2 \\ 1 & \text{for} & -1/2 \le x < 1/2 \\ -1 & \text{for} & 1/2 \le x \le 3/2 \\ 0 & \text{otherwise} \end{cases}$$

with $f \in L_2(\mathbb{R})$ and $f \in L_1(\mathbb{R})$. Find the Fourier transform of f

$$\hat{f}(k) = \int_{-\infty}^{\infty} e^{ikx} f(x) dx.$$

Solution 9. We apply the Maxima program

```
/* integhilbert.mac */
t1: integrate(-exp(%i*k*x),x,-3/2,-1/2); t1: demoivre(t1);
t2: integrate(exp(%i*k*x),x,-1/2,1/2); t2: demoivre(t2);
t3: integrate(-exp(%i*k*x),x,1/2,3/2); t3: demoivre(t3);
t: t1+t2+t3; t: expand(t);
```

The output is

$$\hat{f}(k) = \frac{1}{k}(4\sin(k/2) - 2\sin(3k/2))$$

with $\hat{f}(0) = -1$.

Problem 10. Consider the Hilbert space $L_2(\mathbb{R})$. Find the Fourier transform of the function

$$f(x) = \begin{cases} 1 & \text{if} & -1 \le x \le 0 \\ e^{-x} & \text{if} & x \ge 0 \\ 0 & \text{otherwise} \end{cases}$$

Solution 10. We have

$$\int_{-\infty}^{+\infty} f(x) e^{ikx} dx = \int_{-1}^{0} e^{ikx} dx + \int_{0}^{\infty} e^{(ik-1)x} dx$$

$$= \left. \frac{e^{ikx}}{ik} \right|_{-1}^{0} + \left. \frac{e^{(ik-1)x}}{ik-1} \right|_{0}^{\infty} = \frac{1-e^{-ik}}{ik} - \frac{1}{ik-1}$$

$$= \frac{-1-(ik-1)e^{-ik}}{ik(ik-1)}.$$

Problem 11. Let $a > 0$. Consider the function

$$f_a(x) = \frac{a}{2} \exp(-a|x|).$$

Then $f_a \in L_2(\mathbb{R})$ and $f_a \in L_1(\mathbb{R})$.

(i) Calculate

$$\int_{-\infty}^{\infty} f_a(x)dx.$$

(ii) Find the Fourier transform of f_a. Discuss a large and a small.

Solution 11. (i) We find

$$\int_{\infty}^{\infty} \frac{a}{2} e^{-a|x|} dx = \int_{-\infty}^{0} \frac{a}{2} e^{ax} dx + \int_{0}^{\infty} \frac{a}{2} e^{-ax} dx$$

$$= \left[\frac{ae^{ax}}{2} \right]_{-\infty}^{0} + \left[\frac{ae^{-ax}}{2} \right]_{0}^{\infty} = 1.$$

Thus the integral does not depend on a.

(ii) For the Fourier transform we find

$$\int_{-\infty}^{\infty} \frac{a}{2} e^{ikx} e^{-a|x|} dx = \int_{-\infty}^{0} \frac{a}{2} e^{(a+ik)x} dx + \int_{0}^{\infty} \frac{a}{2} e^{(-a+ik)x} dx$$

$$= \left[\frac{a}{2(a+ik)} e^{(a+ik)x} \right]_{-\infty}^{0} + \left[\frac{a}{2(-a+ik)} e^{(-a+ik)x} \right]_{0}^{\infty}$$

$$= \frac{a}{2(a+ik)} - \frac{a}{2(-a+ik)} = \frac{a^2}{a^2+k^2}.$$

Problem 12. Find the Fourier transform of the *hat function*

$$f(t) = \begin{cases} 1 - |t| & \text{for} \quad -1 < t < 1 \\ 0 & \text{otherwise} \end{cases}$$

Solution 12. We note that

$$\int e^{-i\omega t} dt = -\frac{e^{-i\omega t}}{i\omega}, \qquad \int t e^{-i\omega t} dt = \frac{1}{\omega^2} e^{-i\omega t} - \frac{t}{i\omega} e^{-i\omega t}.$$

Thus the Fourier transform is

$$\hat{f}(\omega) = \left(\frac{\sin(\pi\omega)}{\pi\omega} \right)^2$$

with $f(0) = 1$.

Problem 13. Let $f \in L_1(\mathbb{R})$ and $f \in L_2(\mathbb{R})$ and $\hat{f}$ be the Fourier transform. Let $\hat{\phi}(k)$ be the Fourier transform of $\phi(x) \in S(\mathbb{R})$ (Schwartzian space). Show that

$$\langle f(x), \phi(x) \rangle = \frac{1}{2\pi} \langle \hat{f}(k), \hat{\phi}(k) \rangle.$$

Solution 13. We have

$$\langle f(x), \phi(x) \rangle = \int_{\mathbb{R}} \overline{f}(x)\phi(x)dx = \frac{1}{2\pi} \int_{\mathbb{R}} \overline{f}(x) \left(\int_{\mathbb{R}} \hat{\phi}(k)e^{-ikx}dk \right) dx$$

$$= \frac{1}{2\pi} \int_{\mathbb{R}} \hat{\phi}(k) \left(\int_{\mathbb{R}} \overline{f(x)e^{ikx}}dx \right) dk = \frac{1}{2\pi} \int_{\mathbb{R}} \overline{\hat{f}(k)}\hat{\phi}(k)dk$$

$$= \frac{1}{2\pi} \langle \hat{f}(k), \hat{\phi}(k) \rangle.$$

Problem 14. Let $f \in L_2(\mathbb{R})$ and $f \in L_1(\mathbb{R})$. Assume that $f(x) = f(-x)$. Can we conclude that $\hat{f}(k) = \hat{f}(-k)$?

Solution 14. Yes, we have

$$\hat{f}(-k) = \int_{x=-\infty}^{x=+\infty} e^{-ikx} f(x)dx = \int_{y=+\infty}^{x=-\infty} e^{iky} f(-y)d(-y)$$

$$= -\int_{y=+\infty}^{y=-\infty} e^{iky} f(y)dy = \int_{y=-\infty}^{y=+\infty} e^{iky} f(y)dy$$

$$= \hat{f}(k).$$

Problem 15. Consider the Hilbert space $L_2(\mathbb{R})$. Let Ω be a fixed frequency and

$$\hat{f}(\omega) = \begin{cases} 1 & \text{for} \quad \Omega/2 \leq |\omega| \leq \Omega \\ 0 & \text{otherwise} \end{cases}$$

Find

$$f(t) = \frac{1}{2\pi} \int_{-\infty}^{\infty} \hat{f}(\omega)e^{-i\omega t}d\omega.$$

Solution 15. We obtain

$$f(t) = \frac{1}{2\pi} \int_{-\infty}^{+\infty} \hat{f}(\omega) e^{-i\omega t} d\omega = \frac{1}{2\pi} \int_{-\Omega}^{-\Omega/2} e^{-i\omega t} d\omega + \frac{1}{2\pi} \int_{\Omega/2}^{\Omega} e^{-i\omega t} d\omega$$

$$= \frac{1}{2\pi} \left(\frac{e^{i\Omega t/2}}{-it} - \frac{e^{i\Omega t}}{-it} + \frac{e^{-i\Omega t}}{-it} - \frac{e^{-i\Omega t/2}}{-it} \right)$$

$$= \frac{1}{\pi t} (\sin(\Omega t) - \sin(\Omega t/2)).$$

Note that $f(0) = \Omega/(2\pi)$.

Problem 16. Consider the Hilbert space $L_2(\mathbb{R})$. Let $a > 0$. Define

$$f_a(x) = \begin{cases} \frac{1}{2a} & |x| < a \\ 0 & |x| > a \end{cases}$$

Calculate

$$\int_{\mathbb{R}} f_a(x) dx$$

and the Fourier transform of f_a. Discuss the result in dependence of a.

Solution 16. We find

$$\int_{\mathbb{R}} f_a(x) dx = 1.$$

For the Fourier transform we find

$$\hat{f}_a(k) = \int_{\mathbb{R}} e^{ikx} f_a(x) dx = \frac{1}{2a} \int_{-a}^{a} e^{ikx} dx = \frac{e^{ika}}{ik} - \frac{e^{-ika}}{ik} = \frac{\sin(ka)}{ka}$$

with $\hat{f}_a(0) = 1$.

Problem 17. Consider the Hilbert space $L_2(\mathbb{R})$. Let

$$\hat{\psi}(\omega) = \begin{cases} 1 & \text{if} \quad 1/2 \le |\omega| \le 1 \\ 0 & \text{otherwise} \end{cases}$$

and

$$\hat{\phi}(\omega) = e^{-\alpha|\omega|}, \qquad \alpha > 0.$$

(i) Calculate the inverse Fourier transform of $\hat{\psi}(\omega)$ and $\hat{\phi}(\omega)$, i.e.

$$\psi(t) = \frac{1}{2\pi} \int_{\mathbb{R}} e^{-i\omega t} \hat{\psi}(\omega) d\omega, \qquad \phi(t) = \frac{1}{2\pi} \int_{\mathbb{R}} e^{-i\omega t} \hat{\phi}(\omega) d\omega.$$

(ii) Calculate the scalar product $\langle \psi(t), \phi(t) \rangle$ by utilizing the identity

$$2\pi \langle \psi(t), \phi(t) \rangle \equiv \langle \hat{\psi}(\omega), \hat{\phi}(\omega) \rangle.$$

Solution 17. For the inverse Fourier transform we find

$$\psi(t) = \frac{1}{2\pi} \left(\int_{-1}^{-1/2} e^{-i\omega t} d\omega + \int_{1/2}^{1} e^{-i\omega t} d\omega \right) = \frac{1}{\pi t} (\sin(t) - \sin(t/2))$$

$$\phi(t) = \frac{1}{2\pi} \left(\int_{-\infty}^{0} e^{-i\omega t} e^{\alpha \omega} d\omega + \int_{0}^{\infty} e^{-i\omega t} e^{-\alpha \omega} d\omega \right) = \frac{1}{\pi} \frac{\alpha}{t^2 + \alpha^2}.$$

(ii) We find

$$\langle \psi(t), \phi(t) \rangle = \frac{1}{2\pi} \langle \hat{\psi}(\omega), \hat{\phi}(\omega) \rangle = \frac{1}{2\pi} \left(\int_{-1}^{-1/2} e^{\alpha \omega} d\omega + \int_{1/2}^{1} e^{-\alpha \omega} d\omega \right)$$

$$= \frac{2e^{-3\alpha/2}(e^{\alpha} - e^{\alpha/2})}{\alpha}.$$

Problem 18. Consider the Hilbert space $L_2(\mathbb{R})$ and the function $f \in L_2(\mathbb{R})$

$$f(x) = \begin{cases} 1 & \text{if } |x| < 1 \\ 0 & \text{if } |x| \geq 1 \end{cases}$$

Calculate $f * f$ and verify the *convolution theorem*

$$\widehat{f * f} = \hat{f}\hat{f}.$$

Solution 18. We have

$$(f * f)(x) = \int_{-\infty}^{\infty} f(t) f(t - x) dt.$$

Then for fixed x, the integral is the area common to the rectangle given by f and the rectangle shifted to have its centre at $t = x$. The result is

$$(f * f)(x) = \begin{cases} 2 - |x| & \text{if} & |x| \leq 2 \\ 0 & \text{otherwise} \end{cases}$$

The Fourier transform of f is

$$\hat{f}(k) = 2 \frac{\sin(k)}{k}.$$

It follows that

$$\hat{f}(k)\hat{f}(k) = 4 \frac{\sin^2(k)}{k^2}.$$

Applying the Maxima program

```
/* convolution.mac */
Ff: integrate(exp(%i*k*x),x,-1,+1); Ff: demoivre(Ff); Ff: expand(Ff);
t1: integrate((2+x)*exp(%i*k*x),x,-2,0);
t2: integrate((2-x)*exp(%i*k*x),x,0,2);
t: t1+t2; t: demoivre(t); t: expand(t);
```

we find the Fourier transform of $f * f$ as

$$\frac{2}{k^2}(1 - \cos(2k)) \equiv \frac{4}{k^2}\sin^2(k).$$

Problem 19. Let

$$\hat{f}(\omega) = \begin{cases} (1 - \omega^2) & \text{for } |\omega| \leq 1 \\ 0 & \text{for } |\omega| > 1 \end{cases}$$

Find $f(t)$.

Solution 19. We obtain

$$f(t) = \frac{2}{\pi}\left(\frac{\sin(t) - t\cos(t)}{t^3}\right).$$

Note that $f(0) = 2/(3\pi)$.

Problem 20. Let $a > 0$. Find the Fourier transform of the function $f_a : \mathbb{R} \to \mathbb{R}$

$$f_a(x) = \begin{cases} x/a^2 + 1/a & \text{for } -a \leq x \leq 0 \\ -x/a^2 + 1/a & \text{for } 0 \leq x \leq a \\ 0 & \text{otherwise} \end{cases}$$

Solution 20. Note that

$$\int_{-\infty}^{\infty} f_a(x)dx = \int_{-a}^{a} f_a(x)dx = 1$$

is independent of a. Since the function f_a is symmetric, i.e. $f_a(x) = f_a(-x)$ we have

$$\hat{f}_a(k) = \int_{-\infty}^{\infty} f_a(x)e^{ikx}dx = \int_{-a}^{a} f_a(x)e^{ikx}dx$$

$$= \int_{-a}^{a} f_a(x)\cos(kx)dx$$

$$= 2\int_{0}^{a} f_a(x)\cos(kx)dx.$$

Since
$$\int x \cos(kx)dx = \frac{kx \sin(kx) + \cos(kx)}{k^2}$$

we finally obtain
$$\hat{f}_a(k) = \frac{2}{(ak)^2}(1 - \cos(kx)).$$

Note that $\hat{f}_a(0) = 1$.

Problem 21. Let $a > 0$. Find the Fourier transform of
$$\sqrt{2\pi} f_a(x) + \frac{\sin(ax)}{ax}$$

where f_a is the function with 1 for $|x| \leq a$ and 0 otherwise.

Solution 21. We obtain
$$\sqrt{2\pi}\left(\frac{\sin(ak)}{ak} + \sqrt{2\pi} f_a(k)\right).$$

Problem 22. Consider the *Hermite-Gauss functions*
$$f_n(x) = \frac{2^{1/4}}{\sqrt{2^n n!}} H_n(\sqrt{2\pi}x) \exp(-\pi x^2), \qquad n = 0, 1, 2, \ldots$$

where H_n is the nth *Hermite polynomial*. They form an orthonormal basis in the Hilbert space $L_2(\mathbb{R})$. Do the Fourier transform of the functions form an orthonormal basis in the Hilbert space $L_2(\mathbb{R})$?

Solution 22. Using the *generating function* of the Hermite-Gauss function
$$F(s, x) = e^{-s^2 + 2sx - x^2/2} = \sum_{n=0}^{\infty} \frac{s^n}{n!} f_n(x)$$

where $f_n(x)$ is the n-th order Hermite-Gauss function. The Fourier transform of $F(s, x)$ is
$$\frac{1}{\sqrt{2\pi}} \int_{\mathbb{R}} \exp(-s^2 + 2sx - x^2/2) e^{-ikx} dx$$

$$= \exp(-s^2 - 2isk - k^2) = \sum_{n=0}^{\infty} \frac{(-1)^n s^n}{n!} f_n(k).$$

```
/* intminfinf.mac */
integrate(exp(-s*s+2*s*x-x*x/2)*exp(-%i*k*x),x,minf,inf);
```

This shows that the Fourier transform of $f_n(x)$ is $(-1)^n f_n(k)$. Owing to the phase factor $(-1)^n$ in general a linear combination of Hermite-Gauss functions is not an eigenfunction of the Fourier transform but these function build an orthogonal basis in the Hilbert space $L_2(\mathbb{R})$.

Problem 23. Let $a > 0$. Show that

$$\int_{-\infty}^{+\infty} e^{-i\omega\tau - a\tau^2} d\tau = \sqrt{\frac{\pi}{a}} \exp\left(-\frac{\omega^2}{4a}\right).$$

Solution 23. We have

$$\int_{-\infty}^{\infty} e^{-i\omega\tau - a\tau^2} d\tau = \exp\left(-\frac{\omega^2}{4a}\right) \int_{-\infty}^{+\infty} \exp\left(-\left(\sqrt{a}\tau + i\frac{\omega}{2\sqrt{a}}\right)^2\right) d\tau$$

$$= \frac{1}{\sqrt{a}} \exp\left(-\frac{\omega^2}{4a}\right) \int_{-\infty+ic}^{\infty+ic} e^{-\sigma^2} d\sigma$$

with $c = \omega/(2\sqrt{a})$. One integrates in the complex plane $\sigma = \xi + i\eta$ over the boundary of the rectangle Q with $\sigma_1 = R_0$, $\sigma_2 = R_0 + ic$, $\sigma_3 = -R_0 + ic$, $\sigma_4 = -R_0$ and utilizes *Cauchy's integral theorem*

$$\int_Q e^{-\sigma^2} d\sigma = 0.$$

With

$$\int_0^c e^{-(R_0+i\eta)^2} d\eta \quad \text{and} \quad \int_c^0 e^{-(-R_0+i\eta)^2} d\eta$$

tending to 0 for $R_0 \to \infty$ we obtain

$$\sqrt{\pi} = \lim_{R_0 \to \infty} \int_{-R_0}^{R_0} e^{-\sigma^2} d\sigma = -\lim_{R_0 \to \infty} \int_{R_0+ic}^{-R_0+ic} e^{-\sigma^2} d\sigma$$

and the result follows.

Problem 24. Consider the second order linear differential equation (Airy differential equation)

$$\frac{d^2u(x)}{dx^2} - xu(x) = 0, \quad \lim_{|x| \to \infty} u(x) = 0.$$

Find a solution applying Fourier transform.

Solution 24. Applying the Fourier transform the Airy differential equation provides the first order linear differential equation

$$\frac{d\hat{u}(k)}{dk} = ik^2\hat{u}(k).$$

Integrating $d\hat{u}/\hat{u} = ik^2 dk$ yields

$$\hat{u}(k) = C \exp(ik^3/3)$$

where C is the constant of integration. The inverse Fourier transform provides

$$u(x) = \frac{C}{2\pi} \int_{\mathbb{R}} \exp(i(kx + k^3/3))dk.$$

For $C = 1$ we have the Airy function $\mathrm{Ai}(x)$.

Problem 25. Consider Banach-Gelfand triple $S(\mathbb{R}) \subset L_2(\mathbb{R}) \subset S'(\mathbb{R})$. Consider the second order linear ordinary differential equation

$$-\frac{d^2u(x)}{dx^2} + \gamma^2 u(x) = h(x), \qquad \lim_{|x|\to\infty} u(x) = 0$$

where $\gamma > 0$ and $h \in L_2(\mathbb{R})$. Find a solution applying the Fourier transform.

Solution 25. The *Green function G* for the problem satisfies

$$-\frac{d^2G(x)}{dx^2} + \gamma^2 G(x) = \delta(x - s), \qquad -\infty < x, s < \infty.$$

Then the Fourier transform of the second order differential equation provides

$$k^2\hat{G}(k) + \gamma^2\hat{G} = \frac{e^{-iks}}{\sqrt{2\pi}} \quad \Rightarrow \quad \hat{G}(k) = \frac{e^{-iks}}{\sqrt{2\pi}(k^2 + \gamma^2)}.$$

Taking the inverse Fourier transform yields

$$G(x; s) = \frac{1}{2\gamma}e^{-\gamma|x-s|}$$

and hence a solution follows at

$$u(x) = \int_{\mathbb{R}} h(s)G(x; s)ds = \frac{1}{2\gamma}\int_{\mathbb{R}} h(s)e^{-\gamma|x-s|}ds$$

for a given $h \in L_2(\mathbb{R})$.

Problem 26. Consider the one-dimensional *diffusion equation*

$$\frac{\partial u}{\partial t} = D\frac{\partial^2 u}{\partial x^2}, \quad x \in \mathbb{R}, \ u(x,0) = f(x), \quad \lim_{|x|\to\infty} u(x,t) = 0.$$

Find a solution applying the Fourier transform.

Solution 26. The fundamental solution $E(x,t)$ solves the diffusion equation

$$\frac{\partial E}{\partial t} = D\frac{\partial^2 E}{\partial x^2}$$

with

$$x \in \mathbb{R}, \ E(x,x_0,0) = \delta(x - x_0), \quad \lim_{|x|\to\infty} E(x,x_0,t) = 0.$$

Taking the Fourier transform

$$\hat{E}(k,x_0,t) = \int_{\mathbb{R}} E(x,x_0,t)\exp(-ikx)dx$$

with respect to the space coordinate x of the differential equation we obtain the linear first order ordinary differential equation

$$\frac{d\hat{E}(k,x_0,t)}{dt} = -Dk^2\hat{E}(k,x_0,t).$$

The initial condition of this ordinary differential equation is given by the Fourier transform of $\delta(x - x_0)$ which is given by e^{-ix_0k}. It follows that

$$\hat{E}(k,x_0,t) = \exp(-ix_0k - Dk^2t).$$

Now the inverse Fourier transform provides

$$E(x,x_0,t) = \frac{1}{\sqrt{4\pi Dt}}\exp(-(x - x_0)^2/(4Dt)).$$

Then with $u(x,0) = f(x)$ we obtain

$$u(x,t) = \int_{\mathbb{R}} \frac{f(x_0)}{\sqrt{4\pi Dt}}e^{-(x-x_0)^2/(4Dt)}dx_0.$$

Problem 27. Consider the two-dimensional Laplace equation on the upper half plane $(y > 0)$

$$\frac{\partial^2 u}{\partial x^2} + \frac{\partial^2 u}{\partial y^2} = 0, \quad x \in \mathbb{R}, \ u(x,0) = g(x), \quad \lim_{y\to\infty} u(x,y) = 0$$

and $g \in L_2(\mathbb{R})$. Use the one-dimensional Fourier transform

$$\hat{u}(k,y) = \int_{\mathbb{R}} e^{-ikx} u(x,y) dx$$

to find a solution.

Solution 27. Applying the one-dimensional Fourier transform provides the second order ordinary differential equation

$$-k^2 \hat{u}(k,y) + \frac{d^2 \hat{u}(k,y)}{dy^2} = 0, \quad \hat{u}(k,0) = \hat{g}(k), \quad \lim_{y \to \infty} \hat{u}(k,y) = 0.$$

Taking into account the boundary conditions we obtain the solution

$$\hat{u}(k,y) = Ce^{-|k|y}$$

and utilizing $\hat{u}(k,0) = \hat{g}(k)$ yields

$$\hat{u}(k,y) = \hat{g}(k)e^{-|k|y}.$$

The inverse Fourier transform then provides

$$u(x,y) = g(x) \star \left(\frac{y}{\pi(x^2 + y^2)} \right) = \frac{1}{\pi} \int_{\mathbb{R}} \frac{yg(x_0)}{(x-x_0)^2 + y^2} dx_0$$

where $y/(\pi(x^2 + y^2))$ is the inverse Fourier transform of $\exp(-|k|y)$ and $\star$ denotes the convolution.

Problem 28. Let $f \in L_1(\mathbb{R}) \cap L_2(\mathbb{R})$. The Fourier transform and its inverse is given by

$$\hat{f}(\omega) = \int_{\mathbb{R}} e^{i\omega t} f(t) dt, \quad f(t) = \frac{1}{2\pi} \int_{\mathbb{R}} e^{-i\omega t} \hat{f}(\omega) d\omega.$$

Then with the pairs $(\psi(t), \phi(t))$, $(\hat{\psi}(\omega), \hat{\phi}(\omega))$ we have (*Parseval's equation*)

$$2\pi \langle \psi(t), \phi(t) \rangle = \langle \hat{\psi}(\omega), \hat{\phi}(\omega) \rangle$$

and $2\pi \langle \psi(t), \psi(t) \rangle = \langle \hat{\psi}(\omega), \hat{\psi}(\omega) \rangle$. Let $0 < \Omega < \infty$ be a fixed cutoff frequency and $C(\mathbb{R})$ be the vector space of continuous functions. Then

$$\mathcal{H} = \{ f \in C(\mathbb{R}) : \operatorname{supp}(\hat{f}) \subset [-\Omega, \Omega] \}$$

is a Hilbert space. Show that

$$|f(t)| \leq \sqrt{\frac{\Omega}{\pi}} \|f(t)\|_{L_2(\mathbb{R})}.$$

Consider the *kernel function* K_t given by

$$K_t(\tau) = \frac{\sin(\Omega(\tau - t))}{\pi(\tau - t)}.$$

Show that $\langle f(t), K_t(\tau) \rangle = f(t)$.

Solution 28. We have applying the Cauchy-Schwarz inequality

$$|f(t)| \leq \frac{1}{2\pi} \sqrt{\int_{-\Omega}^{\Omega} 2\Omega |\hat{f}(\omega)|^2 d\omega} = \frac{1}{\pi} \sqrt{\frac{a}{2} \int_{\mathbb{R}} |\hat{f}(\omega)|^2 d\omega} = \sqrt{\frac{a}{\pi}} \|f(t)\|_{L_2(\mathbb{R})}.$$

The Fourier transform of $K_t(\tau)$ is given by

$$\int_{\mathbb{R}} K_t(\tau) e^{i\omega\tau} d\tau = \begin{cases} e^{i\omega t} & \text{if} \quad \omega \in [-\Omega, \Omega] \\ 0 & \text{otherwise} \end{cases}$$

Then we have

$$\langle f(t), K_t(\tau) \rangle = \int_{\mathbb{R}} f(\tau) \overline{K}_t(\tau) d\tau = \frac{1}{2\pi} \int_{-\Omega}^{\Omega} \hat{f}(\omega) e^{-i\omega t} d\omega = f(t).$$

Problem 29. Let $\mathbf{k} \cdot \mathbf{x} = k_1 x_1 + k_2 x_2 + k_3 x_3$, $k = \sqrt{k_1^2 + k_2^2 + k_3^2}$ and $r = \sqrt{x_1^2 + x_2^2 + x_3^2}$. Find

$$\int_{\mathbb{R}^3} \frac{e^{i\mathbf{k}\cdot\mathbf{x}}}{r} d^3 r \quad \text{and} \quad \int_{\mathbb{R}^3} \frac{e^{-r/a}}{r} e^{i\mathbf{k}\cdot\mathbf{x}} d^3 r$$

with $a > 0$.

Solution 29. Utilizing *spherical coordinates*

$$x_1(r, \theta, \phi) = r \cos(\beta) \sin(\alpha), \quad x_2(r, \theta, \phi) = r \cos(\beta) \cos(\alpha), \quad x_3(r, \theta, \phi) = r \sin(\beta)$$

where $-\pi \leq \alpha \leq \pi$ and $-\pi/2 \leq \beta \leq \pi/2$ we obtain

$$\int_{\mathbb{R}^3} \frac{e^{i\mathbf{k}\cdot\mathbf{x}}}{r} d^3 r = \frac{4\pi}{k^2}, \quad \int_{\mathbb{R}^3} \frac{e^{-r/a}}{r} e^{i\mathbf{k}\cdot\mathbf{x}} d^3 r = \frac{4\pi}{k^2 + a^{-2}}.$$

Problem 30. Let V be a metric vector space. A *reproducing kernel* Hilbert space on V is a Hilbert space $\mathcal{H}$ of functions on V such that for each $x \in V$, the point evaluation functional

$$\delta_x(f) := f(x), \qquad f \in \mathcal{H}$$

is continuous. A reproducing kernel Hilbert space $\mathcal{H}$ possesses a unique reproducing kernel K which is a function on $V \times V$ characterized by the properties that for all $f \in \mathcal{H}$ and $x \in V$, $K(x, \cdot) \in \mathcal{H}$ and

$$f(x) = \langle f, K(x, \cdot) \rangle_{\mathcal{H}}$$

where $\langle \cdot, \cdot \rangle_{\mathcal{H}}$ denotes the inner product on $\mathcal{H}$. The reproducing kernel K uniquely determines the reproducing kernel Hilbert space $\mathcal{H}$. The reproducing kernel Hilbert space of a reproducing kernel K is denoted by $\mathcal{H}_K$. The *Paley-Wiener space* is defined by

$$S := \{ f \in C(\mathbb{R}^d) \cap L_2(\mathbb{R}^d) : \operatorname{supp}\hat{f} \subseteq [-\pi, \pi]^d \}$$

is a reproducing kernel Hilbert space. The Fourier transform of $f \in L_1(\mathbb{R}^d)$ is given by

$$\hat{f}(\mathbf{k}) := \frac{1}{(\sqrt{2\pi})^{2d}} \int_{\mathbb{R}^{2d}} f(\mathbf{x}) e^{-i\mathbf{k}\cdot\mathbf{x}} d\mathbf{x}, \qquad \mathbf{k} \in \mathbb{R}^d$$

where $\mathbf{k} \cdot \mathbf{x} := k_1 x_1 + \cdots + k_d x_d$ is the inner product in $\mathbb{R}^d$. The norm on the vector space S inherits from that in $L_2(\mathbb{R}^d)$. Show that the reproducing kernel for the Paley-Wiener space S is the *sinc function*

$$\operatorname{sinc}(\mathbf{x}, \mathbf{y}) := \prod_{j=1}^{d} \frac{\sin(\pi(x_j - y_j))}{\pi(x_j - y_j)}, \qquad \mathbf{x}, \mathbf{y} \in \mathbb{R}^d.$$

Solution 30. We consider the case $d = 1$. The set S is a closed subspace of $C(\mathbb{R}) \cap L_2(\mathbb{R})$ since the Fourier transform $\mathcal{F}$ is a unitary operator and $\mathcal{F}^{-1}(L_2([-\pi, \pi]))$. The Hilbert space $L_2([-\pi, \pi])$ is identified to a closed subspace of $L_2(\mathbb{R})$ by extending to 0 on $\mathbb{R}$ the functions of $L_2([-\pi, \pi])$. By using the inverse Fourier transform, any $f \in S$ can be represented as

$$f(t) = \frac{1}{\sqrt{2\pi}} \int_{-\pi}^{\pi} \hat{f}(k) e^{ikt} dk = \langle \hat{f}, e^{-ikt}/\sqrt{2\pi} \rangle_{L_2([-\pi,\pi])}.$$

Using Cauchy-Schwarz's inequality and Parseval equality

$$\|f\|_{L_2(\mathbb{R})} = \|\hat{f}\|_{L_2([-\pi,\pi])}$$

we have for every $t \in \mathbb{R}$

$$|f| \leq \|\hat{f}\|_{L_2([-\pi,\pi])} \|e^{-ikt}/\sqrt{2\pi}\|_{L_2([-\pi,\pi])} = \|f\|_{L_2(\mathbb{R})}$$

for $f \in S$. Applying Plancherel-Parseval theorem and the inverse Fourier transform

$$\mathcal{F}^{-1}\left(\frac{-iks}{\sqrt{2\pi}} \chi_{[-\pi,\pi]} \right)(t) = \frac{\sin(\pi(t-s))}{\pi(t-s)}, \qquad s \in \mathbb{R}$$

we obtain

$$f(s) = \langle \hat{f}, e^{-ikt}/\sqrt{2\pi} \rangle_{L_2([-\pi,\pi])} = \langle f, \sin(\pi(\cdot - s))/(\pi(\cdot - s)) \rangle_{L_2(\mathbb{R})}, \qquad s \in \mathbb{R}.$$

The higher dimensional case can be proved in an analog manner.

6.3   Supplementary Problems

Problem 1. Consider the six 3×3 permutation matrices P_{123}, P_{132}, P_{213}, P_{231}, P_{312}, P_{321} with P_{123} the 3×3 identity matrix and

$$P_{321} = \begin{pmatrix} 0 & 0 & 1 \\ 0 & 1 & 0 \\ 1 & 0 & 0 \end{pmatrix}.$$

Let F_3 be the 3×3 Fourier matrix

$$F_3 = \frac{1}{\sqrt{3}} \begin{pmatrix} 1 & 1 & 1 \\ 1 & e^{-2\pi i/3} & e^{-4\pi i/3} \\ 1 & e^{-4\pi i/3} & e^{-2\pi i/3} \end{pmatrix}$$

with $F_3^* = F_3^{-1}$. Show that

$$F_3 P_{jk\ell} F_3^* = (1) \oplus U_{jk\ell}$$

where the six matrices 2×2 matrices are unitary and form a group under matrix multiplication. Obviously

$$F_3 P_{123} F_3^* = (1) \oplus \begin{pmatrix} 1 & 0 \\ 0 & 1 \end{pmatrix}.$$

Problem 2. Let $n \geq 2$. *Circulant matrices* $(n \times n)$ are those that are diagonalized by the discrete Fourier transform matrix

$$F_n = \frac{1}{\sqrt{n}} \begin{pmatrix} 1 & 1 & 1 & \cdots & 1 \\ 1 & \omega & \omega^2 & \cdots & \omega^{n-1} \\ 1 & \omega^2 & \omega^4 & \cdots & \omega^{2n-2} \\ \vdots & \vdots & & \ddots & \\ 1 & \omega^{n-1} & \omega^{2n-2} & \cdots & \omega^{(n-1)(n-1)} \end{pmatrix}$$

with $\omega = \exp(-2\pi i/n)$. Hence an $n \times n$ circulant matrix C can be written as

$$C = F_n \Delta U^*$$

where Δ is a diagonal matrix and contains the eigenvalues of C. Let $n = 3$. Then

$$F_3 = \frac{1}{\sqrt{3}} \begin{pmatrix} 1 & 1 & 1 \\ 1 & e^{4\pi i/3} & e^{2\pi i/3} \\ 1 & e^{2\pi i/3} & e^{4\pi i/3} \end{pmatrix}.$$

Let

$$\Delta = \begin{pmatrix} 0 & 0 & 0 \\ 0 & 1 & 0 \\ 0 & 0 & 2 \end{pmatrix}.$$

Find $F_3 C F_3^*$. Let

$$P = \begin{pmatrix} 0 & 1 & 0 \\ 0 & 0 & 1 \\ 1 & 0 & 0 \end{pmatrix}.$$

Find $F_3^* P F_3$.

Problem 3. Let $\sigma > 0$. Show that the Fourier transform of the *Gaussian function*

$$g_\sigma(x) = \frac{1}{\sqrt{2\pi}\sigma} \exp\left(-\frac{x^2}{2\sigma^2}\right)$$

is again a Gaussian function $\hat{g}_\sigma(k) = e^{-\sigma^2 k^2 / 2}$. We have $\int_{-\infty}^{\infty} g_\sigma(x) dx = 1$. Is

$$\int_{-\infty}^{\infty} \hat{g}_{\sigma_k}(k) dk = 1 ?$$

Problem 4. The Hilbert transform $h(t)$ of the function $f(t)$ is the principal value of the convolution of $f(t)$ with the *kernel function* $k(t) = 1/(\pi t)$

$$h(t) = \int_{-\infty}^{\infty} f(s) k(t - s) ds = \int_{-\infty}^{\infty} f(s) \frac{1}{t - s} ds.$$

Let

$$G(\omega) = \int_{-\infty}^{\infty} g(t) \exp(i\omega t) dt$$

be the Fourier transform of g. Show that the Hilbert transform can be written as

$$H(\omega) = F(\omega) K(\omega) = -i\,\mathrm{sgn}(\omega) F(\omega).$$

Problem 5. Let $f \in L_2(\mathbb{R})$ and let us call the Fourier transformed function $\hat{f}$. Is $\hat{f} \in L_2(\mathbb{R})$? What is preserved under the Fourier transform?

Problem 6. Consider the Hilbert space $L_2(\mathbb{R})$. Let $T > 0$. Consider the function in $L_2(\mathbb{R})$

$$f(t) = \begin{cases} A\cos(\Omega t) & \text{for} \quad -T < t < T \\ 0 & \text{otherwise} \end{cases}$$

where A is a positive constant. Calculate the Fourier transform using the identity $\cos(\Omega t) \equiv (e^{i\Omega t} + e^{-i\Omega t})/2$.

Problem 7. Let $N_1, N_2 \in \mathbb{Z}$ with $N_1 \le N_2$ and f be a function of the real variable x such that f possesses a Fourier series expansion over any unit interval in the range $N_1 - 1/2 - \gamma < x < N_2 + 1 - \gamma$. Show that

$$\sum_{\ell=N_1}^{N_2} f(\ell) = \sum_{m=-\infty}^{\infty} \exp(2m\pi i(\gamma - 1/2)) \int_{N_1}^{N_2+1} f(s + \gamma - 1/2)\exp(2m\pi is)ds$$

for any real γ with $|\gamma| < 1/2$.

Problem 8. (i) Let $n \in \mathbb{Z}$ and $\omega_c > 0$. Do the functions

$$\left\{ \phi_n(t) := \frac{1}{\sqrt{\pi}} \frac{\sin(\omega_c t - n\pi)}{(\omega_c t - n\pi)} : n \in \mathbb{Z} \right\}$$

form an orthonormal basis in the Hilbert space $L_2(\mathbb{R})$?
(ii) Show that the Fourier transform of $\phi_n(t)$ is given by

$$\hat{\phi}_n(\omega) = \begin{cases} \frac{\sqrt{\pi}}{\omega_c} \exp(-in\pi\omega/\omega_c) & \text{for } |\omega| \le \omega_c \\ 0 & \text{for } |\omega| > \omega_c \end{cases}$$

Problem 9. Show that the inverse Fourier transform of the symmetric function

$$\hat{f}(\omega) = \begin{cases} 1 & \text{for} \quad \pi < |\omega| < 2\pi \\ 0 & \text{otherwise} \end{cases}$$

is given by

$$f(t) = \frac{\sin(\pi t/2)}{\pi t/2} \cos\left(\frac{3\pi t}{2}\right).$$

Problem 10. Find functions f such that

$$f(\omega) = \int_{-\infty}^{\infty} f(t)e^{i\omega t}dt.$$

Problem 11. Consider a d-dimensional lattice L. Assume that the lattice is finite, cubic and periodic, i.e. $L = \mathbb{Z}_N^d$. Let $\mathbf{x}$ be a lattice site and $|L|$ be the cardinality of the set L. The Fourier transform of a lattice function $f : L \to \mathbb{C}$ is defined as

$$\hat{f}(\mathbf{k}) = \frac{1}{\sqrt{|L|}} \sum_{\mathbf{x} \in L} \exp(-i\mathbf{k} \cdot \mathbf{x}) f(\mathbf{x})$$

for $\mathbf{k} \in B$, where the *Brillouin zone* of a cubic lattice is given by $B = \frac{2\pi}{N} L$. Show that the inverse Fourier transform is given by

$$f(\mathbf{x}) = \frac{1}{\sqrt{|L|}} \sum_{\mathbf{k} \in B} \hat{f}(\mathbf{k}) \exp(i\mathbf{k} \cdot \mathbf{x}).$$

Problem 12. Let

$$\mathbb{C}^+ := \{ z \in \mathbb{C} : \Im(z) > 0 \},$$

$f \in L_2([0, \infty))$ and $k = k_1 + ik_2$ $(k_1, k_2 \in \mathbb{R})$ with $k_2 > 0$. Show that g defined by

$$g(x) = \int_0^\infty f(x) e^{2\pi i k x} dx$$

is analytic in $\mathbb{C}^+$. Note that $\left| e^{2\pi i (k_1 + ik_2)} \right| = e^{-2\pi k_2 x}$.

Problem 13. Let $f \in L_1(\mathbb{R}) \cap L_2(\mathbb{R})$. Show that $\mathcal{F}(\mathcal{F}[f(x)]) = 2\pi f(-x)$.

Problem 14. Let $f \in L_1(\mathbb{R})$ and assume that f is continuously differentiable and

$$\frac{df}{dx} \in L_1(\mathbb{R}).$$

Show that

$$\mathcal{F}\left[\frac{df}{dx}\right] = -ik\mathcal{F}[f].$$

Problem 15. Let

$$\hat{S}(\omega) = \int_{-\infty}^{\infty} S(t) e^{i\omega t} dt, \qquad S(t) = \frac{1}{2\pi} \int_{-\infty}^{\infty} \hat{S}(\omega) e^{-i\omega t} d\omega.$$

Show that (*uncertainty relation*)

$$\Delta_t \cdot \Delta_\omega \geq \frac{1}{2}$$

where

$$\Delta_t^2 := \frac{\int\limits_{-\infty}^{\infty} t^2 |S(t)|^2 dt}{\int\limits_{-\infty}^{\infty} |S(t)|^2 dt}, \qquad \Delta_\omega^2 := \frac{\int\limits_{-\infty}^{\infty} \omega^2 |S(\omega)|^2 d\omega}{\int\limits_{-\infty}^{\infty} |S(\omega)|^2 d\omega}.$$

Problem 16. Let

$$f(t) := \frac{1}{\sqrt{2\pi}\sigma_t} \exp\left(-\frac{t^2}{2\sigma_t^2}\right).$$

Show that

$$\hat{f}(\omega) = \exp\left(-\frac{\omega^2}{2\sigma_\omega^2}\right).$$

Problem 17. Let

$$f(t) = \exp(2\pi i \alpha t)$$

where $\alpha \in \mathbb{R}$. Obviously, $f(t) \notin L_2(\mathbb{R})$. Show that windowed Fourier transform of f with respect to the Gaussian window $g(t) = \exp(-\pi t^2)$ is well defined. Find the windowed Fourier transform.

Problem 18. We consider one space dimension. Let $\hat{p}$, $\hat{x}$ be the momentum and position operators, respectively with

$$\hat{p} = -i\hbar \frac{d}{dx}.$$

The *characteristic operator* is defined as

$$\hat{M}(\hat{x}, \hat{p}, \tau, \phi) := \exp(i(\phi\hat{x} + \tau\hat{p})/\hbar)$$

where τ and ϕ are real parameters with dimensions of position and momentum, respectively. One defines the *characteristic function* as (expectation value)

$$M(\tau, \phi) := \langle\psi| \exp(i(\phi\hat{x} + \tau\hat{p})/\hbar)|\psi\rangle.$$

where $|\psi\rangle$ is a normalized state of the system. Show that

$$M(\tau, \phi) = \langle\psi|e^{i\phi\hat{x}/\hbar}e^{i\tau\hat{p}/\hbar}e^{i\phi\tau/(2\hbar)}|\psi\rangle$$

utilizing that

$$[i\phi\hat{x}/\hbar, i\tau\hat{p}/\hbar] = -\frac{\phi\tau}{\hbar^2}[\hat{x}, \hat{p}] = -\frac{i\phi\tau}{\hbar}I.$$

The *Wigner function* $W(x,p)$ is defined in terms of the characteristic function as

$$W(x,p) = \frac{1}{(2\pi\hbar)^2} \int_{-\infty}^{+\infty} \int_{-\infty}^{+\infty} d\tau d\phi \, e^{-i(\phi x + \tau p)/\hbar} M(\tau, \phi).$$

Show that characteristic function can be expressed as Fourier transform of the Wigner function

$$M(\tau, \phi) = \int_{-\infty}^{+\infty} \int_{-\infty}^{+\infty} dx dp \exp(i(\phi x + \tau p)/\hbar) W(x,p).$$

Problem 19. Let $f, g \in L_2(\mathbb{R}^d)$. One defines the *cross Wigner function*

$$W(f,g)(\mathbf{x}, \boldsymbol{\xi}) := \int_{\mathbb{R}^d} \exp(-2\pi i \boldsymbol{\xi} \cdot \mathbf{y}) f(\mathbf{x} + \mathbf{y}/2) \overline{g(\mathbf{x} - \mathbf{y}/2)} d\mathbf{y}.$$

Let $\alpha, \beta > 0$ and $f(x) = e^{-\alpha x^2/2}$, $g(x) = e^{-\beta x^2/2}$. Find $W(f,g)$. Discuss.

Problem 20. Show that the Fourier transform of the *rectangular window* of size N

$$w_n = \begin{cases} 1 & \text{for} \quad 0 \le n \le N-1 \\ 0 & \text{otherwise} \end{cases}$$

is

$$W(e^{i\omega}) = \frac{\sin(\omega N/2)}{\sin(\omega/2)} e^{-i\omega(N-1)/2}.$$

Problem 21. Consider the Hilbert space $L_2(\mathbb{R}^n)$ and let $\mathcal{F}$ be the Fourier transform operator on $L_2(\mathbb{R}^n)$, i.e. $\tilde{f}(\mathbf{k}) = \mathcal{F}f(\mathbf{x})$. Let $m \in \mathbb{N}_0$. The *Sobolev space* $W^{m,2}(\mathbb{R}^n)$ is defined as the set of all f such that

$$\int_{\mathbb{R}^n} (1 + |\mathbf{k}|^2)^m |\tilde{f}(\mathbf{k})|^2 d^n\mathbf{k} < \infty.$$

Show that $W^{m,2}$ is a Hilbert space with respect to the inner product

$$\langle f, g \rangle_m := \int_{\mathbb{R}^n} (1 + |\mathbf{k}|^2)^m \tilde{f}(\mathbf{k}) \overline{\tilde{g}(\mathbf{k})} d^n\mathbf{k} < \infty$$

with $W^{0,2} = L_2(\mathbb{R}^n)$.

Problem 22. Let $z = x + iy = re^{i\theta}$ $(x, y \in \mathbb{R})$, $r \geq 0$ and

$$D := \{ z \in \mathbb{C} : |z| < 1 \}$$

be the *open unit disk*. Furthermore

$$\Delta u = \frac{\partial^2 u}{\partial x^2} + \frac{\partial^2 u}{\partial y^2} = \frac{\partial^2 u}{\partial r^2} + \frac{1}{r}\frac{\partial u}{\partial r} + \frac{1}{r^2}\frac{\partial^2 u}{\partial \theta^2}$$

where we utilized *polar coordinates*. Assume that $u \in C(\overline{D}) \cap C^2(D)$ and $\Delta u(z) = 0$ for $z \in D$. Let $g = u_{\partial D}$, where ∂D is the boundary of D, i.e.

$$\partial D := \{ z \in \overline{D} : |\dot{z}| = 1 \}$$

and

$$\hat{g}(k) := \frac{1}{2\pi} \int_{-\pi}^{\pi} g(e^{ik\theta}) e^{-ik\theta} \, d\theta.$$

Define

$$\hat{u}(r, k) := \frac{1}{2\pi} \int_{-\pi}^{\pi} u(re^{i\theta}) e^{-ik\theta} \, d\theta.$$

Show that $\hat{u}(r, k)$ satisfies the second order ordinary differential equation

$$\frac{1}{r}\frac{\partial}{\partial r}\left(r\frac{\partial}{\partial r}\hat{u}(r, k) \right) = \frac{1}{r^2}k^2\hat{u}(r, k) \text{ for } r \in (0, 1).$$

Problem 23. Consider the Hilbert space $L_2(\mathbb{R})$. Show that the *Cauchy densities*

$$f_\tau(x) = \frac{\tau}{\pi(\tau^2 + x^2)}$$

define a convolution semigroup on $\mathbb{R}$ utilizing that

$$\int_{\mathbb{R}} f_\tau(x) e^{-ix\zeta} \, dx = e^{-\tau|\zeta|}$$

for all $\zeta \in \mathbb{R}$.

Problem 24. Show that the analytic function $f : \mathbb{R} \to \mathbb{R}$

$$f(x) = \text{sech}(\pi x) \equiv \frac{1}{\cosh(\pi x)} \equiv \frac{2}{e^{\pi x} + e^{-\pi x}}$$

is an element of $L_2(\mathbb{R})$ and $L_1(\mathbb{R})$. Find the Fourier transform of the function.

Chapter 7

Wavelets

7.1 Introduction

The scalar product in $L_2(\mathbb{R})$ is defined as

$$\langle f, g \rangle := \int\limits_{-\infty}^{\infty} f(x)\overline{g(x)}dx.$$

Thus the induced norm is given by $\|f\|_2 := \langle f, f \rangle^{1/2}$, where $f, g \in L_2(\mathbb{R})$. Let $f \in L_2(\mathbb{R})$. We consider $f(2^j x - k)$. The function

$$f(2^j x - k)$$

is obtained from the function $f(x)$ by a binary dilation (i.e. *dilation* by 2^j) and a *dyadic translation* (of $k/2^j$). For any $j, k \in \mathbb{Z}$, we have

$$\|f(2^j x - k)\|_2 = \left(\int\limits_{-\infty}^{\infty} |f(2^j x - k)|^2 dx \right)^{1/2} = 2^{-j/2}\|f\|_2.$$

Hence, if a function $\psi \in L_2(\mathbb{R})$ has unit length, then all of the functions $\psi_{j,k}$, defined by

$$\psi_{j,k}(x) := 2^{j/2}\psi(2^j x - k), \qquad j, k \in \mathbb{Z}$$

also have unit length, i.e.

$$\|\psi_{j,k}\|_2 = \|\psi\|_2 = 1, \qquad j, k \in \mathbb{Z}.$$

Definition. A function $\psi \in L_2(\mathbb{R})$ is called an orthogonal wavelet, if the family $\{\psi_{j,k}\}$, as defined in

$$\psi_{j,k}(x) := 2^{j/2}\psi(2^j x - k), \qquad j, k \in \mathbb{Z}$$

is an orthonormal basis of $L_2(\mathbb{R})$; that is,

$$\langle \psi_{j,k}, \psi_{l,m} \rangle = \delta_{j,l}\delta_{k,m}, \qquad j, k, l, m \in \mathbb{Z}$$

and every $f \in L_2(\mathbb{R})$ can be written as

$$f(x) = \sum_{j,k=-\infty}^{\infty} c_{j,k}\psi_{j,k}(x)$$

where the convergence of the series is in $L_2(\mathbb{R})$, namely

$$\lim_{M_1,N_1,M_2,N_2 \to \infty} \left\| f(x) - \sum_{j=-M_2}^{N_2} \sum_{k=-M_1}^{N_1} c_{j,k}\psi_{j,k}(x) \right\|_2 = 0.$$

We are interested in wavelet functions ψ ([13],[14],[38],[47]) whose binary dilations and dyadic translations are enough to represent all the functions in $L_2(\mathbb{R})$.

The simplest example of an orthogonal wavelet is the *Haar function* ψ_H defined by

$$\psi_H(x) := \begin{cases} 1 & \text{for} & 0 \leq x < \frac{1}{2} \\ -1 & \text{for} & \frac{1}{2} \leq x < 1 \\ 0 & \text{otherwise.} \end{cases}$$

Then

$$\psi_{m,n}(x) := 2^{-m/2}\psi(2^{-m}x - n)$$

where $m, n \in \mathbb{Z}$. Thus $\psi_{m,n}$ is given by

$$\psi_{m,n}(x) = \begin{cases} 1 & \text{for} & 2^m n \leq x < 2^m n + 2^{m-1} \\ -1 & \text{for} & 2^m n + 2^{m-1} \leq x < 2^m n + 2^m \\ 0 & \text{otherwise} \end{cases}$$

Another example is the *Littlewood-Paley orthonormal basis* wavelets [46]. The mother wavelet of this set is

$$L(x) := \frac{1}{\pi x}(\sin(2\pi x) - \sin(\pi x)).$$

Using the definition

$$L_{m,n}(x) := 2^{-m/2}L(2^{-m}x - n), \quad m, n \in \mathbb{Z}$$

we generate an orthonormal set in $L_2(\mathbb{R})$, where $m, n \in \mathbb{Z}$.

The series representation of f

$$f(x) = \sum_{j,k=-\infty}^{\infty} c_{j,k}\psi_{j,k}(x)$$

is called a *wavelet series*. Analogous to the notion of Fourier coefficients, the *wavelet coefficients* $c_{j,k} \in \mathbb{C}$ are given by

$$c_{j,k} = \langle f, \psi_{j,k}\rangle.$$

If we define an integral transform W_ψ on $L_2(\mathbb{R})$ by

$$(W_\psi f)(b,a) := |a|^{-\frac{1}{2}} \int_{-\infty}^{\infty} f(x)\overline{\psi\left(\frac{x-b}{a}\right)}dx, \qquad f \in L_2(\mathbb{R})$$

then the wavelet coefficients can be written as

$$c_{j,k} = (W_\psi f)\left(\frac{k}{2^j}, \frac{1}{2^j}\right).$$

The linear transformation W_ψ is called the *integral wavelet transform* relative to the basic wavelet ψ. Hence, the (j,k)-th wavelet coefficient of f is given by the integral wavelet transformation of f evaluated at the *dyadic position* $b = k/2^j$ with *binary dilation* $a = 2^{-j}$, where the same wavelet ψ is used to generate the wavelet series and to define the integral wavelet transform.

The integral wavelet transform enhances the value of the (integral) Fourier transform $\mathcal{F}$ defined above. The function f has to be reconstructed from the values of $(W_\psi f)(b,a)$. Any formula that expresses every $f \in L_2(\mathbb{R})$ in terms of $(W_\psi f)(b,a)$ will be called an *inverse formula*, and the (kernel) function $\tilde{\psi}$ to be used in this formula will be called a dual of the basic wavelet ψ. The function ψ can be used as a basic wavelet, only if an inversion formula exists. In order to find f from $W_\psi f$, one needs to know the constant

$$C_\psi := \int_{-\infty}^{\infty} \frac{|\hat{\psi}(\omega)|^2}{|\omega|}d\omega < \infty$$

where $\hat{\psi}$ is the Fourier transform of ψ. The finiteness of this constant restricts the class of $L_2(\mathbb{R})$ functions ψ that can be used as basic wavelets in the definition of the integral wavelet transform. In particular, if ψ must also be a window function, then ψ is necessarily in $L_1(\mathbb{R})$, i.e.

$$\int_{-\infty}^{\infty} |\psi(x)|dx < \infty$$

Thus the function $\hat{\psi}$ is a continuous function in $\mathbb{R}$. It follows that $\hat{\psi}$ must vanish at the origin. Thus

$$\int_{-\infty}^{\infty} \psi(x)dx = 0.$$

So, the graph of a basic wavelet ψ is a small wave. With the constant C_ψ, we have the following reconstruction formula

$$f(x) = \frac{1}{C_\psi} \int_{-\infty}^{\infty} \int_{-\infty}^{\infty} \{(W_\psi f)(b,a)\} \left\{ \frac{1}{|a|^{\frac{1}{2}}} \psi \left(\frac{x-b}{a} \right) \right\} \frac{dadb}{a^2}, \qquad f \in L_2(\mathbb{R}).$$

One cannot expect uniqueness of this dual.

Definition. A function $\psi \in L_2(\mathbb{R})$ is called a *dyadic wavelet* if it satisfies the stability condition

$$A \le \sum_{j=-\infty}^{\infty} |\hat{\psi}(2^{-j}\omega)|^2 \le B$$

for almost all $\omega \in \mathbb{R}$ for some constants A and B with $0 < A \le B < \infty$.

The simplest way to build two-dimensional wavelet bases is to use separable products (tensor products) of a one-dimensional wavelet ψ and scaling function ϕ. This provides the following scaling function

$$\Phi(x_1, x_2) = \phi(x_1)\phi(x_2)$$

and there are three wavelets

$$\psi^{(1)}(x_1, x_2) = \phi(x_1)\psi(x_2)$$
$$\psi^{(2)}(x_1, x_2) = \psi(x_1)\phi(x_2)$$
$$\psi^{(3)}(x_1, x_2) = \psi(x_1)\psi(x_2).$$

There are also non-separable two-dimensional wavelets.

7.2 Solved Problems

Problem 1. Consider the function $\phi : \mathbb{R} \to \mathbb{R}$

$$\phi(x) := \begin{cases} 1 & \text{for} \quad x \in [0,1] \\ 0 \text{ otherwise} \end{cases}$$

Find $\psi(x) := \phi(2x) - \phi(2x - 1)$. Calculate

$$\int_{-\infty}^{\infty} \psi(x) dx.$$

Solution 1. We obtain

$$\psi(x) := \begin{cases} 1 & \text{for} \quad x \in [0, 1/2] \\ -1 & \text{for} \quad x \in (1/2, 1] \\ 0 & \text{otherwise} \end{cases}$$

and

$$\int_{-\infty}^{\infty} \psi(x) dx = 0.$$

Problem 2. Consider the Hilbert space $L_2([0,1])$ and the function

$$f(x) = x^2$$

in this Hilbert space. Project the function f onto the subspace of $L_2([0,1])$ spanned by the functions $\phi(x)$, $\psi(x)$, $\psi(2x)$, $\psi(2x - 1)$, where

$$\phi(x) := \begin{cases} 1 & \text{for} \quad 0 \leq x < 1 \\ 0 \text{ otherwise} \end{cases}$$

$$\psi(x) := \begin{cases} 1 & \text{for} \quad 0 \leq x < 1/2 \\ -1 & \text{for} \quad 1/2 \leq x < 1 \\ 0 & \text{otherwise} \end{cases}.$$

This is related to the Haar wavelet expansion of f. The function ϕ is called the *father wavelet* and ψ is called the *mother wavelet*.

Solution 2. We have

$$\langle \phi, f \rangle = \int_0^1 \phi(x) f(x) dx = \int_0^1 x^2 dx = \frac{1}{3}.$$

$$\langle \psi, f \rangle = \int_0^1 \psi(x) f(x) dx = \int_0^{1/2} x^2 dx - \int_{1/2}^1 x^2 dx = -\frac{1}{4}.$$

Problem 3. Consider the function $H \in L_2(\mathbb{R})$

$$H(x) = \begin{cases} 1 & 0 \le x \le 1/2 \\ -1 & 1/2 \le x \le 1 \\ 0 & \text{otherwise} \end{cases}$$

Let

$$H_{m,n}(x) := 2^{-m/2} H(2^{-m} x - n)$$

where $m, n \in \mathbb{Z}$. Show that

$$\langle H_{m,n}(x), H_{k,l}(x) \rangle = \delta_{m,k} \delta_{n,l}, \qquad k, l \in \mathbb{Z}$$

where $\langle \cdot, \cdot \rangle$ denotes the scalar product in the Hilbert space $L_2(\mathbb{R})$. Expand the function

$$f(x) = \exp(-|x|)$$

with respect to $H_{m,n}$. The functions $H_{m,n}$ form an orthonormal basis in the Hilbert space $L_2(\mathbb{R})$.

Solution 3. We have

$$H_{m,n}(x) = 2^{-\frac{m}{2}} H(2^{-m}x - n) = \begin{cases} 2^{-\frac{m}{2}} & 0 \le 2^{-m}x - n \le \frac{1}{2} \\ -2^{-\frac{m}{2}} & \frac{1}{2} \le 2^{-m}x - n \le 1 \\ 0 & \text{otherwise} \end{cases}$$

$$= \begin{cases} 2^{-\frac{m}{2}} & 2^m \le x \le 2^m(n + \frac{1}{2}) \\ -2^{-\frac{m}{2}} & 2^m(n + \frac{1}{2}) \le x \le 2^m(n+1) \\ 0 & \text{otherwise} \end{cases}$$

Thus

$$H_{1,1}(x) = \begin{cases} \frac{1}{\sqrt{2}} & 2 \le x \le 3 \\ -\frac{1}{\sqrt{2}} & 3 \le x \le 4 \\ 0 & \text{otherwise} \end{cases} \qquad H_{1,2}(x) = \begin{cases} \frac{1}{\sqrt{2}} & 4 \le x \le 5 \\ -\frac{1}{\sqrt{2}} & 5 \le x \le 6 \\ 0 & \text{otherwise} \end{cases}$$

$$H_{2,1}(x) = \begin{cases} \frac{1}{2} & 4 \le x \le 6 \\ -\frac{1}{2} & 6 \le x \le 8 \\ 0 & \text{otherwise} \end{cases} \qquad H_{2,2}(x) = \begin{cases} \frac{1}{2} & 8 \le x \le 10 \\ -\frac{1}{2} & 10 \le x \le 12 \\ 0 & \text{otherwise} \end{cases}$$

$$\langle H_{m,n}(x), H_{k,l}(x) \rangle = \int_{-\infty}^{\infty} H_{m,n}(x) H_{k,l}(x) dx := I_{m,n,k,l}.$$

The non-zero intervals are

$$I_{m,n} := (2^m n, 2^m(n+1)), \quad I_{k,l} := (2^k l, 2^k(l+1)).$$

We consider the different cases

(1) $m = k$, $n = l$

$$I_{m,n,k,l} = \int_{2^m n}^{2^m (n+1)} 2^{-m} dx = 1.$$

(2) $m = k$, $n \neq l$ $I_{mnkl} = 0$ since $I_{mn} \cap I_{kl} = \emptyset$.

(3) $m \neq 0$, suppose without loss of generality that $m < k$. Either $I_{m,n} \cap I_{k,l} = (I_{m,n,k,l} = 0)$, or $I_{mn} \subset I_{kl}$ (as shown below). We have the following

$$2^k l \leq 2^m n < 2^k (l + \frac{1}{2}) \Rightarrow 2^{k-m} l \leq n < 2^{k-m} (l + \frac{1}{2})$$

$$\Rightarrow 2^{k-m} l \leq n + 1 \leq 2^{k-m} (l + \frac{1}{2})$$

$$\Rightarrow 2^k l \leq 2^m (n + 1) \leq 2^k (l + \frac{1}{2})$$

$$2^k (l + \frac{1}{2}) \leq 2^m n < 2^k (l + 1) \Rightarrow 2^k (l + \frac{1}{2}) \leq 2^m (n + 1) \leq 2^k (l + 1)$$

$$2^k l < 2^m (n + 1) \leq 2^k (l + \frac{1}{2}) \Rightarrow 2^k l \leq 2^m n \leq 2^k (l + \frac{1}{2})$$

$$2^k (l + \frac{1}{2}) < 2^m (n + 1) \leq 2^k (l + 1) \Rightarrow 2^k (l + \frac{1}{2}) \leq 2^m n \leq 2^k (l + 1)$$

which gives

$$I_{m,n,k,l} = \pm \int_{2^m n}^{2^m (n+\frac{1}{2})} 2^{-\frac{1}{2}(m+k)} dx \mp \int_{2^m (n+\frac{1}{2})}^{2^m (n+1)} 2^{-\frac{1}{2}(m+k)} dx = 0.$$

Thus $I_{m,n,k,l} = \delta_{m,n} \delta_{k,l}$.

$$\langle f(x), H_{mn}(x) \rangle = \int_{-\infty}^{\infty} f(x) H_{mn}(x) dx$$

$$= \int_{2^m n}^{2^m (n+\frac{1}{2})} 2^{-\frac{m}{2}} e^{-|x|} dx - \int_{2^m (n+\frac{1}{2})}^{2^m (n+1)} 2^{-\frac{m}{2}} e^{-|x|} dx$$

$$= 2^{-\frac{m}{2}} \begin{cases} -e^{-x} \Big|_{2^m n}^{2^m (n+\frac{1}{2})} + e^{-x} \Big|_{2^m (n+\frac{1}{2})}^{2^m (n+1)} & n \geq 0 \\ e^x \Big|_{2^m n}^{2^m (n+\frac{1}{2})} - e^x \Big|_{2^m (n+\frac{1}{2})}^{2^m (n+1)} & n < 0 \end{cases}$$

$$= -\frac{n - \delta_{n,0}}{|n| + \delta_{n,0}} 2^{-\frac{m}{2}} (2e^{-2^m |n+\frac{1}{2}|} - e^{-2^m |n|} - e^{-2^m |n+1|}).$$

The expansion is given by

$$f(x) = \sum_{m,n \in \mathbf{Z}} \langle f(x), H_{m,n}(x) \rangle H_{m,n}(x).$$

Problem 4. Consider the Hilbert space $L_2([0,1])$ and the *Haar scaling function* (father wavelet)

$$\phi(x) = \begin{cases} 1 & \text{if} \quad 0 \le x < 1 \\ 0 & \text{otherwise} \end{cases}$$

Let n be a positive integer. We define

$$g_k(x) := \sqrt{n}\phi(nx - k), \qquad k = 0, 1, \ldots, n - 1.$$

(i) Show that the set of functions $\{ g_0, g_1, \ldots, g_{n-1} \}$ is an orthonormal set in the Hilbert space $L_2([0,1])$.

(ii) Let f be a continuous function on the unit interval $[0,1]$. Thus $f \in L_2([0,1])$. Form the projection f_n on the subspace S_n of the Hilbert space $L_2([0,1])$ spanned by $\{ g_0, g_1, \ldots, g_{n-1} \}$, i.e.

$$f_n(x) = \sum_{k=0}^{n-1} \langle f(x), g_k(x) \rangle g_k(x).$$

Show that $f_n(x) \to f(x)$ pointwise in x as $n \to \infty$.

Solution 4. (i) We find

$$\int_0^1 g_k(x) g_\ell(x) dx = 0 \quad \text{if} \quad k \ne \ell$$

since $g_k(x) g_\ell(x) = 0$ for all $x \in (0,1)$. Moreover

$$\int_0^1 g_k^2(x) dx = n \int_{k/n}^{(k+1)/n} dx = 1.$$

(ii) We have

$$f_n(x) = \sum_{k=0}^{n-1} n \int_{k/n}^{(k+1)/n} f(s) ds \, \phi(nx - k).$$

It follows that

$$f_n(x) = n \int_{k/n}^{(k+1)/n} f(s) ds \quad \text{for} \quad k/m \le x < (k+1)/n.$$

Applying the mean value theorem of calculus there is a point $x_{n,k}$ in the interval $(k/n, (k+1)/n)$ such that the integral expression on the right is equal to $f(x_{n,k})$. Therefore

$$f_n(x) = f(x_{n,k}) \quad \text{for} \quad k/n \le x < (k+1)/n.$$

Since f is continuous on the closed bounded set $[0, 1]$, it is uniformly continuous on this set. Thus for any $\epsilon > 0$ there is a $\delta(\epsilon) > 0$ such that $|f(x) - f(x')| < \epsilon$ whenever $|x - x'| < \delta(\epsilon)$. Now given $x \in (0, 1)$ and $\epsilon > 0$, we choose $n > 1/\delta(\epsilon)$. It follows that

$$|f(x) - f_n(x)| = |f(x) - f(x_{n,k})| < \epsilon$$

since $|x - x_{n,k}| < 1/n < \delta(\epsilon)$. Thus $f_n(x) \to f(x)$, uniformly in x as $n \to \infty$.

Problem 5. The *continuous wavelet transform*

$$Wf(a, b) = \frac{1}{a} \int_{-\infty}^{+\infty} f(t)\overline{\psi\left(\frac{t - b}{a}\right)} dt, \qquad (a, b \in \mathbb{R}, \, a > 0)$$

decomposes the function $f \in L_2(\mathbb{R})$ hierarchically in terms of elementary components $\psi((t - b)/a)$. They are obtained from a single *analyzing wavelet* ψ applying *dilations* and *translations*. Here $\bar{\psi}$ denotes the complex conjugate of ψ and a is the scale and b the shift parameter. The function ψ has to be chosen so that it is well localized both in physical and Fourier space. The signal $f(t)$ can be uniquely recovered by the *inverse wavelet transform*

$$f(t) = \frac{1}{C_\psi} \int_{-\infty}^{+\infty} \int_0^{+\infty} Wf(a, b)\psi\left(\frac{t - b}{a}\right) \frac{da}{a} db$$

if $\psi(t)$ (respectively its Fourier transform $\hat{\psi}(\omega)$) satisfies the *admissibility condition*

$$C_\psi = \int_0^{+\infty} \frac{|\hat{\psi}(\omega)|^2}{\omega} d\omega < \infty.$$

Consider the analytic function $\psi(t) = te^{-t^2/2}$. Does ψ satisfy the admissibility condition?

Solution 5. The admissibility condition reduces to

$$\int_{-\infty}^{+\infty} \psi(t) dt = 0$$

for $\psi \in L_1(\mathbb{R})$ which is obviously satisfied for the given ψ.

Problem 6. Consider the Hilbert space $L_2(\mathbb{R})$. Let $\phi \in L_2(\mathbb{R})$ and assume that ϕ satisfies

$$\int_{\mathbb{R}} \phi(t)\overline{\phi(t - k)} dt = \delta_{0,k}.$$

i.e. the integral equals 1 for $k = 0$ and vanishes for $k = 1, 2, \ldots$. Show that for any fixed integer j the functions

$$\phi_{jk}(t) := 2^{j/2}\phi(2^j t - k), \qquad k = 0, \pm 1, \pm 2, \ldots$$

form an orthonormal set.

Solution 6. We have

$$\int_{\mathbb{R}} \phi_{jk}(t)\overline{\phi_{jl}}(t)dt = \int_{\mathbb{R}} 2^{j/2}\phi(2^j t - k)2^{j/2}\overline{\phi(2^j t - l)}dt.$$

Let $2^j t - k = x$. Thus $2^j dt = dx$ and therefore

$$\int_{\mathbb{R}} \phi_{jk}(t)\overline{\phi_{jl}}(t)dt = \int_{\mathbb{R}} \phi(x)\overline{\phi(x + k - l)}dx = \delta_{0,l-k}.$$

When $l = k$ the integral provides 1 otherwise 0. Thus this set forms an orthonormal set.

Problem 7. Consider the Littlewood-Paley orthonormal basis of wavelets. The mother wavelet of this set is

$$L(x) := \frac{1}{\pi x}(\sin(2\pi x) - \sin(\pi x)).$$

Then

$$L_{m,n}(x) = \frac{1}{2^{m/2}}L(2^{-m}x - n), \qquad m, n \in \mathbb{Z}$$

generates an orthonormal basis in the Hilbert space $L_2(\mathbb{R})$. Apply the rule of L'Hospital to find $L(0)$.

Solution 7. We have

$$L(0) = \lim_{x \to 0} \frac{1}{\pi x}(\sin(2\pi x) - \sin(\pi x)) = \frac{1}{\pi}(2\pi \cos(2\pi x) - \pi \cos(\pi x))_{x=0} = 1$$

since $\cos(0) = 1$.

Problem 8. Show that expanding the function

$$f(x) := \begin{cases} \sin(2\pi x) & \text{for} \quad x \in [0, 1] \\ 0 & \text{otherwise} \end{cases}$$

on the Hilbert space of square integrable functions $L_2(\mathbb{R})$ in terms of the *Haar basis*

$$\{ \psi_{j,k}(x) := 2^{j/2}\psi(2^j x - k) \ : \ j, k \in \mathbb{Z} \}$$

where

$$\psi(x) := \begin{cases} -1 & \text{for} & x \in [0, 1/2] \\ 1 & \text{for} & x \in (1/2, 1] \\ 0 & \text{otherwise} \end{cases}$$

yields the expansion

$$f(x) = \frac{1}{2\pi} \sum_{j=0}^{\infty} \sum_{k=0}^{2^j - 1} 2^{\frac{j}{2}} \left(2 \cos \left(\frac{2\pi \left(k + \frac{1}{2}\right)}{2^j} \right) - \cos \left(\frac{2\pi (k+1)}{2^j} \right) - \cos \left(\frac{2\pi k}{2^j} \right) \right)$$

by considering $j \le -1$ and $j \ge 0$ separately.

Solution 8. We have

$$\psi_{j,k}(x) = \begin{cases} -2^{j/2} & \text{for} & 2^j x - k \in [0, 1/2] \\ 2^{j/2} & \text{for} & 2^j x - k \in (1/2, 1] \\ 0 & \text{otherwise} \end{cases}$$

$$= \begin{cases} -2^{j/2} & \text{for} & 2^{-j} k \le x \le 2^{-j}(k + 1/2) \\ 2^{j/2} & \text{for} & 2^{-j}(k + 1/2) < x \le 2^{-j}(k + 1) \\ 0 & \text{otherwise} \end{cases} .$$

We note that

$$\int_{-\infty}^{\infty} (f(x))^2 dx = \int_0^1 \sin^2(2\pi x) dx = \frac{1}{2}$$

so that the function f is indeed in $L_2(\mathbb{R})$. Thus we can perform the expansion. For $j \le -1$ it follows that $2^{-j} \ge 2$ and $k2^{-j}$, $(k + \frac{1}{2})2^{-j}$ and $(k + 1)2^{-j}$ are all integers. Since

$$\int_0^1 f(x) dx = 0$$

we find $\langle f(x), \psi_{j,k}(x) \rangle = 0$ i.e. the intervals for which both functions are nonzero either do not intersect, or

$$[0, 1] \subseteq \left[k2^{-j}, \left(k + \frac{1}{2}\right) 2^{-j} \right]$$

or

$$[0, 1] \subseteq \left[\left(k + \frac{1}{2}\right) 2^{-j}, (k + 1) 2^{-j} \right] .$$

For $j \ge 0$ we consider

$$0 \le k2^{-j} \le 1 \quad \Rightarrow \quad 0 \le k \le 2^j$$

$$0 \le \left(k + \frac{1}{2}\right) 2^{-j} \le 1 \quad \Rightarrow \quad -\frac{1}{2} \le k \le 2^j - \frac{1}{2} \quad \Rightarrow \quad 0 \le k \le 2^j - 1$$

$$0 \le (k+1) 2^{-j} \le 1 \quad \Rightarrow \quad -1 \le k \le 2^j - 1$$

Consequently, unless $k \in \{0, 1, \ldots, 2^j - 1\}$

$$\langle f(x), \psi_{j,k}(x) \rangle = 0.$$

Now we calculate for $j \ge 0$ and $k \in \{0, 1, \ldots, 2^j - 1\}$

$$\langle f(x), \psi_{j,k}(x) \rangle = -2^{\frac{j}{2}} \int_{k2^{-j}}^{(k+\frac{1}{2})2^{-j}} \sin(2\pi x)\, dx + 2^{\frac{j}{2}} \int_{(k+\frac{1}{2})2^{-j}}^{(k+1)2^{-j}} \sin(2\pi x)\, dx$$

$$= \frac{2^{\frac{j}{2}}}{2\pi} \left[2\cos\left(2\pi\left(k+\frac{1}{2}\right)2^{-j}\right) - \cos\left(2\pi k 2^{-j}\right) - \cos\left(2\pi(k+1)2^{-j}\right) \right].$$

Thus the expansion of f is given by

$$f(x) = \sum_{j=0}^{\infty} \sum_{k=0}^{2^j-1} \frac{2^{\frac{j}{2}}}{2\pi} \left[2\cos\left(\frac{2\pi\left(k+\frac{1}{2}\right)}{2^j}\right) - \cos\left(\frac{2\pi(k+1)}{2^j}\right) - \cos\left(\frac{2\pi k}{2^j}\right) \right].$$

Problem 9. Consider the function $\hat{f}(k)$ in the Hilbert space $L_2(\mathbb{R})$

$$\hat{f}(k) = \begin{cases} 1 & \text{if} \quad 1/2 \le |k| \le 1 \\ 0 & \text{otherwise} \end{cases}$$

Then the Fourier transform is

$$f(x) = \frac{1}{2\pi} \int_{\mathbb{R}} e^{-ikx} \hat{f}(k)\, dk = \frac{1}{\pi k}(\sin(k) - \sin(k/2)).$$

Let $m, n \in \mathbb{Z}$. The *translation operator* T_n is given by

$$T_n f(x) := f(x - n)$$

and the *dilation operator* is given by

$$D_{2^m} f(x) := |2^m|^{-1/2} f(x/2^m).$$

Show that the functions $f_{m,n}(x) := D_{2^m} T_n f(x)$, $(m, n \in \mathbb{Z})$ form an orthonormal basis in the Hilbert space $L_2(\mathbb{R})$.

Solution 9. Let $g \in L_2(\mathbb{R})$. We have

$$\sum_{m,n\in\mathbb{Z}} |\langle g, f_{m,n}\rangle|^2 = \frac{1}{4\pi^2} \sum_{m,n\in\mathbb{Z}} |\langle \hat{g}, \hat{f}_{m,n}\rangle|$$

$$= \frac{1}{4\pi^2} \sum_{m,n\in\mathbb{Z}} 2^{-m} \left| \int_{1/2}^{1} \hat{g}(k/2^m)e^{ink}\,dk + \int_{-1}^{-1/2} \hat{g}(k/2^m)e^{ink}\,dk \right|^2$$

$$= \frac{1}{4\pi^2} \sum_{m\in\mathbb{Z}} \left(\int_{1/2^{m+1}}^{1/2^m} |\hat{g}(k)|^2\,dk + \int_{-1/2^m}^{-1/2^{m+1}} |\hat{g}(k)|^2\,dk \right)$$

$$= \frac{1}{4\pi^2} \left(\int_{0}^{\infty} \hat{g}(k)|^2\,dk + \int_{-\infty}^{0} |\hat{g}(k)|^2\,dk \right)$$

$$= \frac{1}{4\pi^2} \int_{-\infty}^{\infty} |\hat{g}(k)|^2\,dk = \int_{\mathbb{R}} |g(x)|^2\,dx = \langle g, g\rangle.$$

Problem 10. Let $f \in L_2(\mathbb{R}^n)$. Consider the following operators

$$T_{\mathbf{y}}f(\mathbf{x}) = f(\mathbf{x} - \mathbf{y}), \qquad \text{translation operator}$$
$$M_{\mathbf{k}}f(\mathbf{x}) = e^{i\mathbf{x}\cdot\mathbf{k}}f(\mathbf{x}), \qquad \text{modulation operator}$$
$$D_s f(\mathbf{x}) = |s|^{-n/2}f(s^{-1}\mathbf{x}), \quad s \in \mathbb{R}\setminus\{0\} \quad \text{dilation operator}$$

where $\mathbf{x}\cdot\mathbf{k} = k_1 x_1 + \cdots + x_n k_n$.
(i) Find $\|T_{\mathbf{y}}f\|$, $\|M_{\mathbf{k}}\|$, $\|D_s f\|$, where $\|\cdot\|$ denotes the norm in $L_2(\mathbb{R}^n)$.
(ii) Find the adjoint operators of these three operators.

Solution 10. (i) These operators are surjections preserving the norm of $L_2(\mathbb{R}^n)$

$$\|T_{\mathbf{y}}f\| = \|f\|, \quad \|M_{\mathbf{k}}f\| = \|f\|, \quad \|D_s f\| = \|f\|.$$

(ii) The adjoint operators are given by their inverse

$$\langle T_{\mathbf{y}}f, g\rangle = \langle f, T_{-\mathbf{y}}g\rangle, \qquad \langle M_{\mathbf{k}}f, g\rangle = \langle f, M_{-\mathbf{k}}g\rangle, \qquad \langle D_s f, g\rangle = \langle f, D_{1/s}g\rangle.$$

Problem 11. Consider the Hilbert space $\mathbb{R}^4$. Given a signal as the column vector

$$\mathbf{x} = (3.0 \ \ 0.5 \ \ 2.0 \ \ 7.0)^T.$$

The *pyramid algorithm* (for *Haar wavelets*) is as follows: The first two entries $(3.0 \ \ 0.5)^T$ in the signal give an average of $(3.0 + 0.5)/2 = 1.75$

and a difference average of $(3.0 - 0.5)/2 = 1.25$. The second two entries
$(2.0 \; 7.0)$ give an average of $(2.0 + 7.0)/2 = 4.5$ and a difference average of
$(2.0 - 7.0)/2 = -2.5$. Thus we end up with a vector

$$(1.75 \; 1.25 \; 4.5 \; -2.5)^T.$$

Now we take the average of 1.75 and 4.5 providing $(1.75 + 4.5)/2 = 3.125$
and the difference average $(1.75 - 4.5)/2 = -1.375$. Thus we end up with
the vector

$$\mathbf{y} = (3.125 \; -1.375 \; 1.25 \; -2.5)^T.$$

(i) Find a 4×4 matrix A such that

$$\mathbf{x} \equiv \begin{pmatrix} 3.0 \\ 0.5 \\ 2.0 \\ 7.0 \end{pmatrix} = A\mathbf{y} \equiv A \begin{pmatrix} 3.125 \\ -1.375 \\ 1.25 \\ -2.5 \end{pmatrix}.$$

(ii) Show that the inverse of A exists. Then find the inverse matrix.

Solution 11. (i) Since we can write

$$\begin{pmatrix} 3.0 \\ 0.5 \\ 2.0 \\ 7.0 \end{pmatrix} = 3.125 \begin{pmatrix} 1 \\ 1 \\ 1 \\ 1 \end{pmatrix} - 1.375 \begin{pmatrix} 1 \\ 1 \\ -1 \\ -1 \end{pmatrix} + 1.25 \begin{pmatrix} 1 \\ -1 \\ 0 \\ 0 \end{pmatrix} - 2.5 \begin{pmatrix} 0 \\ 0 \\ 1 \\ -1 \end{pmatrix}$$

we obtain the matrix

$$A = \begin{pmatrix} 1 & 1 & 1 & 0 \\ 1 & 1 & -1 & 0 \\ 1 & -1 & 0 & 1 \\ 1 & -1 & 0 & -1 \end{pmatrix}.$$

(ii) All the column vectors in the matrix A are nonzero and all the pairwise
scalar products are equal to 0. Thus the column vectors form a basis (not
normalized) in $\mathbb{R}^n$ and the matrix is invertible. The inverse matrix is given
by

$$A^{-1} = \begin{pmatrix} 1/4 & 1/4 & 1/4 & 1/4 \\ 1/4 & 1/4 & -1/4 & -1/4 \\ 1/2 & -1/2 & 0 & 0 \\ 0 & 0 & 1/2 & -1/2 \end{pmatrix}.$$

7.3 Supplementary Problems

Problem 1. (i) Consider the Hilbert space $L_2(\mathbb{R})$ and $\phi \in L_2(\mathbb{R})$. The basic scaling function (father wavelet) satisfies a scaling relation of the form

$$\phi(x) = \sum_{k=0}^{N-1} a_k \phi(2x - k).$$

Show that the *Hilbert transform* of ϕ

$$H(\phi)(y) = \frac{1}{\pi} \int_{\mathbb{R}} \frac{\phi(x)}{x - y} dx$$

is a solution of the same scaling relation. Note that the scaling function ϕ may have compact support, the Hilbert transform has support on the real line and decays as y^{-1}.

(ii) Show that the Hilbert transform of the related mother wavelet ψ is also noncompact and decays like y^{-p-1} where

$$\int_{\mathbb{R}} x^m \psi(x) dx = 0$$

for $m = 0, 1, \ldots, p - 1$.

Problem 2. The *linear spline function* $f : \mathbb{R} \to \mathbb{R}$ is defined as

$$f(x) = \begin{cases} 1 - |x| & -1 \leq x \leq 1 \\ 0 & \text{otherwise} \end{cases}$$

Show that the function generates an orthonormal wavelet basis in the Hilbert space $L_2(\mathbb{R})$ using the technique of multiresolution analysis.

Problem 3. Show that for the *Shannon wavelet* the projection operators are given by

$$(P_j)(x) = \int_{\mathbb{R}} \frac{\sin(2^j \pi (y - x))}{\pi (y - x)} f(y) dy.$$

Problem 4. Consider the analytic function $f : \mathbb{R} \to \mathbb{R}$

$$f(x) = e^{-x^2/2} \cos(x)$$

with $f \in L_2(\mathbb{R})$. Can we use f as a mother wavelet? Study

$$f_{m,n}(x) = \frac{1}{2^{m/2}} f(2^{-m}x - n) \qquad m, n \in \mathbb{Z}.$$

Problem 5. Consider the Hilbert space $L_2(\mathbb{R}^n)$. Show that the three linear operators

$$T_y f(x) := f(x - y) \quad \textit{Translation operator}$$
$$M_\xi f(x) := e^{ix\xi} f(x) \quad \textit{Modulation operator}$$
$$D_\rho f(x) := |\rho|^{-n/2} f(\rho^{-1} x), \quad \rho \in \mathbb{R} \setminus \{0\} \quad \textit{Dilation operator}$$

are unitary on $L_2(\mathbb{R}^n)$. Thus show that

$$\|T_y f\| = \|f\|, \quad \|M_\xi f\| = \|f\|, \quad \|D_\rho f\| = \|f\|.$$

Show that the adjoint operators are given by their inverses

$$\langle T_y f, g \rangle \langle f, T_{-y} g \rangle, \quad \langle M_\xi f, g \rangle = \langle f, M_{-\xi} g \rangle, \quad \langle D_\rho f, g \rangle = \langle f, D_{1/\rho} g \rangle.$$

Problem 6. Let

$$\psi(x) = \frac{1}{2\pi}(x^2 - 1)e^{-x^2/2}.$$

This function is called the *Mexican hat function*. Consider

$$\psi_{m,n}(x) := 2^{-m/2} \psi(2^{-mx-n})$$

where $m, n \in \mathbb{Z}$. Calculate $\langle \psi_{m,n}, \psi_{k,l} \rangle$.

Problem 7. Let

$$f_0(x) = \exp(-x^2/2).$$

We define the *mother wavelets* f_n as

$$f_n(x) = -\frac{d}{dx} f_{n-1}(x), \qquad n = 1, 2, \dots.$$

Show that the family of f_n's obey the *Hermite recursion relation*

$$f_n(x) = x f_{n-1}(x) - (n-1) f_{n-2}(x), \qquad n = 2, 3, \dots.$$

Chapter 8

Holomorphic Entire Functions and Hilbert Space

8.1 Introduction

If a complex-valued function f is differentiable at every point of some open set S of $\mathbb{R}^2$, then f is said to be analytic on S. A function is said to be analytic at a point z_0 if there is a neighbourhood of z_0 on which it is analytic. A functions which is analytic everywhere in the finite complex plane (i.e. everywhere except at ∞) is called an *entire function*. For example z^n ($n \in \mathbb{N}$), $\exp(z)$, $\exp(-z)$, $\cosh(z)$, $\sinh(z)$, $\cos(z)$, $\sin(z)$ are entire functions. An entire function can be represented as a Taylor series which has an infinite radius of convergence.

The *Fock space* $\mathcal{F}$ is the Hilbert space of entire functions with inner product given by

$$\langle f, g \rangle := \frac{1}{\pi} \int_{\mathbb{C}} f(z)\overline{g(z)}e^{-|z|^2}\,dxdy, \qquad z = x + iy$$

where $\mathbb{C}$ denotes the complex numbers, $x, y \in \mathbb{R}$, $z = x + iy$, $\overline{z} = x - iy$, $z\overline{z} = x^2 + y^2$, $z = re^{i\phi}$, $\overline{z} = re^{-i\phi}$, $r \geq 0$. The growth of the functions in the Hilbert space $\mathcal{F}$ is dominated by $\exp(|z|^2/2)$.

Let $\mathbb{C}^{n \times N}$ be the vector space of all $n \times N$ complex matrices and $Z \in \mathbb{C}^{n \times N}$. Then ZZ^* is an $n \times n$ matrix. One defines the *Gaussian measure* $d\mu$ on $\mathbb{C}^{n \cdot N}$ as

$$d\mu(Z) := \frac{1}{\pi^{nN}} \exp(-\text{tr}(ZZ^*))dZ, \quad dZ = \prod_{j=1}^{n} \prod_{k=1}^{N} dX_{jk}dY_{jk}$$

where $\text{tr}(\cdot)$ denotes the trace. Consider the vector space

$$\mathcal{H}(\mathbb{C}^{n \times N}) := \left\{ f : \mathbb{C}^{n \times N} \to \mathbb{C} : f \text{ holomorphic} \int_{\mathbb{C}^{n \times N}} |f(Z)|^2 d\mu(Z) < \infty \right\}$$

297

where $Z = (Z_{jk})$ with $Z_{jk} = X_{jk} + iY_{jk}$ $(j = 1, \ldots, n; \; k = 1, \ldots, N)$. Then $\mathcal{H}(\mathbb{C}^{n \times N})$ is a Hilbert space with respect to the inner product

$$\langle f, g \rangle = \int_{\mathbb{C}^{n \times N}} f(Z)\overline{g(Z)} d\mu(Z).$$

This Hilbert space has a *reproducing kernel* K

$$K(Z, Z') = \exp(\operatorname{tr}(Z(Z')^*)).$$

This means a continuous function $K(Z, Z') : \mathbb{C}^{n \times N} \times \mathbb{C}^{n \times N} \to \mathbb{C}$ such that

$$f(Z) = \int_{\mathbb{C}^{n \times N}} K(Z, Z')f(Z') d\mu(Z')$$

for all $Z \in \mathbb{C}^{n \times N}$ and $f \in \mathcal{F}(\mathbb{C}^{n \times N})$.

Let $\mathbf{z} \in \mathbb{C}^n$. The counterpart of the *Dirac delta function* in the *Bargmann representation* ([5],[6]) is the *reproducing kernel* $K(\mathbf{w}^*, \mathbf{z})$ defined by

$$\widetilde{\phi}(\mathbf{w}^*) = \int_{\mathbb{R}^{2n}} d\mu(\mathbf{z}) \exp(-|\mathbf{z}|^2) K(\mathbf{w}^*, \mathbf{z}) \widetilde{\phi}(\mathbf{z}^*)$$

where $K(\mathbf{w}^*, \mathbf{z}) := \exp(\mathbf{w}^* \cdot \mathbf{z})$. Here $d\mu_N(\mathbf{z})$ is the *Gaussian measure* in the space of N-dimensional variable, namely

$$d\mu_N(\mathbf{z}) = \frac{1}{\pi^N} \exp(-z\overline{z}) d^N z.$$

The *Bargmann transform* ([5],[6],[7]) maps the orthonormal basis $\{h_\alpha\}_{\alpha \in \mathbb{N}^d}$ of the Hilbert space $L_2(\mathbb{R}^d)$ bijectively into the orthonormal basis $\{e_\alpha\}_{\alpha \in \mathbb{N}^d}$ of monomials in $A^2(\mathbb{C}^d)$. Therefore these is a natural way to identify hermite functions series expansion by formal power series expansions. The *Bargmann transform* is defined by

$$(\mathcal{B}_d f)(z) := \frac{1}{\pi^{d/4}} \int_{\mathbb{R}^d} \exp\left(-\frac{1}{2}((z, z) + |y|^2) + 2^{1/2}(z, y)\right) f(y) dy$$

where $f \in L_2(\mathbb{R}^d)$. If $f \in L_2(\mathbb{R}^d)$, then the Bargmann transform $\mathcal{B}_d f$ of f is an entire function. The *Bargmann kernel* is given by

$$\mathcal{K}_d(z, y) := \frac{1}{\pi^{d/4}} \exp\left(-\frac{1}{2}((z, z) + |y|^2) + 2^{1/2}(z, y)\right)$$

where $(z, w) = \sum_{j=1}^{d} z_j w_j$ with $z = (z_1, \ldots, z_d) \in \mathbb{C}^d$, $w = (w_1, \ldots, w_d) \in \mathbb{C}^d$.

Let Ω be a bounded open domain of the complex z-plane ($z = x + iy, x, y \in \mathbb{R}$) and let $H(\Omega)$ be the set of all holomorphic functions f defined on Ω with the norm

$$\|f\| = \left(\int\int_\Omega |f(z)|^2 dxdy \right)^{1/2} < \infty.$$

Then $H(\Omega)$ is a Hilbert space with the scalar product

$$\langle f, g \rangle = \int\int_\Omega f(z)\overline{g}(z)dxdy$$

and $(f + g)(z) = f(z) + g(z)$, $(cf)(z) = cf(z)$ with $c \in \mathbb{C}$.

Consider the class C of all single-valued analytic functions defined over a finite domain D in the complex plane $\mathbb{C}$. Let $z = x + iy$ with $x, y \in \mathbb{R}$. An inner product can be defined as

$$\langle f(z), g(z) \rangle := \int_D f(z)\overline{g}(z)dxdy$$

where $f, g \in C$. This implies a norm

$$\|f(z)\|_D^2 = \int_D |f(z)|^2 dxdy.$$

The class of all f such that $\|f\|_D$ exists provides a Hilbert space $L_2(D)$ with a countable base. Let $\{\phi_j\}_{j=0}^\infty$ be an arbitrary orthonormal basis in the Hilbert space $L_2(D)$ and let $c_0, c_1, c_2, \ldots$ be complex numbers such that

$$\sum_{j=0}^\infty |c_j|^2 < \infty.$$

Then there exists a function $f \in L_2(D)$ such that

$$c_j = \langle f(z), \phi_j(z) \rangle \quad \text{and} \quad \|f\|^2 = \sum_{j=0}^\infty |c_j|^2.$$

8.2   Solved Problems

Problem 1. Let $m, n \in \mathbb{N}_0$ and $z \in \mathbb{C}$. Find

$$I_{m,n} = \frac{1}{\pi} \int_{\mathbb{C}} \exp(-|z|^2) z^n \overline{z}^m d^2 z.$$

Utilize $z = r \exp(i\phi)$.

Solution 1. With $\overline{z} = r \exp(-i\phi)$ we have

$$|z|^2 = r^2, \quad z^n = r^n e^{in\phi}, \quad \overline{z}^m = r^m e^{-im\phi}$$

with $0 \leq r < \infty$, $0 \leq \phi < 2\pi$. Since $dx \wedge dy = r dr \wedge d\phi$ we obtain

$$I_{m,n} = \frac{1}{\pi} \int_{\mathbb{C}} \exp(-|z|^2) z^n \overline{z}^m dx dy = \frac{1}{\pi} \int_{r=0}^{\infty} \int_{\phi=0}^{2\pi} e^{-r^2} r^{n+m+1} e^{i(n-m)\phi} d\phi dr.$$

Hence for $m \neq n$ we have $I_{m,n} = 0$. For $n = m$ we have

$$I_{n,n} = \frac{1}{\pi} \int_0^{\infty} e^{-r^2} r^{2n+1} dr \int_{\phi=0}^{2\pi} d\phi = 2 \int_{r=0}^{\infty} e^{-r^2} r^{2n+1} dr = \Gamma(n+1) = n!.$$

Finally

$$I_{n,m} = \frac{1}{\pi} \int_{\mathbb{C}} \exp(-|z|^2) z^n \overline{z}^m d^2 z = n! \delta_{n,m}.$$

Problem 2. Consider the Bargmann-Fock space which is the space of holomorphic functions $f : \mathbb{C}^n \to \mathbb{C}$ in n complex variables satisfying the square integrability condition

$$\|f\|^2 = \frac{1}{\pi^n} \int_{\mathbb{C}^n} |f(z)|^2 \exp(-|z|^2) dz < \infty$$

where dz denotes the $2n$ dimensional Lebesgue measure on $\mathbb{C}^n$. The norm is implied by a scalar product

$$\langle f(z), g(z) \rangle = \frac{1}{\pi^n} \int_{\mathbb{C}^n} \overline{f}(z) g(z) e^{-|z|^2} dz$$

and one has a Hilbert space. Let $n = 1$, $z, w \in \mathbb{C}$ and $f_w(z) = \exp(\overline{w} \cdot z)$. Find

$$\|f_w\|^2 = \langle f_w(z), f_w(z) \rangle.$$

Hint. With $a > 0$ one has

$$\int_{\mathbb{R}} e^{-(ax^2+bx+c)} dx = \sqrt{\frac{\pi}{a}} e^{(b^2-4ac)/(4a)}.$$

Solution 2. We have

$$f_w(z) = e^{\overline{w}z} \Rightarrow \overline{f}_w(z) = e^{w\overline{z}} \Rightarrow f_w(z)\overline{f}_w(z) = e^{\overline{w}z+w\overline{z}}.$$

With $w = w_1 + iw_2$, $z = x + iy$ ($w_1, w_2, x, y \in \mathbb{R}$) we have

$$\overline{w} \cdot z + w \cdot \overline{z} = 2w_1 x + 2w_2 y.$$

Hence with $\|z\|^2 = x^2 + y^2$ it follows that

$$\langle f_w(z), f_w(z) \rangle = \frac{1}{\pi} \int_{\mathbb{R}} \int_{\mathbb{R}} e^{2w_1 x} e^{-x^2} e^{2w_2 y} e^{-y^2} dx dy$$

$$= e^{w_1^2} e^{w_2^2} = e^{w_1^2 + w_2^2} = e^{|w|^2}.$$

Problem 3. Let Ω be the unit disk. A Hilbert space of analytic functions can be defined by

$$\mathcal{H} := \left\{ f(z) \text{ analytic, } |z| < 1 : \sup_{a<1} \int_{|z|=a} |f(z)|^2 dz < \infty \right\}$$

and the scalar product

$$\langle f, g \rangle := \lim_{a \to 1} \int_{|z|=a} \overline{f(z)} g(z) dz.$$

Let c_n ($n = 0, 1, 2, \dots$) be the coefficients of the power-series expansion of the analytic function f. Find the norm of f.

Solution 3. The norm is

$$\|f\| = \sqrt{\sum_{n=0}^{\infty} |c_n|^2}.$$

Problem 4. Let $\mathbb{C}^n$ denote the complex Euclidean space. Let $\mathbf{z} = (z_1, \dots, z_n) \in \mathbb{C}^n$ and $\mathbf{w} = (w_1, \dots, w_n) \in \mathbb{C}^n$ then the scalar product (inner product) is given by

$$\mathbf{z} \cdot \mathbf{w} := \mathbf{z}\mathbf{w}^* \equiv \mathbf{z}\overline{\mathbf{w}}^T$$

where $\overline{\mathbf{z}} = (\overline{z}_1, \dots, \overline{z}_n)$. Let E_n denote the set of entire functions in $\mathbb{C}^n$ and F_n denote the set of $f \in E_n$ such that

$$\|f\|^2 := \frac{1}{\pi^n} \int_{\mathbb{C}^n} |f(\mathbf{z})|^2 \exp(-|\mathbf{z}|^2) dV$$

is finite. Here dV is the volume element (Lebesgue measure)

$$dV = \prod_{j=1}^{n} dx_j dy_j = \prod_{j=1}^{n} r_j dr_j d\phi_j$$

with $z_j = r_j e^{i\phi_j}$. The norm follows from the scalar product of two functions $f, g \in F_n$

$$\langle f, g \rangle := \frac{1}{\pi^n} \int_{\mathbb{C}^n} f(\mathbf{z})\overline{g(\mathbf{z})} \exp(-|\mathbf{z}|^2) dV.$$

Let

$$\mathbf{z}^m := z_1^{m_1} \cdots z_n^{m_n}$$

where the multiindex m is defined by $m! = m_1! \cdots m_n!$ and $|m| = \sum_{j=1}^{n} m_j$. Find the scalar product $\langle \mathbf{z}^m, \mathbf{z}^p \rangle$.

Solution 4. We obtain

$$\langle \mathbf{z}^m, \mathbf{z}^p \rangle = \langle z_1^{m_1} \cdots z_n^{m_n}, z_1^{p_1} \cdots z_n^{p_n} \rangle = 0 \quad \text{for} \ \ m \neq p$$

and

$$\langle \mathbf{z}^m, \mathbf{z}^m \rangle = \langle z_1^{m_1} \cdots z_n^{m_n}, z_1^{m_1} \cdots z_n^{m_n} \rangle m! = m_1! \cdots m_n!.$$

Thus the *monomials*

$$\frac{1}{\sqrt{m!}} \mathbf{z}^m$$

are orthonormal in this Hilbert space.

Problem 5. Consider the *Fock space* $\mathcal{F}$. Let $f, g \in \mathcal{F}$ with Taylor expansions

$$f(z) = \sum_{j=0}^{\infty} a_j z^j, \qquad g(z) = \sum_{j=0}^{\infty} b_j z^j.$$

(i) Find $\langle f, g \rangle$ and $\|f\|^2$.
(ii) Consider the special that $f(z) = \sin(z)$ and $g(z) = \cos(z)$. Calculate $\langle f, g \rangle$.
(iii) Let

$$\mathcal{K}(z, w) := e^{z\overline{w}}, \qquad z, w \in \mathbb{C}.$$

Calculate $\langle f(z), \mathcal{K}(z, w) \rangle$.

Solution 5. (i) We obtain

$$\langle f, g \rangle = \sum_{j=0}^{\infty} a_j \bar{b}_j j!$$

and

$$\|f\|^2 = \sum_{j=0}^{\infty} |a_j|^2 j!.$$

(ii) Since

$$\sin(z) = \sin(x)\cosh(y) + i\cos(x)\sinh(y), \quad \cos(\bar{z}) = \cos(x)\cosh(y) + i\sin(x)\sinh(y)$$

and

$$\sin(-x) = -\sin(x), \quad \cos(-x) = \cos(x), \quad \sinh(-y) = -\sinh(y), \quad \cosh(-y) = \cosh(y)$$

we obtain $\langle \sin(z), \cos(z) \rangle = 0$.

(iii) We find

$$\langle f(z), e^{z\bar{w}} \rangle = f(w).$$

Hence the *Dirac delta function* in the Fock Hilbert space is the exponential function.

Problem 6. Let $z \in \mathbb{C}$ and $z = x + iy$ $(x, y \in \mathbb{R})$. Consider the entire functions

$$f(z) = \sin(z), \qquad g(z) = \cos(z).$$

Find the scalar products $\langle f, g \rangle$, $\langle f, f \rangle$, $\langle g, g \rangle$ with

$$\langle f, g \rangle = \frac{1}{\pi} \int_{\mathbb{C}} f(z)\bar{g}(z)e^{-|z|^2}\, dx dy.$$

Note that

$$\int_{\mathbb{R}} \exp(-x^2)dx = \sqrt{\pi}$$

$$\int_{\mathbb{R}} \sin^2(x)e^{-x^2}\, dx = \frac{\sqrt{\pi}}{2}(1 - e^{-1})$$

$$\int_{\mathbb{R}} \cos^2(x)e^{-x^2}\, dx = \frac{\sqrt{\pi}}{2}(1 + e^{-1})$$

$$\int_{\mathbb{R}} \sinh^2(x)e^{-x^2}\, dx = \frac{\sqrt{\pi}}{2}(e - 1)$$

$$\int_{\mathbb{R}} \cosh^2(x)e^{-x^2}\, dx = \frac{\sqrt{\pi}}{2}(e + 1)$$

and $\overline{f}(z) = f(\overline{z})$, $\overline{g}(z) = g(\overline{z})$.

Solution 6. We have

$$f(z) = \sin(z) = \sin(x + iy) = \sin(x)\cosh(y) + i\cos(x)\sinh(y)$$
$$f(\overline{z}) = \sin(\overline{z}) = \sin(x - iy) = \sin(x)\cosh(y) - i\cos(x)\sinh(y)$$
$$g(z) = \cos(z) = \cos(x + iy) = \cos(x)\cosh(y) - i\sin(x)\sinh(y)$$
$$g(\overline{z}) = \cos(\overline{z}) = \cos(x - iy) = \cos(x)\cosh(y) + i\sin(x)\sinh(y).$$

Now

$$f(z)g(\overline{z}) = \sin(x)\cos(x) + i\sinh(y)\cosh(y) = \frac{1}{2}\sin(2x) + \frac{i}{2}\sinh(2y).$$

Note that $\sin(-2x) = -\sin(2x)$, $\sinh(-2y) = -\sinh(2y)$ and

$$\exp(-x^2 - y^2) \mapsto \exp(-x^2 - y^2)$$

under the map $x \mapsto -x$, $y \mapsto -y$. Thus $\langle \cos(z), \sin(z) \rangle = 0$. Now

$$f(z)f(\overline{z}) = \sin^2(x)\cosh^2(y) + \cos^2(x)\sinh^2(y) = \sin^2(x) + \sinh^2(y).$$

Utilizing the integrals given above we find

$$\langle \sin(z), \sin(z) \rangle = \frac{1}{\pi} \int_{\mathbb{R}} \int_{\mathbb{R}} (\sin^2(x) + \sinh^2(y)) e^{-(x^2 + y^2)} dx dy = \frac{1}{2}(e - e^{-1}).$$

For g we have

$$g(z)g(\overline{z}) = \cos^2(x) + \sinh^2(y).$$

Utilizing the integrals given above we find

$$\langle \cos(z), \cos(z) \rangle = \frac{1}{\pi} \int_{\mathbb{R}} \int_{\mathbb{R}} (\cos^2(x) + \sinh^2(y)) e^{-(x^2 + y^2)} dx dy = \frac{1}{2}(e + e^{-1}).$$

Problem 7. Consider Fourier series and analytic (harmonic) functions on the disc

$$\mathbb{D} := \{ z \in \mathbb{C} : |z| \leq 1 \}.$$

A Fourier series can be viewed as the boundary values of a *Laurent series*

$$\sum_{n=-\infty}^{\infty} c_n z^n.$$

Suppose we are given a function f on $\partial\mathbb{D} = \mathbb{T}$. Find the harmonic extension u of f into $\mathbb{D}$. This means

$$\Delta u = 0 \quad \text{and} \quad u = f \quad \text{on} \quad \partial\mathbb{D} = \mathbb{T}$$

where $\Delta := \partial^2/\partial x^2 + \partial^2/\partial y^2$.

Solution 7. Since $\Delta z^n = 0$ and $\Delta \bar{z}^n = 0$ for every integer $n \geq 0$, we define

$$u(z) = \sum_{n=0}^{\infty} \hat{f}(n)z^n + \sum_{n=-\infty}^{-1} \hat{f}(n)\bar{z}^{|n|}$$

which formally satisfies

$$u(e^{2\pi i\theta}) = \sum_{n=-\infty}^{\infty} \hat{f}(n)e^{2\pi in\theta} = f(\theta).$$

Inserting $z = re^{2\pi i\theta}$ and

$$\hat{f}(n) = \int_0^1 e^{-2\pi in\phi} f(\phi)d\phi$$

into $u(z)$ yields

$$u(re^{2\pi i\theta}) = \int_{\mathbb{T}} \sum_{n=-\infty}^{\infty} r^{|n|} e^{2\pi in(\theta-\phi)} f(\phi)d\phi.$$

Problem 8. Let $\hat{b}^\dagger$ be a *Bose creation operator*. Show that the eigenvalue equation

$$\hat{b}^\dagger|\lambda\rangle = \lambda|\lambda\rangle$$

implies that $|\lambda\rangle = 0$, i.e. the point spectrum is empty.

Solution 8. Projecting both sides of the eigenvalue equation onto a coherent state $\langle z|$ with

$$\langle z|\hat{b}^\dagger = \langle z|\bar{z}$$

yields

$$(\bar{z} - \lambda)\langle z|\lambda\rangle = 0.$$

Taking into account that $\langle z|\lambda\rangle$ is an entire function of $\bar{z}$ we find that $\langle z|\lambda\rangle = 0$, i.e. $|\lambda\rangle = 0$.

Problem 9. The Hardy-Lebesgue class $H - L_2$ consists of the set of all functions f which are *holomorphic* in the *unit disk* $\{ z : |z| < 1 \}$ of the complex z-plane such that

$$\sup_{0<r<1} \left(\int_0^{2\pi} |f(re^{i\theta})|^2 d\theta \right) < \infty.$$

Let $f \in H - L_2$ and

$$f(z) = \sum_{j=0}^{\infty} c_j z^j$$

a Taylor expansion of f. Find

$$F(r) = \frac{1}{2\pi} \int_0^{2\pi} |f(re^{i\theta})|^2 d\theta.$$

Solution 9. We have

$$F(r) = \frac{1}{2\pi} \int_0^{2\pi} |f(re^{i\theta})|^2 d\theta = \frac{1}{2\pi} \sum_{j,k=0}^{\infty} \int_0^{2\pi} c_j \bar{c}_k r^{j+k} e^{i(j-k)\theta} d\theta$$

$$= \sum_{j=0}^{\infty} |c_j|^2 r^{2n}.$$

The function F is monotone increasing with r, where $0 < r < 1$ and F is bounded from above. It follows that

$$\|f\| = \sup_{0<r<1} \left(\frac{1}{2\pi} \left(\int_0^{2\pi} |f(re^{i\theta})|^2 d\theta \right) \right)^{1/2} = \left(\sum_{j=0}^{\infty} |c_j|^2 \right)^{1/2}$$

with $(c_0, c_1, c_2, \dots)$ be an element of the Hilbert space $\ell_2(\mathbb{N}_0)$.

Problem 10. Let $R > 0$ and consider the circular domain

$$D := \{ z \,:\, |z| < R \}$$

and

$$|z\rangle_D = \sum_{j=0}^{\infty} \phi_j(z) |j\rangle$$

where $|j\rangle$ $(j = 0, 1, 2, \dots)$ are the number states with

$$\langle j|k \rangle = \delta_{j,k}, \quad \sum_{j=0}^{\infty} |j\rangle\langle j| = I$$

and

$$\phi_j(z) = \frac{\sqrt{(j+1)/\pi}}{R^{j+1}} z^j.$$

Find the scalar product $\langle z'|z\rangle_D$ and the square of the norm $\||z\rangle_D\|^2$..

Solution 10. Let

$$|z\rangle_D = \sum_{j=0}^{\infty} \frac{\sqrt{(j+1)/\pi}}{R^{j+1}} z^j |j\rangle, \quad |z'\rangle_D = \sum_{k=0}^{\infty} \frac{\sqrt{(k+1)/\pi}}{R^{k+1}} z'^k |k\rangle.$$

With $\langle k, j \rangle = \delta_{j,k}$ we obtain the scalar product

$$\langle z'|z\rangle_D = \sum_{j=0}^{\infty} \frac{(j+1)}{\pi} \frac{\overline{(z')^j}(z^j)}{R^{2(j+1)}} = \frac{R^2}{\pi(R^2 - \overline{z}'z)^2}.$$

Then the square of the norm is given by

$$\| |z\rangle_D \|^2 = \frac{R^2}{\pi(R^2 - |z|^2)^2}.$$

Problem 11. Let $z = x_1 + ix_2$, $\overline{z} = x_1 - ix_2$ with $x_1, x_2 \in \mathbb{R}$ and dimension length. One has

$$\frac{\partial}{\partial z} = \frac{1}{2}\left(\frac{\partial}{\partial x_1} - i\frac{\partial}{\partial x_2}\right), \quad \frac{\partial}{\partial \overline{z}} = \frac{1}{2}\left(\frac{\partial}{\partial x_1} + i\frac{\partial}{\partial x_2}\right).$$

Consider the *Bose annihilation operator* and *Bose creation operator*

$$\hat{b} = \frac{z}{2\ell} + \ell\frac{\partial}{\partial \overline{z}}, \quad \hat{b}^\dagger = \frac{\overline{z}}{2\ell} - \ell\frac{\partial}{\partial z}$$

with the constant $\ell > 0$ of dimension length. Then $\hat{b}|0\rangle = 0|0\rangle$ maps to

$$\hat{b}e^{-z\overline{z}/(2\ell^2)} \equiv \left(\frac{z}{2\ell} + \ell\frac{\partial}{\partial \overline{z}}\right)e^{-z\overline{z}/(2\ell^2)} = 0e^{-z\overline{z}/(2\ell^2)} = 0$$

with

$$\frac{\partial}{\partial \overline{z}}(z\overline{z}) = z.$$

Note that $\exp(-z\overline{z}/(2\ell^2)) \in L_2(\mathbb{R}^2)$. Find the commutator $[\hat{b}, \hat{b}^\dagger]$.

Solution 11. We have

$$\begin{aligned}
[\hat{b}, \hat{b}^\dagger]f(z, \overline{z}) &= [z/(2\ell) + \ell\partial/\partial\overline{z}, \overline{z}/(2\ell) - \ell\partial/\partial z]f(z, \overline{z}) \\
&= [z/(2\ell), -\ell\partial/\partial z]f(z, \overline{z}) + [\ell\partial/\partial\overline{z}, \overline{z}/(2\ell)]f(z, \overline{z}) \\
&= [z/2, -\partial/\partial z]f(z, \overline{z}) + [\partial/\partial\overline{z}, \overline{z}/2]f(z, \overline{z}) \\
&= f(z, \overline{z}).
\end{aligned}$$

Thus $[\hat{b}, \hat{b}^\dagger] = I$.

Problem 12. Let $\hat{b}^\dagger$, $\hat{b}$ be Bose creation and annihilation operators, respectively. Let $|n\rangle$ be the number state

$$|n\rangle = \frac{1}{\sqrt{n!}}(\hat{b}^\dagger)^n|0\rangle, \quad n = 0, 1, 2, \ldots$$

with $\langle 0|0\rangle = 1$, $\langle n|n\rangle = 1$. Then

$$\langle n|\hat{b}^\dagger\hat{b}|n\rangle = n, \qquad \langle n|(\hat{b}^\dagger + \hat{b})|n\rangle = 0.$$

Do the calculation in the Bargmann space (entire functions)

$$\hat{b}^\dagger \leftrightarrow \zeta, \quad \hat{b} \leftrightarrow \frac{d}{d\zeta}, \quad (\hat{b}^\dagger)^n|0\rangle \leftrightarrow \zeta^n \quad (n = 0, 1, 2, \ldots)$$

i.e. $|0\rangle \leftrightarrow 1$. The scalar product in the Bargmann space is

$$\langle f(\zeta)|g(\zeta)\rangle = \frac{1}{\pi}\int_{\mathbb{C}} \exp(-|\zeta|^2)f(\overline{\zeta})g(\zeta)d\Re(\zeta)d\Im(\zeta).$$

Solution 12. We have $\hat{b}^\dagger\hat{b} \leftrightarrow \zeta d/d\zeta$. Then with $\zeta = re^{i\phi}$, $\overline{\zeta} = re^{-i\phi}$, $\zeta\overline{\zeta} = r^2$ we have

$$\zeta\frac{d}{d\zeta}\frac{1}{\sqrt{n!}}\zeta^n = \frac{n\zeta}{\sqrt{n!}}\zeta^{n-1} = \frac{\sqrt{n}}{\sqrt{(n-1)!}}\zeta^n$$

and

$$\frac{1}{\sqrt{n!}}\overline{\zeta}^n \frac{\sqrt{n}}{\sqrt{(n-1)!}}\zeta^n = \frac{1}{(n-1)!}(\zeta\overline{\zeta})^n = \frac{1}{(n-1)!}(r^2)^n.$$

The scalar product is given by

$$\frac{1}{\pi}\int_{\mathbb{C}} e^{-|\zeta|^2}\frac{1}{(n-1)!}(\zeta\overline{\zeta})^n d\Re(\zeta)d\Im(\zeta) = \frac{1}{\pi}\frac{1}{(n-1)!}\int_{\phi=0}^{2\pi}\int_{r=0}^{\infty} e^{-r^2}(r^2)^n r\,dr\,d\phi$$

$$= \frac{1}{\pi}\frac{2\pi}{(n-1)!}\int_{r=0}^{\infty} e^{-r^2}r^{2n+1}dr$$

$$= \frac{2}{(n-1)!}\frac{n!}{2} = n.$$

We have with $\zeta\overline{\zeta} = r^2$

$$\frac{1}{\sqrt{n!}}\overline{\zeta}^n\left(\zeta + \frac{d}{d\zeta}\right)\frac{1}{\sqrt{n!}}\zeta^n = \frac{1}{n!}\overline{\zeta}^n(\zeta^{n+1} + n\zeta^{n-1})$$

$$= \frac{1}{n!}(\overline{\zeta}\zeta)^n\zeta + n(\overline{\zeta}\zeta)^{n-1}\overline{\zeta}$$

$$= \frac{1}{n!}((r^2)^n re^{i\phi} + n(r^2)^{n-1}re^{-i\phi}).$$

Since

$$\int_{\phi=0}^{2\pi} e^{i\phi}d\phi = 0, \qquad \int_{\phi=0}^{2\pi} e^{-i\phi}d\phi = 0.$$

Hence we find

$$\frac{1}{\pi} \int_{\phi=0}^{2\pi} \int_{r=0}^{\infty} e^{r^2} \left(\frac{1}{n!}((r^2)^n re^{i\phi} + n(r^2)^{n-1}re^{-i\phi}) \right) rdrd\phi = 0.$$

Problem 13. Let $\hat{b}^\dagger$, $\hat{b}$ be Bose creation and annihilation operators, respectively. To solve the eigenvalue problem

$$(\hat{b}^\dagger\hat{b} + \frac{1}{2}I + \gamma(\hat{b} + \hat{b}^\dagger))|\psi\rangle = E|\psi\rangle$$

find the solution of the first order ordinary differential equation

$$\left(\zeta\frac{d}{d\zeta} + \frac{1}{2} + \gamma\left(\frac{d}{d\zeta} + \zeta\right) \right) w(\zeta) = Ew(\zeta)$$

with the condition that $w(\zeta)$ is an entire function.

Solution 13. The differential equation can be written as

$$\frac{dw(\zeta)}{d\zeta} = \frac{E - 1/2 - \gamma\zeta}{\zeta + \gamma}w(\zeta) \equiv \frac{E - 1/2}{\zeta + \gamma}w(\zeta) - \frac{\gamma\zeta}{\zeta + \gamma}w(\zeta).$$

It follows that

$$\ln(w(\zeta)) = (E - 1/2) \int \frac{1}{\zeta + \gamma}d\zeta - \gamma \int \frac{\zeta}{\zeta + \gamma}d\zeta.$$

Then

$$\ln(w(\zeta)) = (E - 1/2 + \gamma^2)\ln(\zeta + \gamma) - \gamma\zeta + C$$

and

$$w(\zeta) = (\zeta + \gamma)^{E+\gamma^2-1/2}e^{-\gamma\zeta}e^C$$

With $w(\zeta)$ to be entire we have $E + \gamma^2 - 1/2 = n$ ($n = 0, 1, 2, \ldots$) or $E_n = n - \gamma^2 + 1/2$. Hence

$$w_n(\zeta) = (\zeta + \gamma)^n e^{-\gamma\zeta}e^C$$

where $e^C = K$ is determined by the normalization condition

$$1 = \frac{1}{\pi} \int_\mathbb{C} e^{-|\zeta|^2}\overline{w_n(\overline{\zeta})}w_n(\zeta)d\Re(\zeta)d\Im(\zeta).$$

8.3   Supplementary Problems

Problem 1. Let $\mathbb{E}$ be the exterior of the unit disc

$$\{\, z \in \mathbb{C} \,:\, |z| > 1\,\}$$

and $\mathbb{T}$ the unit circle

$$\{\, z \in \mathbb{C} \,:\, |z| = 1\,\}.$$

Let $\mathcal{H}_2(\mathbb{E})$ be the *Hardy space* of square integrable functions on $\mathbb{T}$, analytic in the region $\mathbb{E}$. The inner product for $f(z), g(z) \in \mathcal{H}_2(\mathbb{E})$ is defined by

$$\langle f, g \rangle = \frac{1}{2\pi} \int_{-\pi}^{\pi} f(e^{i\omega})^* g(e^{i\omega}) d\omega = \frac{1}{2\pi i} \oint_{\mathbb{T}} f^*(1/z^*) g(z) \frac{dz}{z}.$$

Let $f(z) = z^2$ and $g(z) = z + 1$. Find the scalar product $\langle f, g \rangle$.

Problem 2. Show that for $z \in \mathbb{C}$ one has

$$\frac{\sin(\pi z)}{\pi z} = \int_{-1/2}^{1/2} e^{2\pi i z t} dt.$$

Problem 3. Show that for every $t \in \mathbb{R}$ the function

$$z \mapsto \exp(2\pi i z t)$$

is an *entire function* of $z \in \mathbb{C}$.

Problem 4. Consider the Hilbert space $L_2(\mathbb{R}^d)$ and $f \in L_2(\mathbb{R}^d)$. The *Bargmann transform* is defined by

$$(B_d f)(z) := \frac{1}{\pi^{d/4}} \int_{\mathbb{R}^d} \exp\left(-\frac{1}{2}((\mathbf{z}, \mathbf{z}) + |\mathbf{y}|^2) + \sqrt{2}(\mathbf{z}, \mathbf{y}) \right) f(\mathbf{y}) d\mathbf{y}$$

where $\mathbf{z} \in \mathbb{C}^d$ and

$$(\mathbf{z}, \mathbf{w}) := \sum_{j=1}^{d} z_j w_j.$$

The Bargmann transform $f \mapsto B_d f$ is a bijective and isometric map from $L_2(\mathbb{R}^d)$ to the Hilbert space $A^2(\mathbb{C})$. The Bargmann transform maps the orthonormal basis $\{\phi_\alpha\}_{\alpha \in \mathbb{N}^d}$ in $L_2(\mathbb{R}^d)$ bijectively into the orthonormal basis $\{e_\alpha\}_{\alpha \in \mathbb{N}^d}$ of the monomials in the Hilbert space $A^2(\mathbb{C}^d)$.

(i) Let $d = 1$, $k > 0$ and $f(x) = e^{-k|x|} \in L_2(\mathbb{R})$. Find the Bargmann transform of f.

(ii) Let $d = 1$, $k > 0$ and $g(x) = e^{-k^2 x^2} \in L_2(\mathbb{R})$. Find the Bargmann transform of g.

(iii) Let $d = 1$, $k > 0$ and $h(x) = \cos(kx)e^{-k^2 x^2}$. Find the Bargmann transform of h.

Problem 5. The *confluent hypergeometric function* of the first kind $F(\alpha, \gamma; z)$ depends on two complex parameters α, γ and one complex argument z and is defined for $\gamma \neq 0, -1, -2, \ldots$ by

$$F(\alpha, \gamma; z) = 1 + \sum_{n=1}^{\infty} \frac{(\alpha)_n z^n}{(\gamma)_n n!}$$

where

$$(\alpha)_n := \alpha(\alpha + 1) \cdots (\alpha + n - 1).$$

Show that this series converges for all finite z and F is an entire function of z for fixed parameters α, γ.

Problem 6. Show that the functions

$$z^2 e^{-z}, \quad \frac{\sin(z)}{z}, \quad \sinh(z^2), \quad \frac{1 - \cos(z)}{z}, \quad \frac{\sin(\sqrt{z})}{\sqrt{z}}, \quad \cos(\sqrt{z})$$

are entire.

Problem 7. (i) Show that $f : \mathbb{C} \to \mathbb{C}$

$$f(z) = \frac{e^z - 1}{z} = 1 + \frac{z}{2!} + \frac{z^2}{3!} + \frac{z^3}{4!} + \cdots$$

is an entire function. Let

$$A = \begin{pmatrix} 0 & 1 \\ 1 & 0 \end{pmatrix}.$$

Find $f(A)$.

(ii) Is $g : \mathbb{C} \to \mathbb{C}$

$$g(z) = \frac{e^{-z} - 1}{z}$$

an entire function? Find $g(A)$.

Problem 8. Show that the 2-dimensional complex δ-function can be written as $(w \in \mathbb{C})$

$$\delta^{(2)}(z) = \frac{1}{\pi^2} \int_{\mathbb{C}} d^2 w \exp(w^* z - z^* w) = \frac{1}{\pi^2} \int_{\mathbb{C}} d^2 w \exp(i(w^* z + z^* w)).$$

Problem 9. Consider the *Lie group*

$$SU(1,1) = \left\{ \begin{pmatrix} \alpha & \beta \\ \bar{\beta} & \bar{\alpha} \end{pmatrix} \ |\alpha|^2 - |\beta|^2 = 1 \right\}.$$

The elements of this Lie group act as analytic automorphism of the disk

$$\Omega := \{ \, |z| < 1 \, \}$$

under

$$z \to zg = \frac{\bar{\alpha} z + \beta}{\bar{\beta} z + \alpha}$$

where $(zg)h = z(gh)$. Let $n \geq 2$. We define

$$\mathcal{H}_n := \{ \, f(z) \text{ analytic in } \Omega, \ \|f\|^2 = \int_{\Omega} |f(z)|^2 (1 - |z|^2)^{n-2} dx dy < \infty \, \}$$

and

$$U_n(g) f(z) := \frac{1}{(\bar{\beta} z + \alpha)^n} f((\bar{\alpha} z + \beta)/(\bar{\beta} z + \alpha)).$$

Then $\mathcal{H}_n$ is a Hilbert space, i.e. the analytic functions in

$$L_2(\Omega, (1 - |z|^2)^{n-2} dx dy)$$

form a closed subspace. U_n is a representation, i.e.

$$U_n(gh) = U_n(g) U_n(h)$$

and $U_n(e) = I$, where e is the identity element in $SU(1,1)$ (2×2 unit matrix). Show that

$$\frac{1}{(1 - |z|^2)^2} dx \wedge dy$$

is invariant $z \to zg$. Note that an group element of $SU(1,1)$ can be written as

$$\begin{pmatrix} \cosh(\tau/2) e^{-i\alpha} & \sinh(\tau/2) e^{-i\beta} \\ \sinh(\tau/2) e^{i\beta} & \cosh(\tau/2) e^{i\alpha} \end{pmatrix}.$$

Chapter 9

Generalized Functions

9.1 Introduction

The vector space $D(\mathbb{R}^n)$ denotes all infinitely-differentiable functions in $\mathbb{R}$ with compact support. The functions $\phi \in D(\mathbb{R}^n)$ are called test functions. Each linear continuous functional over the vector space of test functions $D(\mathbb{R}^n)$ is called a generalized function.

Instead of considering the space $D(\mathbb{R}^n)$ for the test functions we can also consider the vector space $S(\mathbb{R}^n)$. The vector space $S(\mathbb{R}^n)$ is the set of all infinitely differentiable functions which decrease as $|\mathbf{x}| \to \infty$, together with all their derivatives, faster than any power of $|\mathbf{x}|^{-1}$. These functions are called test functions. Each linear continuous functional over the vector space $S(\mathbb{R}^n)$ is called a generalized function (of slow growth). This vector space of linear continuous functional is denoted by $S'(\mathbb{R}^n)$. Note that $S(\mathbb{R}^n) \subset L_2(\mathbb{R}^n)$ and the vector space $S(\mathbb{R}^n)$ is dense in the Hilbert space $L_2(\mathbb{R}^n)$ and

$$S(\mathbb{R}^n) \subset L_2(\mathbb{R}^n) \subset S'(\mathbb{R}^n)$$

is a Banach-Gelfand triple ([26],[27],[36],[37],[78],[86]).

The *delta function* δ is defined by

$$(\delta(\mathbf{x}), \phi(\mathbf{x})) := \phi(\mathbf{0})$$
$$(\delta(\mathbf{x} - \mathbf{x}_0), \phi(\mathbf{x})) := \phi(\mathbf{x}_0).$$

The delta function δ is a singular generalized function.

Let ϕ be a test function. Then the derivative of a generalized function T is

313

defined as

$$\left(\frac{\partial T}{\partial x_j}, \phi\right) := -\left(T, \frac{\partial \phi}{\partial x_j}\right).$$

Let ϕ be a test function, i.e. $\phi \in S(\mathbb{R}^n)$ and T be a generalized function. Then the *Fourier transform* $F(T)$ of the generalized function T is defined as

$$(F(T), \psi) := (2\pi)^n (T, \phi)$$

where

$$\psi(\mathbf{k}) \equiv F[\phi(\mathbf{x})] = \int_{\mathbb{R}^n} \phi(\mathbf{x}) \exp(i\mathbf{k} \cdot \mathbf{x}) d\mathbf{x}$$

with $\mathbf{k} \cdot \mathbf{x} = \sum_{j=1}^n k_j x_j$. The Fourier transform operation is continuous from $S'(\mathbb{R}^n)$ to $S'(\mathbb{R}^n)$.

In the sense of generalized functions we have

$$\sum_{j=1}^{\infty} \cos(jx) = -\frac{1}{2} + \pi \sum_{n \in \mathbb{Z}} \delta(x - 2\pi n)$$

$$\sum_{j=1}^{\infty} j \sin(jx) = -\pi \sum_{n \in \mathbb{Z}} \delta'(x - 2\pi n)$$

$$\sum_{j=1}^{\infty} j^2 \cos(jx) = -\pi \sum_{n \in \mathbb{Z}} \delta''(x - 2\pi n)$$

$$1 + \sum_{k=1}^{\infty} e^{ikx} + \sum_{k=1}^{\infty} e^{-ikx} = 2\pi \sum_{n \in \mathbb{Z}} \delta(x - 2\pi n)$$

or

$$\frac{1}{2\pi} \sum_{k \in \mathbb{Z}} e^{ikx} = \sum_{n \in \mathbb{Z}} \delta(x - 2\pi n).$$

In the sense of generalized functions we have the Fourier transforms

$$F[\exp(kx)] = 2\pi\delta(s - ik)$$
$$F[\sin(kx)] = i\pi(\delta(s - k) - \delta(s + k))$$
$$F[\cos(kx)] = \pi(\delta(s - k) + \delta(s + k))$$
$$F[\sinh(kx)] = \pi(\delta(s - ik) - \delta(s + ik))$$
$$F[\cosh(kx)] = \pi(\delta(s - ik) + \delta(s + ik)).$$

The Fourier transform of a polynomial $p(x)$ in the sense of generalized functions is

$$F(p(x)) = 2\pi p(-id/ds)\delta(s).$$

Furthermore

$$F(p(d/dx)\delta(x)) = p(-is).$$

An entire holomorphic function $f(\mathbf{z}) = f(z_1, \ldots, z_n)$ of n complex variables $z_j = x_j + jy_j$ $(x_j, y_j \in \mathbb{R})$ is the Fourier-Laplace transform of a distribution $T \in \mathcal{E}(\mathbb{R}^n)'$ if and only if for some positive constants B, N and C

$$|f(\mathbf{z})| \leq C(1 - |\mathbf{z}|)^N \exp(B|\Im(\mathbf{z})|).$$

Let $\{\phi_j(x)\}_{j=0}^\infty$ be an orthonormal basis in $L_2(\mathbb{R})$. Then

$$K(x, x') := \sum_{j=0}^\infty \phi_j(x)\overline{\phi_j(x')} = \delta(x - x').$$

The delta function $\delta(x - x')$ is a *reproducing function*. However it is not continuous or analytic.

In the sense of generalized function we have

$$\frac{1}{x - i\epsilon} = P\left(\frac{1}{x}\right) + \pi i\delta(x)$$

where P denotes the *Cauchy principal value*, ϵ is a positive constant and x is real and

$$\frac{1}{x - i\epsilon} = i \int_0^\infty \exp(-is(x - i\epsilon))ds.$$

There is a *Banach-Gelfand triple* isomorphism between

$$(S(\mathbb{R}), L_2(\mathbb{R}), S'(\mathbb{R})) \quad \text{and} \quad (\ell_1(\mathbb{N}_0), \ell_2(\mathbb{N}_0), \ell_\infty(\mathbb{N}_0)).$$

The δ function in *spherical coordinates* in $\mathbb{R}^3$

$$\mathbf{x}_1 = r\sin(\theta)\cos(\phi), \quad \mathbf{x}_2 = r\sin(\theta)\sin(\phi), \quad \mathbf{x}_3 = r\cos(\theta)$$

is given by

$$\delta(\mathbf{r} - \mathbf{r}_0) = \frac{1}{r^2\sin(\theta)}\delta(r - r_0)\delta(\theta - \theta_0)\delta(\phi - \phi_0)$$

and the delta function over the spherical angles θ and ϕ can be written as sum over *spherical harmonics* $Y_{\ell,m}$ $(\ell = 0, 1, 2, \ldots; m = -\ell, -\ell+1, \ldots, +\ell)$

$$\delta(\cos(\theta) - \cos(\theta'))\delta(\phi - \phi') = \frac{1}{\sin(\theta)}\delta(\theta - \theta')\delta(\phi - \phi')$$

$$= \sum_{\ell=0}^\infty \sum_{m=-\ell}^{+\ell} Y_{\ell,m}(\theta, \phi)Y_{\ell,m}^*(\theta', \phi').$$

9.2   Solved Problems

Problem 1. Consider the function $H : \mathbb{R} \to \mathbb{R}$

$$H(x) := \begin{cases} 1 & 0 \leq x \leq 1/2 \\ -1 & 1/2 \leq x \leq 1 \\ 0 & \text{otherwise} \end{cases}$$

(i) Find the derivative of H in the sense of generalized functions. Obviously H can be considered as a regular functional

$$\int_{\mathbb{R}} H(x)\phi(x)dx.$$

(ii) Find the Fourier transform of H. Draw a picture of the Fourier transform.

Solution 1. (i) We have

$$\left(\frac{dH}{dx}, \phi\right) := -\left(H, \frac{d\phi}{dx}\right) = -\int_0^{\frac{1}{2}} \frac{d\phi}{dx}dx + \int_{\frac{1}{2}}^1 \frac{d\phi}{dx}dx$$
$$= -\phi(1/2) + \phi(0) + \phi(1) - \phi(1/2)$$
$$= (\delta(x) + \delta(x-1) - 2\delta(x-1/2), \phi).$$

(ii) For the Fourier transform we find

$$\int_{-\infty}^{\infty} e^{ikx} H(x)dx = \int_0^{1/2} e^{ikx}dx - \int_{1/2}^1 e^{ikx}dx = \frac{e^{ikx}}{ik}\bigg|_0^{1/2} - \frac{e^{ikx}}{ik}\bigg|_{1/2}^1$$
$$= \frac{1}{ik}(2e^{ik/2} - 1 - e^{ik}) = \frac{e^{ik/2}}{ik}(2 - (e^{ik/2} + e^{-ik/2}))$$
$$= \frac{2ie^{ik/2}}{k}(\cos(k/2) - 1).$$

Problem 2. Let $a > 0$. We define

$$f_a(x) := \begin{cases} \dfrac{1}{2a} & \text{for} \quad |x| \leq a \\ 0 & \text{for} \quad |x| > a. \end{cases}$$

(i) Find

$$\int_{\mathbb{R}} f_a(x)dx.$$

(ii) Calculate the derivative of f_a in the sense of generalized functions.

(iii) What happens if $a \to +0$?

Solution 2. (i) We obtain

$$\int_{\mathbb{R}} f_a(x)dx = 1$$

which is independent of a.

(ii) Using the definition of the derivative we find

$$\left(\frac{df_a(x)}{dx}, \phi(x) \right) = - \left(f_a(x), \frac{d\phi(x)}{dx} \right).$$

Thus

$$\left(\frac{df_a(x)}{dx}, \phi(x) \right) = - \int_{\mathbb{R}} f_a(x) \frac{d\phi}{dx} dx = -\frac{1}{2a} \int_{-a}^{a} \frac{d\phi}{dx} dx.$$

Finally

$$\left(\frac{df_a(x)}{dx}, \phi(x) \right) = \frac{1}{2a} \left(-\phi(a) + \phi(-a) \right) = \frac{1}{2a} ((\delta(x+a), \phi(x)) - (\delta(x-a), \phi(x))).$$

We can write

$$\frac{df_a(x)}{dx} = \frac{1}{2a} \left(\delta(x+a) - \delta(x-a) \right)$$

where δ denotes the *delta function*.

(iii) In the sense of generalized functions the function f_a tends to the delta function for $a \to +0$.

Problem 3. Show that

$$\frac{1}{2\pi} \sum_{k=-\infty}^{\infty} e^{ikx} = \sum_{k=-\infty}^{\infty} \delta(x - 2k\pi)$$

in the sense of generalized functions. Hint. Expand the 2π periodic function

$$f(x) = \frac{1}{2} - \frac{x}{2\pi}$$

into a Fourier series and consider the Hilbert space $L_2([-\pi, \pi])$. An orthonormal basis in the Hilbert space $L_2([-\pi, \pi])$ is given by

$$\phi_k(x) = \frac{1}{2\pi} e^{ikx}, \quad k \in \mathbb{Z}.$$

Solution 3. The expansion of the 2π periodic function f provides

$$f(x) = -\frac{i}{2\pi} \sum_{\substack{k=-\infty \\ k \neq 0}}^{k=\infty} \frac{1}{k} e^{ikx}.$$

In the sense of generalized functions we can now differentiate term by term and find

$$\frac{df}{dx} = -\frac{1}{2\pi} + \sum_{k=-\infty}^{k=\infty} \delta(x - 2k\pi) = \frac{1}{2\pi} \sum_{\substack{k=-\infty \\ k \neq 0}}^{\infty} e^{ikx}.$$

Problem 4. An orthonormal basis in the Hilbert space $L_2([-\pi, \pi])$ is given by

$$B = \left\{ \phi_k(x) = \frac{1}{\sqrt{2\pi}} e^{ikx} \; : \; k \in \mathbb{Z} \right\}.$$

Find the sum

$$\sum_{k \in \mathbb{Z}} \phi_k(x') \phi_k^*(x).$$

Solution 4. We have in the sense of generalized functions

$$\sum_{k \in \mathbb{Z}} \phi_k(x') \phi_k^*(x) = \frac{1}{2\pi} \sum_{k \in \mathbb{Z}} e^{ikx'} e^{-ikx} = \frac{1}{2\pi} \sum_{k \in \mathbb{Z}} e^{ik(x'-x)} = \delta(x' - x).$$

Problem 5. Let $f : \mathbb{R} \to \mathbb{R}$ be a continuous function. Show that

$$\sum_{n=0}^{\infty} f(n + 1/2) = \sum_{\ell=-\infty}^{\infty} (-1)^\ell \int_{x=0}^{\infty} f(x) e^{2\pi i \ell x} dx.$$

Solution 5. Note that in the sense of generalized function

$$\delta(x) = \frac{1}{2\pi} \int_0^{\infty} e^{-ikx} dk$$

i.e.

$$\delta(x - (n + 1/2)) = \frac{1}{2\pi} \int_0^{\infty} e^{-ik(x-(n+1/2))} dx = \frac{1}{2\pi} e^{ikn} e^{ik/2} \int_0^{\infty} e^{-ikx} dx.$$

It follows that

$$\sum_{n=0}^{\infty} f(n+1/2) = \sum_{n=0}^{\infty} \int_{x=0}^{\infty} f(x)\delta(x-(n+1/2))dx$$

$$= \int_{x=0}^{\infty} dx f(x) \int_{-\infty}^{\infty} \frac{e^{2\pi iqx}}{2i\sin(\pi q)} dq$$

$$= \sum_{\ell=-\infty}^{\infty} (-1)^\ell \int_{x=0}^{\infty} f(x) e^{2\pi i\ell x} dx.$$

Problem 6. Let $C^m([a,b])$ be the vector space of m-times differentiable functions and the m-th derivative is continuous over the interval $[a,b]$ ($b > a$). We define an inner product (scalar product) of such two functions f and g as

$$\langle f,g\rangle_m := \int_a^b \left(fg + \frac{df}{dx}\frac{dg}{dx} + \cdots + \frac{d^m f}{dx^m}\frac{d^m g}{dx^m} \right) dx.$$

Given (Legendre polynomials)

$$f(x) = \frac{1}{2}(3x^2 - 1), \qquad g(x) = \frac{1}{2}(5x^3 - 3x)$$

and the interval $[-1,1]$, i.e. $a = -1$ and $b = 1$. Show that f and g are orthogonal with respect to the inner product $\langle f,g\rangle_0$. Are they orthogonal with respect to $\langle f,g\rangle_1$?

Solution 6. Straightforward calculation yields

$$\langle f,g\rangle_0 = \frac{1}{4}\int_{-1}^1 (3x^2 - 1)(5x^3 - 3x)dx = \frac{1}{4}\int_{-1}^1 (15x^5 - 14x^3 + 3x)dx = 0.$$

We have

$$\frac{df}{dx} = 3x, \qquad \frac{dg}{dx} = \frac{15}{2}x^2 - \frac{3}{2}.$$

Then since $\langle f,g\rangle_0 = 0$

$$\langle f,g\rangle_1 = \int_{-1}^1 3x \left(\frac{15}{2}x^2 - \frac{3}{2} \right) dx = \int_{-1}^1 \left(\frac{45}{2}x^3 - \frac{9}{2}x \right) dx = 0.$$

Problem 7. Let $a > 0$. Find the derivative of $f_a(x) = \exp(-a|x|)$ in the sense of generalized functions.

Solution 7. Using integration by parts we find

$$\left(\frac{d}{dx}e^{-a|x|}, \phi(x)\right) = -\left(e^{-a|x|}, \frac{d\phi}{dx}\right) = -\int_{-\infty}^{\infty} e^{-a|x|}\frac{d\phi}{dx}dx$$

$$= -\int_{-\infty}^{0} e^{ax}\frac{d\phi}{dx}dx - \int_{0}^{\infty} e^{-ax}\frac{d\phi}{dx}dx$$

$$= -\phi(0) + \int_{-\infty}^{0} ae^{ax}\phi(x)dx + \phi(0) - \int_{0}^{\infty} ae^{-ax}\phi(x)dx$$

$$= a\left(\int_{-\infty}^{0} e^{ax}\phi(x)dx - \int_{0}^{\infty} e^{-ax}\phi(x)dx\right)$$

$$= -a(e^{-a|x|}\mathrm{sgn}(x), \phi(x)).$$

Hence in the sense of generalized functions we have

$$\frac{d}{dx}e^{-a|x|} = -a\mathrm{sgn}(x)e^{-a|x|}.$$

Problem 8. Let $r^2 = \sum_{j=1}^{3} x_j^2$ and $\phi(\mathbf{x}) = e^{-r^2}$, i.e. $\phi \in S(\mathbb{R}^3)$. Find

$$\left(\frac{1}{r}, \phi(\mathbf{x})\right) = \int_{\mathbb{R}^3} \frac{1}{r}\phi(\mathbf{x})dx_1 dx_2 dx_3.$$

Solution 8. Introducing *spherical coordinates* we have

$$\int_{\mathbb{R}^3} \frac{1}{r}\phi(\mathbf{x})dx_1 dx_2 dx_3 = \int_{r=0}^{\infty}\int_{\theta=0}^{\pi}\int_{\phi=0}^{2\pi} \frac{1}{r}e^{-r^2}r^2 \sin(\theta)d\theta dr d\phi$$

$$= 4\pi \int_{r=0}^{\infty} re^{-r^2} dr.$$

Using the substitution $u = r^2$, $du = 2rdr$ we have $dr = du/(2r)$. Hence

$$\int_{\mathbb{R}^3} \frac{1}{r}\phi(\mathbf{x})dx_1 dx_2 dx_3 = 4\pi \int_{u=0}^{\infty} re^{-u}\frac{du}{2r} = 2\pi \int_{u=0}^{\infty} e^{-u}du = 2\pi.$$

Problem 9. Let T be the generalized function generated by $f : \mathbb{R} \to \mathbb{R}$, $f(x) = 1$. Find the Fourier transform.

Solution 9. We have

$$(F(T), \psi) := 2\pi(1, \phi) = 2\pi \int_{\mathbb{R}} \phi(x)dx.$$

Since

$$\psi(k) = \int_{\mathbb{R}} \phi(x)e^{ikx}\,dx$$

we obtain

$$\psi(0) = \int_{\mathbb{R}} \phi(x)\,dx.$$

Consequently

$$(F(T), \psi) = 2\pi\psi(0) = 2\pi(\delta(k), \psi(k)).$$

Hence $F[1] = 2\pi\delta$.

Problem 10. Consider the function $f : \mathbb{R} \to \mathbb{R}$

$$f(x) = \begin{cases} 1 & \text{for} \quad 0 \le x \le 1 \\ 0 & \text{otherwise} \end{cases}$$

Find the derivative of f in the sense of generalized functions. Then apply the Fourier transform in the sense of generalized function.

Solution 10. We have

$$(df/dx, \phi) = -(f, d\phi/dx) = -\int_{\mathbb{R}} f(x)\frac{d\phi}{dx}\,dx$$

$$= -\int_0^1 \frac{d\phi}{dx}\,dx = -\phi(0) + \phi(0)$$

$$= -(\delta(x-1), \phi(x)) + (\delta(x), \phi(x)).$$

The Fourier transform of $-\delta(x-1) + \delta(x)$ in the sense of generalized functions is $-e^{ik} + 1$.

Problem 11. Let $r^2 = \sum_{j=1}^3 x_j^2$ and $\phi(\mathbf{x}) \in S(\mathbb{R}^3)$. Assume that ϕ only depends on r. Let

$$\Delta := \frac{1}{r^2}\frac{\partial}{\partial r}\left(r^2\frac{\partial}{\partial r}\right).$$

Find

$$\left(\frac{1}{r}, \Delta\phi\right) = \int_{\mathbb{R}^3} \frac{1}{r}\Delta\phi(r)\,dx_1\,dx_2\,dx_3.$$

Solution 11. We utilize integration by parts and that $\phi(\infty) = 0$. Introducing spherical coordinates we have

$$\left(\frac{1}{r}, \Delta\phi\right) = \int_{r=0}^{\infty} \int_{\theta=0}^{\pi} \int_{\phi=0}^{2\pi} \frac{1}{r} \frac{1}{r^2} \left(\frac{\partial}{\partial r} r^2 \frac{\partial}{\partial r} \phi(r)\right) r^2 \sin(\theta) dr d\theta d\phi$$

$$= 4\pi \int_{r=0}^{\infty} \frac{1}{r} \left(\frac{d}{dr} r^2 \frac{d}{dr} \phi(r)\right) dr$$

$$= 4\pi \int_{0}^{\infty} \left(2\frac{d}{dr}\phi(r)\right) dr + 4\pi \int_{0}^{\infty} \left(r \frac{d^2}{dr^2}\phi(r)\right) dr$$

$$= -8\pi\phi(0) + 4\pi \int_{0}^{\infty} \left(r \frac{d^2}{dr^2}\phi(r)\right) dr$$

$$= -8\pi\phi(0) - 4\pi \int_{0}^{\infty} \left(\frac{d}{dr}\phi\right) dr = -8\pi\phi(0) + 4\pi\phi(0)$$

$$= -4\pi\phi(0) = -4\pi(\delta, \phi).$$

So formally we can write $\Delta \frac{1}{r} = -4\pi\delta$.

Problem 12. Find the derivative of $f : \mathbb{R} \to \mathbb{R}$, $f(x) = |x|$ in the sense of generalized functions.

Solution 12. We have

$$\left(\frac{df}{dx}, \phi\right) = -\left(f, \frac{d\phi}{dx}\right) = -\int_{-\infty}^{\infty} |x| \frac{d\phi}{dx} dx = \int_{-\infty}^{0} x \frac{d\phi}{dx} dx - \int_{0}^{\infty} x \frac{d\phi}{dx} dx$$

$$= x\phi(x)\Big|_{-\infty}^{0} - \int_{-\infty}^{0} \phi(x) dx - x\phi(x)\Big|_{0}^{\infty} + \int_{0}^{\infty} \phi(x) dx$$

$$= -\int_{-\infty}^{0} \phi(x) dx + \int_{0}^{\infty} \phi(x) dx = (\text{sgn}(x), \phi(x))$$

where $\text{sgn}(x)$ is the *sign function*

$$\text{sgn}(x) := \begin{cases} 1 & x > 0 \\ 0 & x = 0 \\ -1 & x < 0 \end{cases}$$

Problem 13. Let δ be the delta function. Find $x\delta'(x)$.

Solution 13. We have

$$(x\delta'(x), \phi(x)) = (\delta'(x), x\phi(x)) = (\delta, -d(x\phi(x))/dx)$$

$$= (\delta, -\phi(x) - x d\phi(x)/dx)$$

$$= -\phi(0) = -(\delta(x), \phi(x)).$$

Hence $x\delta'(x) = -\delta(x)$.

Problem 14. Find the solution of the equation $x^2 f(x) = 0$ in the sense of generalized functions.

Solution 14. Since $(\delta(x), x^2\phi(x)) = 0$ and
$$(\delta'(x), x^2\phi(x)) = (\delta(x), -2x\phi(x) + x^2 d\phi(x)/dx) = 0$$
we obtain the solution
$$f(x) = C_0\delta(x) + C_1\delta'(x)$$
where C_0 and C_1 are arbitrary constants.

Problem 15. Find the solution of the first order ordinary differential equation
$$x\frac{du}{dx} + u(x) = 0$$
in the sense of generalized functions.

Solution 15. Obviously $1/x$ is a solution of the differential equation. The delta function δ is also a solution of the differential equation since $x\delta'(x) = -\delta(x)$. Hence the solution in the sense of generalized function is the linear combination
$$u(x) = \frac{C_0}{x} + C_1\delta(x)$$
where C_0 and C_1 are arbitrary constants.

Problem 16. Find the first three derivatives of the function $f : \mathbb{R} \to \mathbb{R}$ defined by
$$f(x) = e^{-|x|}$$
in the sense of generalized functions. Note that $e^{-|x|} \in L_2(\mathbb{R})$ and f is continuous.

Solution 16. Using integration by parts we have for the first derivative
$$\left(\frac{df}{dx}, \phi\right) = -\left(f, \frac{d\phi}{dx}\right) = -\int_{-\infty}^{0} e^x \frac{d\phi}{dx} dx - \int_{0}^{\infty} e^{-x} \frac{d\phi}{dx} dx$$
$$= \int_{-\infty}^{0} e^x \phi(x) dx - \int_{0}^{\infty} e^{-x} \phi(x) dx.$$

Using the *sign function*

$$\text{sign}(x) = \begin{cases} 1 & \text{for } x \geq 0 \\ -1 & \text{for } x < 0 \end{cases}$$

we have in the sense of generalized functions

$$g(x) = \frac{df}{dx} = -\text{sign}(x)e^{-|x|}.$$

For the second order derivative using integration by parts we find

$$\left(\frac{dg}{dx}, \phi\right) = -\left(g, \frac{d\phi}{dx}\right) = -\int_{\mathbb{R}} g(x)\frac{d\phi}{dx}dx = \int_{\mathbb{R}} \text{sign}(x)e^{-|x|}\frac{d\phi}{dx}dx$$

$$= -\phi(0) + \int_{-\infty}^{0} e^{x}\phi(x)dx - \phi(0) + \int_{0}^{\infty} e^{-x}\phi(x)dx$$

$$= -2\phi(0) + (e^{-|x|}, \phi)$$

$$= -2(\delta, \phi) + (e^{-|x|}, \phi).$$

Thus in sense of generalized function we have the derivative

$$h(x) = -2\delta + e^{-|x|}.$$

Using the result from the first derivative we obtain for the third derivative

$$-2\phi'(0) - \text{sign}(x)e^{-|x|} = 2\delta' - \text{sign}(x)e^{-|x|}.$$

Problem 17. The *Sobolev space* of order m, denoted by $H^m(\Omega)$, is defined to be the space consisting of those functions in the Hilbert space $L_2(\Omega)$ that, together with all their weak partial derivatives up to and including those of order m, belong to the Hilbert space $L_2(\Omega)$, i.e.

$$H^m(\Omega) := \{ u : D^\alpha u \in L_2(\Omega) \text{ for all } \alpha \text{ such that } |\alpha| \leq m \}.$$

We consider real-valued functions only, and make $H^m(\Omega)$ an inner product space by introducing the Sobolev inner product $\langle \cdot, \cdot \rangle_{H^m}$ defined by

$$\langle u, v \rangle_{H^m} := \int_{\Omega} \sum_{|\alpha| \leq m} (D^\alpha u)(D^\alpha v)dx \quad \text{for} \quad u, v \in H^m(\Omega).$$

This inner product generates the Sobolev norm $\| \cdot \|_{H^m}$ defined by

$$\|u\|_{H^m}^2 = \langle u, u \rangle_{H^m} = \int_{\Omega} \sum_{|\alpha| \leq m} (D^\alpha u)^2 dx.$$

Thus $H^0(\Omega) = L_2(\Omega)$. We can write

$$\langle u, v \rangle = \sum_{|\alpha| \le m} \langle D^\alpha u, D^\alpha v \rangle_{L_2(\Omega)}.$$

In other words the Sobolev inner product $\langle u, v \rangle_{H^m(\Omega)}$ is equal to the sum of the $L_2(\Omega)$ inner products of $D^\alpha u$ and $D^\alpha v$ over all α such that $|\alpha| \le m$.
(i) Consider the domain $\Omega = (0, 2)$ and the function

$$u(x) = \begin{cases} x^2 & 0 < x \le 1 \\ 2x^2 - 2x + 1 & 1 < x < 2. \end{cases}$$

Obviously $u \in L_2(\Omega)$. Find the Sobolev space to which u belongs.
(ii) Find the norm of u.

Solution 17. The derivative of u in the ordinary sense exists and is given by

$$\frac{du}{dx} = \begin{cases} 2x & 0 < x \le 1 \\ 4x - 2 & 1 < x < 2. \end{cases}$$

This function is continuous but not differentiable in the ordinary sense and it belong to the Hilbert space $L_2(\Omega)$. The derivative of du/dx in the sense of generalized functions is given by

$$\frac{d^2u}{dx^2} = \begin{cases} 2 & 0 < x \le 1 \\ 4 & 1 < x < 2. \end{cases}$$

The function is not continuous. Thus next derivative in the sense of generalized functions yields a delta function $2\delta(x - 1)$ which is not an element of the Hilbert space $L_2(\Omega)$. Thus $u \in H^2(0, 2)$.
(ii) The square of norm is

$$\|u\|_{H^2}^2 = \int_0^2 \left| u^2 + \left(\frac{du}{dx} \right)^2 + \left(\frac{d^2u}{dx^2} \right)^2 \right| dx = 71.37.$$

Problem 18. Let $x \in \mathbb{R}$ and $m \ge 1$. Show that the general solution if the equation

$$x^m u(x) = 0$$

in the sense of generalized function is given by

$$u(x) = \sum_{j=0}^{m-1} c_j \delta^{(j)}(x)$$

where c_j are arbitrary constants.

Solution 18. For all $\phi \in S(\mathbb{R})$ we have

$$(x^m \delta^{(j)}, \phi(x)) = (\delta^{(j)}, x^m \phi(x)) = (-1)^j (\delta, (x^m \phi(x))^{(j)})$$
$$= (-1)^j (x^m \phi(x))^{(j)} \Big|_{x=0} = 0.$$

Problem 19. Consider the one-dimensional wave equation

$$\frac{\partial^2 u}{\partial x_0^2} = \frac{\partial^2 u}{\partial x_1^2}, \quad -\infty < x < \infty.$$

Show that the fundamental solution is given by

$$E(x_0, x_1) = \begin{cases} \frac{1}{2} \text{ for } |x_1| < x_0 \\ 0 \text{ for } |x_1| > x_0 \end{cases} \equiv \frac{1}{2} \theta(x_0 - |x_1|)$$

for the Cauchy problem.

Solution 19. For a fixed $x_0 > 0$ we have

$$\frac{\partial E(x_0, x_1)}{\partial x_1} = \frac{1}{2} \delta(x_1 + x_0) - \frac{1}{2} \delta(x_1 - x_0)$$

$$\frac{\partial^2 E(x_0, x_1)}{\partial x_1^2} = \frac{1}{2} \delta'(x_1 + x_0) - \frac{1}{2} \delta'(x_1 - x_0).$$

For a fixed x_1 we find

$$\frac{\partial E(x_0, x_1)}{\partial x_0} = \frac{1}{2} \delta(x_1 + x_0) - \frac{1}{2} \delta(x_1 - x_0)$$

$$\frac{\partial^2 E(x_0, x_1)}{\partial x_0^2} = \frac{1}{2} \delta'(x_1 + x_0) - \frac{1}{2} \delta'(x_1 - x_0)$$

so that the equation is satisfied. Furthermore

$$\frac{\partial E(x_0, x_1)}{\partial x_0} \Big|_{x_0=0} = \delta(x_1).$$

Problem 20. Let $a > 0$ and let $f_a : \mathbb{R} \to \mathbb{R}$ be given by

$$f_a(x) = \begin{cases} x/a^2 + 1/a \text{ for } -a \leq x \leq 0 \\ -x/a^2 + 1/a \text{ for } 0 \leq x \leq a \end{cases}$$

The function f_a generates regular functional. Find the derivative of f_a in the sense of generalized functions.

Solution 20. Let ϕ be a test function. Thus

$$\frac{d}{dx}(f_a(x), \phi(x)) = \left(f_a(x), -\frac{d\phi(x)}{dx}\right)$$

$$= -\int_{-\infty}^{\infty} f_a(x)\frac{d\phi(x)}{dx}dx = -\int_{-a}^{a} f_a(x)\frac{d\phi(x)}{dx}dx$$

$$= -\int_{-a}^{0} \left(\frac{x}{a^2} + \frac{1}{a}\right)\frac{d\phi(x)}{dx}dx - \int_{0}^{a} \left(-\frac{x}{a^2} + \frac{1}{a}\right)\frac{d\phi(x)}{dx}dx.$$

Using integration by parts we arrive at

$$\frac{d}{dx}(f_a(x), \phi(x)) = \frac{1}{a^2}(\Theta(x+a) + \Theta(x-a) - 2\Theta(x), \phi(x))$$

where Θ is the step function.

Problem 21. Consider a one-dimensional lattice (chain) with lattice constant a. Let k be the sum over the first *Brillouin zone* we have

$$\frac{1}{N}\sum_{k\in 1.BZ} F(\epsilon(k)) \rightarrow \frac{a}{2\pi}\int_{-\pi/a}^{\pi/a} F(\epsilon(k))dk = G$$

where $\epsilon(k) = \epsilon_0 - 2\epsilon_1 \cos(ka)$. Using the identity

$$\int_{-\infty}^{\infty} \delta(E - \epsilon(k))F(E)dE \equiv F(\epsilon(k))$$

we can write

$$G = \frac{a}{2\pi}\int_{-\infty}^{\infty} F(E)\left(\int_{-\pi/a}^{\pi/a} \delta(E - \epsilon(k))dk\right)dE.$$

Calculate

$$g(E) = \int_{-\pi/a}^{\pi/a} \delta(E - \epsilon(k))dk$$

where $g(E)$ is called the density of states.

Solution 21. Since $\epsilon(k) = \epsilon_0 - 2\epsilon_1 \cos(ka)$ where $\epsilon_1 > 0$ we have

$$g(E) = \int_{-\pi/a}^{\pi/a} \delta(E - \epsilon_0 + 2\epsilon_1 \cos(ka))dk.$$

Setting

$$\bar{E} := \frac{E - \epsilon_0}{2\epsilon_1}$$

we obtain

$$g(E) = \int_{-\pi/a}^{\pi/a} \delta(2\epsilon_1(\bar{E} + \cos(ka)))dk = \frac{1}{2\epsilon_1} \int_{-\pi/a}^{\pi/a} \delta(\bar{E} + \cos(ka))dk.$$

The substitution $u = \cos(ka)$, $du = -a\sin(ka)dk$ provides

$$g(E) = \frac{1}{\epsilon_1} \int_{-1}^{1} \frac{\delta(\bar{E} + u)}{\sqrt{1 - u^2}} du.$$

Therefore

$$g(E) = \begin{cases} \dfrac{2}{\sqrt{4\epsilon_1^2 - (\bar{E} - \epsilon_0)^2}} & \text{for} \quad \bar{E} \in [-1, 1] \\ 0 & \text{otherwise} \end{cases}$$

Problem 22. Give two interpretations of the series of derivatives of δ functions

$$f(k) = 2\pi \sum_{n=0}^{\infty} c_n(-1)^n \delta^{(n)}(k). \tag{1}$$

Solution 22. Let ϕ be a suitable test function. Then by integrating term-by-term with this suitable test function we have

$$(f(k), \phi(k)) = 2\pi \sum_{n=0}^{\infty} c_n(-1)^n (\phi(k), \delta^{(n)}) = 2\pi \sum_{n=0}^{\infty} c_n \phi^{(n)}(0). \tag{2}$$

The class of test functions ϕ must be chosen so that this series converges. Another interpretation is to consider (1) as the term-by-term Fourier transform of the function

$$g(x) = \sum_{n=0}^{\infty} c_n(-ix)^n. \tag{3}$$

Then formally we can write

$$f(k) = \int_{-\infty}^{\infty} dx e^{ikx} \sum_{n=0}^{\infty} c_n(-ix)^n. \tag{4}$$

This interpretation is invalid unless the series (3) converges for all real values of x.

Problem 23. Let $a > 0$. Show that

$$\delta(x^2 - a^2) = \frac{1}{2a}(\delta(x - a) + \delta(x + a)).$$

Note that $x^2 - a^2 \equiv (x - a)(x + a)$.

Solution 23. Let $u \geq 0$. Setting $x = \sqrt{u}$ we have $dx = 1/(2\sqrt{u})du$ and

$$\int_0^\infty \delta(x^2 - a^2)\phi(x)dx = \int_0^\infty \delta(u - a^2)\frac{\phi(\sqrt{u})}{2\sqrt{u}}du = \frac{\phi(|a|)}{2|a|}.$$

Setting $x = -\sqrt{u}$ we have $dx = -1/(2\sqrt{u})du$ and

$$\int_{-\infty}^0 \delta(x^2 - a^2)\phi(x)dx = \int_{-\infty}^0 \delta(u - a^2)\frac{\phi(-\sqrt{u})}{2\sqrt{u}}du = \frac{\phi(-|a|)}{2|a|}.$$

Addition of the two equations yields

$$\int_{-\infty}^\infty \delta(x^2 - a^2)dx = \frac{1}{2|a|}(\phi(|a|) + \phi(-|a|)).$$

Problem 24. Find the Fourier transform in the sense of generalized functions of $1 + \sqrt{2\pi}\delta(x)$.

Solution 24. Since the Fourier transform of 1 is $2\pi\delta(k)$ and the Fourier transform of $\delta(x)$ is 1 we obtain

$$\sqrt{2\pi}(1 + \sqrt{2\pi}\delta(k)).$$

Problem 25. Let f be a differentiable function with a simple zero at $x = a$ such that $f(x = a) = 0$ and $df(x = a)/dx \neq 0$. Let g be a differentiable function with a simple zero at $x = b \neq a$ such that $g(x = b) = 0$ and $dg(x = b)/dx \neq 0$. Find $\delta(f(x)g(x))$.

Solution 25. We obtain

$$\delta(f(x)g(x)) = \frac{1}{|f'(a)g(a)|}\delta(x - a) + \frac{1}{|f(b)g'(b)|}\delta(x - b)$$

where $'$ denotes differentiation.

Problem 26. Find the derivative of $\ln(|x|)$ in the sense of generalized functions.

Solution 26. Let $\phi \in S(\mathbb{R})$. Then

$$\left(\frac{d\ln(|x|)}{dx}, \phi(x)\right) = (\ln(|x|), -d\phi(x)/dx) = -\int_{\mathbb{R}} \ln(|x|)\frac{d\phi(x)}{dx}dx$$

$$= -\lim_{\epsilon \to 0} \int_{|x|>\epsilon} \ln(|x|)\frac{d\phi(x)}{dx}dx$$

$$= -\lim_{\epsilon \to 0} \left(\ln(|x|)\phi(x)|_{-\infty}^{-\epsilon} + \ln(|x|)\phi(x)|_{\epsilon}^{\infty} - \int_{|x|>\epsilon} \frac{\phi(x)}{x}dx\right)$$

$$= \lim_{\epsilon \to 0} \int_{|x|>\epsilon} \frac{\phi(x)}{x}dx.$$

The limit is in the sense of the *Cauchy mean value.*

Problem 27. Let $\epsilon > 0$. Show that

$$f_\epsilon(x - c) = \frac{1}{\sqrt{\pi\epsilon}} \exp\left(-\frac{(x-c)^2}{\epsilon}\right)$$

tends to $\delta(x - c)$ in the sense of generalized function if $\epsilon \to 0_+$.

Solution 27. Let $\beta > \alpha$, $c \in (\alpha, \beta)$ and

$$\lim_{\epsilon \to 0} \frac{1}{\sqrt{\pi\epsilon}} \int_{x=\alpha}^{x=\beta} \exp\left(-\frac{(x-c)^2}{\epsilon}\right) dx.$$

Setting $s := (x - c)/\sqrt{\epsilon}$ we have $ds = (1/\sqrt{\epsilon})dx$ and

$$\lim_{\epsilon \to 0} \frac{1}{\sqrt{\pi}} \int_{s=(\alpha-c)/\sqrt{\epsilon}}^{s=(\beta-a)/\sqrt{\epsilon}} \exp(-s^2)ds = \frac{1}{\sqrt{\pi}} \int_{s=-\infty}^{s=\infty} \exp(-s^2)ds = 1.$$

Problem 28. Let $p \in [0, 1]$ and

$$\rho(x) = \frac{1}{2}pe^{-|x|} + (1 - p)\delta(x).$$

Then $\rho(x) \geq 0$. Find

$$\int_{\mathbb{R}} \rho(x)dx.$$

Solution 28. Since

$$\int_{\mathbb{R}} \frac{1}{2}pe^{-|x|}dx = \frac{1}{2}p\int_{\mathbb{R}} e^{-|x|}dx = p$$

we find in the sense of generalized functions

$$\int_{\mathbb{R}} \rho(x)dx = 1.$$

Problem 29. Show that in the sense of generalized functions

$$\exp\left(-a\frac{\partial}{\partial x}\right)\delta(x) = \delta(x - a).$$

Solution 29. The operator $\exp(-a\partial/\partial x)$ is the *translation operator*. Let $\phi \in S(\mathbb{R})$. Then

$$\exp(-a\partial/\partial x)\phi(x) = \phi(x - a).$$

Problem 30. (i) Show that the Fourier transform in the sense of generalized function of the *Dirac comb*

$$\sum_{n\in\mathbb{Z}} \delta(x - n)$$

is again a Dirac comb.
(ii) The two-dimensional *Dirac comb* is defined by

$$C(x_1, x_1) := \sum_{m\in\mathbb{Z}}\sum_{n\in\mathbb{Z}} \delta(x_1 - m)\delta(x_2 - n).$$

Find the Fourier transform of C in the sense of generalized functions.

Solution 30. (i) The Fourier transform of the delta function is 1. The Dirac comb consists of infinite many delta function shifted by n. Hence the Fourier transform in the sense of generalized functions is given by

$$\mathcal{F}\left(\sum_{n=-\infty}^{+\infty} \delta(x - n)\right)(k) = \sum_{n=-\infty}^{+\infty} e^{-2\pi ink}.$$

As generalized function $\exp(-2\pi ink)$ acts on a test function ϕ by integrating against it. From the definition of a (classical) Fourier transform, this yields the Fourier transform of ϕ evaluated at $n \in \mathbb{Z}$. Hence the Fourier transform of the Dirac comb acts on ϕ by summing the values of ϕ's Fourier transform over all integers. By the *Poisson summation formula* this is the same as summing the values of ϕ over all integers which is the same as applying Dirac's comb to ϕ.

Problem 31. Let $T > 0$. Consider the sequence of functions

$$f_n(t) = \frac{1}{n!} \frac{n}{T} \left(\frac{nt}{T} \right)^n \exp(-nt/T)$$

where $n = 1, 2, \ldots$. Find $f_n(t)$ for $n \to \infty$ in the sense of generalized functions.

Solution 31. We obtain $f_\infty(t) = \delta(t - T)$.

Problem 32. What charge distribution $\rho(r)$ does the spherical symmetric potential

$$V(r) = \frac{e^{-\mu r}}{r}$$

give? For $r \neq 0$ *Poisson's equation* in spherical coordinates is given by

$$\Delta V(\mathbf{r}) = \frac{1}{r} \frac{d^2}{dr^2} (rV(\mathbf{r})) + R(\theta, \phi)V(\mathbf{r}) = -4\pi\rho(\mathbf{r})$$

where $R(\theta, \phi)$ is the differential operator depending on the angles θ, ϕ.

Solution 32. Since $V(r)$ is spherical symmetric we have $R(\theta, \phi)V(r) = 0$ and we arrive at

$$\rho(r) = -\frac{1}{4\pi r} \frac{d^2}{dr^2} e^{-\mu r} = -\frac{\mu^2 e^{-\mu r}}{4\pi r}$$

for $r \neq 0$. For $r \to 0$ we have $V(r) \to \frac{1}{r}$. Since

$$\Delta \frac{1}{r} = -4\pi\delta(r)$$

we obtain the charge distribution

$$\rho(r) = \delta(r) - \frac{\mu^2}{4\pi} \frac{e^{-\mu r}}{r}.$$

Problem 33. Consider the Hilbert space $L_2([-1, 1])$. The *Legendre polynomials* are given by

$$P_0 = 1, \quad P_n(x) = \frac{1}{2^n n!} \frac{d^n}{dx^n} (x^2 - 1)^n$$

where $n = 1, 2, \ldots$. They satisfy

$$\int_{-1}^{+1} P_m(x) P_n(x) dx = \frac{2}{2n + 1} \delta_{m,n}, \quad n, m = 0, 1, 2, \ldots$$

Let

$$\delta(x) = \sum_{j'=0}^{\infty} d_{j'} P_{j'}(x).$$

Find the expansion coefficients $d_{j'}$.

Solution 33. We have

$$P_j(0) = \int_{-1}^{+1} \delta(x) P_j(x) dx = \int_{-1}^{+1} \sum_{j'=0}^{\infty} d_{j'} P_{j'}(x) P_j(x) dx = \sum_{j'=0}^{\infty} d_{j'} \frac{2}{2j'+1} \delta_{j',j}$$

$$= \frac{2}{2j+1} d_j.$$

Thus

$$d_j = \frac{2j+1}{2} P_j(0)$$

and hence

$$\delta(x) = \sum_{j=0}^{\infty} \frac{2j+1}{2} P_j(0) P_j(x).$$

Utilize this result to show that

$$\Theta(x) = \frac{1}{2} + \sum_{j=1}^{\infty} \frac{P_{j-1}(0) - P_{j+1}(0)}{2} P_j(x).$$

Problem 34. Let $\mathbf{f} : \mathbb{R}^n \to \mathbb{R}^n$ be an analytic function. Consider the map

$$\mathbf{x}_j = \mathbf{f}(\mathbf{x}_{j-1}) = \cdots = \mathbf{f}(\mathbf{x}_0).$$

To study the evolution of phase-space distributions, we can introduce the evolution operator $U(\mathbf{x}', \mathbf{x}, j)$ such that any initial phase-space distribution $\rho(\mathbf{x}, 0)$ evolves into

$$\rho(\mathbf{x}'; j) = \int_{\Omega} U(\mathbf{x}', \mathbf{x}; j) \rho(\mathbf{x}, 0) d\mathbf{x}$$

where Ω is the phase space area. Find $U(\mathbf{x}', \mathbf{x}; j)$.

Solution 34. The operator $U(\mathbf{x}', \mathbf{x}; j)$ is the unitary *Frobenius-Perron operator*

$$U(\mathbf{x}', \mathbf{x}; j) = \delta(\mathbf{x}' - f^{(j)}(\mathbf{x}))$$

where δ is the delta function.

Problem 35. Find the sense of generalized functions

$$\lim_{\tau \to 0} \frac{1}{\tau}(\delta(t + 2\tau) - 2\delta(t + \tau) + \delta(t)).$$

Solution 35. With $\phi \in S(\mathbb{R})$ we have

$$\left(\frac{\delta(t + 2\tau) - 2\delta(t + \tau) + \delta(t)}{\tau}, \phi(t)\right) = \frac{1}{\tau}(\phi(-2\tau) - 2\phi(-\tau) + \phi(0))$$

$$= 2\tau\phi''(-\theta_1 2\tau) - \tau\phi''(-\theta_2\tau)$$

with $\theta_1, \theta_2, 0 < \theta_1, \theta_2 < 1$. Hence for $\tau \to 0$ we obtain the zero distribution.

Problem 36. Let T be a linear operator in a linear topological space L. A linear functional F on ϕ, such that

$$(TF, \phi) = \lambda(F, \phi)$$

for every element $\phi \in L$, is called a *generalized eigenvector* of the operator T, corresponding to the eigenvalue λ, where

$$(TF, \phi) = (TF, \phi) = F(T^*\phi).$$

The topological space under consideration is $L = S(\mathbb{R}^n)$. Let us consider two examples.
(i) Consider the operator $-id/dx$ defined on $S(\mathbb{R}) \subset L_2(\mathbb{R})$. Find the generalized eigenvector.
(ii) Consider the eigenvalue equation $xf(x) = \lambda f(x)$. Find the generalized eigenvector.

Solution 36. (i) The eigenvalue equation is takes the form

$$-i\frac{df(x)}{dx} = \lambda f(x).$$

The solution is given by

$$f(x) = e^{i\lambda x}.$$

However

$$e^{i\lambda x} \notin S(\mathbb{R}), \qquad e^{i\lambda x} \notin L_2(\mathbb{R}).$$

Consider the differential operator $-id/dx$ in the space $S'(\mathbb{R})$. Then obviously we find that

$$f(x) = e^{i\lambda x}$$

is a generalized eigenvector and $\lambda \in \mathbb{R}$, i.e. the spectrum is the real axis. The generalized function given by the function $e^{i\lambda x}$ is

$$(e^{i\lambda x}, \phi(x)) = \int_{\mathbb{R}} e^{-i\lambda x} \phi(x) dx.$$

If we consider complex valued test functions and generalized functions, then we have

$$c(T, \phi) = (\bar{c}T, \phi) = (T, c\phi)$$

and

$$(f, \phi) = \int_{\mathbb{R}^n} \bar{f}\phi d\mathbf{x}$$

where f is a regular functional and $c \in \mathbb{C}$.

(ii) There is no solution in the Hilbert space $L_2(\mathbb{R})$. However there is a solution in the sense of generalized functions

$$(x\delta(x - \lambda), \phi(x)) = (\delta(x - \lambda), x\phi(x)) = \lambda(\delta(x - \lambda), \phi(x))$$

where $\lambda \in \mathbb{R}$. The "eigenfunction" is the delta function. Here we used the fact that $(xT, \phi) = (T, x\phi)$.

Problem 37. Show that

$$\delta(\sin(x)) = \sum_{n=-\infty}^{\infty} \frac{\delta(x - n\pi)}{|\cos(n\pi)|} = \sum_{n=-\infty}^{\infty} \delta(x - n\pi).$$

Solution 37. We utilize the identity

$$\delta(f(x)) \equiv \sum_m \frac{1}{|f'(x_m)|} \delta(x - x_m)$$

where the sum over m runs over all zeros of f and $f'(x_m) \equiv df(x = x_m)/dx$. Since $f(x) = \sin(x)$ we have $df(x)/dx = \cos(x)$ and $\cos(mx) = 1$ for m even and -1 for m odd.

Problem 38. Consider the function $f : \mathbb{R} \to \mathbb{R}$

$$f(x) = \begin{cases} 1 & \text{for} \quad 0 \leq x \leq 1 \\ 0 & \text{otherwise} \end{cases}$$

(i) Find the derivative in the sense of generalized functions. Then apply the Fourier transform.

(ii) Find the Fourier transform of f and then calculate the derivative in the sense of generalized functions.

Solution 38. (i) For the derivative in the sense of generalized functions we have ($\phi \in S(\mathbb{R})$)

$$(df(x)/dx, \phi(x)) = -(f(x), d\phi(x)/dx) = -\int_{-\infty}^{\infty} f(x) \frac{d\phi}{dx} dx$$

$$= -\int_0^1 \frac{d\phi}{dx} dx = -\phi(x)|_0^1 = -\phi(1) + \phi(0)$$

$$= -(\delta(x-1), \phi(x)) + (\delta(x), \phi(x)).$$

The Fourier transform of $-\delta(x-1) + \delta(x)$ in the sense of generalized functions is $-e^{ik} + 1$.

(ii) The Fourier transform of f is

$$\int_{-\infty}^{\infty} f(x) e^{ikx} dx = \int_0^1 e^{ikx} dx = \frac{e^{ikx}}{ik} \Big|_0^1 = \frac{e^{ik}}{ik} - \frac{1}{ik} = \frac{e^{ik} - 1}{ik}.$$

The derivative in the sense of generalized functions of $(e^{ik} - 1)/(ik)$ is

$$\frac{e^{ik}(k+i) - i}{k^2}.$$

Problem 39. Given a function (signal) $f(\mathbf{t}) = f(t_1, t_2, \ldots, t_n) \in L_2(\mathbb{R}^n)$ of n real variables $\mathbf{t} = (t_1, t_2, \ldots, t_n)$. We define the *symplectic tomogram* associated with the square integrable function f

$$w(\mathbf{X}, \boldsymbol{\mu}, \boldsymbol{\nu}) = \prod_{k=1}^n \frac{1}{2\pi|\nu_k|} \left| \int_{\mathbb{R}^n} dt_1 dt_2 \cdots dt_n f(\mathbf{t}) \exp\left(\sum_{j=1}^n \left(\frac{i\mu_j}{2\nu_j} t_j^2 - \frac{iX_j}{\nu_j} t_j \right) \right) \right|^2$$

where ($\nu_j \neq 0$ for $j = 1, 2, \ldots, n$)

$$\mathbf{X} = (X_1, X_2, \ldots, X_n), \quad \boldsymbol{\mu} = (\mu_1, \mu_2, \ldots, \mu_n), \quad \boldsymbol{\nu} = (\nu_1, \nu_2, \ldots, \nu_n).$$

(i) Prove the equality

$$\int_{\mathbb{R}^n} w(\mathbf{X}, \boldsymbol{\mu}, \boldsymbol{\nu}) d\mathbf{X} = \int_{\mathbb{R}^n} |f(\mathbf{t})|^2 d\mathbf{t} \tag{1}$$

for the special case $n = 1$. The tomogram is the probability distribution function of the random variable $\mathbf{X}$. This probability distribution function depends on $2n$ extra real parameters $\boldsymbol{\mu}$ and $\boldsymbol{\nu}$.

(ii) The map of the function $f(\mathbf{t})$ onto the tomogram $w(\mathbf{X}, \boldsymbol{\mu}, \boldsymbol{\nu})$ is invertible. The square integrable function $f(\mathbf{t})$ can be associated to the density matrix

$$\rho_f(\mathbf{t}, \mathbf{t}') = f(\mathbf{t}) f(\mathbf{t}').$$

This density matrix can be mapped onto the *Ville-Wigner function*

$$W(\mathbf{q}, \mathbf{p}) = \int_{\mathbb{R}^n} \rho_f \left(\mathbf{q} + \frac{\mathbf{u}}{2}, \mathbf{q} - \frac{\mathbf{u}}{2} \right) e^{-i\mathbf{p} \cdot \mathbf{u}} d\mathbf{u}.$$

Show that this map is invertible.

(iii) How is the tomogram $w(\mathbf{X}, \boldsymbol{\mu}, \boldsymbol{\nu})$ related to the Ville-Wigner function?

(iv) Show that the Ville-Wigner function can be reconstructed from the function $w(\mathbf{X}, \boldsymbol{\mu}, \boldsymbol{\nu})$.

(v) Show that the density matrix $f(\mathbf{t}) f^*(\mathbf{t}')$ can be found from $w(\mathbf{X}, \boldsymbol{\mu}, \boldsymbol{\nu})$.

Solution 39. (i) Since

$$\frac{1}{2\pi\nu} \left| \int_{\mathbb{R}} dt f(t) e^{i\mu t^2/(2\nu) - iXt/\nu} \right|^2 =$$

$$\frac{1}{2\pi\nu} \int_{\mathbb{R}} \int_{\mathbb{R}} dt dt' f(t) f^*(t') e^{i\mu(t^2 - (t')^2)/(2\nu)} e^{iX(t'-t)/\nu}$$

and in the sense of generalized functions ($\nu \neq 0$)

$$\int_{\mathbb{R}} e^{iX(t'-t)/\nu} dX = 2\pi\delta((t'-t)/\nu) = 2\pi\nu\delta(t'-t)$$

we find, applying the property of the delta function, that equality (1) holds.

(ii) Since we have a Fourier transform we inverse transform is given by

$$\rho(\mathbf{t}, \mathbf{t}') = \frac{1}{(2\pi)^n} \int_{\mathbb{R}^n} W \left(\frac{\mathbf{t} + \mathbf{t}'}{2}, \mathbf{p} \right) e^{i\mathbf{p} \cdot (\mathbf{t} - \mathbf{t}')} d\mathbf{p}.$$

(iii) Using a delta function we can write

$$w(\mathbf{X}, \boldsymbol{\mu}, \boldsymbol{\nu}) = \int_{\mathbb{R}^n} W(\mathbf{q}, \mathbf{q}) \prod_{k=1}^{n} \delta(X_k - \mu_k q_k - \nu_k p_k) \frac{dp_k dq_k}{2\pi}.$$

(iv) The Ville-Wigner function can be reconstructed from the function $w(\mathbf{X}, \boldsymbol{\mu}, \boldsymbol{\nu})$ via

$$W(\mathbf{p}, \mathbf{q}) = \frac{1}{(2\pi)^n} \int_{\mathbb{R}^n} w(\mathbf{X}, \boldsymbol{\mu}, \boldsymbol{\nu}) \prod_{k=1}^{n} e^{i(X_k - \mu_k q_k - \nu_k p_k)} dX_k d\mu_k d\nu_k.$$

(v) The *density matrix* $f(\mathbf{t}) f^*(\mathbf{t}')$ can be found from $w(\mathbf{X}, \boldsymbol{\mu}, \boldsymbol{\nu})$ as

$$f(\mathbf{t}) f^*(\mathbf{t}') = \frac{1}{(2\pi)^n} \int_{\mathbb{R}^n} w(\mathbf{X}, \boldsymbol{\mu}, \mathbf{t} - \mathbf{t}') \prod_{k=1}^{n} \exp \left(i \left(X_k - \mu_k \frac{t_k + t_k'}{2} \right) \right) dX_k d\mu_k.$$

9.3   Supplementary Problems

Problem 1. Given $f \in L_2(\mathbb{R})$. Does there exist $g \in L_2(\mathbb{R})$ such that

$$\langle f, d\phi/dx \rangle = -\langle g, \phi \rangle$$

for all $\phi \in S(\mathbb{R})$. Note that $S(\mathbb{R})$ is dense in $L_2(\mathbb{R})$.

Problem 2. (i) Let $a \neq 0$. Show that

$$\delta(ax + b) = \frac{1}{|a|}\delta(x + b/a).$$

(ii) Let $b > a$. Show that

$$\delta((x - a)(x - b)) = \frac{1}{b - a}(\delta(x - a) + \delta(x - b)).$$

(iii) Show that

$$\int_0^\infty \cos(\omega t)dt = \pi\delta(\omega).$$

(iv) Show that

$$\delta(x - x') = \frac{1}{\pi}\left(1 + 2\sum_{k=1}^\infty \cos(kx)\cos(kx')\right).$$

(v) Show that

$$\delta\left(t - \frac{x}{c}\right) = \frac{1}{2\pi}\int_{-\infty}^\infty e^{-i\omega(t - x/c)}d\omega.$$

(vi) Let $\phi \in S(\mathbb{R})$. Show that in the sense of generalized functions

$$\lim_{\zeta \to \infty}\frac{\sin(\zeta x)}{x}\phi(x)dx = \pi\phi(0).$$

Problem 3. Show that in the sense of generalized functions

$$\delta(x) = \lim_{n\to\infty} \frac{1}{\pi} \frac{n}{1 + n^2 x^2}$$

$$\delta(x) = \lim_{n\to\infty} n \exp(-\pi n^2 x^2)$$

$$\delta(x) = \lim_{\epsilon\to 0} \frac{1}{\pi} \frac{\epsilon}{\epsilon^2 + x^2}$$

$$\delta(x) = \frac{1}{2} \lim_{\epsilon\to 0} \frac{1}{\epsilon} e^{-|x|/\epsilon}$$

$$\delta(x) = \frac{1}{\pi} \lim_{\epsilon\to\infty} \epsilon \frac{\sin^2(\epsilon x)}{(\epsilon x)^2}$$

$$\delta(x) = \frac{1}{4} \lim_{\epsilon\to 0} \frac{1}{\epsilon} \left(1 + \frac{|x|}{\epsilon}\right) e^{-|x|/\epsilon}$$

$$\delta(x) = \lim_{n\to\infty} \frac{2}{\pi} \frac{n}{e^{nx} + e^{-nx}}.$$

Problem 4. (i) Let $a > 0$. Show that

$$\sum_{m=-\infty}^{\infty} \exp(i2\pi m(x+q)/a) \equiv a \sum_{k=-\infty}^{\infty} \delta(x+q-ka).$$

(ii) Let $x \in \mathbb{R}$ and $\alpha > 0$. Show that

$$\sum_{n=-\infty}^{\infty} \delta(x - n\alpha) \equiv \frac{1}{\alpha} \sum_{n=-\infty}^{\infty} \exp(2\pi inx/\alpha).$$

(iii) Let $p > 0$. Show that

$$\frac{1}{2\pi} \sum_{m\in\mathbb{Z}} e^{i(m+1/2)\phi} \cos(p|m+1/2|x) = \frac{1}{2p} \left(\delta(x - \phi/p) + \delta(x + \phi/p)\right).$$

(iv) Show that

$$\delta(x - x') = \frac{1}{2\pi} + \frac{1}{\pi} \sum_{n=1}^{\infty} (\cos(nx)\cos(nx') + \sin(nx)\sin(nx')).$$

Problem 5. (i) Let $\epsilon \geq 0$. Consider the $C^\infty(\mathbb{R})$ function

$$\rho_\epsilon(x) = \begin{cases} C\epsilon^{-1} \exp(-\epsilon^2/(\epsilon^2 - x^2)) & \text{for } |x| < \epsilon \\ 0 & \text{for } |x| \geq \epsilon \end{cases}$$

where

$$C^{-1} = \int_{-1}^{+1} \exp(\exp(-1/(1 - x^2)))dx.$$

Show that for $x = 0$ the function ρ_ϵ has a maximum. Show that ρ_ϵ is equal to 0 for $x = \pm\epsilon$. Show that

$$\int_{\mathbb{R}} \rho_\epsilon(x)dx = 1.$$

Show that in the sense of generalized function

$$\lim_{\epsilon \to +0} \rho_\epsilon = \delta(x).$$

(ii) Let $\epsilon > 0$ and

$$f_\epsilon(\mathbf{x}) = \begin{cases} C_\epsilon \exp\left(-\dfrac{\epsilon^2}{\epsilon^2 - |\mathbf{x}|^2}\right) & |\mathbf{x}| < \epsilon \\ 0 & |\mathbf{x}| \geq \epsilon \end{cases}$$

where $|\mathbf{x}| = \sqrt{x_1^2 + x_2^2 + \cdots + x_n^2}$. We choose C_ϵ so that

$$\int_{\mathbb{R}^n} f_\epsilon(\mathbf{x})d\mathbf{x} = 1.$$

Show that

$$\lim_{\epsilon \to 0} f_\epsilon(\mathbf{x}) = \delta(\mathbf{x})$$

in the sense of generalized functions.

Problem 6. (i) Consider the Hilbert space $L_2([0, \infty))$. The *Laguerre polynomials* are defined as

$$L_n(x) = e^x \frac{d^n}{dx^n}(e^{-x}x^n), \quad n = 0, 1, 2, \ldots$$

For the Hilbert space $L_2([0, \infty))$ we have the basis

$$B = \{ e^{-x/2}L_n(x) : n = 0, 1, 2, \ldots \}.$$

Let $a \in \mathbb{R}$. Show that

$$\delta(x - a) = e^{-(x+a)/2} \sum_{k=0}^{\infty} L_k(x)L_k(a).$$

(ii) Let H_k ($k = 0, 1, 2, \ldots$) be the *Hermite polynomials*. Show that

$$\delta(x - a) = \frac{e^{-(x^2+a^2)/2}}{\sqrt{\pi}} \sum_{k=0}^{\infty} \frac{H_k(x)H_k(a)}{2^k k!}.$$

(iii) Show that the sum

$$\frac{1}{2} \sum_{\ell=0}^{\infty} (2\ell + 1) P_\ell(x) P_\ell(y)$$

of *Legendre polynomials* P_ℓ is given by the Dirac delta function $\delta(y - x)$ for $-1 \leq x \leq +1$ and $-1 \leq y \leq +1$.

Problem 7. Let $\ell = 0, 1, 2, \ldots$ and $m = -\ell, -\ell + 1, \ldots, +\ell$. Show that

$$\delta(\cos(\theta_1) - \cos(\theta_2))\delta(\phi_1 - \phi_2) = \sum_{\ell=0}^{\infty} \sum_{m=-\ell}^{\ell} Y_{\ell,m}(\theta_1, \phi_1) Y_{\ell,m}^*(\theta_2, \phi_2).$$

Problem 8. (i) Calculate the first and second derivatives in the sense of generalized function of

$$f(x) = \begin{cases} 0 & x < 0 \\ 4x(1 - x) & 0 \leq x \leq 1 \\ 0 & x > 1 \end{cases}$$

(ii) Consider the generalized function $f(x) = |\cos(x)|$. Find the derivative in the sense of generalized functions.
(iii) Consider the generalized function

$$f(x) := \begin{cases} \cos(x) & \text{for} \quad x \in [0, 2\pi) \\ 0 & \text{otherwise} \end{cases}$$

Find the first and second derivatives of f in the sense of generalized functions.

Problem 9. Find the first and second derivatives of the function

$$f(x) := \begin{cases} x \exp(-cx^2) & \text{for } x \geq 0 \\ 0 & \text{for } x < 0 \end{cases}$$

in the sense of generalized function. Note that the function is continuous.

Problem 10. (i) Let $r^2 = x_1^2 + x_2^2$ and $r'^2 = x_1'^2 + x_2'^2$. Show that

$$\delta(x_1, x_2) = \frac{1}{2\pi r} \delta(r).$$

Show that

$$\delta(x_1 - x_1', x_2 - x_2') = \frac{1}{r} \delta(r - r')\delta(\theta - \theta'), \quad r' > 0$$

where we applied polar coordinates.

(ii) Let $r^2 = x_1^2 + x_2^2 + x_3^2$. Show that

$$\delta(x_1 - x_1', x_2 - x_2', x_3 - x_3') = \frac{1}{r\sin(\theta)}\delta(r - r')\delta(\theta - \theta')\delta(\phi - \phi')$$

where we applied spherical coordinates.

Problem 11. Consider the linear partial differential equation

$$\frac{\partial u}{\partial t} + c\frac{\partial u}{\partial x} = 0.$$

Show that $u(x, t) = f(x - ct)$ is a solution for any one-dimensional generalized function.

Problem 12. Let $a > 0$. Consider

$$f_a(x) = \begin{cases} 0 & \text{for} & x < -a \\ \exp(-\exp(x/(x^2 - a^2))) & \text{for} & -a \le x \le +a \\ 1 & \text{for} & a < x \end{cases}$$

Is the function analytic? Is the function an element of $C^\infty(\mathbb{R})$? Find

$$\delta_a(x) = \frac{d}{dx}f_a(x).$$

Then consider $a \to 0$.

Problem 13. Show that

$$\delta(\mathbf{x} - \mathbf{x}') = \lim_{\alpha \to \beta}\left(\frac{2\pi}{\beta - \alpha}\right)^{3/2}\exp\left(-\frac{\alpha\beta}{2(\beta - \alpha)}(\mathbf{x} - \mathbf{x}')^2\right).$$

Problem 14. (i) Show that the integral representation of the *step function* is given by

$$\theta(t - t') = \lim_{\epsilon \to 0}\frac{-i}{2\pi}\int_{-\infty}^{\infty}d\omega\frac{e^{-i\omega(t-t')}}{\omega + i\epsilon}, \quad \epsilon > 0.$$

(ii) Show that

$$\theta(x - a) = \int_{-\infty}^{\infty}du\int_{-\infty}^{\infty}\frac{d\tau}{2\pi}\exp(iu(\tau - x)).$$

Problem 15. Show that (distributional identity on $L_1(\mathbb{R})$)

$$\frac{1}{\pi^2} \int_{\mathbb{R}} \frac{1}{(t-x)(s-x)} dx = \delta(t-s)$$

where the integral is evaluated in the principal value sense.

Problem 16. (i) Let $c > 0$. Show that an integral representation of the delta function is given by

$$\delta(x) = \frac{1}{2\pi i} \int_{c-i\infty}^{c+i\infty} e^{tx} dt$$

where the path of the t-integration can be closed to the right or left.
(ii) Show that in the sense of generalized functions

$$\delta(x) = \frac{1}{2\pi} \int_{-\infty}^{+\infty} \exp(ikx) dk, \qquad \theta(x) = \int_0^\infty \delta(\lambda - x) d\lambda.$$

Problem 17. Let $f : \mathbb{R} \to \mathbb{R}$ be a differentiable function and x_0 a root of f, i.e. $f(x_0) = 0$. Show that

$$\delta'(f(x)) = \frac{1}{(f'(x_0))^2} \left(\delta'(x - x_0) + \frac{f''(x_0)}{f'(x_0)} \delta(x - x_0) \right).$$

Problem 18. The *Airy function* of the first kind can be defined as

$$Ai(x) = \frac{1}{\pi} \lim_{c \to \infty} \int_0^c \cos(t^3/3 + xt) dt.$$

The Airy function of the first kind satisfies the second order linear differential equation $d^2y/dx^2 = xy$.
(i) Show that in the sense of generalized functions

$$\int_{-\infty}^\infty Ai(t + x) Ai(t + y) dt = \delta(x - y).$$

Show that $Ai(x) \in L_2(\mathbb{R})$. On the real axis $Ai(x)$ has an infinite number of zeros, all of which are negative.
(ii) Show that

$$\lim_{t \to 0} \frac{1}{(3t)^{1/3}} Ai(x/(3t)^{1/3}) = \delta(x).$$

(iii) Consider in the complex domain $d^2w/dz^2 = zw$. Show that under the transformation

$$W = \frac{1}{w} \frac{dw}{dz}$$

where w is any nontrivial solution of $d^2w/dz^2 = zw$ we obtain

$$\frac{dW}{dz} + W^2 = z.$$

(iv) Let $a \in \mathbb{R}$. Show that

$$\delta(x - a) = \int_{-\infty}^{+\infty} Ai(s - x)Ai(s - a)ds$$

where Ai is the *Airy function*.

Problem 19. Consider the first order ordinary differential equation

$$x\frac{du}{dx} = 0.$$

Show that two linear independent solutions are given by $u_1(x) = 1$, $u_2(x) = \theta(x)$, where θ is the step function.

Problem 20. The *static potential* V of a point charge q on the axis at $x_3 = 0$ interior to an infinitely long grounded cylinder (radius r_0), with perfectly reflecting walls, is given by

$$V(r, x_3) = \frac{2q}{r_0^2} \sum_{j=1}^{\infty} \frac{e^{-k_j|x_3|} J_0(k_j r)}{k_j(J_1(k_j r_0))^2}$$

where J_0 and J_1 are the zeroth and first order *Bessel functions*, respectively. Show that $V(r, x_3)$ contains the proper singularity at $x_3 = 0$ with the Dirac delta function representation

$$\delta(r) = \frac{1}{\pi} \sum_{j=1}^{\infty} \frac{J_0(k_j r)}{r_0^2(J_1(k_j r_0))^2}$$

with $2\pi \int_0^{r_0} \delta(r)r dr = 1$, $J_0(k_j r_0) = 0$.

Problem 21. Show that in the sense of generalized functions

$$\delta(q - q')\delta(p - p') = \frac{1}{(2\pi)^2} \int_{\mathbb{R}} \int_{\mathbb{R}} d\chi d\eta e^{i\chi q + i\eta p - i\chi q' - i\eta p'}.$$

Problem 22. (i) Consider the one-dimensional *diffusion equation*

$$\frac{\partial u}{\partial t} = D\frac{\partial^2 u}{\partial x^2}, \quad x > 0, T > 0$$

$$u(0, t) = 0, \quad 0 < t; \quad \lim_{x \to \infty} u(x, t) = 1, \quad 0 < t; \quad u(x, 0) = 1.$$

Show that

$$u(x,t) = \frac{2}{\sqrt{\pi}} \int_0^{x/\sqrt{4Dt}} e^{-s^2} ds$$

is an analytic solution of the one-dimensional diffusion equation.
(ii) Show that

$$u(x,t) = \frac{1}{\sqrt{4\pi Dt}} \exp\left(-\frac{(x-x_0)^2}{4Dt}\right)$$

satisfies the one-dimensional diffusion equation

$$\frac{\partial u}{\partial t} = D \frac{\partial^2 u}{\partial x^2}$$

with the initial condition $u(x,0) = \delta(x - x_0)$.

Problem 23. Let J_0 be the *Bessel functions.* given by

$$J_0(x) = 1 - \frac{x^2}{2^2} + \frac{x^4}{2^2 \cdot 4^2} - \frac{x^6}{2^2 \cdot 4^2 \cdot 6^2} + \cdots$$

Show that

$$\delta(x)\delta(y) = \frac{1}{2\pi} \int_0^\infty k J_0(k\sqrt{(x^2+y^2)})dk$$

in the sense of generalized functions.

Problem 24. Let $0 < x_0 < 1$. Consider the ordinary second order differential equation

$$-\frac{d^2u}{dx^2} = \delta(x - x_0), \quad 0 < x < 1$$

with the boundary conditions $u(0) = u(1) = 0$. Show that a solution is given by

$$u(x) = \begin{cases} (1 - x_0)x \text{ for } 0 < x < x_0 \\ x_0(1 - x) \text{ for } x_0 < x < 1. \end{cases}$$

Problem 25. (i) Let $\alpha \in [0,1)$. Show that

$$\int_0^\infty x^{\alpha-1} P\left(\frac{1}{1-x^2}\right) dx = \frac{\pi}{2} \cot(\pi\alpha/2).$$

(ii) Show that in the sense of generalized functions

$$\sum_{n=1}^\infty \sin(nx) \equiv \frac{1}{2} \cot\left(\frac{x}{2}\right).$$

Problem 26. Show that the Fourier transform of $\cosh(at)$ in the sense of generalized functions is given by $\pi(\delta(\omega - ia) + \delta(\omega + ia))$.

Problem 27. Let ω be a frequency. The *cal function* and *sal function* are defined by

$$\mathrm{cal}_\omega(t) := \mathrm{sgn}(\cos(\omega t)), \qquad \mathrm{sal}_\omega(t) := \mathrm{sgn}(\sin(\omega t)).$$

Find the derivative in the sense of generalized function.

Problem 28. Consider the Schwartzian space $S(\mathbb{R})$, the Hilbert space $L_2(\mathbb{R})$ and the space $S'(\mathbb{R})$ and the operator $\hat{p} = -i\hbar d/dx$ on $S(\mathbb{R})$. The eigenvalue equation takes the form

$$-i\hbar \frac{df(x)}{dx} = \lambda f(x).$$

Show that the solution is given by $f(x) = \exp(i\lambda x/\hbar)$ with λ dimension of momentum. Show that

$$e^{i\lambda x/\hbar} \notin S(\mathbb{R}), \qquad e^{i\lambda x/\hbar} \notin L_2(\mathbb{R}).$$

Consider the differential operator $\hat{p} = -i\hbar d/dx$ in the vector space $S'(\mathbb{R})$. Show that $f(x) = \exp(i\lambda x/\hbar)$ is a generalized eigenvector and $\lambda \in \mathbb{R}$ (dimension momentum).

Problem 29. Consider the eigenvalue equation $xf(x) = \mu f(x)$. Show that there is not solution in the Hilbert space $L_2(\mathbb{R})$. Show that there is a solution in the sense of generalized functions. Apply ($\phi \in S(\mathbb{R})$)

$$(x\delta(x - \mu), \phi(x)) = (\delta(x - \mu), x\phi(x)) = \mu(\delta(x - \mu), \phi(x))$$

where $\mu \in \mathbb{R}$.

Problem 30. Let $a_1, a_2, \ldots, a_n \neq 0$. Show that

$$\frac{1}{a_1 a_2 \cdots a_n} = (n-1)! \int_0^1 d\epsilon_1 \cdots \int_0^1 d\epsilon_n \frac{\delta(1 - \sum_{j=1}^n \epsilon_j)}{(\sum_{j=1}^n \epsilon_j a_j)^n}.$$

Problem 31. Let $n \in \mathbb{N}_0$. Show that in the sense of generalized functions

$$\int_0^\infty \exp(i\omega\alpha)\omega^n d\omega = i^{n+1} n! \alpha^{-n-1} + (-i)^n \pi \delta^{(n)}(\alpha).$$

Problem 32. Let $\alpha, \beta \in \mathbb{R}$ and $\alpha \neq 0$. Let $\mathcal{L}$ be the *Laplace transform*. Show that

$$\mathcal{L}(\delta(\alpha t + \beta)) = \frac{1}{|\alpha|} \exp\left(\frac{\beta s}{\alpha}\right), \quad s \in \mathbb{C}.$$

Problem 33. The *Bessel function* of the first kind of order n is defined by

$$J_m(t) := \sum_{n=0}^{\infty} \frac{(-1)^n}{n!(n+m)!} \left(\frac{t}{2}\right)^{2n+m}, \quad m = 0, 1, 2, \ldots$$

Utilizing the properties

$$\frac{dJ_0(t)}{dt} = -J_1(t), \quad J_{m+1}(t) = J_{m-1}(t) - 2\frac{dJ_m}{dt}, \quad m \in \mathbb{N}$$

show that

$$\mathcal{L}(I_m(t)H(t)) = \frac{(\sqrt{1+s^2} - s)^m}{\sqrt{1+s^2}} \quad \text{for } \Re(s) > 0.$$

Problem 34. The *Hilbert transform* is defined by

$$H[f(x)] := \frac{1}{\pi} P \int_{\mathbb{R}} \frac{f(x)}{x' - x} dx'.$$

Show that

$$H(e^{ikx}) = i(\operatorname{sgn}(k))e^{ikx}.$$

Problem 35. Let $z = x + iy$, $\bar{z} = x - iy$ with $x, y \in \mathbb{R}$. Then

$$\frac{\partial}{\partial \bar{z}} = \frac{\partial}{\partial x} + i\frac{\partial}{\partial y}, \quad \frac{\partial}{\partial z} = \frac{\partial}{\partial x} - i\frac{\partial}{\partial y}.$$

Show that the fundamental solution of

$$\frac{\partial E(x, y)}{\partial \bar{z}} = \delta(x, y)$$

in the sense of generalized functions is given by

$$E(x, y) = \frac{1}{\pi z}.$$

Problem 36. (i) Show that

$$\left(\frac{\partial^2}{\partial x_1^2} + \frac{\partial^2}{\partial x_2^2}\right) \ln(|\mathbf{x}|) = 2\pi\delta(\mathbf{x})$$

in the sense of generalized functions.

(ii) Show that

$$\left(\frac{\partial^2}{\partial x_1^2} + \frac{\partial^2}{\partial x_2^2} + \frac{\partial^2}{\partial x_3^2}\right) \frac{1}{|\mathbf{x}|} = -4\pi\delta(\mathbf{x})$$

in the sense of generalized functions.

(iii) Show that

$$E_1(\mathbf{x}) = -\frac{e^{ik|\mathbf{x}|}}{4\pi|\mathbf{x}|}, \qquad E_2(\mathbf{x}) = -\frac{e^{-ik|\mathbf{x}|}}{4\pi|\mathbf{x}|}$$

satisfy

$$\left(\frac{\partial^2}{\partial x_1^2} + \frac{\partial^2}{\partial x_2^2} + \frac{\partial^2}{\partial x_3^2}\right) E(\mathbf{x}) + k^2 E(\mathbf{x}) = \delta(\mathbf{x})$$

in the sense of generalized functions.

(iv) Show that

$$E(\mathbf{x}, t) = \frac{\Theta(t)}{(2a\sqrt{\pi t})^n} \exp(-|\mathbf{x}|^2/(4a^2 t))$$

satisfies

$$\left(\frac{\partial}{\partial t} - a^2 \Delta\right) E = \delta(\mathbf{x}, t)$$

in the sense of generalized functions.

(v) Show that

$$E(x, t) = \frac{1}{2a}\Theta(at - |x|)$$

satisfies

$$\left(\frac{\partial^2}{\partial t^2} - a^2 \frac{\partial^2}{\partial x^2}\right) E(x, t) = 0$$

in the sense of generalized functions.

(vi) Show that

$$\left(\frac{\partial^2}{\partial x_1^2} + \frac{\partial^2}{\partial x_2^2} + \frac{\partial^2}{\partial x_3^2} - \frac{m^2 c^2}{\hbar^2}\right) \frac{\exp(-mcr/\hbar)}{r} = -4\pi\delta(\mathbf{x})$$

with $r^2 = x_1^2 + x_2^2 + x_3^2$. Note that

$$\left(\frac{\partial^2}{\partial x_1^2} + \frac{\partial^2}{\partial x_2^2} + \frac{\partial^2}{\partial x_3^2}\right) v = \frac{1}{r^2}\frac{\partial}{\partial r}\left(r^2\frac{\partial v}{\partial r}\right) + \frac{1}{r^2\sin(\theta)}\frac{\partial}{\partial\theta}\left(\sin(\theta)\frac{\partial v}{\partial\theta}\right)$$
$$+ \frac{1}{r^2\sin^2(\theta)}\frac{\partial^2 v}{\partial\phi^2}.$$

Problem 37. Prove the identity in the sense of generalized function

$$f(y)\frac{\partial}{\partial y}\delta(x-y) \equiv -f(x)\frac{\partial}{\partial x}\delta(x-y) + \frac{df(y)}{dy}\delta(x-y).$$

Problem 38. Show that in the sense of generalized function

$$F\left(\frac{1}{z}\right) = \frac{2\pi i}{w}$$

where F is the Fourier transform and $w = w_1 + iw_2$ with $w_1, w_2 \in \mathbb{R}$.

Problem 39. (i) Consider the nonlinear differential equation

$$3u\frac{du}{dx} = 2\frac{du}{dx}\frac{d^2u}{dx^2} + u\frac{d^3u}{dx^3}.$$

Show that $u(x) = e^{-|x|}$ is a solution in the sense of generalized function.
(ii) Consider the nonlinear partial differential equation

$$\frac{\partial u}{\partial t} - \frac{\partial^3 u}{\partial x^2\partial t} + 3u\frac{\partial u}{\partial x} = 2\frac{\partial u}{\partial x}\frac{\partial^2 u}{\partial x^2} + u\frac{\partial^3 u}{\partial x^3}.$$

Show that $u(x,t) = c\exp(-|x - ct|)$ (*peakon*) is a solution in the sense of generalized functions.

Problem 40. Let A, B be $n \times n$ matrices. Show that

$$e^{A+B} = \int_0^\infty d\alpha_1 e^{\alpha_1 A}\delta(1 - \alpha_1) + \int_0^\infty\int_0^\infty d\alpha_1 d\alpha_2 e^{\alpha_1 A}Be^{\alpha_2 A}\delta(1 - \alpha_1 - \alpha_2)$$
$$+ \int_0^\infty\int_0^\infty\int_0^\infty e^{\alpha_1 A}Be^{\alpha_2 A}Be^{\alpha_3 A}\delta(1 - \alpha_1 - \alpha_2 - \alpha_3) + \cdots$$

Chapter 10

Quantum Mechanics

10.1 Introduction

In quantum theory the system is described by a Hamilton operator $\hat{H}$ which is in most cases a self-adjoint operator.

We consider first the finite dimensional Hilbert spaces and then infinite dimensional Hilbert space ([10],[11],[21],[23],[25],[28],[31],[41],[48],[52], [54],[57],[66],[79]). For finite dimensional Hilbert spaces the Hamilton operator $\hat{H}$ in most case is a hermitian matrix.

The states ψ evolve in time according to the Schrödinger equation

$$i\hbar\frac{\partial\psi}{\partial t} = \hat{H}\psi$$

where $\hat{H}$ is a given self-adjoint operator which specifies the dynamics of the quantum system. The formal solution takes the form

$$\psi(t) = \exp(-i\hat{H}t/\hbar)\psi(0)$$

where $\psi(0) = \psi(t=0)$. The probability for finding $\psi(t)$ in the initial state is

$$|\langle\psi(t)|\psi(0)\rangle|^2.$$

In non-relativistic quantum dynamics we describe the quantum system by the time-dependent Schrödinger equation

$$i\hbar\frac{\partial\psi(t,q)}{\partial t} = -\frac{\hbar^2}{2m}\Delta\psi(t,q) + V(q)\psi(t,q).$$

Hence we have a linear partial differential equation. Under reasonable regularity and growth assumptions on the potential $V(\cdot) : \mathbb{R}^d \to \mathbb{R}$ the Hamilton operator

$$\hat{H} = -\frac{\hbar^2}{2m}\Delta + V(q)$$

is a self-adjoint operator in $L_2(\mathbb{R}^d)$. The solution can be written as

$$\psi(t, q) = \exp(-i\hat{H}t/\hbar)\psi_0(q).$$

Let $\hat{H}$ be the Hamilton operator and $\hat{B}$ an observable. Then $\hat{B}$ satisfy the *Heisenberg equation of motion*

$$-i\hbar\frac{d\hat{B}(t)}{dt} = [\hat{H}, \hat{B}](t)$$

where we assume that the Hamilton operator does not depend explicitly on t. The formal solution of the initial value problem is

$$\hat{B}(t) = \exp(i\hat{H}/\hbar)\hat{B}\exp(-i\hat{H}t/\hbar)$$

with $\hat{B}(0) = \hat{B}$.

A density operator ρ is an operator in a Hilbert space $\mathcal{H}$ of the form

$$\rho := \sum_{j=1}^{\infty} w_j\Pi(\phi_j) = \text{norm} \lim_{N\to\infty} \sum_{j=1}^{N} w_j\Pi(\phi_j)$$

where $\{\Pi(\phi_j)\}$ are the projection operators for an orthonormal sequence $\{\phi_j\}$ of vectors and $\{w_j\}$ is a sequence of non-negative numbers whose sum is unity. A density operator is bounded and positive.

The *Liouville-von Neumann equation* for the density matrix ρ is given by

$$\frac{\partial\rho}{\partial t} = \frac{1}{i\hbar}[\hat{H}, \rho](t)$$

where $\hat{H}$ is the Hamilton operator. The equation has the formal solution

$$\rho(t) = e^{-i\hat{H}t/\hbar}\rho(0)e^{i\hat{H}t/\hbar}.$$

10.2   Solved Problems

Problem 1. Do the states

$$|\psi_+(\theta,\phi)\rangle = \begin{pmatrix} \cos(\theta/2) \\ e^{i\phi}\sin(\theta/2) \end{pmatrix}, \quad |\psi_-(\theta,\phi)\rangle = \begin{pmatrix} \sin(\theta/2) \\ -e^{i\phi}\cos(\theta/2) \end{pmatrix}$$

form an orthonormal basis in the Hilbert space $\mathbb{C}^2$? Find the density matrices

$$|\psi_+(\theta,\phi)\rangle\langle\psi_+(\theta,\phi)|, \quad |\psi_-(\theta,\phi)\rangle\langle\psi_-(\theta,\phi)|.$$

Solution 1. We have

$$\langle\psi_+(\theta,\phi)|\psi_+(\theta,\phi)\rangle = 1, \quad \langle\psi_-(\theta,\phi)|\psi_-(\theta,\phi)\rangle = 1, \quad \langle\psi_+(\theta,\phi)|\psi_-(\theta,\phi)\rangle = 0.$$

Hence we have an orthonormal basis in $\mathbb{C}^2$. We obtain the density matrices

$$|\psi_+(\theta,\phi)\rangle\langle\psi_+(\theta,\phi)| = \begin{pmatrix} \cos^2(\theta/2) & e^{-i\phi}\cos(\theta/2)\sin(\theta/2) \\ e^{i\phi}\cos(\theta/2)\sin(\theta/2) & \sin^2(\theta/2) \end{pmatrix}$$

$$|\psi_-(\theta,\phi)\rangle\langle\psi_-(\theta,\phi)| = \begin{pmatrix} \sin^2(\theta/2) & -e^{-i\phi}\cos(\theta/2)\sin(\theta/2) \\ -e^{i\phi}\cos(\theta/2)\sin(\theta/2) & \cos^2(\theta/2) \end{pmatrix}.$$

Problem 2. Consider the Hilbert space $\mathbb{C}^3$. Let $\hat{H} = \hbar\omega S_1^{(1)}$ be a Hamilton operator, where $S_1^{(1)}$ is the hermitian spin-1 matrix

$$S_1^{(1)} := \frac{1}{\sqrt{2}}\begin{pmatrix} 0 & 1 & 0 \\ 1 & 0 & 1 \\ 0 & 1 & 0 \end{pmatrix}$$

and ω is the frequency.
(i) Find the *spectral decomposition* of $S_1^{(1)}$.
(ii) Find the unitary 3×3 matrix $\exp(-i\hat{H}t/\hbar)$.
(iii) Find the normalized state $\exp(-i\hat{H}t/\hbar)\psi(0)$, where $\psi(0) = (1,1,1)^T/\sqrt{3}$.
(iv) Find the probability $|\langle\psi(t)|\psi(0)\rangle|^2$ of finding $\psi(t)$ in the initial state $\psi(0)$.

Solution 2. (i) The eigenvalues of the hermitian matrix $S_1^{(1)}$ are

$$\lambda_1 = -1, \quad \lambda_2 = 0, \quad \lambda_3 = +1$$

with the corresponding normalized eigenvectors

$$\mathbf{v}_1 = \frac{1}{2}\begin{pmatrix} 1 \\ -\sqrt{2} \\ 1 \end{pmatrix}, \quad \mathbf{v}_2 = \frac{1}{\sqrt{2}}\begin{pmatrix} 1 \\ 0 \\ -1 \end{pmatrix}, \quad \mathbf{v}_3 = \frac{1}{2}\begin{pmatrix} 1 \\ \sqrt{2} \\ 1 \end{pmatrix}.$$

These eigenvectors form an orthonormal basis in the Hilbert space $\mathbb{C}^3$. Then

$$S_1^{(1)} = \lambda_1 \mathbf{v}_1 \mathbf{v}_1^* + \lambda_2 \mathbf{v}_2 \mathbf{v}_2^* + \lambda_3 \mathbf{v}_3 \mathbf{v}_3^* = -\mathbf{v}_1 \mathbf{v}_1^* + \mathbf{v}_3 \mathbf{v}_3^*.$$

(ii) Utilizing the spectral decomposition of $S_1^{(1)}$ and

$$(\mathbf{v}_1 \mathbf{v}_1^*)^2 = \mathbf{v}_1 \mathbf{v}_1^*, \quad (\mathbf{v}_2 \mathbf{v}_2^*)^2 = \mathbf{v}_2 \mathbf{v}_2^*, \quad (\mathbf{v}_3 \mathbf{v}_3^*)^2 = \mathbf{v}_3 \mathbf{v}_3^*$$

we obtain

$$\exp(-i\hat{H}t/\hbar) = I_3 - i\sin(\omega t)S_1^{(1)} + (\cos(\omega t) - 1)(S_1^{(1)})^2.$$

Thus we obtain the unitary matrix

$$\exp(-i\hat{H}t/\hbar) = \begin{pmatrix} \frac{1}{2}\cos(\omega t) + \frac{1}{2} & -\frac{i}{\sqrt{2}}\sin(\omega t) & \frac{1}{2}\cos(\omega t) - \frac{1}{2} \\ -\frac{i}{\sqrt{2}}\sin(\omega t) & \cos(\omega t) & -\frac{i}{\sqrt{2}}\sin(\omega t) \\ \frac{1}{2}\cos(\omega t) - \frac{1}{2} & -\frac{i}{\sqrt{2}}\sin(\omega t) & \frac{1}{2}\cos(\omega t) + \frac{1}{2} \end{pmatrix}.$$

(iii) We obtain the normalized vector

$$\psi(t) = \exp(-i\hat{H}t/\hbar)\psi(0) = \frac{1}{\sqrt{3}}\begin{pmatrix} \cos(\omega t) - \frac{i}{\sqrt{2}}\sin(\omega t) \\ \cos(\omega t) - i\sqrt{2}\sin(\omega t) \\ \cos(\omega t) - \frac{i}{\sqrt{2}}\sin(\omega t) \end{pmatrix}.$$

(iv) Thus the probability of finding $\psi(t)$ in the initial state $\psi(0)$ is

$$|\langle \psi(t)|\psi(0)\rangle|^2 = \frac{1}{9}|3\cos(\omega t) - i2\sqrt{2}\sin(\omega t)|^2 = \frac{1}{9}(9\cos^2(\omega t) + 8\sin^2(\omega t))$$

$$= 1 - \frac{1}{9}\sin^2(\omega t).$$

Problem 3. Let $s = 1/2, 1, 3/2, 2, 5/2, \dots$ be the *spin*. Consider the Hilbert space $\mathbb{C}^{2s+1}$. The *spin matrices* $S_1^{(s)}$, $S_2^{(s)}$, $S_3^{(s)}$ satisfy the commutation relations

$$[S_1^{(s)}, S_2^{(s)}] = iS_3^{(s)}, \quad [S_2^{(s)}, S_3^{(s)}] = iS_1^{(s)}, \quad [S_3^{(s)}, S_1^{(s)}] = iS_2^{(s)}.$$

One defines

$$S_+^{(s)} := S_1^{(s)} + iS_2^{(s)}, \quad S_-^{(s)} := S_1^{(s)} - iS_2^{(s)}.$$

Then we have the commutation relations

$$[S_+^{(s)}, S_-^{(s)}] = 2S_3^{(s)}, \quad [S_3^{(s)}, S_+^{(s)}] = S_+^{(s)}, \quad [S_3^{(s)}, S_-^{(s)}] = -S_-^{(s)}.$$

Let $S_-^{(s)}$ be the spin matrices for spin 1/2, 1, 3/2, 2, i.e.

$$S_-^{(1/2)} = \begin{pmatrix} 0 & 0 \\ 1 & 0 \end{pmatrix}, \quad S_-^{(1)} = \begin{pmatrix} 0 & 0 & 0 \\ \sqrt{2} & 0 & 0 \\ 0 & \sqrt{2} & 0 \end{pmatrix},$$

$$S_-^{(3/2)} = \begin{pmatrix} 0 & 0 & 0 & 0 \\ \sqrt{3} & 0 & 0 & 0 \\ 0 & 2 & 0 & 0 \\ 0 & 0 & \sqrt{3} & 0 \end{pmatrix}, \quad S_-^{(2)} = \begin{pmatrix} 0 & 0 & 0 & 0 & 0 \\ 2 & 0 & 0 & 0 & 0 \\ 0 & \sqrt{6} & 0 & 0 & 0 \\ 0 & 0 & \sqrt{6} & 0 & 0 \\ 0 & 0 & 0 & 2 & 0 \end{pmatrix}.$$

Let $z \in \mathbb{C}$. Find $\exp(zS_-^{(s)})$ for $s = 1/2, 1, 3/2, 2$. Find the state in the Hilbert space $\mathbb{C}^{2s+1}$ for $s = 1/2$, $s = 1$, $s = 3/2$, $s = 2$

$$\exp(zS_-^{(s)})|0\rangle \quad \text{with} \quad |0\rangle \equiv |s, m = s\rangle = \begin{pmatrix} 1 \\ 0 \\ 0 \\ \vdots \\ 0 \end{pmatrix} \in \mathbb{C}^{2s+1}$$

where $m = -s, -s+1, \ldots, s$. Normalize the vectors. These vectors are called *spin coherent states*.

Solution 3. For spin-$\frac{1}{2}$ we have

$$\exp(zS_-^{(1/2)}) = \begin{pmatrix} 1 & 0 \\ z & 1 \end{pmatrix} \quad \Rightarrow \quad \exp(zS_-^{(1/2)}) \begin{pmatrix} 1 \\ 0 \end{pmatrix} = \begin{pmatrix} 1 \\ z \end{pmatrix}.$$

Normalizing the vector we obtain

$$\frac{1}{\sqrt{1 + z\bar{z}}} \begin{pmatrix} 1 \\ z \end{pmatrix}.$$

For spin-1 we have

$$\exp(zS_-^{(1)}) = \begin{pmatrix} 1 & 0 & 0 \\ \sqrt{2}z & 1 & 0 \\ z^2 & \sqrt{2}z & 1 \end{pmatrix} \quad \Rightarrow \quad \exp(zS_-^{(1)}) \begin{pmatrix} 1 \\ 0 \\ 0 \end{pmatrix} = \begin{pmatrix} 1 \\ \sqrt{2}z \\ z^2 \end{pmatrix}.$$

The normalized vector is

$$\frac{1}{\sqrt{1 + 2z\overline{z} + (z\overline{z})^2}} \begin{pmatrix} 1 \\ \sqrt{2}z \\ z^2 \end{pmatrix}.$$

For spin-$\frac{3}{2}$ we have

$$\exp(zS_-^{(3/2)}) = \begin{pmatrix} 1 & 0 & 0 & 0 \\ \sqrt{3}z & 1 & 0 & 0 \\ \sqrt{3}z^2 & 2z & 1 & 0 \\ z^3 & \sqrt{3}z^2 & \sqrt{3}z & 1 \end{pmatrix} \Rightarrow \exp(zS_-^{(3/2)}) \begin{pmatrix} 1 \\ 0 \\ 0 \\ 0 \end{pmatrix} = \begin{pmatrix} 1 \\ \sqrt{3}z \\ \sqrt{3}z^2 \\ z^3 \end{pmatrix}.$$

The normalized vector is

$$\frac{1}{\sqrt{1 + 3z\overline{z} + 3(z\overline{z})^2 + (z\overline{z})^3}} \begin{pmatrix} 1 \\ \sqrt{3}z \\ \sqrt{3}z^2 \\ z^3 \end{pmatrix}.$$

For spin-2 we have

$$\exp(zS_-^{(2)}) = \begin{pmatrix} 1 & 0 & 0 & 0 & 0 \\ 2z & 1 & 0 & 0 & 0 \\ \sqrt{6}z^2 & \sqrt{6}z & 1 & 0 & 0 \\ 2z^3 & 3z^2 & \sqrt{6}z & 1 & 0 \\ z^4 & 2z^3 & \sqrt{6}z^2 & 2z & 1 \end{pmatrix}$$

and

$$\exp(zS_-^{(2)}) \begin{pmatrix} 1 \\ 0 \\ 0 \\ 0 \\ 0 \end{pmatrix} = \begin{pmatrix} 1 \\ 2z \\ \sqrt{6}z^2 \\ 2z^3 \\ z^4 \end{pmatrix}.$$

The normalized vector is

$$\frac{1}{\sqrt{1 + 4z\overline{z} + 6(z\overline{z})^2 + 4(z\overline{z})^3 + (z\overline{z})^4}} \begin{pmatrix} 1 \\ 2z \\ \sqrt{6}z^2 \\ 2z^3 \\ z^4 \end{pmatrix}.$$

Problem 4. Let $z = x + iy$, $\bar{z} = x - iy$ $(x, y \in \mathbb{R})$. Consider the Hilbert space $\mathbb{C}^2$ and the normalized spin-$\frac{1}{2}$ coherent state

$$|z\rangle = \frac{1}{\sqrt{1 + |z|^2}} \begin{pmatrix} 1 \\ z \end{pmatrix} \quad \Rightarrow \quad \langle z| = \frac{1}{\sqrt{1 + |z|^2}} \begin{pmatrix} 1 & \bar{z} \end{pmatrix}.$$

Consider the function

$$f(z, \bar{z}) = \frac{1}{\sqrt{1 + |z|^2}} \begin{pmatrix} 1 & \bar{z} \end{pmatrix} \begin{pmatrix} 0 & 1 \\ 1 & 0 \end{pmatrix} \frac{1}{\sqrt{1 + |z|^2}} \begin{pmatrix} 1 \\ z \end{pmatrix}.$$

Find the minima and maxima of the function. Discuss.

Solution 4. We have

$$f(z, \bar{z}) = \frac{1}{1 + |z|^2}(z + \bar{z}) \quad \Rightarrow \quad f(x, y) = \frac{2x}{1 + x^2 + y^2}.$$

Now

$$\frac{\partial f}{\partial x} = \frac{2(1 - x^2 + y^2)}{(1 + x^2 + y^2)^2} = 0, \qquad \frac{\partial f}{\partial y} = \frac{-4xy}{(1 + x^2 + y^2)^2} = 0$$

provides the two nonlinear equations $1 - x^2 + y^2 = 0$, $xy = 0$ with $x, y \in \mathbb{R}$. Hence we find that $y = 0$ and $x = \pm 1$. It follows that

$$f(x = +1, y = 0) = 1, \qquad f(x = -1, y = 0) = -1$$

which are the eigenvalues of the matrix $\sigma_1 = \begin{pmatrix} 0 & 1 \\ 1 & 0 \end{pmatrix}$. Explain.

Problem 5. Consider the Hilbert space $\mathbb{C}^4$. Let σ_1, σ_2, σ_3 be the Pauli spin matrices. Find the eigenvalues and normalized eigenvectors of the Hamilton operator

$$\hat{K} \equiv \frac{\hat{H}}{\hbar\omega} = \sigma_1 \otimes \sigma_1 + \sigma_2 \otimes \sigma_2 + \sigma_3 \otimes \sigma_3.$$

Solution 5. We obtain the hermitian 4×4 matrix

$$\sigma_1 \otimes \sigma_1 + \sigma_2 \otimes \sigma_2 + \sigma_3 \otimes \sigma_3 = \begin{pmatrix} 1 & 0 & 0 & 0 \\ 0 & -1 & 2 & 0 \\ 0 & 2 & -1 & 0 \\ 0 & 0 & 0 & 1 \end{pmatrix} \equiv (1) \oplus \begin{pmatrix} -1 & 2 \\ 2 & -1 \end{pmatrix} \oplus (1)$$

with the eigenvalues $\lambda_1 = +1$, $\lambda_2 = +1$, $\lambda_3 = -3$, $\lambda_4 = +1$. The corresponding normalized eigenvectors are

$$\mathbf{v}_1 = \begin{pmatrix} 1 \\ 0 \\ 0 \\ 0 \end{pmatrix}, \quad \mathbf{v}_2 = \frac{1}{\sqrt{2}} \begin{pmatrix} 0 \\ 1 \\ 1 \\ 0 \end{pmatrix}, \quad \mathbf{v}_3 = \frac{1}{\sqrt{2}} \begin{pmatrix} 0 \\ 1 \\ -1 \\ 0 \end{pmatrix}, \quad \mathbf{v}_4 = \begin{pmatrix} 0 \\ 0 \\ 0 \\ 1 \end{pmatrix}.$$

The normalized eigenvectors form an orthonormal basis in the Hilbert space $\mathbb{C}^4$ with the vectors $\mathbf{v}_2$ and $\mathbf{v}_3$ are *Bell states*.

Problem 6. Consider the Hilbert space $\mathbb{C}^4$. Find the eigenvalues and normalized eigenvectors of the Hamilton operator

$$\hat{K} \equiv \frac{\hat{H}}{\hbar\omega} = \sigma_1 \otimes \sigma_2 + \sigma_2 \otimes \sigma_3 + \sigma_3 \otimes \sigma_1.$$

Solution 6. We obtain the hermitian matrix

$$\sigma_1 \otimes \sigma_2 + \sigma_2 \otimes \sigma_3 + \sigma_3 \otimes \sigma_1 = \begin{pmatrix} 0 & 1 & -i & -i \\ 1 & 0 & i & i \\ i & -i & 0 & -1 \\ i & -i & -1 & 0 \end{pmatrix}$$

with the eigenvalues $\lambda_1 = -3$, $\lambda_2 = +1$, $\lambda_3 = +1$, $\lambda_4 = +1$ and the corresponding normalized eigenvectors

$$\mathbf{v}_1 = \frac{1}{2} \begin{pmatrix} 1 \\ -1 \\ -i \\ -i \end{pmatrix}, \quad \mathbf{v}_2 = \frac{1}{\sqrt{2}} \begin{pmatrix} 1 \\ 0 \\ 0 \\ i \end{pmatrix}, \quad \mathbf{v}_3 = \frac{1}{\sqrt{2}} \begin{pmatrix} 0 \\ 1 \\ 0 \\ -i \end{pmatrix}, \quad \mathbf{v}_4 = \frac{1}{\sqrt{2}} \begin{pmatrix} 0 \\ 0 \\ 1 \\ -1 \end{pmatrix}.$$

The normalized eigenvectors form a basis in the Hilbert space $\mathbb{C}^4$.

Problem 7. Consider the Hilbert space $\mathbb{C}^4$. Find the eigenvalues and normalized eigenvectors of the hermitian 4×4 matrices

$$\frac{\hat{H}}{\hbar\omega} = \sigma_3 \otimes \sigma_1, \qquad \frac{\hat{K}}{\hbar\omega} = \sigma_1 \otimes \sigma_3.$$

Solution 7. The eigenvalues of $\sigma_3 \otimes \sigma_1$ and $\sigma_1 \otimes \sigma_3$ are 1 (2×) and -1 (2×). Since the normalized eigenvectors of σ_1 are

$$\frac{1}{\sqrt{2}} \begin{pmatrix} 1 \\ 1 \end{pmatrix}, \quad \frac{1}{\sqrt{2}} \begin{pmatrix} 1 \\ -1 \end{pmatrix}$$

and the normalized eigenvectors of σ_3 are

$$\begin{pmatrix} 1 \\ 0 \end{pmatrix}, \quad \begin{pmatrix} 0 \\ 1 \end{pmatrix}$$

we find the normalized eigenvectors of $\sigma_3 \otimes \sigma_1$ as

$$\begin{pmatrix} 1 \\ 0 \end{pmatrix} \otimes \frac{1}{\sqrt{2}} \begin{pmatrix} 1 \\ 1 \end{pmatrix}, \quad \begin{pmatrix} 1 \\ 0 \end{pmatrix} \otimes \frac{1}{\sqrt{2}} \begin{pmatrix} 1 \\ -1 \end{pmatrix},$$

$$\begin{pmatrix} 0 \\ 1 \end{pmatrix} \otimes \frac{1}{\sqrt{2}} \begin{pmatrix} 1 \\ 1 \end{pmatrix}, \quad \begin{pmatrix} 0 \\ 1 \end{pmatrix} \otimes \frac{1}{\sqrt{2}} \begin{pmatrix} 1 \\ -1 \end{pmatrix}$$

and the normalized eigenvectors of $\sigma_1 \otimes \sigma_3$ as

$$\frac{1}{\sqrt{2}} \begin{pmatrix} 1 \\ 1 \end{pmatrix} \otimes \begin{pmatrix} 1 \\ 0 \end{pmatrix}, \quad \frac{1}{\sqrt{2}} \begin{pmatrix} 1 \\ 1 \end{pmatrix} \otimes \begin{pmatrix} 0 \\ 1 \end{pmatrix},$$

$$\frac{1}{\sqrt{2}} \begin{pmatrix} 1 \\ -1 \end{pmatrix} \otimes \begin{pmatrix} 1 \\ 0 \end{pmatrix}, \quad \frac{1}{\sqrt{2}} \begin{pmatrix} 1 \\ -1 \end{pmatrix} \otimes \begin{pmatrix} 0 \\ 1 \end{pmatrix}.$$

Since the eigenvalues are degenerate we can also form linear combinations of the eigenvectors. It follows that $\sigma_3 \otimes \sigma_1$, $\sigma_1 \otimes \sigma_3$ admit the normalized eigenvector

$$\mathbf{v} = \frac{1}{2} \begin{pmatrix} 1 & -1 & 1 & 1 \end{pmatrix}^T$$

with the eigenvalues $+1$ and -1, respectively.

Problem 8. Consider the Hilbert space $\mathbb{C}^n$. Let A, B be $n \times n$ hermitian matrices and let ρ be an $n \times n$ *density matrix* (i.e. ρ is positive semidefinite and $\mathrm{tr}(\rho) = 1$). The *expectation values* $\langle A \rangle$, $\langle B \rangle$ are defined as

$$\langle A\rho \rangle := \mathrm{tr}(A\rho), \qquad \langle B\rho \rangle := \mathrm{tr}(B\rho).$$

The *uncertainty relation* is

$$(\langle A^2 \rangle - \langle A \rangle^2)(\langle B^2 \rangle - \langle B \rangle^2) \geq \frac{1}{4} |\langle [A, B] \rangle|^2$$

where $[\cdot, \cdot]$ denotes the commutator. A stronger version is

$$(\langle A^2 \rangle - \langle A \rangle^2)(\langle B^2 \rangle - \langle B \rangle^2) \geq \left| \frac{1}{2} \langle [A, B]_+ \rangle - \langle A \rangle \langle B \rangle \right|^2 + \frac{1}{4} |\langle [A, B] \rangle|^2$$

where $[\cdot, \cdot]_+$ denotes the anti-commutator. Let $A = \sigma_3$, $B = \sigma_1$ and the density matrix

$$\rho = \begin{pmatrix} 1/2 & 0 \\ 0 & 1/2 \end{pmatrix}.$$

Find the left and right-hand sides of the inequalities.

Solution 8. We have $\sigma_3^2 = \sigma_1^2 = I_2$. Hence $\langle A^2 \rangle = \langle B^2 \rangle = 1$. Furthermore $\langle A \rangle = \langle B \rangle = 0$. Hence

$$(\langle A^2 \rangle - \langle A \rangle^2)(\langle B^2 \rangle - \langle B \rangle^2) = 1.$$

Now $[\sigma_3, \sigma_1] = \begin{pmatrix} 0 & 2 \\ -2 & 0 \end{pmatrix}$ and $\mathrm{tr}([\sigma_3, \sigma_1]) = 0$. Thus for the first uncertainty relation we have $1 > 0$. Since $[\sigma_3, \sigma_1]_+ = 0_2$ the second inequality provides the same result.

Problem 9. Consider the Hilbert space $\mathbb{C}^4$ and the *permutation matrix*

$$P_{4321} = \begin{pmatrix} 0 & 0 & 0 & 1 \\ 0 & 0 & 1 & 0 \\ 1 & 0 & 0 & 0 \\ 0 & 1 & 0 & 0 \end{pmatrix}$$

with the eigenvalues $\lambda_1 = +1$, $\lambda_2 = -1$, $\lambda_3 = +i$, $\lambda_4 = -i$ and the corresponding normalized eigenvectors

$$\mathbf{v}_1 = \frac{1}{2}\begin{pmatrix} 1 \\ 1 \\ 1 \\ 1 \end{pmatrix}, \quad \mathbf{v}_2 = \frac{1}{2}\begin{pmatrix} 1 \\ 1 \\ -1 \\ -1 \end{pmatrix}, \quad \mathbf{v}_3 = \frac{1}{2}\begin{pmatrix} 1 \\ -1 \\ -i \\ i \end{pmatrix}, \quad \mathbf{v}_4 = \frac{1}{2}\begin{pmatrix} 1 \\ -1 \\ i \\ -i \end{pmatrix}$$

which form an orthonormal basis in $\mathbb{C}^4$. Let $|0\rangle$, $|1\rangle$ be the standard basis in $\mathbb{C}^2$

$$|0\rangle = \begin{pmatrix} 1 \\ 0 \end{pmatrix}, \quad |1\rangle = \begin{pmatrix} 0 \\ 1 \end{pmatrix}.$$

Then

$$P_{4321}(|0\rangle \otimes |0\rangle) = P_{4321} \begin{pmatrix} 1 \\ 0 \\ 0 \\ 0 \end{pmatrix} = \begin{pmatrix} 0 \\ 0 \\ 1 \\ 0 \end{pmatrix} \equiv |1\rangle \otimes |0\rangle$$

$$P_{4321}(|0\rangle \otimes |1\rangle) = P_{4321} \begin{pmatrix} 0 \\ 1 \\ 0 \\ 0 \end{pmatrix} = \begin{pmatrix} 0 \\ 0 \\ 0 \\ 1 \end{pmatrix} \equiv |1\rangle \otimes |1\rangle$$

$$P_{4321}(|1\rangle \otimes |0\rangle) = P_{4321} \begin{pmatrix} 0 \\ 0 \\ 1 \\ 0 \end{pmatrix} = \begin{pmatrix} 0 \\ 1 \\ 0 \\ 0 \end{pmatrix} \equiv |0\rangle \otimes |1\rangle$$

$$P_{4321}(|1\rangle \otimes |1\rangle) = P_{4321} \begin{pmatrix} 0 \\ 0 \\ 0 \\ 1 \end{pmatrix} = \begin{pmatrix} 1 \\ 0 \\ 0 \\ 0 \end{pmatrix} \equiv |0\rangle \otimes |0\rangle.$$

Thus we have the mapping

$$|0\rangle \otimes |0\rangle \mapsto |1\rangle \otimes |0\rangle, \quad |0\rangle \otimes |1\rangle \mapsto |1\rangle \otimes |1\rangle,$$

$$|1\rangle \otimes |0\rangle \mapsto |0\rangle \otimes |1\rangle, \quad |1\rangle \otimes |1\rangle \mapsto |0\rangle \otimes |0\rangle.$$

Find the corresponding *boolean map* $\mathbf{f} : \{0,1\}^2 \to \{0,1\}^2$.

Solution 9. Let $x_1, x_2, y_1, y_2 \in \{0,1\}$. The *truth table* is

```
|x1|x2| ||y1|y2|
 0 |0  |||1 |0 |
 0 |1  |||1 |1 |
 1 |0  |||0 |1 |
 1 |1  |||0 |0 |
```

Applying *sum of products* we obtain

$$f_1(x_1, x_2) = \overline{x}_1 \cdot \overline{x}_2 + \overline{x}_1 \cdot x_2 = \overline{x}_1 \cdot (\overline{x}_2 + x_2) = \overline{x}_1$$

$$f_2(x_1, x_2) = \overline{x}_1 \cdot x_2 + x_2 \cdot \overline{x}_1 = x_1 \oplus x_2$$

where $\cdot$ is the AND operation, $+$ the OR operation and $\overline{(\cdot)}$ the NOT operation. It follows that

$$|x_1\rangle \otimes |x_2\rangle \mapsto |\overline{x}_1\rangle \otimes |x_1 \oplus x_2\rangle$$

or $P_{4321}(|x_1\rangle \otimes |x_2\rangle) = |\overline{x}_1\rangle \otimes |x_1 \oplus x_2\rangle$.

Problem 10. Let $|0\rangle$, $|1\rangle$ be the standard basis in the Hilbert space $\mathbb{C}^2$, i.e.

$$|0\rangle = \begin{pmatrix} 1 \\ 0 \end{pmatrix}, \quad |1\rangle = \begin{pmatrix} 0 \\ 1 \end{pmatrix},$$

$x_1, x_2 \in \{0, 1\}$ and $\oplus$ the *XOR operation* with $0 \oplus 0 = 0$, $0 \oplus 1 = 1$, $1 \oplus 0 = 1$, $1 \oplus 1 = 0$. Find the 4×4 permutation matrix P such that

$$|x_1\rangle \otimes |x_2\rangle \mapsto |x_1\rangle \otimes |x_1 \oplus x_2\rangle$$

i.e. $P(|x_1\rangle \otimes |x_2\rangle) = |x_1\rangle \otimes |x_1 \oplus x_2\rangle$.

Solution 10. We have

$$x_1 = 0, x_2 = 0, x_1 \oplus x_2 = 0 \ \Rightarrow \ |0\rangle \otimes |0\rangle \mapsto |0\rangle \otimes |0\rangle$$
$$x_1 = 0, x_2 = 1, x_1 \oplus x_2 = 1 \ \Rightarrow \ |0\rangle \otimes |1\rangle \mapsto |0\rangle \otimes |1\rangle$$
$$x_1 = 1, x_2 = 0, x_1 \oplus x_2 = 1 \ \Rightarrow \ |1\rangle \otimes |0\rangle \mapsto |1\rangle \otimes |1\rangle$$
$$x_1 = 1, x_2 = 1, x_1 \oplus x_2 = 0 \ \Rightarrow \ |1\rangle \otimes |1\rangle \mapsto |1\rangle \otimes |0\rangle.$$

Hence the permutation matrix P is

$$P = \begin{pmatrix} 1 & 0 & 0 & 0 \\ 0 & 1 & 0 & 0 \\ 0 & 0 & 0 & 1 \\ 0 & 0 & 1 & 0 \end{pmatrix}.$$

Problem 11. Let A, H be $n \times n$ hermitian matrices, where H plays the role of the Hamilton operator. The *Heisenberg equation of motion* is given by

$$\frac{dA(t)}{dt} = \frac{i}{\hbar}[H, A(t)]$$

with $A = A(t = 0) = A(0)$. Let E_j $(j = 1, 2, \ldots, n^2)$ be an orthonormal basis in the Hilbert space $\mathcal{H}$ of the $n \times n$ matrices with scalar product

$$\langle X, Y \rangle := \text{tr}(XY^*), \qquad X, Y \in \mathcal{H}.$$

Now $A(t)$ can be expanded using this orthonormal basis as

$$A(t) = \sum_{j=1}^{n^2} c_j(t) E_j$$

and H can be expanded as

$$H = \sum_{j=1}^{n^2} h_j E_j.$$

Find the time evolution for the coefficients $c_j(t)$, i.e. dc_j/dt, where $j = 1, 2, \ldots, n^2$.

Solution 11. We have

$$\frac{dA(t)}{dt} = \sum_{j=1}^{n^2} \frac{dc_j(t)}{dt} E_j.$$

Inserting this equation and the expansion for H into the Heisenberg equation of motion we arrive at

$$\sum_{j=1}^{n^2} \frac{dc_j(t)}{dt} E_j = \frac{i}{\hbar} \sum_{k=1}^{n^2} \sum_{j=1}^{n^2} h_k c_j(t) [E_k, E_j].$$

Taking the scalar product of the left and right-hand sides of this equation with E_ℓ gives

$$\sum_{j=1}^{n^2} \frac{dc_j(t)}{dt} \mathrm{tr}(E_j E_\ell^*) = \frac{i}{\hbar} \sum_{k=1}^{n^2} \sum_{j=1}^{n^2} h_k c_j(t) \mathrm{tr}(([E_k, E_j]) E_\ell^*)$$

where $\ell = 1, 2, \ldots, n^2$. Since $\mathrm{tr}(E_j E_\ell^*) = \delta_{j,\ell}$ we obtain

$$\frac{dc_\ell}{dt} = \frac{i}{\hbar} \sum_{k=1}^{n^2} \sum_{j=1}^{n^2} h_k c_j(t) \mathrm{tr}(E_k E_j E_\ell^* - E_j E_k E_\ell^*)$$

where $\ell = 1, 2, \ldots, n^2$.

Problem 12. Let $S_1^{(s)}$, $S_2^{(s)}$, $S_3^{(s)}$ be the $(2s+1) \times (2s+1)$ *spin matrices* for spin $s = 1/2$, $s = 1$, $s = 3/2$, $s = 2$, $\ldots$. They obey the commutation relations

$$[S_1^{(s)}, S_2^{(s)}] = iS_3^{(s)}, \quad [S_2^{(s)}, S_3^{(s)}] = iS_1^{(s)}, \quad [S_3^{(s)}, S_1^{(s)}] = iS_2^{(s)}.$$

Thus we consider the Hilbert space $\mathbb{C}^{2s+1}$. Let $\mathbf{a} \in \mathbb{R}^3$, $\mathbf{b} \in \mathbb{R}^3$ and $\mathbf{S} = (S_1, S_2, S_3)$. Let $\times$ be the *vector product* and $\cdot$ be the scalar product.
(i) Show that $\mathbf{S} \cdot (\mathbf{S} \times \mathbf{a}) + (\mathbf{S} \times \mathbf{a}) \cdot \mathbf{S} \equiv 0_{2s+1}$.
(ii) Show that $(\mathbf{S} \cdot \mathbf{a})(\mathbf{S} \cdot \mathbf{b}) - (\mathbf{S} \cdot \mathbf{b})(\mathbf{S} \cdot \mathbf{a}) = i\mathbf{S} \cdot (\mathbf{a} \times \mathbf{b})$.

Solution 12. (i) From

$$\mathbf{S} \times \mathbf{a} = \begin{pmatrix} S_2 a_3 - S_3 a_2 \\ S_3 a_1 - S_1 a_3 \\ S_1 a_2 - S_2 a_1 \end{pmatrix}$$

and

$$\mathbf{S} \cdot (\mathbf{S} \times \mathbf{a}) = S_1 S_2 a_3 - S_1 S_3 a_2 + S_2 S_3 a_1 - S_2 S_1 a_3 + S_3 S_1 a_2 - S_3 S_2 a_1$$

$$(\mathbf{S} \times \mathbf{a}) \cdot \mathbf{S} = S_2 S_1 a_3 - S_3 S_1 a_2 + S_3 S_2 a_1 - S_1 S_2 a_3 + S_1 S_3 a_2 - S_2 S_3 a_1$$

the identity follows.

(ii) From

$$\mathbf{a} \times \mathbf{b} = \begin{pmatrix} a_2 b_3 - a_3 b_2 \\ a_3 b_1 - a_1 b_3 \\ a_1 b_2 - a_2 b_1 \end{pmatrix}$$

$$\mathbf{S} \cdot (\mathbf{a} \times \mathbf{b}) = S_1 (a_2 b_3 - a_3 b_2) + S_2 (a_3 b_1 - a_1 b_3) + S_3 (a_1 b_2 - a_2 b_1)$$

$$\mathbf{S} \cdot \mathbf{a} = a_1 S_1 + a_2 S_2 + a_3 S_3, \quad \mathbf{S} \cdot \mathbf{b} = b_1 S_1 + b_2 S_2 + b_3 S_3$$

and the commutation relations $[S_1, S_2] = iS_3$, $[S_2, S_3] = iS_1$, $[S_3, S_1] = iS_2$ the identity follows.

Problem 13. Consider the Hilbert spaces $\mathbb{C}^{2s+1}$ with $s = 1/2, 1, 3/2, 2, \dots$. Let $S_1^{(s)}$, $S_2^{(s)}$, $S_3^{(s)}$ be the $(2s+1) \times (2s+1)$ *spin matrices* with the *commutation relations*

$$[S_1^{(s)}, S_2^{(s)}] = iS_3^{(s)}, \quad [S_2^{(s)}, S_3^{(s)}] = iS_1^{(s)}, \quad [S_3^{(s)}, S_1^{(s)}] = iS_2^{(s)}.$$

The *spin coherent states* are defined as

$$|z\rangle = \frac{1}{(1 + |z|^2)^s} \exp(z S_-^{(s)}) |0\rangle$$

with $z \in \mathbb{C}$,

$$S_-^{(s)} := S_1^{(s)} - iS_2^{(s)}, \quad S_+^{(s)} := S_1^{(s)} + iS_2^{(s)}$$

and

$$|0\rangle = \begin{pmatrix} 1 \\ 0 \\ \vdots \\ 0 \end{pmatrix} \in \mathbb{C}^{2s+1}.$$

Hence $S_3^{(s)}|0\rangle = s|0\rangle$ (eigenvalue equation). The normalized states $|z\rangle$ are complete, but not orthogonal.

(i) Find the scalar product $\langle w|z\rangle$ with $|w\rangle$ be a spin coherent state.

(ii) Find

$$\frac{1}{\pi} \int_C \frac{2s+1}{(1+|z|^2)^2} |z\rangle\langle z| dz$$

where

$$\frac{1}{\pi} \frac{2s+1}{(1+|z|^2)^2}$$

is the *weight factor*.

(iii) Find $\langle w|S_3^{(s)}|z\rangle$, $\langle w|S_+^{(s)}|z\rangle$, $\langle w|S_-^{(s)}|z\rangle$.

Solution 13. (i) We find

$$\langle w|z\rangle = \frac{(1+\overline{w}z)^{2s}}{(1+|w|^2)^s(1+|z|^2)^s}.$$

(ii) We obtain the identity matrix

$$\frac{1}{\pi} \int_C \frac{2s+1}{(1+|z|^2)^2} |z\rangle\langle z| dz = I_{2s+1}.$$

(iii) We obtain

$$\langle w|S_3^{(s)}|z\rangle = s\frac{1-\overline{w}z}{1+\overline{w}z}\langle w|z\rangle,$$

$$\langle w|S_+^{(s)}|z\rangle = 2s\frac{z}{1+\overline{w}z}\langle w|z\rangle, \qquad \langle w|S_-^{(s)}|z\rangle = 2s\frac{\overline{w}}{1+\overline{w}z}\langle w|z\rangle.$$

Problem 14. Let $n \geq 2$. Consider the Hilbert space $\mathbb{C}^n$. An $n \times n$ *density matrix* ρ is defined a positive semidefinite matrix ($\rho \geq 0$) with $\text{tr}(\rho) = 1$. Consider $n = 3$ and the *group* G of the 3×3 permutation matrices

$$P_0 = I_3 = \begin{pmatrix} 1 & 0 & 0 \\ 0 & 1 & 0 \\ 0 & 0 & 1 \end{pmatrix}, \quad P_1 = \begin{pmatrix} 1 & 0 & 0 \\ 0 & 0 & 1 \\ 0 & 1 & 0 \end{pmatrix}, \quad P_2 = \begin{pmatrix} 0 & 1 & 0 \\ 1 & 0 & 0 \\ 0 & 0 & 1 \end{pmatrix},$$

$$P_3 = \begin{pmatrix} 0 & 1 & 0 \\ 0 & 0 & 1 \\ 1 & 0 & 0 \end{pmatrix}, \quad P_4 = \begin{pmatrix} 0 & 0 & 1 \\ 1 & 0 & 0 \\ 0 & 1 & 0 \end{pmatrix}, \quad P_5 = \begin{pmatrix} 0 & 0 & 1 \\ 0 & 1 & 0 \\ 1 & 0 & 0 \end{pmatrix}.$$

Hence $|G| = 6$. Show that

$$\rho := \frac{1}{|G|} \sum_{j=0}^{5} P_j$$

is a *density matrix*.

Solution 14. We obtain

$$\rho = \frac{1}{3} \begin{pmatrix} 1 & 1 & 1 \\ 1 & 1 & 1 \\ 1 & 1 & 1 \end{pmatrix}.$$

The matrix is hermitian and eigenvalues are 1 and 0 (twice). Furthermore $\text{tr}(\rho) = 1$. This result can be extended to the set of $n \times n$ permutation matrices.

Problem 15. Consider the Hilbert space $\mathbb{C}^2$ and the Hamilton operator

$$\hat{H} = \hbar\omega\sigma_2 \equiv \hbar\omega \begin{pmatrix} 0 & -i \\ i & 0 \end{pmatrix}.$$

Then

$$U(t) = \exp(-i\hat{H}t/\hbar) = \exp\left(\omega t \begin{pmatrix} 0 & -1 \\ 1 & 0 \end{pmatrix}\right) = \begin{pmatrix} \cos(\omega t) & -\sin(\omega t) \\ \sin(\omega t) & \cos(\omega t) \end{pmatrix}.$$

Let $a = \omega t$. Find the *Cayley transform* of

$$A(a) = \begin{pmatrix} 0 & -ai \\ ai & 0 \end{pmatrix}.$$

Solution 15. Utilizing the Maxima program

```
/* CayleyTrans.mac */
I2: matrix([1,0],[0,1]);
sig2: matrix([0,-%i],[%i,0]);
A: a*sig2; Ap: A + %i*I2; Am: A - %i*I2;
Api: invert(Ap); Api: ratsimp(Api);
UC: Am . Api; UC: ratsimp(UC);
UCT: transpose(UC); UCT: ratsimp(UCT);
F: UC . UCT; F: ratsimp(F);
```

we obtain

$$A + iI_2 = \begin{pmatrix} i & -ai \\ ai & i \end{pmatrix} \quad \Rightarrow \quad (A + iI_2)^{-1} = \frac{i}{1+a^2} \begin{pmatrix} -1 & -a \\ a & -1 \end{pmatrix}.$$

It follows that

$$(A - iI_2)(A + iI_2)^{-1} = \frac{1}{1+a^2} \begin{pmatrix} a^2 - 1 & -2a \\ 2a & a^2 - 1 \end{pmatrix}.$$

Problem 16. Consider the Hilbert space $L_2(\mathbb{R})$ and the one-dimensional Schrödinger equation (eigenvalue equation)

$$\left(-\frac{d^2}{dx^2} + V(x) \right) u(x) = Eu(x)$$

where the potential V is given by

$$V(x) = x^2 + \frac{ax^2}{1+bx^2}$$

and $b > 0$. Insert the ansatz $u(x) = e^{-x^2/2}v(x)$ and find the differential equation for v. Discuss. Make a polynomial ansatz for v.

Solution 16. We find

$$(1 + bx^2) \left(\frac{d^2v}{dx^2} - 2x\frac{dv}{dx} \right) + (E - 1 + x^2(Eb - b - a))v = 0.$$

At $x = 0$ we have an ordinary point and at $x = \pm\infty$ there is an irregular singular point. Thus the second differential equation admits a convergent series solution for v about $x = 0$. Since $u \in L_2(\mathbb{R})$ the series must be truncated to satisfy this boundary condition. Since $V(x) = V(-x)$ we can look at an even- and odd-parity solutions for u. Consider

$$u(x) = A(1 + gx^2) \exp(-x^2/2).$$

The function u is a solution of the differential equation if

$$E = 1 - g, \qquad \lambda = -4g - 2g^2.$$

Problem 17. Let $a > 0$. Consider the Hilbert space $L_2([-a, a])$ and the Hamilton operator

$$\hat{H} = -\frac{\hbar^2}{2m}\frac{d^2}{dx^2} + V(x)$$

where

$$V(\dot{x}) = \begin{cases} 0 & \text{for} \quad |x| \le a \\ \infty & \text{otherwise} \end{cases}$$

Solve the Schrödinger equation, where the initial function $\psi(t = 0) = \phi(x)$ is given by

$$\phi(x) = \begin{cases} x/a^2 + 1/a & \text{for } -a \le x \le 0 \\ -x/a^2 + 1/a & \text{for } \; 0 \le x \le a \end{cases}$$

Normalize ϕ. Calculate the probability to find the particle in the state

$$\chi(x) = \frac{1}{\sqrt{a}} \sin\left(\frac{\pi x}{a}\right)$$

after time t. A basis in the Hilbert space $L_2([-a, a])$ is given by

$$\left\{ \frac{1}{\sqrt{a}} \sin\left(\frac{n\pi x}{a}\right), \frac{1}{\sqrt{a}} \cos\left(\frac{(n - 1/2)\pi x}{a}\right) \quad n = 1, 2, \dots \right\}.$$

Solution 17. We have

$$\psi(t) = \exp(-i\hat{H}t/\hbar)\psi(0)$$

where $\psi(0) = \phi(x)$. Let $\{ f_n(x) : n = 1, 2, \dots \}$ be the basis given above. Then ϕ can be expanded as

$$\phi(x) = \sum_{n=1}^{\infty} \langle \phi, f_n \rangle f_n(x).$$

Since $V(x) = V(-x)$ and ϕ is an even function we only need

$$u_n^+(x) = \frac{1}{\sqrt{a}} \cos\left(\frac{(n - 1/2)\pi x}{a}\right)$$

for the expansion. Now

$$\hat{H} \frac{1}{\sqrt{a}} \cos\left(\frac{(n-1)\pi x}{a}\right) = E_n^+ \frac{1}{\sqrt{a}} \cos\left(\frac{(n-1)\pi x}{a}\right)$$

where the eigenvalue E_n^+ is given by

$$E_n^+ = \frac{(n - 1/2)^2 \pi^2 \hbar^2}{2ma^2}.$$

Now from

$$\phi(x) = \sum_{n=1}^{\infty} \langle \phi(x), u_n^+(x) \rangle u_n^+(x)$$

we obtain

$$\psi(t) = \exp(-i\hat{H}t/\hbar) \sum_{n=1}^{\infty} \langle \phi(x), u_n^+(x) \rangle u_n^+(x) = \sum_{n=1}^{\infty} \langle \phi(x), u_n^+(x) \rangle e^{-i\hat{H}t/\hbar} u_n^+(x).$$

From $\exp(-i\hat{H}t/\hbar)u_n^+(x) = \exp(-iE_n^+t/\hbar)u_n^+(x)$ it follows that

$$\psi(t) = \sum_{n=1}^{\infty} \langle \phi(x), u_n^+(x) \rangle \exp(-iE_n^+t/\hbar)u_n^+(x).$$

Since $\chi(x)$ is an odd function, we have $\langle \psi(t), \chi(x) \rangle = 0$ and therefore

$$P = |\langle \psi(t), \chi(x) \rangle|^2 = 0.$$

Problem 18. Let $a > 0$ with dimension length. Consider the Hilbert space $L_2([-a, a])$ and the Hamilton operator

$$\hat{H} = \frac{\hat{p}^2}{2m} + V(x), \quad \hat{p} = -i\hbar\frac{d}{dx}$$

with the potential V given by $V(x) = 0$ if $|x| < a$ and $V(x) = +\infty$ if $|x| > a$. Find the eigenvalues and eigenfunctions of $\hat{H}$.

Solution 18. An orthonormal basis in the Hilbert space $L_2([-a, a])$ is given by

$$\phi_n(x) = \left\{ \frac{1}{2a} e^{i\pi nx/a} : n \in \mathbb{Z} \right\}.$$

Noting that the boundary conditions for the wave function is $u(a) = 0$, $u(-a) = 0$ and $V(x) = V(-x)$ we obtain the eigenvectors

$$u_n(x) = \begin{cases} 0 & |x| > a \\ \frac{1}{\sqrt{a}} \cos(((n+1)\pi x)/(2a)) & -a < x < a \end{cases} \quad n = 0, 2, 4, \ldots$$

$$u_n(x) = \begin{cases} 0 & |x| > a \\ \frac{1}{\sqrt{a}} \sin(((n+1)\pi x)/(2a)) & -a < x < a \end{cases} \quad n = 1, 3, 5, \ldots$$

with the corresponding eigenvalues

$$E_n = \frac{\pi^2\hbar^2}{8ma^2}(n+1)^2, \quad n \in \mathbb{N}_0.$$

For the classical case we have a free particle in the square well potential with elastic reflection on the boundaries at a and $-a$. With the Hamilton function

$$H(p, x) = \frac{p^2}{2m} + V(x)$$

we have the *Hamilton equations of motion*

$$\frac{dp}{dt} = -\frac{\partial H}{\partial x} = 0, \qquad \frac{dx}{dt} = \frac{\partial H}{\partial p} = \frac{p}{m}$$

with the solution inside the square well potential

$$p(t) = p_0, \qquad x(t) = \frac{p_0 t}{m} + x_0$$

together with the elastic reflection on the boundaries.

Problem 19. Consider the problem of a free particle in a one-dimensional box $[-a, a]$. The underlying Hilbert space is $L_2([-a, a])$. An orthonormal basis in $L_2([-a, a])$ is given by

$$\mathcal{B} = \{ u_k^{(+)}(q), u_k^{(-)}(q) \ : \ k \in \mathbb{N} \}$$

where

$$u_k^{(+)} = \frac{1}{\sqrt{a}} \cos\left(\frac{(k-1/2)\pi q}{a} \right), \qquad u_k^{(-)} = \frac{1}{\sqrt{a}} \sin\left(\frac{k\pi q}{a} \right).$$

The formal solution of the initial value problem of the Schrödinger equation $i\hbar \frac{\partial \psi}{\partial t} = \hat{H}\psi$ is given by

$$\psi(t) = \exp(-i\hat{H}t/\hbar)\psi(0).$$

Let

$$\psi(q, 0) = \frac{1}{\sqrt{a}} \sin(\pi q/a), \qquad \phi(q) = \frac{1}{\sqrt{a}} \sin(\pi q/a).$$

Find $\exp(-i\hat{H}t/\hbar)$ and the probability $P = |\langle \phi, \psi(t) \rangle|^2$.

Solution 19. Since $\hat{H} = -(\hbar^2/(2m))d^2/dq^2$ we have

$$\psi(q, t) = e^{-i\hat{H}t/\hbar}\psi(q, 0) = \sum_{j=0}^{\infty} \frac{\left(-\frac{it\hbar}{2m}\frac{d^2}{dq^2} \right)^j}{j!} \frac{1}{\sqrt{a}} \sin\left(\frac{\pi q}{a} \right)$$

$$= \frac{1}{\sqrt{a}} \sin\left(\frac{\pi q}{a} \right) \left(\sum_{j=0}^{\infty} \frac{\left(\frac{it\hbar\pi^2}{2ma^2} \right)^j}{j!} \right) = \frac{1}{\sqrt{a}} \sin\left(\frac{\pi q}{a} \right) \exp\left(\frac{it\hbar\pi^2}{2ma^2} \right).$$

Furthermore

$$\langle \phi, \psi(q, t) \rangle = \int_{-a}^{a} \frac{1}{a} \sin^2\left(\frac{\pi q}{a} \right) \exp\left(-\frac{it\hbar\pi^2}{2ma^2} \right) dq$$

$$= \int_{-a}^{a} \frac{1}{a}\left(\frac{1}{2} - \frac{1}{2}\cos\left(\frac{2\pi q}{a} \right) \right) \exp\left(-\frac{it\hbar\pi^2}{2ma^2} \right) dq$$

$$= \left(\frac{q}{2a} - \frac{1}{4\pi} \sin\left(\frac{2\pi q}{a} \right) \right) \exp\left(-\frac{it\hbar\pi^2}{2ma^2} \right) \Bigg|_{-a}^{a}$$

$$= \exp\left(-\frac{it\hbar\pi^2}{2ma^2} \right).$$

Problem 20. Consider the one-dimensional time-dependent Hamilton operator

$$\hat{H} = -\frac{\hbar^2}{2m}\frac{d^2}{dx^2} + D(1 - e^{-\alpha x})^2 + eEx\cos(\omega t)$$

where $\alpha > 0$. So the third term is a driving force. Find the *quantum Liouville equation* for this Hamilton operator.

Solution 20. From the definition of the *Wigner function* we find that

$$\dot{\rho}(p,q,t) = \frac{1}{\pi\hbar}\int dx e^{2ipx/\hbar}[\dot{\psi}^*(q+x)\psi(q-x) + \psi^*(q+x)\dot{\psi}(q-x)].$$

Using the Schrödinger equation we arrive at

$$\dot{\rho}(p,q,t) = \frac{1}{\pi\hbar}\int dx e^{2ipx/\hbar}[(\hat{H}\psi^*)(q+x)\psi(q-x) - \psi^*(q+x)(\hat{H}\psi)(q-x)].$$

This expression is linear in $\hat{H}$. Thus each part of the Hamilton operator can be considered separately. The kinetic part $\hat{H} = -(\hbar^2/2m)d^2/dx^2$ becomes, after integration by parts

$$\dot{\rho} = -\frac{p}{m}\frac{\partial\rho}{\partial q}$$

which is the classical Liouville operator for a free particle. The next contribution is from the exponential potential and are both of the form $Ce^{-\beta x}$. The expression for $\dot{\rho}$ can be written compactly if it is assumed that ρ can be analytically continued. We obtain

$$\dot{\rho} = e^{-\beta q}\frac{i}{\hbar}[\rho(q,p+i\beta\hbar/2) - \rho(q,p-i\beta\hbar/2)].$$

For the linear driving force we find

$$\dot{\rho} = eE\cos(\omega t)\frac{\partial\rho}{\partial p}.$$

Thus the quantum Liouville equation for the Hamilton operator $\hat{H}$ is given by

$$\dot{\rho} = -\frac{p}{m}\frac{\partial\rho}{\partial q} + eE\cos(\omega t)\frac{\partial\rho}{\partial p} + 2D\alpha e^{-\alpha q}\frac{i}{\hbar}[\rho(q,p+i\alpha\hbar/2) - \rho(q,p-i\alpha\hbar/2)]$$

$$- 2D\alpha e^{-2\alpha q}\frac{i}{\hbar}[\rho(q,p+i\alpha h) - \rho(q,p-i\alpha\hbar)].$$

Problem 21. Let $c > 0$. Consider the Schrödinger equation

$$-\frac{\hbar^2}{2m}\frac{d^2u}{dx^2} + c\delta^{(n)}(x)u = Eu$$

where $\delta^{(n)}$ ($n = 0, 1, 2, \dots$) denotes the n-th derivative of the delta function. Derive the joining conditions on the wave function u.

Solution 21. Let $\epsilon > 0$. We define

$$\overline{f}(0) := \frac{1}{2}(f(0^+) + f(0^-)).$$

For functions f that are discontinuous at the origin the expression

$$\int_{-\epsilon}^{\epsilon} \delta(x)f(x)dx$$

is in general not well-defined. If we assume that the delta function is the limit of a sequence of even functions, then

$$\int_{-\epsilon}^{\epsilon} \delta(x)f(x)dx = \overline{f}(0).$$

This we assume in the following. Integrating from $-\epsilon$ to ϵ the eigenvalue equation yields

$$-\frac{\hbar^2}{2m}(u'(\epsilon) - u'(-\epsilon)) + c\int_{-\epsilon}^{\epsilon}\left(\left(\frac{d}{dx}\right)^n\delta(x)\right)u(x)dx = E\int_{-\epsilon}^{\epsilon}u(x)dx.$$

Using integration by parts n times and using the fact that all boundary terms vanish yields

$$\int_{-\epsilon}^{\epsilon}\left(\left(\frac{d}{dx}\right)^n\delta(x)\right)u(x)dx = (-1)^n\int_{-\epsilon}^{\epsilon}\delta(x)\left(\left(\frac{d}{dx}\right)^n u(x)\right)dx$$

$$= (-1)^n\overline{u}^{(n)}(0).$$

In the limit $\epsilon \to 0$ we have

$$\lim_{\epsilon \to 0}\int_{-\epsilon}^{\epsilon}u(\epsilon)dx = 0.$$

Thus we have the first boundary condition

$$\Delta u' = (-1)^n\frac{2mc}{\hbar^2}\overline{u}^{(n)}(0).$$

Furthermore integrating the eigenvalue equation from $-L$ (L positive) to x yields

$$-\frac{\hbar^2}{2m}(u'(x) - u'(-L)) + c\int_{-L}^{x}\left(\left(\frac{d}{ds}\right)^n\delta(s)\right)u(s)ds = E\int_{-L}^{x}u(s)ds. \quad (1)$$

We integrate by parts and find

$$\int_{-L}^{x} \left(\left(\frac{d}{ds} \right)^n \delta(s) \right) u(s) ds = \delta^{(n-1)}(x)u(x) - \delta^{(n-2)}(x)u'(x)$$
$$+ \delta^{(n-3)}(x)u^{(2)}(x) - \cdots + (-1)^{n-1}\delta(x)u^{(n-1)}(x) + (-1)^n\theta(x)\overline{u}^{(n)}(0)$$

where now the upper boundary terms are nonzero. Integrating (1) from $-\epsilon$ to ϵ and taking the limit $\epsilon \to 0$ we obtain

$$-\frac{\hbar^2}{2m}\Delta u + c \left(\int_{-\epsilon}^{\epsilon} \delta^{(n-1)}(x)u(x)dx - \int_{-\epsilon}^{\epsilon} \delta^{(n-2)}(x)u'(x)dx \right.$$
$$\left. + \int_{-\epsilon}^{\epsilon} \delta^{(n-3)}(x)u^{(2)}(x)dx - \cdots + (-1)^{n-1} \int_{-\epsilon}^{\epsilon} \delta(x)u^{(n-1)}(x)dx \right) = 0.$$

Using once more integration by parts we obtain for the expression in the parenthesis $(-1)^{n-1}n\overline{\psi}^{(n-1)}(0)$. Thus we have the second boundary condition

$$\Delta u = (-1)^{n-1}\frac{2mc}{\hbar^2}n\overline{u}^{(n-1)}(0).$$

For $n = 0$ we have

$$\Delta u = 0, \qquad \Delta u' = \frac{2mc}{\hbar^2}u(0)$$

and for $n = 1$ we have

$$\Delta u = \frac{2mc}{\hbar^2}\overline{u}(0), \qquad \Delta u' = -\frac{2mc}{\hbar^2}\overline{u}'(0).$$

Problem 22. Consider the Schrödinger equation

$$i\hbar\frac{\partial\psi}{\partial t} = \left(-\frac{1}{2m}\Delta + V(x) \right) \psi.$$

Find the partial differential equation for the density $\rho := \psi^*\psi$.

Solution 22. First we calculate

$$\frac{\partial\rho}{\partial t} = \frac{\partial}{\partial t}(\psi\psi^*) = \frac{\partial\psi}{\partial t}\psi^* + \psi\frac{\partial\psi^*}{\partial t}.$$

Inserting the Schrödinger equation and

$$-i\hbar\frac{\partial\psi^*}{\partial t} = \left(-\frac{1}{2m}\Delta + V(\mathbf{x}) \right) \psi^*$$

yields

$$\frac{\partial\rho}{\partial t} = -\frac{1}{2im\hbar}\Delta(\psi\psi^*) + \frac{1}{2im\hbar}\psi\Delta\psi^*.$$

Problem 23. Let $\epsilon > 0$. Consider the Schrödinger eigenvalue equation

$$\left(-\frac{d^2}{dx^2} + 2\epsilon\delta(x) \right) u(x, \epsilon) = E(\epsilon)u(x, \epsilon)$$

with the boundary conditions $u(\pm 1, \epsilon) = 0$. Here ϵ is the coupling constant and determines the penetrability of the potential barrier. Find the eigenfunctions and the eigenvalues.

Solution 23. The eigenfunctions are given by

$$u(x, \epsilon) = \begin{cases} c_- \sin(k(\epsilon)x + 1) & \text{for } x < 0 \\ c_+ \sin(k(\epsilon)x - 1) & \text{for } x > 0 \end{cases}$$

The eigenvalues are $E(\epsilon) = k^2(\epsilon)$. Connecting the left and right branches of the eigenfunctions at $x = 0$ yields

$$\frac{d}{dx}\ln(u(x, \epsilon))\bigg|_{x=-0}^{x=+0} = 2\epsilon.$$

Using $E(\epsilon) = k^2(\epsilon)$ we find the transcendental equation for $E(\epsilon)$

$$\sqrt{E(\epsilon)}\cot(\sqrt{E(\epsilon)}) = -\epsilon.$$

This equation has an infinite number of solutions for any real value of ϵ. This describes only the even part of the discrete spectrum. The eigenfunctions of the odd states have nodes at $x = 0$. Thus the corresponding energy levels do not depend on ϵ.

Problem 24. Let $a > 0$ with dimension length and $n_1, n_2 \in \mathbb{N}$. Consider the two-dimensional *Helmholtz equation*

$$\left(\frac{\partial^2}{\partial x_1^2} + \frac{\partial^2}{\partial x_2^2} \right) u(x_1, x_2) + k^2 u(x_1, x_2) = 0.$$

Consider the ansatz

$$u(x_1, x_2) = \sin(n_1\pi x_1/a)\sin(n_2\pi x_2/a) - (-1)^{n_1+n_2}\sin(n_2\pi x_1/a)\sin(n_1\pi x_2/a).$$

(i) Find k^2.
(ii) Find $u(0, x_2)$, $u(x_1, 0)$, $u(x_1, a - x_1)$. Discuss.

Solution 24. (i) Since

$$\frac{\partial^2 u}{\partial x_1^2} + \frac{\partial^2 u}{\partial x_2^2} = -\frac{\pi^2}{a^2}(n_1^2 + n_2^2)u$$

we obtain

$$k^2 = \frac{\pi^2}{a^2}(n_1^2 + n_2^2) \;\Rightarrow\; k = \frac{\pi}{a}\sqrt{n_1^2 + n_2^2}.$$

(ii) We have $u(0, x_2) = 0$, $u(x_1, 0) = 0$. Utilizing that $\cos(n_1\pi) = (-1)^{n_1}$, $\cos(n_2\pi) = (-1)^{n_2}$ we have

$$\cos(n_1\pi) - (-1)^{n_1+n_2}\cos(n_2\pi) = 0.$$

It follows that $u(x_1, a - x_1) = 0$. Hence we found a solution of Helmholtz equation for the domain

$$D := \{\,(x_1, x_2) : x_1 \geq 0,\, x_2 \geq 0,\, x_1 + x_2 \leq a\,\}.$$

The function u vanishes at the boundary ∂D of D. We consider the Hilbert space $L_2(D)$. Note that

$$\int_{x_2=0}^{x_2=a} \int_{x_1=0}^{x_1=a-x_2} dx_1 dx_2 = \int_{x_2=0}^{x_2=a} (a - x_2)dx_2 = \frac{a^2}{2}.$$

Problem 25. Consider the Hamilton operator of the one-dimensional harmonic oscillator

$$\hat{H} = -\frac{\hbar}{2m}\frac{d^2}{dx^2} + \frac{1}{2}m\omega^2 x^2$$

and the operators

$$L_- := -\frac{1}{2}\frac{d^2}{dx^2}, \quad L_+ := \frac{1}{2}x^2, \quad L_3 := \frac{1}{2}x\frac{d}{dx} + \frac{1}{4}.$$

These operators satisfy the commutation relation of the simple Lie algebra $su(1,1)$, namely

$$[L_+, L_-] = 2L_3, \quad [L_3, L_+] = L_+, \quad [L_3, L_-] = -L_-.$$

Consider the one-dimensional harmonic oscillator in a gravitational field. The Hamilton operator is

$$\hat{H} = -\frac{\hbar^2}{2m}\frac{d^2}{dx^2} + \frac{1}{2}m\omega^2 x^2 + mgx$$

$$\equiv -\frac{\hbar^2}{2m}\frac{d^2}{dx^2} + \frac{1}{2}m\omega^2(x + g/\omega^2)^2 - \frac{mg^2}{2\omega^2}.$$

Find $\exp(-i\hat{H}t/\hbar)$.

Solution 25. Using the *entanglement techniques* for finite dimensional Lie algebras we obtain

$$\exp(-i\hat{H}t/\hbar) = \exp\left(\frac{img^2t}{2\hbar\omega^2}\right)\exp\left(-\frac{im\omega}{2\hbar}\tan(\omega t)(x+g/\omega^2)^2\right)$$
$$\times \exp\left(-\frac{i\hbar}{2m\omega}\sin(\omega t)\left(-\frac{1}{2}\frac{d^2}{dx^2}\right)\right)$$
$$\times \exp\left(-\ln(\cos(\omega t))\left((x+g/\omega^2)\frac{d}{dx}+\frac{1}{2}\right)\right).$$

Problem 26. Consider the linear operators $\hat{E}$, $\hat{x}$, $\hat{p}$ defined by

$$\hat{E} = i\hbar\frac{\partial}{\partial t}$$
$$\hat{p} = -i\hbar\alpha\cos(\omega t)\frac{\partial}{\partial x} + i\hbar\frac{\omega x\sin(\omega t)}{\alpha c^2}\frac{\partial}{\partial t} + \frac{m\omega\sin(\omega t)}{\alpha}x$$
$$\hat{x} = \frac{\cos(\omega t)}{\alpha} + i\hbar\frac{\alpha\sin(\omega t)}{m\omega}\frac{\partial}{\partial x} + i\hbar\frac{x\cos(\omega t)}{\alpha mc^2}\frac{\partial}{\partial t}$$

where α is the dimensionless quantity $\alpha = \sqrt{1+\omega^2x^2/c^2}$. Find the commutators $[\hat{E},\hat{x}]$, $[\hat{E},\hat{p}]$, $[\hat{x},\hat{p}]$.

Solution 26. The operators $\hat{E}$, $\hat{x}$, $\hat{p}$ satisfy the commutation relations

$$[\hat{E},\hat{x}] = -i\frac{\hbar}{m}\hat{p}, \quad [\hat{E},\hat{p}] = im\omega^2\hbar\hat{x}, \quad [\hat{x},\hat{p}] = i\hbar\left(1+\frac{1}{mc^2}\hat{E}\right).$$

Hence $\hat{E}$, $\hat{x}$, $\hat{p}$, $i\hbar$ form a basis of a four-dimensional *Lie algebra*. Is the Lie algebra semi-simple?

Problem 27. Consider the function $u : \mathbb{R} \to \mathbb{R}$

$$u(x) = e^{-k|x|}, \quad k > 0$$

which is an element of the Hilbert space $L_2(\mathbb{R})$. The function is continuous, but not differentiable at $x = 0$. Find the derivatives du/dx, d^2u/dx^2 in the sense of generalized functions. Show that u satisfies the second order ordinary differential equation (eigenvalue equation)

$$\left(-\frac{1}{2k}\cdot\frac{d^2}{dx^2} - \delta(x)\right)u(x) = \frac{1}{2}ku(x)$$

in the sense of generalized functions.

Solution 27. With $\phi \in S(\mathbb{R})$ (note that $S(\mathbb{R})$ is dense in $L_2(\mathbb{R})$) and applying integration by parts we have

$$\left(\frac{du}{dx}, \phi\right) = -\left(u, \frac{d\phi}{dx}\right) = -\int_{\mathbb{R}} u(x) \frac{d\phi}{dx} dx$$

$$= -\int_{-\infty}^{0} e^{kx} \frac{d\phi}{dx} dx - \int_{0}^{\infty} e^{-kx} \frac{d\phi}{dx} dx$$

$$= -(\phi(0) - k \int_{\infty}^{0} e^{kx} \phi(x) dx - (-\phi(0) + k \int_{0}^{\infty} e^{-kx} \phi(x) dx)$$

$$= k(\int_{-\infty}^{0} e^{kx} \phi(x) dx - \int_{0}^{\infty} e^{-kx} \phi(x) dx)$$

$$= -k(\text{sign}(x) e^{-k|x|}, \phi(x))$$

where

$$\text{sign}(x) = \begin{cases} 1 & \text{for } x \geq 0 \\ -1 & \text{for } x < 0 \end{cases}$$

We set now

$$v(x) = -k\text{sign}(x) e^{-k|x|}.$$

For the second order derivative we find by applying integration by parts

$$\left(\frac{dv}{dx}, \phi(x)\right) = -\left(v, \frac{d\phi}{dx}\right) = -\int_{\mathbb{R}} v(x) \frac{d\phi}{dx} dx = k \int_{\mathbb{R}} \text{sign}(x) e^{-k|x|} \frac{d\phi}{dx} dx$$

$$= k\left(-\int_{-\infty}^{0} e^{kx} \frac{d\phi}{dx} dx + \int_{0}^{\infty} e^{-kx} \frac{d\phi}{dx} dx\right)$$

$$= k(-\phi(0) + k(e^{-k|x|}, \phi(x)) - \phi(0))$$

$$= -2k\phi(0) + k^2(e^{-k|x|}, \phi(x))$$

$$= -2k(\delta(x), \phi(x)) + k^2(e^{-k|x|}, \phi(x)).$$

Thus we can write the second derivative as $-2k\delta(x) + k^2 e^{-k|x|}$.

Problem 28. Consider the eigenvalue problem

$$\left(-\frac{\hbar^2}{2m} \cdot \frac{d^2}{dx^2} - \gamma \hbar^2 \delta(x)\right) u(x) = E u(x)$$

with $\gamma > 0$. Then there is a bound state (see previous problem)

$$u_B(x) = \sqrt{m\gamma} e^{-m\gamma|x|} \in L_2(\mathbb{R})$$

with eigenvalue $E_B = -\frac{1}{2} m \gamma^2 \hbar^2$. Hence $E_B < 0$. Study the case $E \geq 0$.

Solution 28. First we note that

$$(e^{ikx}\delta(x), \phi(x)) = \phi(0).$$

For the case $E \geq 0$ one finds two degenerate stationary scattering states. The first one is a plane wave moving in the positive x direction with $k = (2mE)^{1/2}/\hbar$ (k wave vector) and scattered waves moving from the origin and the second scattering state is a plane wave moving in the negative x direction together with scattered waves. Note that $u_{-k}(x) = u_k(-x)$ $k > 0$. Hence we have ($k \in (-\infty, +\infty)$)

$$u_k(x) = \frac{1}{\sqrt{2\pi}}(e^{ikx} + im\gamma(|k| - im\gamma(|k| - im\gamma)^{-1})\exp(i|k| \cdot |x|)$$

and the *completeness relation* is

$$\int_{\mathbb{R}} u_k(x)u_k^*(x')dk + u_B(x)u_B^*(x') = \delta(x - x').$$

Problem 29. Let $L > 0$ with dimension length. Consider the Hilbert space $L_2([0, L] \times [0, L] \times [0, L])$ and the eigenvalue problem

$$-\frac{\hbar^2}{2m}\left(\frac{\partial^2}{\partial x_1^2} + \frac{\partial^2}{\partial x_2^2} + \frac{\partial^2}{\partial x_3^2}\right)u_{\mathbf{n}}(\mathbf{x}) = E_{\mathbf{n}}u_{\mathbf{n}}(\mathbf{x}) \quad \mathbf{n} = (n_1, n_2, n_3).$$

Periodic boundary conditions are imposed for the three-dimensional box of side length L. Find the eigenfunctions and the corresponding energy eigenvalues.

Solution 29. Obviously the eigenfunctions are

$$u_{\mathbf{n}}(\mathbf{x}) = \frac{1}{\sqrt{L^3}}\exp\left(\frac{i}{\hbar}(p_1 x_1 + p_2 x_2 + p_3 x_3)\right)$$

with $\mathbf{n} = (n_1, n_2, n_3)$ $(n_1, n_2, n_3 \in \mathbb{Z})$

$$p_1 = \frac{2\pi\hbar n_1}{L}, \quad p_2 = \frac{2\pi\hbar n_2}{L}, \quad p_3 = \frac{2\pi\hbar n_3}{L}.$$

Then the energy eigenvalues are

$$E_{\mathbf{n}} = \frac{1}{2m}(p_1^2 + p_2^2 + p_3^2) = \frac{2\pi^2\hbar^2}{mL^2}(n_1^2 + n_2^2 + n_3^2).$$

Problem 30. Let $L > 0$. Consider the Hilbert space $L_2([0, L] \times [0, L])$ and the Hamilton operator

$$\hat{H} = \frac{\hat{p}^2}{2m} + V(\mathbf{x}) = -\frac{\hbar^2}{2m}\left(\frac{\partial^2}{\partial x_1^2} + \frac{\partial^2}{\partial x_2^2}\right) + V(\mathbf{x})$$

with

$$V(\mathbf{x}) = \begin{cases} 0 & \text{for} \quad 0 \leq x_1 \leq L, \ 0 \leq x_2 \leq L \\ \infty & \text{otherwise} \end{cases}$$

Let $n_1, n_2 \in \mathbb{N}$. Consider the normalized state

$$u(x_1, x_2) = \frac{2}{L} \sin\left(\frac{n_1 \pi x_1}{L}\right) \sin\left(\frac{n_2 \pi x_2}{L}\right)$$

with $u(0,0) = 0$, $u(L,0) = 0$, $u(0,L) = 0$, $u(L,L) = 0$.
(i) Find $\hat{H}u$ and thus the eigenvalues.
(ii) Calculate

$$P(x_2) = \int_0^L |u(x_1, x_2)|^2 dx_1.$$

Solution 30. (i) We obtain

$$\hat{H}u = -\frac{\hbar^2}{2m}\left(-\frac{2}{L}\frac{n_1^2\pi^2}{L^2} - \frac{2}{L}\frac{n_2^2\pi}{L^2}\right)\sin\left(\frac{n_1\pi x_1}{L}\right)\sin\left(\frac{n_2\pi x_2}{L}\right)$$

$$= E_{n_1,n_2}\frac{2}{L}\sin\left(\frac{n_1\pi x_1}{L}\right)\sin\left(\frac{n_2\pi x_2}{L}\right).$$

Hence

$$E_{n_1,n_2} = \frac{\hbar^2}{2m}\cdot\frac{\pi^2}{L^2}(n_1^2 + n_2^2).$$

(ii) Note that

$$\int_0^L \sin^2\left(\frac{n_1\pi x_1}{L}\right)dx_1 = \frac{L}{2}.$$

Hence we find

$$P(x_2) = \int_0^L |u(x_1, x_2)|^2 dx_1 = \frac{4}{L^2}\int_0^L \sin^2\left(\frac{n_1\pi x_1}{L}\right)\sin^2\left(\frac{n_2\pi x_2}{L}\right)dx_1$$

$$= \frac{2}{L}\sin^2\left(\frac{n_2\pi x_2}{L}\right).$$

Problem 31. Consider the Banach Gelfand triple $(S(\mathbb{R}^2), L_2(\mathbb{R}^2), S'(\mathbb{R}^2))$ and the Hamilton operator $\hat{H}$ expressed in *polar coordinates*

$$\hat{H} = \frac{\mathbf{p}^2}{2m} = -\frac{\hbar^2}{2m}\left(\frac{\partial^2}{\partial x_1^2} + \frac{\partial^2}{\partial x_2^2}\right) = -\frac{\hbar^2}{2m}\left(\frac{\partial^2}{\partial r^2} + \frac{1}{r}\frac{\partial}{\partial r} + \frac{1}{r^2}\frac{\partial^2}{\partial \phi^2}\right).$$

Consider the eigenvalue problem $\hat{H}u(r,\phi) = Eu(r,\phi)$

$$-\frac{\hbar^2}{2m}\left(r^2\frac{\partial^2}{\partial r^2} + r\frac{\partial}{\partial r} + \frac{\partial^2}{\partial \phi^2}\right)u(r,\phi) = r^2 Eu(r,\phi).$$

Show that the solution is given by

$$u(r,\phi) = i^n J_n(kr)e^{in\phi}, \quad n = 0, \pm1, \pm2, \ldots$$

with $E = k^2\hbar^2/(2m)$, where $J_n(kr)$ denotes the *Bessel function* of first order.

Solution 31. We have

$$\frac{\partial u}{\partial \phi} = i^n J_n(kr)e^{in\phi}in, \quad \frac{\partial^2 u}{\partial \phi^2} = -i^n n^2 J_n(kr)e^{in\phi}.$$

It follows that

$$-\frac{\hbar^2}{2m}\left(r^2\frac{\partial^2}{\partial r^2} + r\frac{\partial}{\partial r} - n^2 + (kr)^2\right)J_n(kr) = 0.$$

Setting $x = kr$ we obtain the *Bessel differential equation*

$$\left(x^2\frac{d^2}{dx^2} + x\frac{d}{dx} - n^2 + x^2\right)J_n(x) = 0.$$

Utilizing the identities

$$x\frac{dJ_n(x)}{dx} = xJ_{n-1}(x) - nJ_n(x), \quad x\frac{dJ_n(x)}{dx} = nJ_n(x) - xJ_{n+1}(x)$$

we can show that $J_n(x)$ satisfies the differential equation.

Problem 32. Let $\gamma > 0$ with dimension *meter*$^{-1}$. Consider the one-dimensional eigenvalue problem

$$\frac{d^2u}{dx^2} + (k^2 + 2\gamma^2\text{sech}^2(\gamma x))u = 0$$

with $k^2 = 2mE/\hbar^2$. Note that $\text{sech}^2(\gamma x) \equiv 1/\cosh^2(\gamma x)$. Show that

$$u(x) = a\left(\cos(kx + \delta) - \frac{\gamma}{k}\sin(kx + \delta)\frac{\sinh(\gamma x)}{\cosh(\gamma x)}\right)$$

is the general solution with a, δ be arbitrary real constants.

Solution 32. Applying the Maxima program

```
/* diffeigen.mac */
u: a*(cos(k*x+d)-(g/k)*sin(k*x+d)*sinh(g*x)/cosh(g*x));
ud: diff(u,x); udd: diff(ud,x);
R: udd + k*k*u + (2*g*g)/(cosh(g*x)*cosh(g*x))*u;
R: trigsimp(R);
```

shows that $R = 0$. Let ℓ be an length. Impose the boundary conditions $u(\ell/2) = u(-\ell/2)$ and $du(\ell/2)/dx = du(-\ell/2)/dx$.

Problem 33. Consider the eigenvalue problem

$$-\frac{\hbar^2}{2m}\frac{d^2u(x)}{dx^2}u(x) + V(x)u(x) = Eu(x).$$

We set

$$\widetilde{V}(x) = \frac{2mV(x)}{\hbar^2}, \quad \widetilde{E} = \frac{2mE}{\hbar^2}$$

and write

$$-\frac{d^2u(x)}{dx^2} + \widetilde{V}(x)u(x) = \widetilde{E}u(x).$$

Hence the dimension of $\widetilde{E}$, $\widetilde{V}(x)$ is $meter^{-2}$.

(i) Let $u(x) = \exp(-ax^4/4)$, where $a > 0$ with dimension $1/meter^4$. Then $u(x) \in L_2(\mathbb{R})$. Find the corresponding potential $\widetilde{V}$ and the eigenvalue $\widetilde{E}$.

(ii) Let $a > 0$, $b > 0$ and $u(x) = \exp(-ax^4/4 - bx^2/2) \in L_2(\mathbb{R})$ with b of dimension $1/meter^2$. Find the corresponding potential $\widetilde{V}$ and the eigenvalue $\widetilde{E}$.

(iii) Study the case $u(x) = \exp(-f(x))$ with f a smooth function $f : \mathbb{R} \to \mathbb{R}$ and $\exp(-f(x)) \in L_2(\mathbb{R})$. Hence $f(x) \to \infty$ for $x \to \pm\infty$.

Solution 33. (i) Inserting $u(x) = \exp(-ax^4/4)$ into the differential equation yields

$$3ax^2 - a^2x^6 + \widetilde{V}(x) = \widetilde{E}.$$

Hence $\widetilde{E}(a) = 0$ and $\widetilde{V}(x) = a^2x^6 - 3ax^2$.

(ii) Inserting $u(x) = \exp(-ax^4/4 - bx^2/2)$ into the differential equation yields

$$3ax^2 + b - a^2x^6 - 2abx^4 - b^2x^2 + \widetilde{V} = \widetilde{E}$$

with the solution $\widetilde{E}(a, b) = b$ and $\widetilde{V} = a^2x^6 + 2abx^4 + (b^2 - 3a)x^2$.

(iii) Inserting $u(x) = \exp(-f(x))$ into the differential equation yields

$$\frac{d^2f}{dx^2} - \left(\frac{df}{dx}\right)^2 + \widetilde{V}(x) = \widetilde{E}$$

where we have the constraint that $\widetilde{E}$ must be real number. This equation can be considered from two points of view. If f is given, then $\widetilde{V}$ and $\widetilde{E}$ can easily be determined. The other case is if $\widetilde{V}$ is given. Then we have to solve the second order differential equation. Introduction $g = df/dx$ we obtain the system

$$\frac{df}{dx} = g, \quad \frac{dg}{dx} = g^2 - \widetilde{V} + \widetilde{E}.$$

The second equation can be linearized with

$$g(x) = -\frac{1}{h(x)}\frac{dh}{dx}, \quad h(x) = \exp(-\int^x g(s)ds).$$

We obtain the second order linear differential equation

$$\frac{d^2h}{dx^2} = (\widetilde{V}(x) - \widetilde{E})h.$$

Whether or not we can solve this equation depends on $\widetilde{V}$. The cases discussed above are included as special case. Another example is

$$\widetilde{V}(x) = -\cosh(x) + \sinh^2(x) + 4c^2x^2 + 2cx\sinh(x)$$

where $f(x) = \cosh(x) + cx^2$ and $E = 2c$ with $c \geq 0$.

Problem 34. Consider the Banach-Gelfand triple $S(\mathbb{R}) \subset L_2(\mathbb{R}) \subset S'(\mathbb{R})$. Let $f \in S(\mathbb{R})$. Find the commutator

$$[-d/dx, x + d/dx]f(x).$$

Let

$$\phi_0(x) = \frac{1}{\sqrt{2\pi}}\exp(-x^2/2)$$

and

$$\phi_n(x) = \phi_0(x)\psi_n(x), \quad \psi_n(x) = \frac{1}{\sqrt{2^n n!}}H_n(x/\sqrt{2})$$

where H_n are the Hermite polynomials. Let $T_+ = -d/dx$, $T_- = x + d/dx$. Find $T_+\phi_n(x)$, $T_-\phi_n(x)$ $(n = 0, 1, 2, \dots)$.

Solution 34. We obtain

$$[-d/dx, x + d/dx]f(x) = -f(x) - x\frac{df(x)}{dx} - \frac{d^2f(x)}{dx^2} + x\frac{df(x)}{dx} + \frac{d^2f(x)}{dx^2}$$
$$= -f(x)$$

and

$$T_+\phi_n(x) = \sqrt{n+1}\phi_{n+1}(x), \quad n = 0, 1, 2, \ldots$$

$$T_-\phi_n(x) = \sqrt{n}\phi_{n-1}(x), \quad n = 1, 2, \ldots$$

with $T_-\phi_0(x) = 0$.

Problem 35. Let $\hat{b}^\dagger$, $\hat{b}$ be Bose creation and annihilation operators and $|\beta\rangle$ be a coherent state, where β is a complex number which determines the *average field amplitude*. We have

$$\hat{b}|\beta\rangle = \beta|\beta\rangle.$$

i.e. the spectrum of $\hat{b}$ is the whole complex plane. Find $\langle\beta|\hat{b}^\dagger\hat{b}|\beta\rangle$.

Solution 35. Since $\langle\beta|\hat{b}^\dagger = \langle\beta|\bar{\beta}$ we have $\langle\beta|\hat{b}^\dagger\hat{b}|\beta\rangle = \bar{\beta}\beta$.

Problem 36. Let $z = x_1 + ix_2$, $\bar{z} = x_1 - ix_2$ ($x_1, x_2 \in \mathbb{R}$) with dimension length. We have

$$\frac{\partial}{\partial z} := \frac{1}{2}\left(\frac{\partial}{\partial x_1} - i\frac{\partial}{\partial x_2}\right), \quad \frac{\partial}{\partial \bar{z}} := \frac{1}{2}\left(\frac{\partial}{\partial x_1} + i\frac{\partial}{\partial x_2}\right).$$

Consider the *Bose operators*

$$\hat{b} := \frac{z}{2\ell} + \ell\frac{\partial}{\partial \bar{z}}, \quad \hat{b}^\dagger := \frac{\bar{z}}{2\ell} - \ell\frac{\partial}{\partial z}$$

with the nonzero constant of dimension length. Find the commutator $[\hat{b}, \hat{b}^\dagger]$.

Solution 36. We have

$$[\hat{b}, \hat{b}^\dagger]f(x_1, x_2) = \left[\frac{z}{2\ell} + \ell\frac{\partial}{\partial \bar{z}}, \frac{\bar{z}}{2\ell} - \ell\frac{\partial}{\partial z}\right]f(x_1, x_2)$$

$$= \left[\frac{z}{2\ell}, -\ell\frac{\partial}{\partial z}\right]f(x_1, x_2) + \left[\ell\frac{\partial}{\partial \bar{z}}, \frac{\bar{z}}{2\ell}\right]f(x_1, x_2)$$

$$= \left[\frac{z}{2}, -\frac{\partial}{\partial z}\right]f(x_1, x_2) + \left[\frac{\partial}{\partial \bar{z}}, \frac{\bar{z}}{2}\right]f(x_1, x_2)$$

$$= f.$$

Hence $[\hat{b}, \hat{b}^\dagger] = 1$.

Problem 37. Consider the Hamilton operator for the one-dimensional harmonic oscillator

$$\hat{H} = -\frac{\hbar^2}{2m}\frac{d^2}{dx^2} + \frac{m\omega^2}{2}\hat{x}^2.$$

We set

$$\hat{x} = \sqrt{\frac{\hbar}{2m\omega}}(\hat{b} + \hat{b}^\dagger), \quad \hat{p} = -i\hbar\frac{d}{dx} = -i\sqrt{\frac{m\hbar\omega}{2}}(\hat{b} - \hat{b}^\dagger)$$

Find $\exp(-i\hat{H}t/\hbar)$.

Solution 37. We obtain

$$\exp(-i\hat{H}t/\hbar) = \exp(-i\omega t(\hat{b}^\dagger\hat{b} + I/2)) = \exp(-i\omega t\hat{b}^\dagger\hat{b})\exp(-i\omega t/2).$$

Problem 38. (i) Let $\epsilon \in \mathbb{R}$. Find

$$f_c(\epsilon) = e^{-i\epsilon\hat{b}^\dagger\hat{b}}\hat{b}^\dagger e^{i\epsilon\hat{b}^\dagger\hat{b}}, \quad f_a(\epsilon) = e^{-i\epsilon\hat{b}^\dagger\hat{b}}\hat{b}e^{i\epsilon\hat{b}^\dagger\hat{b}}.$$

(ii) Then find the 2×2 matrix $A(\epsilon)$ such that

$$\begin{pmatrix} e^{-i\epsilon\hat{b}^\dagger\hat{b}}(\hat{b}^\dagger + \hat{b})e^{i\epsilon\hat{b}^\dagger\hat{b}} \\ e^{-i\epsilon\hat{b}^\dagger\hat{b}}(i\hat{b}^\dagger - i\hat{b})e^{i\epsilon\hat{b}^\dagger\hat{b}} \end{pmatrix} = A(\epsilon) \begin{pmatrix} \hat{b}^\dagger + \hat{b} \\ i\hat{b}^\dagger - i\hat{b} \end{pmatrix}.$$

Solution 38. (i) Applying *parameter differentiation* and $\hat{b}\hat{b}^\dagger = I + \hat{b}^\dagger\hat{b}$ we obtain the differential equation

$$\frac{df_c(\epsilon)}{d\epsilon} = -if_c(\epsilon)$$

with the initial condition $f_c(0) = \hat{b}^\dagger$. The solution is $f_c(\epsilon) = e^{-i\epsilon}\hat{b}^\dagger$. Analogously for f_a we find the differential equation $df_a/d\epsilon = if_a$ with the initial condition $f_a(0) = \hat{b}$. Hence $f_a(\epsilon) = e^{i\epsilon}\hat{b}$.
(ii) Since

$$e^{-i\epsilon\hat{b}^\dagger\hat{b}}(\hat{b}^\dagger + \hat{b})e^{i\epsilon\hat{b}^\dagger\hat{b}} = e^{-i\epsilon}\hat{b}^\dagger + e^{i\epsilon}\hat{b}$$
$$e^{-i\epsilon\hat{b}^\dagger\hat{b}}(i\hat{b}^\dagger - i\hat{b})e^{i\epsilon\hat{b}^\dagger\hat{b}} = ie^{-i\epsilon}\hat{b}^\dagger - ie^{i\epsilon}\hat{b}$$

we obtain

$$\begin{pmatrix} e^{-i\epsilon}\hat{b}^\dagger + e^{i\epsilon}\hat{b} \\ ie^{-i\epsilon}\hat{b}^\dagger - ie^{i\epsilon}\hat{b} \end{pmatrix} = \begin{pmatrix} \cos(\epsilon) & -\sin(\epsilon) \\ \sin(\epsilon) & \cos(\epsilon) \end{pmatrix} \begin{pmatrix} \hat{b}^\dagger + \hat{b} \\ i\hat{b}^\dagger - i\hat{b} \end{pmatrix}.$$

Thus the matrix is a *rotation matrix* and an element of the compact Lie group $SO(2)$.

Problem 39. Let $\hat{b}_1^\dagger$, $\hat{b}_2^\dagger$, $\hat{b}_3^\dagger$, $\hat{b}_1$, $\hat{b}_2$, $\hat{b}_3$ be Bose creation and annihilation operators, respectively. The commutation relations are

$$[\hat{b}_j, \hat{b}_k] = [\hat{b}_j^\dagger, \hat{b}_k^\dagger] = 0, \quad j, k = 1, 2, 3$$

$$[\hat{b}_j, \hat{b}_k^\dagger] = \delta_{j,k} I, \quad j, k = 1, 2, 3.$$

(i) The number operators are $\hat{N}_j = \hat{b}_j^\dagger \hat{b}_j$ $(j = 1, 2, 3)$ and the total number operator is

$$\hat{N} = \hat{N}_1 + \hat{N}_2 + \hat{N}_3.$$

Find the commutators $[\hat{N}, \hat{b}_k]$, $[\hat{N}, \hat{b}_k^\dagger]$ for $j = 1, 2, 3$.
(ii) Consider the vacuum state $|000\rangle \equiv |0\rangle \otimes |0\rangle \otimes |0\rangle$ with $\hat{b}_j|000\rangle = 0|000\rangle$ $(j = 1, 2, 3)$ and the number states

$$|n_1, n_2, n_3\rangle \equiv |n_1\rangle \otimes |n_2\rangle \otimes |n_3\rangle = \frac{1}{\sqrt{n_1! n_2! n_3!}} (\hat{b}_1^\dagger)^{n_1} (\hat{b}_2^\dagger)^{n_2} (\hat{b}_3^\dagger)^{n_3} |0, 0, 0\rangle$$

with $\langle m_1, m_2, m_3 | n_1, n_2, n_3 \rangle = \delta_{m_1, n_1} \delta_{m_2, n_2} \delta_{m_3, n_3}$ and the *completeness relation*

$$\sum_{n_1=0}^{\infty} \sum_{n_2=0}^{\infty} \sum_{n_3=0}^{\infty} |n_1, n_2, n_3\rangle\langle n_1, n_2, n_3| = I.$$

Find $\hat{N}_j|n_1, n_2, n_3\rangle$, $\hat{N}|n_1, n_2, n_3\rangle$ and $\hat{b}_1|n_1, n_2, n_3\rangle$, $\hat{b}_1^\dagger|n_1, n_2, n_3\rangle$.
(iii) Find the *isospin operators* I_j $(j = 1, 2, 3)$ defined by

$$I_j = \frac{1}{i} \sum_{k=1}^{3} \sum_{\ell=1}^{3} \epsilon_{jk\ell} \hat{b}_k^\dagger \hat{b}_\ell$$

with $\epsilon_{123} = +1$, $\epsilon_{132} = -1$, $\epsilon_{213} = -1$, $\epsilon_{231} = +1$, $\epsilon_{312} = +1$, $\epsilon_{321} = -1$ and 0 otherwise. Find the commutators

$$[I_j, I_k], \quad (j, k = 1, 2, 3)$$

(iv) Find $I_1^2 + I_2^2 + I_3^2$. Let $\hat{K} := \hat{b}_1^2 + \hat{b}_2^2 + \hat{b}_3^2$. Find $\hat{K}^\dagger$ and $\hat{K}^\dagger \hat{K}$. Find

$$[\hat{K}, \hat{K}^\dagger], \quad [I_j, \hat{K}], \quad [\hat{N}, \hat{K}], \quad [\hat{I}_j, \hat{K}^\dagger], \quad [\hat{N}, \hat{K}^\dagger].$$

Find $[\hat{b}_j, \hat{K}^\dagger]$, $[\hat{K}, \hat{b}_j^\dagger]$.

Solution 39. (i) We obtain

$$[\hat{N}_j, \hat{b}_k] = -\delta_{j,k} \hat{b}_j, \quad [\hat{N}_j, \hat{b}_k^\dagger] = \delta_{j,k} \hat{b}_j^\dagger \quad j, k = 1, 2, 3$$

(ii) We have

$$\hat{N}_j|n_1, n_2, n_3\rangle = n_j|n_1, n_2, n_3\rangle, \quad \hat{N}|n_1, n_2, n_3\rangle = (n_1 + n_2 + n_3)|n_1, n_2, n_3\rangle.$$

(iii) The three *isospin operators* are

$$I_1 = \frac{1}{i}(\hat{b}_2^\dagger \hat{b}_3 - \hat{b}_3^\dagger \hat{b}_2), \quad I_2 = \frac{1}{i}(\hat{b}_3^\dagger \hat{b}_1 - \hat{b}_1^\dagger \hat{b}_3), \quad I_3 = \frac{1}{i}(\hat{b}_1^\dagger \hat{b}_2 - \hat{b}_2^\dagger \hat{b}_1)$$

with the commutation relations

$$[I_j, I_k] = i\sum_{\ell=1}^{3} \epsilon_{jk\ell} I_\ell = i(\epsilon_{jk1} I_1 + \epsilon_{jk2} I_2 + \epsilon_{jk3} I_3).$$

Hence

$$[I_1, I_2] = i(\epsilon_{121} I_1 + \epsilon_{122} I_2 + \epsilon_{123} I_3) = iI_3$$

and $[I_2, I_3] = iI_1$, $[I_3, I_1] = iI_2$. We have a basis of a simple Lie algebra.

(iv) With $[\hat{b}_j^2, (\hat{b}_j^\dagger)^2] = 4\hat{b}_j^\dagger \hat{b}_j + 2I$ $(j = 1, 2, 3)$ we have

$$[\hat{K}, \hat{K}^\dagger] = 4\hat{N} + 6I, \quad \hat{I} = I_1 + I_2^2 + I_3^2 = \hat{N}^2 + \hat{N} - \hat{K}^\dagger \hat{K}$$

and

$$[\hat{N}, \hat{K}] = -2\hat{K}, \quad [\hat{N}, \hat{K}^\dagger] = 2\hat{K}^\dagger, \quad [\hat{K}, \hat{K}^\dagger] = 4\hat{N} + 6I.$$

Problem 40. Consider the *uncertainty relation*

$$\left(\langle\psi|\hat{A}^2|\psi\rangle - \langle\psi|\hat{A}|\psi\rangle^2\right)\left(\langle\psi|\hat{B}^2|\psi\rangle - \langle\psi|\hat{B}|\psi\rangle^2\right) \geq \frac{1}{4}|\langle\psi|[\hat{A}, \hat{B}]|\psi\rangle|^2$$

where $\hat{A}$ and $\hat{B}$ are observables, $[\hat{A}, \hat{B}]$ denotes the commutator and $|\psi\rangle$ is a normalized state. Let

$$|\psi\rangle = |n\rangle, \quad \hat{A} = \frac{1}{\sqrt{2}}(i\hat{b} - i\hat{b}^\dagger), \quad \hat{B} = \frac{1}{\sqrt{2}}(\hat{b} + \hat{b}^\dagger)$$

where $|n\rangle$ is a *number state*. Find the left and right-hand sides of the inequality. Discuss.

Solution 40. For the commutator we find

$$\frac{1}{2}[i\hat{b} - i\hat{b}^\dagger, \hat{b} + \hat{b}^\dagger] = \frac{1}{2}([i\hat{b}, \hat{b}^\dagger] - [i\hat{b}^\dagger, \hat{b}]) = iI$$

where I is the identity operator. Then with $\langle n|n\rangle = 1$ we obtain for the right-hand side

$$\frac{1}{4}|\langle n|iI|n\rangle|^2 = \frac{1}{4}.$$

For the left-hand side we have $\langle n|\hat{A}|n\rangle = 0$, $\langle n|\hat{B}|n\rangle = 0$. Now

$$\hat{A}^2 = \frac{1}{2}(-(\hat{b}^\dagger)^2 - \hat{b}^2 + I + 2\hat{b}^\dagger\hat{b}) \quad \hat{B}^2 = \frac{1}{2}(\hat{b}^2 + (\hat{b}^\dagger)^2 + I + 2\hat{b}^\dagger\hat{b}).$$

Then

$$\langle n|\hat{A}^2|n\rangle = \frac{1}{2}(1+2n) \qquad \langle n|\hat{B}^2|n\rangle = \frac{1}{2}(1+2n)$$

and the inequality is given by

$$\frac{1}{4}(1+2n)^2 \geq \frac{1}{4}.$$

If $n = 0$ we have an equality.

Problem 41. Consider the *uncertainty relation*

$$\left(\langle\psi|\hat{A}^2|\psi\rangle - \langle\psi|\hat{A}|\psi\rangle^2\right)\left(\langle\psi|\hat{B}^2|\psi\rangle - \langle\psi|\hat{B}|\psi\rangle^2\right) \geq \frac{1}{4}|\langle\psi|[\hat{A},\hat{B}]|\psi\rangle|^2$$

where $\hat{A}$ and $\hat{B}$ are observables, $[\hat{A},\hat{B}]$ denotes the commutator and $|\psi\rangle$ is a normalized state. Let

$$|\psi\rangle = |\beta\rangle, \quad \hat{A} = \frac{1}{\sqrt{2}}(i\hat{b} - i\hat{b}^\dagger), \quad \hat{B} = \frac{1}{\sqrt{2}}(\hat{b} + \hat{b}^\dagger)$$

where $|\beta\rangle$ is a *coherent state*. Find the left and right-hand sides of the inequality. Discuss.

Solution 41. For the commutator we find

$$\frac{1}{2}[i\hat{b} - i\hat{b}^\dagger, \hat{b} + \hat{b}^\dagger] = \frac{1}{2}([i\hat{b}, \hat{b}^\dagger] - [i\hat{b}^\dagger, \hat{b}]) = iI$$

where I is the identity operator. Then with $\langle\beta|\beta\rangle = 1$ we obtain for the right-hand side

$$\frac{1}{4}|\langle\beta|iI|\beta\rangle|^2 = \frac{1}{4}.$$

Now $\hat{b}|\beta\rangle = \beta|\beta\rangle$ and $\langle\beta|\hat{b}^\dagger = \langle\beta|\overline{\beta}$. It follows that

$$\langle\beta|\hat{A}|\beta\rangle^2 = \frac{1}{2}\langle\beta|(i\hat{b} - i\hat{b}^\dagger)|\beta\rangle^2 = \frac{1}{2}(\langle\beta|i\hat{b}|\beta\rangle - \langle\beta|i\hat{b}^\dagger|\beta\rangle)^2$$

$$= \frac{1}{2}(-\beta^2 - \overline{\beta}^2 + 2\beta\overline{\beta})$$

and

$$\langle\beta|\hat{B}|\beta\rangle^2 = \frac{1}{2}\langle\beta|(\hat{b} + \hat{b}^\dagger)|\beta\rangle^2 = \frac{1}{2}(\langle\beta|\hat{b}|\beta\rangle + \langle\beta|\hat{b}^\dagger|\beta\rangle)^2 = \frac{1}{2}(\beta^2 + \overline{\beta}^2 + 2\beta\overline{\beta}).$$

Furthermore

$$\langle\beta|\hat{A}^2|\beta\rangle = \frac{1}{2}(1 - \beta^2 - \overline{\beta}^2 + 2\beta\overline{\beta})$$

$$\langle\beta|\hat{B}^2|\beta\rangle = \frac{1}{2}(1 + \beta^2 + \overline{\beta}^2 + 2\beta\overline{\beta}).$$

Hence

$$\langle\beta|\hat{A}^2|\beta\rangle - \langle\beta|\hat{A}|\beta\rangle^2 = \frac{1}{2}, \quad \langle\beta|\hat{B}^2|\beta\rangle - \langle\beta|\hat{B}|\beta\rangle^2 = \frac{1}{2}.$$

It follows that the uncertainty relation is an equality in the present case.

Problem 42. Consider the *uncertainty relation*

$$\left(\langle\psi|\hat{A}^2|\psi\rangle - \langle\psi|\hat{A}|\psi\rangle^2\right)\left(\langle\psi|\hat{B}^2|\psi\rangle - \langle\psi|\hat{B}|\psi\rangle^2\right) \geq \frac{1}{4}|\langle\psi|[\hat{A},\hat{B}]|\psi\rangle|^2$$

where $\hat{A}$ and $\hat{B}$ are observables, $[\hat{A},\hat{B}]$ denotes the commutator and $|\psi\rangle$ is a normalized state. Let

$$|\psi\rangle = |\zeta\rangle, \quad \hat{A} = \frac{1}{\sqrt{2}}(i\hat{b} - i\hat{b}^\dagger), \quad \hat{B} = \frac{1}{\sqrt{2}}(\hat{b} + \hat{b}^\dagger)$$

where $|\zeta\rangle$ is a *squeezed state*. Find the left and right-hand sides of the inequality. Discuss.

Solution 42. With a squeezed state $|\zeta\rangle$ ($\zeta \in \mathbb{C}$) given by

$$|\zeta\rangle = S(\zeta)|0\rangle, \quad S(\zeta) := \exp\left(-\frac{\zeta}{2}(\hat{b}^\dagger)^2 + \frac{\zeta^*}{2}\hat{b}^2\right)$$

where $S(\zeta)$ is the *squeezing operator* and $\zeta = se^{i\theta}$ with $s \geq 0$. With

$$\langle\zeta|\hat{b}|\zeta\rangle = 0 \qquad \langle\zeta|\hat{b}^\dagger|\zeta\rangle = 0$$

one obtains

$$\frac{1}{4}(\cosh^2(2s) - \cos^2(\theta)\sinh^2(2s)) \geq \frac{1}{4}.$$

Problem 43. Consider the simple Lie algebra $su(1,1)$ with the basis K_+, K_-, K_0 and the commutation relations

$$[K_+, K_-] = -2K_0, \quad [K_0, K_+] = K_+, \quad [K_0, K_-] = -K_-.$$

A representation with Bose operators would be

$$K_+ = \hat{b}_1^\dagger\hat{b}_2^\dagger, \quad K_- = \hat{b}_1\hat{b}_2, \quad K_0 = \frac{1}{2}(\hat{b}_1^\dagger\hat{b}_1 + \hat{b}_2^\dagger\hat{b}_2 + I).$$

Let ϵ be a parameter. Calculate

$$f_{+,0}(\epsilon) = e^{\epsilon K_+}K_0e^{-\epsilon K_+}, \quad f_{+,-}(\epsilon) = e^{\epsilon K_+}K_-e^{-\epsilon K_+}$$

$$f_{-,+}(\epsilon) = e^{\epsilon K_-}K_+e^{-\epsilon K_-}, \quad f_{-,0}(\epsilon) = e^{\epsilon K_-}K_0e^{-\epsilon K_-}$$

$$f_{0,+}(\epsilon) = e^{\epsilon K_0}K_+e^{-\epsilon K_0}, \quad f_{0,-}(\epsilon) = e^{\epsilon K_0}K_-e^{-\epsilon K_0}$$

applying *parameter differentiation*. The initial values are

$$f_{+,0}(0) = K_0, \quad f_{+,-}(0) = K_-, \quad f_{-,+}(0) = K_+,$$

$$f_{-,0}(0) = K_0, \quad f_{0,+}(0) = K_+, \quad f_{0,-}(0) = K_-.$$

Solution 43. We obtain

$$\frac{df_{+,0}}{d\epsilon} = e^{\epsilon K_+}[K_+, K_0]e^{-\epsilon K_+} = -K_+$$

$$\frac{df_{+,-}}{d\epsilon} = e^{\epsilon K_+}[K_+, K_-]e^{-\epsilon K_+} = -2f_{+,0}$$

$$\frac{df_{-,+}}{d\epsilon} = e^{\epsilon K_-}[K_-, K_+]e^{-\epsilon K_-} = 2f_{-,0}$$

$$\frac{df_{-,0}}{d\epsilon} = e^{\epsilon K_-}[K_-, K_0]e^{-\epsilon K_-} = K_-$$

$$\frac{df_{0,+}}{d\epsilon} = e^{\epsilon K_0}[K_0, K_+]e^{-\epsilon K_0} = f_{0,+}$$

$$\frac{df_{0,-}}{d\epsilon} = e^{\epsilon K_0}[K_0, K_-]e^{-\epsilon K_0} = -f_{0,-}.$$

The solutions of the first and fourth equations are

$$f_{+,0}(\epsilon) = -K_+\epsilon + K_0, \quad f_{-,0}(\epsilon) = K_-\epsilon + K_0.$$

Using this result we can solve the second and third equations and find

$$f_{+,-}(\epsilon) = K_+\epsilon^2 - 2K_0\epsilon + K_-, \quad f_{-,+}(\epsilon) = K_-\epsilon^2 + 2K_0\epsilon + K_+.$$

The fifth and sixth equations can be integrated directly

$$f_{0,+}(\epsilon) = K_+e^\epsilon, \quad f_{0,-}(\epsilon) = K_-e^{-\epsilon}.$$

Problem 44. Consider the spin-$\frac{1}{2}$ matrices

$$S_1^{(1/2)} = \frac{1}{2}\begin{pmatrix} 0 & 1 \\ 1 & 0 \end{pmatrix}, \quad S_2^{(1/2)} = \frac{1}{2}\begin{pmatrix} 0 & -i \\ i & 0 \end{pmatrix}, \quad S_3^{(1/2)} = \frac{1}{2}\begin{pmatrix} 1 & 0 \\ 0 & -1 \end{pmatrix}$$

with the commutation relations

$$[S_1^{(1/2)}, S_2^{(1/2)}] = iS_3^{(1/2)}, \quad [S_2^{(1/2)}, S_3^{(1/2)}] = iS_1^{(1/2)}, \quad [S_3^{(1/2)}, S_1^{(1/2)}] = iS_2^{(1/2)}.$$

Let $\hat{b}_1^\dagger \equiv \hat{b}^\dagger \otimes I$, $\hat{b}_2^\dagger \equiv I \otimes \hat{b}^\dagger$, $\hat{b}_1 = \hat{b} \otimes I$, $\hat{b}_2 = I \otimes \hat{b}$ be Bose creation and annihilation operators, respectively. The *commutation relations* are

$$[\hat{b}_j, \hat{b}_k^\dagger] = \delta_{j,k}(I \otimes I), \quad [\hat{b}_j, \hat{b}_k] = [\hat{b}_j^\dagger, \hat{b}_k^\dagger] = 0, \quad j, k = 1, 2.$$

We define the three operators

$$J_k := \begin{pmatrix} \hat{b}_1^\dagger & \hat{b}_2^\dagger \end{pmatrix} S_k^{(1/2)} \begin{pmatrix} \hat{b}_1 \\ \hat{b}_2 \end{pmatrix}, \quad k = 1, 2, 3.$$

Then

$$J_1 = \frac{1}{2}(\hat{b}_1^\dagger \hat{b}_2 + \hat{b}_2^\dagger \hat{b}_1), \quad J_2 = \frac{1}{2}(-i\hat{b}_1^\dagger \hat{b}_2 + i\hat{b}_2^\dagger \hat{b}_1), \quad J_3 = \frac{1}{2}(\hat{b}_1^\dagger \hat{b}_1 - \hat{b}_2^\dagger \hat{b}_2)$$

with the commutation relations

$$[J_1, J_2] = iJ_3, \quad [J_2, J_3] = iJ_1, \quad [J_3, J_1] = iJ_2.$$

Let $|0, 0\rangle \equiv |0\rangle \otimes |0\rangle$ be the vacuum state, i.e. $\hat{b}_1|0, 0\rangle = 0|0, 0\rangle$, $\hat{b}_2|0, 0\rangle = 0|0, 0\rangle$. Let $j = 0, 1/2, 1, 3/2, 2, \ldots$ and $m = -j, -j + 1, \ldots, j - 1, j$. Then one can introduce the normalized states

$$|j, m\rangle = \frac{(\hat{b}_1^\dagger)^{j+m}(\hat{b}_2^\dagger)^{j-m}}{((j + m)!(j - m)!)^{1/2}}|0, 0\rangle$$

with $\langle j', m'|j, m\rangle = \delta_{j,j'}\delta_{m,m'}$.

(i) Let $j = 1/2$. Find the orthonormal basis $|1/2, 1/2\rangle$, $|1/2, -1/2\rangle$. Hence for $j = 1/2$ we have a two-dimensional Hilbert space. Find the matrix representation of J_1, J_2, J_3.

(ii) Let $j = 1$. Find the orthonormal basis $|1, 1\rangle$, $|1, 0\rangle$, $|1, -1\rangle$. Hence for $j = 1$ we have a three-dimensional Hilbert space. Find the matrix representation of J_1, J_2, J_3.

(iii) Let $j = 3/2$. Find the orthonormal basis $|3/2, 3/2\rangle$, $|3/2, 1/2\rangle$, $|3/2, -1/2\rangle$, $|3/2, -3/2\rangle$. Hence for $j = 3/2$ we have a four-dimensional Hilbert space. Find the matrix representation of J_1, J_2, J_3.

Solution 44. (i) We obtain the orthonormal basis

$$|1/2, 1/2\rangle = \hat{b}_1^\dagger|0, 0\rangle, \quad |1/2, -1/2\rangle = \hat{b}_2^\dagger|0, 0\rangle.$$

Since

$$J_1|1/2, 1/2\rangle = \frac{1}{2}\hat{b}_2^\dagger|0, 0\rangle = \frac{1}{2}|1/2, -1/2\rangle$$

$$J_1|1/2, -1/2\rangle = \frac{1}{2}\hat{b}_1^\dagger|0, 0\rangle = \frac{1}{2}|1/2, 1/2\rangle$$

we obtain

$$\langle 1/2, 1/2|J_1|1/2, 1/2\rangle = 0, \quad \langle 1/2, 1/2|J_1|1/2, -1/2\rangle = 1/2,$$

$$\langle 1/2, -1/2|J_1|1/2, 1/2\rangle = 1/2, \quad \langle 1/2, -1/2|J_1|1/2, -1/2\rangle = 0.$$

Hence the matrix representation is

$$J_1 \mapsto S_1^{(1/2)} = \frac{1}{2} \begin{pmatrix} 0 & 1 \\ 1 & 0 \end{pmatrix}.$$

Since

$$J_2|1/2, 1/2\rangle = \frac{i}{2}\hat{b}_2^\dagger|0,0\rangle = \frac{i}{2}|1/2, -1/2\rangle$$

$$J_2|1/2, -1/2\rangle = -\frac{i}{2}\hat{b}_1^\dagger|0,0\rangle = -\frac{i}{2}|1/2, 1/2\rangle$$

we obtain

$$\langle 1/2, 1/2|J_2|1/2, 1/2\rangle = 0, \quad \langle 1/2, 1/2|J_2|1/2, -1/2\rangle = -i/2,$$

$$\langle 1/2, -1/2|J_2|1/2, 1/2\rangle = i/2, \quad \langle 1/2, -1/2|J_2|1/2, -1/2\rangle = 0.$$

Hence the matrix representation is

$$J_2 \mapsto S_2^{(1/2)} = \frac{1}{2} \begin{pmatrix} 0 & -i \\ i & 0 \end{pmatrix}.$$

Since

$$J_3|1/2, 1/2\rangle = \frac{1}{2}\hat{b}_1^\dagger|00\rangle = \frac{1}{2}|1/2, 1/2\rangle$$

$$J_3|1/2, -1/2\rangle = -\frac{1}{2}\hat{b}_2^\dagger|00\rangle = -\frac{1}{2}|1/2, 1/2\rangle$$

we obtain

$$\langle 1/2, 1/2|J_3|1/2, 1/2\rangle = 1/2, \quad \langle 1/2, 1/2|J_1|1/2, -1/2\rangle = 0,$$

$$\langle 1/2, -1/2|J_1|1/2, 1/2\rangle = 0, \quad \langle 1/2, -1/2|J_1|1/2, -1/2\rangle = -1/2.$$

Hence the matrix representation is

$$J_3 \mapsto S_3^{(1/2)} = \frac{1}{2} \begin{pmatrix} 1 & 0 \\ 0 & -1 \end{pmatrix}.$$

(ii) Let $j = 1$. Then the orthonormal basis is three dimensional, i.e.

$$|1, 1\rangle = \frac{(\hat{b}_1^\dagger)^2}{\sqrt{2}}|0,0\rangle, \quad |1, 0\rangle = \hat{b}_1^\dagger\hat{b}_2^\dagger|0,0\rangle, \quad |1, -1\rangle = \frac{(\hat{b}_2^\dagger)^2}{\sqrt{2}}|0,0\rangle.$$

From

$$J_1|1, 1\rangle = \frac{1}{\sqrt{2}}\hat{b}_1^\dagger\hat{b}_2^\dagger|0,0\rangle = \frac{1}{\sqrt{2}}|1,0\rangle$$

$$J_1|1, 0\rangle = \frac{1}{2}((\hat{b}_1^\dagger)^2 + (\hat{b}_2^\dagger)^2)|0,0\rangle = \frac{1}{\sqrt{2}}(|1,1\rangle + |1,-1\rangle)$$

$$J_1|1, -1\rangle = \frac{1}{\sqrt{2}}\hat{b}_1^\dagger\hat{b}_2^\dagger|1,0\rangle$$

we obtain the matrix representation

$$J_1 \mapsto S_1^{(1)} = \frac{1}{\sqrt{2}} \begin{pmatrix} 0 & 1 & 0 \\ 1 & 0 & 1 \\ 0 & 1 & 0 \end{pmatrix}.$$

Since

$$J_2|1,1\rangle = \frac{i}{\sqrt{2}} \hat{b}_1^\dagger \hat{b}_2^\dagger |00\rangle = \frac{i}{\sqrt{2}}|1,0\rangle$$

$$J_2|1,0\rangle = \frac{i}{2}(-(\hat{b}_1^\dagger)^2 + (\hat{b}_2^\dagger)^2)|0,0\rangle = \frac{i}{\sqrt{2}}(-|1,1\rangle + |1,-1\rangle)$$

$$J_2|1,-1\rangle = -\frac{i}{\sqrt{2}} \hat{b}_1^\dagger \hat{b}_2^\dagger |0,0\rangle = -\frac{i}{\sqrt{2}}|1,0\rangle$$

we obtain the matrix representation

$$J_2 \mapsto S_2^{(1)} = \frac{1}{\sqrt{2}} \begin{pmatrix} 0 & -i & 0 \\ i & 0 & -i \\ 0 & i & 0 \end{pmatrix}.$$

Since

$$J_3|1,1\rangle = \frac{1}{\sqrt{2}}(\hat{b}_1^\dagger)^2|0,0\rangle = \frac{1}{2}|1,1\rangle$$

$$J_3|1,0\rangle = 0|1,0\rangle$$

$$J_3|1,-1\rangle = -\frac{1}{\sqrt{2}}(\hat{b}_2^\dagger)^2|0,0\rangle = -|1,-1\rangle$$

we obtain the matrix representation is

$$J_3 \mapsto S_3^{(1)} = \begin{pmatrix} 1 & 0 & 0 \\ 0 & 0 & 0 \\ 0 & 0 & -1 \end{pmatrix}.$$

(iii) Let $j = 3/2$. Then the orthonormal basis is four dimensional, i.e.

$$|3/2, 3/2\rangle = \frac{(\hat{b}_1^\dagger)^3}{\sqrt{3!}}|0,0\rangle, \quad |3/2, 1/2\rangle = \frac{(\hat{b}_1^\dagger)^2 \hat{b}_2^\dagger}{\sqrt{2!}}|0,0\rangle,$$

$$|3/2, -1/2\rangle = \frac{\hat{b}_1^\dagger (\hat{b}_2^\dagger)^2}{\sqrt{2!}}|0,0\rangle, \quad |3/2, -3/2\rangle = \frac{(\hat{b}_2^\dagger)^3}{\sqrt{3!}}|0,0\rangle.$$

Since

$$J_3|3/2, 3/2\rangle = \frac{1}{2} \frac{1}{\sqrt{3!}} \hat{b}_1^\dagger \hat{b}_1 \hat{b}_1^\dagger \hat{b}_1^\dagger \hat{b}_1^\dagger |0,0\rangle = \frac{1}{2} \frac{3}{\sqrt{3!}} \hat{b}_1^\dagger \hat{b}_1^\dagger \hat{b}_1^\dagger |0,0\rangle = \frac{3}{2}|3/2, 3/2\rangle.$$

For spin $s = \frac{3}{2}$ we obtain the matrix representations

$$S_1^{(3/2)} = \frac{1}{2}\begin{pmatrix} 0 & \sqrt{3} & 0 & 0 \\ \sqrt{3} & 0 & 2 & 0 \\ 0 & 2 & 0 & \sqrt{3} \\ 0 & 0 & \sqrt{3} & 0 \end{pmatrix}, \quad S_2^{(3/2)} = \frac{1}{2}\begin{pmatrix} 0 & -i\sqrt{3} & 0 & 0 \\ i\sqrt{3} & 0 & -2i & 0 \\ 0 & 2i & 0 & -i\sqrt{3} \\ 0 & 0 & i\sqrt{3} & 0 \end{pmatrix},$$

$$S_3^{(3/2)} = \begin{pmatrix} 3/2 & 0 & 0 & 0 \\ 0 & 1/2 & 0 & 0 \\ 0 & 0 & -1/2 & 0 \\ 0 & 0 & 0 & -3/2 \end{pmatrix}.$$

Problem 45. Let $\alpha \in \mathbb{R}$ and $|n\rangle$ $(n = 0, 1, 2, \dots)$ be the number states. Find $\exp(\alpha(\hat{b}^\dagger \otimes \hat{b} - \hat{b} \otimes \hat{b}^\dagger))(|1\rangle \otimes |0\rangle)$.

Solution 45. We have

$$(\hat{b}^\dagger \otimes \hat{b} - \hat{b} \otimes \hat{b}^\dagger)(|1\rangle \otimes |0\rangle) = -|0\rangle \otimes |1\rangle$$
$$(\hat{b}^\dagger \otimes b - b \otimes b^\dagger)(-|0\rangle \otimes |1\rangle) = -|1\rangle \otimes |0\rangle$$
$$(b^\dagger \otimes \hat{b} - \hat{b} \otimes \hat{b}^\dagger)(-|1\rangle \otimes |0\rangle) = |0\rangle \otimes |1\rangle$$
$$(\hat{b}^\dagger \otimes \hat{b} - \hat{b} \otimes \hat{b}^\dagger)(|0\rangle \otimes |1\rangle) = |1\rangle \otimes |0\rangle.$$

Therefore we find

$$\exp(\alpha(\hat{b}^\dagger \otimes \hat{b} - \hat{b} \otimes \hat{b}^\dagger)) = |1\rangle \otimes |0\rangle \left(1 - \frac{\alpha^2}{2!} + \frac{\alpha^4}{4!} + \cdots\right)$$
$$+ |0\rangle \otimes |1\rangle \left(-\alpha + \frac{\alpha^3}{3!} - \frac{\alpha^5}{5!} + \cdots\right)$$
$$= \cos(\alpha)(|1\rangle \otimes |0\rangle) - \sin(\alpha)(|0\rangle \otimes |1\rangle).$$

If $\alpha = \pi/2$, then

$$\exp(\alpha(\hat{b}^\dagger \otimes \hat{b} - \hat{b} \otimes \hat{b}^\dagger))(|1\rangle \otimes |0\rangle) = -|0\rangle \otimes |1\rangle.$$

If $\alpha = 3\pi/2$, then

$$\exp(\alpha(\hat{b}^\dagger \otimes \hat{b} - \hat{b} \otimes \hat{b}^\dagger))(|1\rangle \otimes |0\rangle) = |0\rangle \otimes |1\rangle.$$

Problem 46. Let $\hat{b}_1^\dagger = \hat{b}^\dagger \otimes I$, $\hat{b}_2^\dagger = I \otimes \hat{b}^\dagger$, $\hat{b}_1 = \hat{b} \otimes I$, $\hat{b}_2 = I \otimes \hat{b}$ be Bose creation and annihilation operators and I be the identity operator. The three operators

$$\hat{T}_+ = \hat{b}_1^\dagger \hat{b}_2^\dagger, \quad \hat{T}_- = \hat{b}_2 \hat{b}_1, \quad \hat{T}_0 = \frac{1}{2}(\hat{b}_1^\dagger \hat{b}_1 + \hat{b}_2^\dagger \hat{b}_2 + I \otimes I)$$

form a basis of the Lie algebra $su(1,1)$.

(i) Find the commutators $[\hat{T}_+, \hat{T}_-]$, $[\hat{T}_0, \hat{T}_+]$, $[\hat{T}_0, \hat{T}_-]$.

(ii) Find the *Casimir operator* $\hat{T}^2$ given by $\hat{T}^2 = \hat{T}_0^2 - \frac{1}{2}(\hat{T}_+\hat{T}_- + \hat{T}_-\hat{T}_+)$.

(iii) Find the eigenstates of $\hat{T}_0$ and $\hat{T}$ given by $\hat{T} = \frac{1}{2}(-\hat{b}_1^\dagger\hat{b}_1 + \hat{b}_2^\dagger\hat{b}_2 + I \otimes I)$ with $|0,0\rangle \equiv |0\rangle \otimes |0\rangle$ be the *vacuum state* with $\hat{b}_1|0,0\rangle = 0|0,0\rangle$, $\hat{b}_2|0,0\rangle = 0|0,0\rangle$.

Solution 46. (i) Using $[\hat{b}, \hat{b}^\dagger] = I$ and $(\hat{b}^\dagger \otimes I)(I \otimes \hat{b}) = \hat{b}^\dagger \otimes \hat{b}$ we have

$$[\hat{T}_+, \hat{T}_-] = -2\hat{T}_0, \quad [\hat{T}_0, \hat{T}_+] = \hat{T}_+, \quad [\hat{T}_0, \hat{T}_-] = -\hat{T}_-.$$

(ii) Applying $[\hat{b}_j, \hat{b}_k^\dagger] = \delta_{j,k}I \otimes I$ we obtain for the Casimir operator

$$\hat{T}^2 = \hat{T}(\hat{T} - I \otimes I)$$

where

$$\hat{T} = \frac{1}{2}(-\hat{b}_1^\dagger\hat{b}_1 + \hat{b}_2^\dagger\hat{b}_2 + I \otimes I)$$

with $[\hat{T}, \hat{T}_0] = 0$, $[\hat{T}, \hat{T}_+] = 0$, $[\hat{T}, \hat{T}_-] = 0$.

(iii) With $j_0 - j = 0, 1, 2, \dots$ and $j_0 + j_1 - 1 = 0, 1, 2, \dots$ the normalized eigenstates $|j, j_0\rangle$ are given by

$$|j, j_0\rangle = \frac{1}{\sqrt{(j_0 - j)!(j_0 + j - 1)!}}(\hat{b}_1^\dagger)^{j_0-j}(\hat{b}_2^\dagger)^{j_0+j-1}|0,0\rangle.$$

Since $j_0 - j = 0, 1, 2, \dots$ and $j_0 + j - 1 = 0, 1, 2, \dots$ the eigenstates $|j, j_0\rangle$ can be treated separately for the following two cases

$$(a) \ j = 1/2, 1, 3/2, 2, \dots, \infty, \quad j_0 = j, j+1, j+2, j+3, \dots, \infty$$

for a given j and

$$(b) \ j = 0, -1/2, -1, -3/2, \dots, -\infty, \quad j_0 = -j+1, -j+2, -j+3, \dots, \infty$$

for a given j.

Problem 47. Consider the two-dimensional *Laplace operator*

$$\Delta := \left(\frac{\partial^2}{\partial x_1^2} + \frac{\partial^2}{\partial x_2^2}\right) \equiv \frac{1}{r}\frac{\partial}{\partial r}\left(r\frac{\partial}{\partial r}\right) + \frac{1}{r^2}\frac{\partial^2}{\partial \phi^2}$$

with $x_1(r, \phi) = r\cos(\phi)$ and $x_2(r, \phi) = r\sin(\phi)$ and the Hilbert space $L_2(\mathbb{R}^2)$ with $\mathbb{C} \cong \mathbb{R}^2$. The eigenvalue equation of the two-dimensional *hydrogen atom* [10] is given by

$$\left(-\frac{\hbar^2}{2m}\Delta - \frac{k}{r}\right)u(r, \phi) = Eu(r, \phi).$$

With $a_0 = \hbar^2/(km)$ (*Bohr radius*), $\lambda = 8/a_0$, $\eta^4 = -8mE/\hbar^2$ we can write

$$(4r\Delta + \lambda)u(r,\phi) = \eta^4 r u(r,\phi).$$

Introducing the *complex coordinates* (ξ, ξ^*) via

$$x_1 + ix_2 = 2\xi^2, \quad x_1 - ix_2 = 2(\xi^*)^2$$

with $r = 2\xi\xi^*$, $\phi = \arctan(x_2/x_1) = $ phase of $(\xi) - $ phase of (ξ^*). Note that

$$\xi = \sqrt{r/2}\exp(i\phi/2), \quad \xi^* = \sqrt{r/2}\exp(-i\phi/2).$$

Then the eigenvalue equation can be written as

$$-\frac{1}{\eta^2}\frac{\partial^2 \tilde{u}(\xi,\xi^*)}{\partial\xi^*\partial\xi} + \eta^2\xi\xi^*\tilde{u}(\xi,\xi^*) = \frac{\lambda}{2\eta^2}\tilde{u}(\xi,\xi^*).$$

We define the Bose operators

$$\hat{b}_+ := \frac{1}{\sqrt{2}\eta}\left(\frac{\partial}{\partial\xi} + \eta^2\xi^*\right), \quad \hat{b}_+^\dagger := \frac{1}{\sqrt{2}\eta}\left(-\frac{\partial}{\partial\xi^*} + \eta^2\xi\right)$$

$$\hat{b}_- := \frac{1}{\sqrt{2}\eta}\left(\frac{\partial}{\partial\xi^*} + \eta^2\xi\right), \quad \hat{b}_-^\dagger := \frac{1}{\sqrt{2}\eta}\left(-\frac{\partial}{\partial\xi} + \eta^2\xi^*\right).$$

Find the commutation relation and the eigenvalues.

Solution 47. We obtain

$$[\hat{b}_+^\dagger, \hat{b}_+^\dagger] = [\hat{b}_-^\dagger, \hat{b}_-^\dagger] = [\hat{b}_+, \hat{b}_+] = [\hat{b}_-, \hat{b}_-] = \hat{0}$$

and

$$[\hat{b}_+, \hat{b}_+^\dagger] = I, \quad [\hat{b}_-, \hat{b}_-^\dagger] = I, \quad [\hat{b}_-, \hat{b}_+^\dagger] = \hat{0}, \quad [\hat{b}_+, \hat{b}_-^\dagger] = \hat{0}.$$

Hence we have Bose creation and annihilation operators. The eigenvalue equation takes the form

$$(\hat{b}_+^\dagger\hat{b}_+ + \hat{b}_-^\dagger\hat{b}_- + 1)\tilde{u} = \frac{\lambda}{2\eta^2}\tilde{u}.$$

Now $\hat{b}_+^\dagger\hat{b}_+$, $\hat{b}_-^\dagger\hat{b}_-$ represent number operators with the eigenvalues $n_+ = 0, 1, 2, \ldots$ and $n_- = 0, 1, 2, \ldots$, respectively. It follows that

$$n_+ + n_- + 1 = \frac{\lambda}{2\eta^2}$$

with η^2 containing the square root of the energy E. We also have

$$\hat{b}_+^\dagger\hat{b}_+ - \hat{b}_-^\dagger\hat{b}_- = \xi\frac{\partial}{\partial\xi} - \xi^*\frac{\partial}{\partial\xi^*}.$$

Problem 48. Let $\hat{b}^\dagger$, $\hat{b}$ be Bose creation and annihilation operators and I the identity operator. Let

$$\hat{A} = \hat{b}^\dagger \otimes \hat{b}, \qquad \hat{B} = \hat{b} \otimes \hat{b}^\dagger.$$

Then

$$[\hat{A}, \hat{B}] = \hat{b}^\dagger \hat{b} \otimes I - I \otimes \hat{b}^\dagger \hat{b} \equiv \hat{C}.$$

Now $[\hat{A}, \hat{B}] = \hat{C}$, $[\hat{A}, \hat{C}] = -2\hat{A}$, $[\hat{B}, \hat{C}] = 2\hat{B}$. Hence $\hat{A}$, $\hat{B}$, $\hat{C}$ form a basis of a semi-simple *Lie algebra*. Find

$$f_{AB}(\epsilon) = e^{\epsilon A} B e^{-\epsilon A}, \quad f_{BA}(\epsilon) = e^{\epsilon B} A e^{-\epsilon B},$$

$$f_{BC}(\epsilon) = e^{\epsilon B} C e^{-\epsilon B}, \quad f_{CB}(\epsilon) = e^{\epsilon C} B e^{-\epsilon C},$$

$$f_{CA}(\epsilon) = e^{\epsilon C} A e^{-\epsilon C}, \quad f_{AC}(\epsilon) = e^{\epsilon A} C e^{-\epsilon A}$$

using *parameter differentiation* and solving the initial value problem of the system of ordinary differential equations.

Solution 48. We obtain

$$\frac{df_{AB}}{d\epsilon} = e^{\epsilon A}[A, B]e^{-\epsilon A} = e^{\epsilon A} C e^{-\epsilon A} = f_{AC}$$

$$\frac{df_{BA}}{d\epsilon} = e^{\epsilon B}[B, A]e^{-\epsilon B} = -e^{\epsilon B} C e^{-\epsilon B} = -f_{BC}$$

$$\frac{df_{BC}}{d\epsilon} = e^{\epsilon B}[B, C]e^{-\epsilon B} = 2B$$

$$\frac{df_{CB}}{d\epsilon} = e^{\epsilon C}[C, B]e^{-\epsilon C} = -2e^{\epsilon C} B e^{-\epsilon C} = -2f_{CB}$$

$$\frac{df_{CA}}{d\epsilon} = e^{\epsilon C}[C, A]e^{-\epsilon C} = 2e^{\epsilon C} A e^{-\epsilon C} = 2f_{CA}$$

$$\frac{df_{AC}}{d\epsilon} = e^{\epsilon A}[A, C]e^{-\epsilon A} = -2A$$

integrating the system of differential equations with the initial conditions

$$f_{AB}(0) = B, \qquad f_{BA}(0) = A, \qquad f_{BC}(0) = C,$$

$$f_{CB}(0) = B, \qquad f_{CA}(0) = A, \qquad f_{AC}(0) = C$$

we obtain

$$f_{AB}(\epsilon) = -A\epsilon^2 + C\epsilon + B, \quad f_{BA}(\epsilon) = B\epsilon^2 + C\epsilon + A, \quad f_{BC}(\epsilon) = 2B\epsilon + C,$$

$$f_{CB}(\epsilon) = e^{-2\epsilon}B, \quad f_{CA}(\epsilon) = e^{2\epsilon}A, \quad f_{AC}(\epsilon) = -2A\epsilon + C.$$

Problem 49. The operators $\hat{b}^\dagger$, $\hat{b}$, I_B, $\hat{b}^\dagger\hat{b}$ form a Lie algebra. Note that the operators $b^\dagger$, b, $b^\dagger b$ are unbounded. For the matrix representation of $b^\dagger b$ we have $\hat{b}^\dagger\hat{b} = \text{diag}(0, 1, 2, \dots)$. If σ_j is one of the Pauli spin matrices. Find the commutator $[\hat{b} \otimes \sigma_j, \hat{b}^\dagger \otimes \sigma_j]$.

Solution 49. With $\sigma_j^2 = I_2$ for $j = 1, 2, 3$ we obtain

$$[\hat{b} \otimes \sigma_j, \hat{b}^\dagger \otimes \sigma_j] = I_B \otimes I_2.$$

Problem 50. Consider the *momentum operator* $\hat{P}_j = -i\hbar\partial/\partial q_j$ and the *position operator* $\hat{Q}_k$ with $j, k = 1, 2, 3$. Then we have the commutation relation

$$[\hat{Q}_k, \hat{P}_j] = i\hbar\delta_{j,k}I$$

where I is the identity operator. Let S be an $n \times n$ matrix over $\mathbb{C}$ with $S^2 = I_n$. Find the commutator $[\hat{Q}_k \otimes S, \hat{P}_j \otimes S]$.

Solution 50. We obtain

$$[\hat{Q}_k \otimes S, \hat{P}_j \otimes S] = Q_k P_j \otimes I_n - P_j Q_k \otimes I_n = [Q_k, P_j] \otimes I_n = i\hbar\delta_{j,k}I \otimes I_n.$$

Problem 51. (i) Let $\hat{b}^\dagger$, $\hat{b}$ be Bose creation and annihilation operators, respectively. Consider the operators

$$J_1 = \frac{i}{4}(\hat{b}^\dagger\hat{b}^\dagger - \hat{b}\hat{b}), \quad J_2 = \frac{1}{4}(\hat{b}^\dagger\hat{b}^\dagger + \hat{b}\hat{b}), \quad J_3 = \frac{1}{4}(\hat{b}^\dagger\hat{b} + \hat{b}\hat{b}^\dagger).$$

Find the commutators $[J_1, J_2]$, $[J_2, J_3]$, $[J_3, J_1]$.
(ii) Let $\hat{c}^\dagger$, $\hat{c}$ be Fermi creation and annihilation operators, respectively. Consider the operators

$$K_1 = \frac{1}{2}(\hat{c}^\dagger + \hat{c}), \quad K_2 = \frac{1}{2i}(\hat{c}^\dagger - \hat{c}), \quad K_3 = \frac{1}{2}(\hat{c}^\dagger\hat{c} - \hat{c}\hat{c}^\dagger).$$

Find the commutators $[K_1, K_2]$, $[K_2, K_3]$, $[K_3, K_1]$. Note that $[\hat{c}, \hat{c}^\dagger]_+ = I$.

Solution 51. (i) We obtain $[J_1, J_2] = -iJ_3$, $[J_2, J_3] = iJ_1$, $[J_3, J_1] = iJ_2$.

(ii) We obtain $[K_1, K_2] = iK_3$, $[K_2, K_3] = iK_1$, $[K_3, K_1] = iK_2$.

Problem 52. (i) Let $z \in \mathbb{C}$ and $\hat{c}^\dagger$ be a Fermi creation operator. Find $e^{z\hat{c}^\dagger}|0\rangle$ with $|0\rangle$ be the vacuum state, i.e. $\hat{c}|0\rangle = 0|0\rangle$ and $\langle 0|0\rangle = 1$.
(ii) Let $z \in \mathbb{C}$ and $c_\uparrow^\dagger$, $c_\downarrow^\dagger$ be Fermi operators with spin down and spin up, respectively. Find $e^{z\hat{c}_\uparrow^\dagger}|0\rangle$, $e^{z\hat{c}_\downarrow^\dagger}|0\rangle$.

Solution 52. (i) Since $(\hat{c}^\dagger)^2 = 0$ we obtain

$$e^{z\hat{c}^\dagger} = I + z\hat{c}^\dagger \quad \Rightarrow \quad e^{z\hat{c}^\dagger}|0\rangle = |0\rangle + z\hat{c}^\dagger|0\rangle.$$

Normalizing this state with $\langle 0|\hat{c}^\dagger = \langle 0|0$ and

$$(\langle 0| + \langle 0|\hat{c}\bar{z})(|0\rangle + z\hat{c}^\dagger|0\rangle) = 1 + z\bar{z}$$

provides the normalized state

$$\frac{1}{\sqrt{1+z\bar{z}}}(|0\rangle + z\hat{c}^\dagger|0\rangle).$$

(ii) We have

$$e^{z\hat{c}_\uparrow^\dagger}|0\rangle = |0\rangle + z\hat{c}_\uparrow^\dagger|0\rangle, \quad e^{z\hat{c}_\downarrow^\dagger}|0\rangle = |0\rangle + z\hat{c}_\downarrow^\dagger|0\rangle.$$

Find

$$e^{z\hat{c}_\uparrow^\dagger\hat{c}_\downarrow^\dagger}|0\rangle, \quad e^{z(\hat{c}_\uparrow^\dagger+\hat{c}_\downarrow^\dagger)}|0\rangle.$$

Problem 53. Consider the Fermi creation and annihilation operators $\hat{c}_j^\dagger$, $\hat{c}_j$ $(j = 1, 2, 3, 4)$, respectively and the states

$$\hat{c}_1^\dagger\hat{c}_2^\dagger|0\rangle, \quad \hat{c}_1^\dagger\hat{c}_3^\dagger|0\rangle, \quad \hat{c}_1^\dagger\hat{c}_4^\dagger|0\rangle, \quad \hat{c}_2^\dagger\hat{c}_3^\dagger|0\rangle, \quad \hat{c}_2^\dagger\hat{c}_4^\dagger|0\rangle, \quad \hat{c}_3^\dagger\hat{c}_4^\dagger|0\rangle$$

which is a basis in a six-dimensional Hilbert space (standard basis for two Fermions). The dual basis is given by

$$\langle 0|\hat{c}_2\hat{c}_1, \quad \langle 0|\hat{c}_3\hat{c}_1, \quad \langle 0|\hat{c}_4\hat{c}_1, \quad \langle 0|\hat{c}_3\hat{c}_2, \quad \langle 0|\hat{c}_4\hat{c}_2, \quad \langle 0|\hat{c}_4\hat{c}_3.$$

Consider the six normalized states

$$|\psi_1\rangle = \frac{1}{\sqrt{2}}(\hat{c}_1^\dagger\hat{c}_2^\dagger + \hat{c}_3^\dagger\hat{c}_4^\dagger)|0\rangle, \qquad |\psi_2\rangle = \frac{1}{\sqrt{2}}(\hat{c}_1^\dagger\hat{c}_2^\dagger - \hat{c}_3^\dagger\hat{c}_4^\dagger|0\rangle$$

$$|\psi_3\rangle = \frac{1}{\sqrt{2}}(\hat{c}_1^\dagger\hat{c}_3^\dagger + \hat{c}_2^\dagger\hat{c}_4^\dagger)|0\rangle, \qquad |\psi_4\rangle = \frac{1}{\sqrt{2}}(\hat{c}_1^\dagger\hat{c}_3^\dagger - \hat{c}_2^\dagger\hat{c}_4^\dagger)|0\rangle$$

$$|\psi_5\rangle = \frac{1}{\sqrt{2}}(\hat{c}_1^\dagger\hat{c}_4^\dagger + \hat{c}_2^\dagger\hat{c}_3^\dagger)|0\rangle, \qquad |\psi_6\rangle = \frac{1}{\sqrt{2}}(\hat{c}_1^\dagger\hat{c}_4^\dagger - \hat{c}_2^\dagger\hat{c}_3^\dagger)|0\rangle.$$

Find the matrix representation in the Hilbert space $\mathbb{C}^6$ of these normalized states with the standard basis. Can these six vectors in the Hilbert space be written as a Kronecker product of a vector in $\mathbb{C}^2$ and a vector in $\mathbb{C}^3$? Can these six vectors in the Hilbert space be written as a Kronecker product of a vector in $\mathbb{C}^3$ and a vector in $\mathbb{C}^2$? Let $\hat{N}$ be the *number operator*

$$\hat{N} = \sum_{j=1}^{4} \hat{c}_j^\dagger \hat{c}_j.$$

Find $\hat{N}|\psi_j\rangle$ for $j = 1, 2, 3, 4$.

Solution 53. For the state $|\psi\rangle$ we have

$$\langle 0|\hat{c}_2\hat{c}_1|\psi_1\rangle = \frac{1}{\sqrt{2}}, \quad \langle 0|\hat{c}_3\hat{c}_1|\psi_1\rangle = 0, \quad \langle 0|\hat{c}_4\hat{c}_1|\psi_1\rangle = 0,$$

$$\langle 0|\hat{c}_4\hat{c}_1|\psi_1\rangle = 0, \quad \langle 0|\hat{c}_4\hat{c}_2|\psi_1\rangle = 0, \quad \langle 0|\hat{c}_4\hat{c}_3|\psi_1\rangle = 0.$$

Analogously we calculate the representation for the other states and find

$$|v_1\rangle = \frac{1}{\sqrt{2}}\begin{pmatrix} 1 \\ 0 \\ 0 \\ 0 \\ 0 \\ 1 \end{pmatrix}, \quad |v_2\rangle = \frac{1}{\sqrt{2}}\begin{pmatrix} 1 \\ 0 \\ 0 \\ 0 \\ 0 \\ -1 \end{pmatrix}, \quad |v_3\rangle = \frac{1}{\sqrt{2}}\begin{pmatrix} 0 \\ 1 \\ 0 \\ 0 \\ 1 \\ 0 \end{pmatrix}$$

$$|v_4\rangle = \frac{1}{\sqrt{2}}\begin{pmatrix} 0 \\ 1 \\ 0 \\ 0 \\ -1 \\ 0 \end{pmatrix}, \quad |v_5\rangle = \frac{1}{\sqrt{2}}\begin{pmatrix} 0 \\ 0 \\ 1 \\ 1 \\ 0 \\ 0 \end{pmatrix}, \quad |v_6\rangle = \frac{1}{\sqrt{2}}\begin{pmatrix} 0 \\ 0 \\ 1 \\ -1 \\ 0 \\ 0 \end{pmatrix}.$$

Obviously these six vectors form an orthonormal basis in $\mathbb{C}^6$. The vectors $|v_1\rangle$, $|v_2\rangle$ can neither be written as Kronecker product of vectors in $\mathbb{C}^2$ and $\mathbb{C}^3$ or a Kronecker product of vectors in $\mathbb{C}^3$ and $\mathbb{C}^2$. For the vectors $|v_3\rangle$, $|v_4\rangle$ we have

$$|v_3\rangle = \frac{1}{\sqrt{2}}\begin{pmatrix} 1 \\ 1 \end{pmatrix} \otimes \begin{pmatrix} 0 \\ 1 \\ 0 \end{pmatrix}, \quad |v_4\rangle = \frac{1}{\sqrt{2}}\begin{pmatrix} 1 \\ -1 \end{pmatrix} \otimes \begin{pmatrix} 0 \\ 1 \\ 0 \end{pmatrix}.$$

However $|v_3\rangle$, $|v_4\rangle$ cannot be written as a Kronecker product of vectors in $\mathbb{C}^3$ and $\mathbb{C}^2$. For the vectors $|v_5\rangle$, $|v_6\rangle$ we have

$$|v_5\rangle = \begin{pmatrix} 0 \\ 1 \\ 0 \end{pmatrix} \otimes \frac{1}{\sqrt{2}} \begin{pmatrix} 1 \\ 1 \end{pmatrix}, \quad |v_6\rangle = \begin{pmatrix} 0 \\ 1 \\ 0 \end{pmatrix} \otimes \frac{1}{\sqrt{2}} \begin{pmatrix} 1 \\ -1 \end{pmatrix}.$$

However $|v_5\rangle$ and $|v_6\rangle$ cannot be written as a Kronecker products of a vector in $\mathbb{C}^2$ and $\mathbb{C}^3$. With the number operator we obtain the eigenvalue equation

$$\hat{N}|\psi_j\rangle = 2|\psi_j\rangle.$$

Problem 54. Let $\hat{c}_1^\dagger$, $\hat{c}_2^\dagger$, $\hat{c}_1$, $\hat{c}_2$ be Fermi creation and annihilation operators. The states are given by

$$|0\rangle, \quad \hat{c}_1^\dagger|0\rangle, \quad \hat{c}_2^\dagger|0\rangle, \quad \hat{c}_1^\dagger\hat{c}_2^\dagger|0\rangle.$$

So we have a four-dimensional Hilbert space. Let $\hat{K}_1 = \hat{c}_1^\dagger\hat{c}_2$, $\hat{K}_2 = \hat{c}_2^\dagger\hat{c}_1$.
(i) Find $\hat{K}_1^2$, $\hat{K}_2^2$.
(ii) Find the commutator $[\hat{K}_1, \hat{K}_2]$ and the anti-commutator $[\hat{K}_1, \hat{K}_2]_+$.
(iii) Find $\hat{K}_1 - \hat{K}_2$, $(\hat{K}_1 - \hat{K}_2)^2$, $(\hat{K}_1 - \hat{K}_2)^3$, $(\hat{K}_1 - \hat{K}_2)^4$.
(iv) Let $\epsilon \in \mathbb{R}$. Find $\exp(\epsilon(\hat{K}_1 - \hat{K}_2))$.
(v) We define

$$\hat{K}_3 := [K_1, K_2] = \hat{c}_1^\dagger\hat{c}_1 - \hat{c}_2^\dagger\hat{c}_2.$$

The operators $\hat{K}_1$, $\hat{K}_2$, $\hat{K}_3$ form a basis of a Lie algebra. Consider the equation (*disentanglement*)

$$e^{\epsilon(\hat{K}_1 - \hat{K}_2)} = e^{f_1(\epsilon)\hat{K}_1} + e^{f_3(\epsilon)\hat{K}_3} + e^{f_2(\epsilon)\hat{K}_2}.$$

Find $f_1(\epsilon)$, $f_2(\epsilon)$, $f_3(\epsilon)$ by *parameter differentiation* and solving the initial value problem of the resulting system of differential equation for f_1, f_2, f_3.

Solution 54. (i) With $\hat{c}_j^\dagger\hat{c}_j^\dagger = 0$ $\hat{c}_j\hat{c}_j = 0$ we obtain $\hat{K}_1^2 = 0$, $\hat{K}_2^2 = 0$.
(ii) The commutator is given by

$$[\hat{c}_1^\dagger\hat{c}_2, \hat{c}_2^\dagger\hat{c}_1] = \hat{c}_1^\dagger\hat{c}_1 - \hat{c}_2^\dagger\hat{c}_2.$$

We set $\hat{K}_3 = \hat{c}_1^\dagger\hat{c}_1 - \hat{c}_2^\dagger\hat{c}_2$. The operators $\hat{K}_1$, $\hat{K}_2$, $\hat{K}_3$ form a basis of a Lie algebras. The anti-commutator is given by

$$[\hat{c}_1^\dagger\hat{c}_2, \hat{c}_2^\dagger\hat{c}_1]_+ = \hat{c}_1^\dagger\hat{c}_1 + \hat{c}_2^\dagger\hat{c}_2 - 2\hat{c}_1^\dagger\hat{c}_1\hat{c}_2^\dagger\hat{c}_2.$$

(iii) We have

$$\hat{K}_1 - \hat{K}_2 = \hat{c}_1^\dagger\hat{c}_2 - \hat{c}_2^\dagger\hat{c}_1, \quad (\hat{K}_1 - \hat{K}_2)^2 = -[\hat{K}_1, \hat{K}_2],$$

$$(\hat{K}_1 - \hat{K}_2)^3 = -(\hat{K}_1 - \hat{K}_2), \quad (\hat{K}_1 - \hat{K}_2)^4 = [\hat{K}_1, \hat{K}_2]_+.$$

Utilizing the results from (iii) and (ii) we obtain

$$\exp(\epsilon(\hat{K}_1 - \hat{K}_2)) = I + (\hat{K}_2 - \hat{K}_1)\left(\epsilon - \frac{\epsilon^3}{3!} + \frac{\epsilon^5}{5!} - \cdots\right)$$
$$+ [\hat{K}_1, \hat{K}_2]_+\left(-\frac{\epsilon^2}{2!} + \frac{\epsilon^4}{4!} - \frac{\epsilon^6}{6!} + \cdots\right)$$
$$= I + (\hat{K}_1 - \hat{K}_2)\sin(\epsilon) + [\hat{K}_1, \hat{K}_2]_+(\cos(\epsilon) - 1).$$

(v) With

$$\frac{d}{d\epsilon}e^{\epsilon(\hat{K}_1 - \hat{K}_2)}(\hat{K}_1 - \hat{K}_2)$$

one finds

$$e^{\epsilon(\hat{K}_1 - \hat{K}_2)}(\hat{K}_1 - \hat{K}_2) = e^{f_1\hat{K}_1}\frac{df_1}{d\epsilon}\hat{K}_1 e^{f_3\hat{K}_3}e^{f_2\hat{K}_2} + e^{f_1\hat{K}_1}e^{f_3\hat{K}_3}\frac{df_3}{d\epsilon}\hat{K}_3 e^{f_2\hat{K}_2}$$
$$+ e^{f_1\hat{K}_1}e^{f_3\hat{K}_3}e^{f_2\hat{K}_2}\frac{df_2}{d\epsilon}\hat{K}_2$$

From

$$e^{-\epsilon(\hat{K}_1 - \hat{K}_2)} = e^{-f_2\hat{K}_2}e^{-f_3\hat{K}_3}e^{-f_1\hat{K}_1}$$

and multiplication it follows that

$$\hat{K}_1 - \hat{K}_2 = \frac{df_1}{d\epsilon}\hat{K}_1 + \frac{df_3}{d\epsilon}e^{f_1K_1}K_3 e^{-f_1K_1} + \frac{df_2}{d\epsilon}e^{f_1K_1}e^{f_3K_3}K_2 e^{-f_3K_3}e^{-f_1\hat{K}_1}$$

$$\hat{K}_1 - \hat{K}_2 = \frac{df_1}{d\epsilon}\hat{K}_1 + \frac{df_3}{d\epsilon}(\hat{K}_3 - 2f_1K_1) + \frac{df_2}{d\epsilon}e^{f_1\hat{K}_1}\hat{K}_2 e^{-2f_3}e^{-f_1\hat{K}_1}$$

$$\hat{K}_1 - \hat{K}_2 = \frac{df_1}{d\epsilon} + \frac{df_3}{d\epsilon}(\hat{K}_3 - 2f_1\hat{K}_1) + \frac{df_2}{d\epsilon}e^{-2f_3}e^{f_1\hat{K}_1}\hat{K}_2 e^{-f_1\hat{K}_1}$$

$$\hat{K}_1 - \hat{K}_2 = \frac{df_1}{d\epsilon}K_1 + \frac{df_3}{d\epsilon}(K_3 - 2f_1K_1) + \frac{df_2}{d\epsilon}e^{-2f_3}(K_2 + f_1K_3 - f_1^2K_1).$$

Comparing coefficients with respect to the basis $\hat{K}_1$, $\hat{K}_2$, $\hat{K}_3$ of the Lie algebra

$$0 = K_1\left(1 - \frac{df_1}{d\epsilon} + 2f_1\frac{df_3}{d\epsilon} + f_1^2\frac{df_2}{d\epsilon}e^{-2f_3}\right)$$
$$+ K_2\left(-1 - \frac{df_2}{d\epsilon}e^{-2f_3}\right) + K_3\left(-\frac{df_3}{d\epsilon} - f_1\frac{df_2}{d\epsilon}e^{-2f_3}\right)$$

we obtain the autonomous system of ordinary differential equations (ordering $\hat{K}_2$, $\hat{K}_3$, $\hat{K}_1$)

$$\frac{df_2}{d\epsilon} = -e^{2f_3}, \quad \frac{df_3}{d\epsilon} = -f_1\frac{df_2}{d\epsilon}e^{-2f_3} = f_1$$

$$\frac{df_1}{d\epsilon} = 1 + 2f_1\frac{df_3}{d\epsilon} + f_1^2\frac{df_2}{d\epsilon}e^{-2f_3} = 1 + f_1^2.$$

Hence we have the system

$$\frac{df_1}{d\epsilon} = 1 + f_1^2, \quad \frac{df_2}{d\epsilon} = -e^{2f_3}, \quad \frac{df_3}{d\epsilon} = f_1$$

with the initial values $f_1(0) = f_2(0) = f_3(0) = 0$. We first integrate the differential equation for f_1 then for f_3 and finally for f_2. We obtain

$$f_1(\epsilon) = \tan(\epsilon), \quad f_3(\epsilon) = -\ln(\cos(\epsilon)), \quad f_2(\epsilon) = -\tan(\epsilon).$$

Note that $d(\tan(\epsilon))/d\epsilon = 1/\cos^2(\epsilon)$. So we obtain

$$e^{f_1\hat{K}_1}e^{f_3\hat{K}_3}e^{f_2\hat{K}_2} = I + \hat{n}_1(e^{f_2} - 1) + \hat{n}_2(e^{-f_3} - 1) + \hat{n}_1\hat{n}_2(2 - e^{f_3} - e^{-f_3})$$
$$+ f_2\hat{K}_2e^{-f_3} + f_1\hat{K}_1e^{-f_3} + f_1f_2e^{-f_3}(\hat{n}_1 - \hat{n}_1\hat{n}_2)$$

where $\hat{n}_1 = \hat{c}_1^\dagger\hat{c}_1$, $\hat{n}_2 = \hat{c}_2^\dagger\hat{c}_2$.

10.3 Supplementary Problems

Problem 1. Let $S_1^{(1/2)}$, $S_2^{(1/2)}$, $S_3^{(1/2)}$ be the spin-$\frac{1}{2}$ matrices

$$S_1^{(1/2)} = \frac{1}{2}\begin{pmatrix} 0 & 1 \\ 1 & 0 \end{pmatrix}, \quad S_2^{(1/2)} = \frac{1}{2}\begin{pmatrix} 0 & -i \\ i & 0 \end{pmatrix}, \quad S_3^{(1/2)} = \frac{1}{2}\begin{pmatrix} 1 & 0 \\ 0 & -1 \end{pmatrix}$$

and σ_1, σ_2, σ_3 be the Pauli spin matrices

$$\sigma_1 = \begin{pmatrix} 0 & 1 \\ 1 & 0 \end{pmatrix}, \quad \sigma_2 = \begin{pmatrix} 0 & -i \\ i & 0 \end{pmatrix}, \quad \sigma_3 = \begin{pmatrix} 1 & 0 \\ 0 & -1 \end{pmatrix}.$$

(i) Find the eigenvalues and normalized eigenvectors of the hermitian 4×4 matrix

$$\hat{K}_1 = \frac{\hat{H}_1}{\hbar\omega} = S_1 \otimes S_1 + S_2 \otimes S_2 + S_3 \otimes S_3.$$

(ii) Find the eigenvalues and normalized eigenvectors of the hermitian 4×4 matrix

$$\hat{K}_2 = \frac{\hat{H}_2}{\hbar\omega} = S_1 \otimes S_2 + S_2 \otimes S_3 + S_3 \otimes S_1.$$

(iii) Find the eigenvalues and normalized eigenvectors of the hermitian 8×8 matrix

$$\hat{K}_3 = \frac{\hat{H}_3}{\hbar\omega} = S_1 \otimes S_2 \otimes S_3 + S_3 \otimes S_1 \otimes S_2 + S_2 \otimes S_3 \otimes S_1.$$

(iv) Find the eigenvalues and normalized eigenvectors of the hermitian 8×8 matrix

$$\hat{K}_4 = \frac{\hat{H}_4}{\hbar\omega} = S_1 \otimes S_3 \otimes S_2 + S_2 \otimes S_1 \otimes S_3 + S_3 \otimes S_2 \otimes S_1.$$

(v) Find the spectrum of the Hamilton operators

$$\hat{H}_1 = \hbar\omega_1 \sigma_1 \otimes \sigma_1 + \hbar\omega_2(\sigma_3 \otimes I_2 + I_2 \otimes \sigma_3)$$

$$\hat{H}_2 = \hbar\omega_1 \sigma_3 \otimes \sigma_3 + \hbar\omega_2(\sigma_1 \otimes I_2 + I_2 \otimes \sigma_1).$$

Problem 2. Consider the spin matrices $S_1^{(s)}$, $S_2^{(s)}$, $S_3^{(s)}$ satisfying the *commutation relations*

$$[S_1^{(s)}, S_2^{(s)}] = iS_3^{(s)}, \quad [S_2^{(s)}, S_3^{(s)}] = iS_1^{(s)}, \quad [S_3^{(s)}, S_1^{(s)}] = iS_2^{(s)}$$

with the spin $s = 1/2, 1, 3/2, 2, \ldots$. One defines

$$S_+^{(s)} := S_1^{(s)} + iS_2^{(s)}, \qquad S_-^{(s)} := S_1^{(s)} - iS_2^{(s)}.$$

Solve the eigenvalue problem for the Hamilton operator

$$\hat{H} = \hbar\omega_1(S_3^{(s)} \otimes S_3^{(s)}) + \hbar\omega_2(S_+^{(s)} \otimes S_-^{(s)} + S_-^{(s)} \otimes S_+^{(s)})$$

for $s = 1/2$, $s = 1$, $s = 3/2$, $s = 2$.

Problem 3. Let

$$S_1^{(1/2)} = \frac{1}{2}\begin{pmatrix} 0 & 1 \\ 1 & 0 \end{pmatrix}, \quad S_2^{(1/2)} = \frac{1}{2}\begin{pmatrix} 0 & -i \\ i & 0 \end{pmatrix}, \quad S_3^{(1/2)} = \frac{1}{2}\begin{pmatrix} 1 & 0 \\ 0 & -1 \end{pmatrix}$$

be the spin-$\frac{1}{2}$ matrices. Consider the 16×16 hermitian matrix (open end boundary conditions) with trace equal to 0

$$\begin{aligned}
\hat{K}_1 = \frac{\hat{H}_1}{\hbar\omega} = {} & S_{1,1}S_{2,1} + S_{1,2}S_{2,2} + S_{1,3}S_{2,3} + S_{2,1}S_{3,1} + S_{2,2}S_{3,2} + S_{2,3}S_{3,3} \\
& + S_{3,1}S_{4,1} + S_{3,2}S_{4,2} + S_{3,3}S_{4,3}
\end{aligned}$$

where the first index indicates the lattice site and the second index the spin number, i.e.

$$\begin{aligned}
S_{1,j} = S_j \otimes I_2 \otimes I_2 \otimes I_2, \qquad & S_{2,j} = I_2 \otimes S_j \otimes I_2 \otimes I_2 \\
S_{3,j} = I_2 \otimes I_2 \otimes S_j \otimes I_2, \qquad & S_{4,j} = I_2 \otimes I_2 \otimes I_2 \otimes S_j
\end{aligned}$$

where $j = 1, 2, 3$. Let F_{16} be the 16×16 Fourier matrix. Find $F_{16}\hat{K}_1 F_{16}^*$. Consider the 16×16 hermitian matrix (cyclic boundary conditions) with trace equal to 0

$$\begin{aligned}
\hat{K}_2 = \frac{\hat{H}_1}{\hbar\omega} = {} & S_{1,1}S_{2,1} + S_{1,2}S_{2,2} + S_{1,3}S_{2,3} + S_{2,1}S_{3,1} + S_{2,2}S_{3,2} + S_{2,3}S_{3,3} \\
& + S_{3,1}S_{4,1} + S_{3,2}S_{4,2} + S_{3,3}S_{4,3} + S_{4,1}S_{1,1} + S_{4,2}S_{1,2} + S_{4,3}S_{1,3}.
\end{aligned}$$

Let F_{16} be the 16×16 Fourier matrix. Find $F_{16}\hat{K}_2 F_{16}^*$.

Problem 4. Consider the Hilbert space $\mathcal{H} = \mathcal{H}_1 \otimes \mathcal{H}_2 = \mathbb{C}^4$ with $\mathcal{H}_1 = \mathcal{H}_2 = \mathbb{C}^2$. Consider the density matrix (pure state)

$$\rho = |\psi\rangle\langle\psi| = \frac{1}{2}\begin{pmatrix} 1 & 0 & 0 & 1 \\ 0 & 0 & 0 & 0 \\ 0 & 0 & 0 & 0 \\ 1 & 0 & 0 & 1 \end{pmatrix}.$$

Show that

$$\text{tr}_{\mathcal{H}_1}(|\psi\rangle\langle\psi|) = \frac{1}{2}I_2, \quad \text{tr}_{\mathcal{H}_2}(|\psi\rangle\langle\psi|) = \frac{1}{2}I_2.$$

Problem 5. Let A, B be $n \times n$ hermitian matrices and $\mathbf{v}_1$, $\mathbf{v}_2$, ..., $\mathbf{v}_n$ be the normalized eigenvectors of A which should form an orthonormal basis in the Hilbert space $\mathbb{C}^n$ and analogously $\mathbf{u}_1$, $\mathbf{u}_2$, ..., $\mathbf{u}_n$ are the normalized eigenvectors of B which should form an orthonormal basis in the Hilbert space $\mathbb{C}^n$. The quantity

$$\max_{j,k=1,\ldots,n} |\langle \mathbf{v}_j | \mathbf{u}_k \rangle|^2$$

plays a role for *entropic inequalities*. Let $n = 4$. Find this quantity for the two 4×4 hermitian and invertible matrices

$$A = \frac{1}{\sqrt{2}} \begin{pmatrix} 1 & 0 & 0 & 1 \\ 0 & 1 & 1 & 0 \\ 0 & 1 & -1 & 0 \\ 1 & 0 & 0 & -1 \end{pmatrix}, \quad B = \frac{1}{\sqrt{2}} \begin{pmatrix} 1 & 1 \\ 1 & -1 \end{pmatrix} \otimes \frac{1}{\sqrt{2}} \begin{pmatrix} 0 & 1 \\ 1 & 0 \end{pmatrix}.$$

Problem 6. Consider two finite dimensional Hilbert spaces $\mathcal{H}_A$ and $\mathcal{H}_B$ with $\dim(\mathcal{H}_A) = d_A$, $\dim(\mathcal{H}_B) = d_B$. The *Schmidt decomposition theorem* tells us that any bi-partite normalized pure state $|\Psi\rangle \in \mathcal{H}_A \otimes \mathcal{H}_B$ can be written in terms of some orthonormal bases $|\psi_k\rangle_A \in \mathcal{H}_A$ ($k = 0, 1, \ldots, d_A - 1$), $|\phi_\ell\rangle_B$ ($\ell = 0, 1, \ldots, d_B - 1$)

$$|\Psi\rangle = \sum_{k=0}^{\min(d_A-1,d_B-1)} \lambda_k |\psi_k\rangle \otimes |\phi_k\rangle.$$

The positive coefficients λ_k satisfy

$$\sum_{k=0}^{\min(d_A-1,d_B-1)} \lambda_k = 1.$$

The orthonormal bases $|\psi_k\rangle_A$ and $|\phi_\ell\rangle_B$ dependent on the given normalized state $|\Psi\rangle$. Let $|j\rangle_A$, $|\ell\rangle_B$ ($j = 0, 1, \ldots, d_A - 1; \ell = 0, 1, \ldots, d_B - 1$) be state independent orthonormal bases. Then there are $d_A \times d_A$ and $d_B \times d_B$ unitary matrices $U_A(\alpha)$ and $U_B(\beta)$ such that

$$|\psi_j\rangle_A = U_A(\alpha)|j\rangle_A, \quad |\phi_k\rangle_B = U_B(\beta)|k\rangle_B$$

where α, β are parameters to specify the unitary matrices. The quantum entanglement of the joint state $|\Psi\rangle$ resides only in the coefficients λ_k and not in the parameters α and β. Find the Schmidt decomposition of the normalized vector in $\mathbb{C}^4$

$$|\Psi\rangle = \frac{1}{2}\begin{pmatrix} 1 \\ -1 \\ 1 \\ 1 \end{pmatrix}.$$

Problem 7. Let $n \geq 2$. Consider the Hilbert space $\mathbb{C}^n$. Let ρ be the *state* (*density matrix*) of a quantum system, a positive definite hermitian matrix with $\mathrm{tr}(\rho) = 1$. The mean value functional is defined as

$$\langle \cdot \rangle := \mathrm{tr}(\rho \cdot).$$

For two $n \times n$ hermitian matrices A and B, the *variance* is defined as

$$(\Delta A)^2 := \langle A^2 \rangle - \langle A \rangle^2, \quad (\Delta B)^2 := \langle B^2 \rangle - \langle B \rangle^2.$$

One has the inequalities

$$(\Delta A)(\Delta B) \geq |\langle AB \rangle - \langle A \rangle \langle B \rangle| \geq \frac{1}{2}|\langle [A, B] \rangle|.$$

The inequality

$$(\Delta A)(\Delta B) \geq |\langle AB \rangle - \langle A \rangle \langle B \rangle|$$

is due to *Schrödinger*. The inequality

$$(\Delta A)(\Delta B) \geq \frac{1}{2}\langle [A, B] \rangle|$$

is due to *Robertson*. Show that one can write

$$\langle AB \rangle = \mathrm{tr}(\sqrt{\rho} AB \sqrt{\rho}).$$

Show that for $n \times n$ hermitian matrices A, B

$$|\langle AB \rangle|^2 = \frac{1}{4}\langle [A, B]_+ \rangle + \langle [A, B] \rangle|^2 \equiv \frac{1}{4}|\langle [A, B]_+ \rangle|^2 + \frac{1}{4}|\langle [A, B] \rangle|^2$$

where $[.,.]_+$ denotes the anti-commutator.

Let $n = 2$ and $A = \sigma_1$, $B = \sigma_2$ and $\rho = \begin{pmatrix} 1/4 & 0 \\ 0 & 3/4 \end{pmatrix}$. Find ΔA, ΔB, $\langle AB \rangle$ $\langle A \rangle$, $\langle B \rangle$ and show that the inequalities of Schrödinger and Robertson are satisfied.

Problem 8. Consider the Hamilton operator $\hat{H} = \hbar\omega K$ acting in the finite dimensional Hilbert space $\mathbb{C}^n$, where K is a hermitian matrix. Let $|\psi\rangle$ be a normalized vector in $\mathbb{C}^n$. We select

$$|\psi\rangle = \frac{1}{\sqrt{n}}(1\ 1\ \ldots\ 1\ 1)^T$$

since this normalized vector has a nonzero overlap with the exact ground state $|\phi_0\rangle$, i.e.

$$|\langle\psi|\phi_0\rangle|^2 \neq 0.$$

We set $\langle H\rangle := \langle\psi|\hat{H}|\psi\rangle$. An approximation of the lowest eigenvalue E_0 can be found from

$$E_0 = I_1 - \frac{I_2^2}{I_3} - \left(\frac{1}{I_3}\right)\frac{(I_2 I_4 - I_3^2)^2}{I_3 I_5 - I_4^2} - \cdots$$

where

$$I_1 = \langle\hat{H}\rangle$$
$$I_2 = \langle\hat{H}^2\rangle - \langle\hat{H}\rangle^2$$
$$I_3 = \langle\hat{H}^3\rangle - 3\langle\hat{H}^2\rangle\langle\hat{H}\rangle + 2\langle\hat{H}\rangle^3$$
$$I_4 = \langle\hat{H}^4\rangle - 4\langle\hat{H}^3\rangle\langle\hat{H}\rangle - 3\langle\hat{H}^2\rangle^2 + 12\langle\hat{H}^2\rangle\langle\hat{H}\rangle^2 - 6\langle\hat{H}\rangle^4$$
$$I_5 = \langle\hat{H}^5\rangle - 5\langle\hat{H}^4\rangle\langle\hat{H}\rangle - 10\langle\hat{H}^3\rangle\langle\hat{H}^2\rangle + 20\langle\hat{H}^3\rangle\langle\hat{H}\rangle^2$$
$$+ 30\langle\hat{H}^2\rangle^2\langle\hat{H}\rangle - 60\langle\hat{H}^2\rangle\langle\hat{H}\rangle^3 + 24\langle\hat{H}\rangle^5.$$

Apply it to the Hamilton operator in $\mathbb{C}^8$

$$\frac{\hat{H}}{\hbar\omega} = \sigma_1 \otimes \sigma_2 \otimes I_2 + I_2 \otimes \sigma_1 \otimes \sigma_2 + \sigma_2 \otimes I_2 \otimes \sigma_1.$$

Compare with the exact ground state.

Problem 9. Let $\ell = 0, 1/2, 1, 3/2, 2, \ldots$ and $m = -\ell, -\ell+1, \ldots, \ell-1, \ell$. We set

$$I_\ell = \{0, 1/2, 1, 3/2, 2, \ldots\}, \qquad S_\ell = \{-\ell, -\ell+1, \ldots, \ell-1, \ell\}.$$

Consider the Hilbert space $\mathcal{H}$ consisting of the tensor sum of the Hilbert subspaces $\mathcal{H}_\ell$

$$\mathcal{H} = \bigoplus_{\ell=0}^{\infty} \mathcal{H}_\ell$$

where the Hilbert subspaces are unitary subspaces of dimension $2\ell+1$ and basis vectors $|\ell, m\rangle$. The basis vectors are normalized, i.e.

$$\langle\ell', m'|\ell, m\rangle = \delta_{\ell,\ell'}\delta_{m,m'}$$

and (*completeness relation*)

$$\sum_{\ell=0}^{\infty} \sum_{m=-\ell}^{m=+\ell} |\ell, m\rangle\langle\ell, m| = I.$$

In $\mathcal{H}$ we consider the linear operators J_3, J_+, J_- defined by

$$J_3|\ell, m\rangle = m|\ell, m\rangle \quad \text{eigenvalue equation}$$
$$J_+|\ell, m\rangle = f(\ell, m)|\ell, m+1\rangle$$
$$J_+|\ell, \ell\rangle = 0|\ell, \ell\rangle$$
$$J_- = (J_+)^\dagger$$

where $m \in S_\ell$, $\ell \in I_\ell$ and $f(\ell, m)$ is a real *entire function* satisfying

$$f(\ell, \ell) = 0, \quad f(\ell, -\ell - 1) = 0.$$

Show that

$$[J_3, J_+] = J_+, \quad [J_3, J_-] = -J_-, \quad J_3^n J_+ = J_+(J_3 + I)^n, \quad J_3^n J_- = J_-(J_3 - I)^n.$$

Problem 10. Consider the one-dimensional eigenvalue problem $\hat{H}u(x) = Eu(x)$ with

$$\hat{H} = \frac{\hat{p}^2}{2m} + V(x), \quad V(x) = D\tanh^2(x/R)$$

with $R > 0$. Show that the energy spectrum $E_n(D, R)$ is given by

$$E_n(D, R) = \frac{\hbar^2}{mR^2}\left(-\frac{1}{8} + \frac{n+1/2}{2}\sqrt{8mDR^2/\hbar^2 + 1} - \frac{(n+1/2)^2}{2}\right)$$

where $n = 0, 1, 2, \ldots$.

Problem 11. Consider the non-relativistic hydrogen atom, where a_0 is the Bohr radius and $a = a_0/Z$. The Schrödinger-Coulomb Green function $G(\mathbf{r}_1, \mathbf{r}_2; E)$ corresponding to the energy variable E is the solution of the partial differential equation

$$\left(-\frac{\hbar^2}{2m}\nabla_1^2 - \frac{\hbar^2}{amr_1} - E\right)G(\mathbf{r}_1, \mathbf{r}_2; E) = \delta(\mathbf{r}_1 - \mathbf{r}_2)$$

with the appropriate boundary conditions. Show that expanding G in terms of spherical harmonics $Y_{\ell,m}$

$$G(\mathbf{r}_1, \mathbf{r}_2; E) = \sum_{\ell=0}^{\infty} \sum_{m=-\ell}^{\ell} g_\ell(r_1, r_2; E)Y_{\ell,m}(\theta_1, \phi_1)Y_{\ell,m}^*(\theta_2, \phi_2)$$

we find for the radial part g_ℓ of the Schrödinger-Coulomb Green function

$$\left(\frac{1}{r_1^2}\frac{d}{dr_1}\left(r_1^2\frac{d}{dr_1}\right) - \frac{\ell(\ell+1)}{r_1^2} + \frac{2}{ar_1} - \frac{1}{\nu^2 a^2}\right)g_\ell(r_1, r_2; \nu) = -\frac{2m}{\hbar^2}\frac{\delta(r_1 - r_2)}{r_1 r_2}$$

where $\nu^2 a^2 := -\hbar^2/(2mE)$. Hint. Utilize the identity

$$\delta(\mathbf{r}_1 - \mathbf{r}_2) = \frac{\delta(r_1 - r_2)}{r_1 r_2}\sum_{\ell=0}^\infty\sum_{m=-\ell}^\ell Y_{\ell,m}(\theta_1, \phi_1)Y_{\ell,m}^*(\theta_2, \phi_2).$$

Problem 12. Consider the Hamilton operator

$$\hat{H} = \frac{1}{2m}(\hat{p}_{x_1}^2 + \hat{p}_{x_2}^2) + \frac{1}{2}m\Omega(x_1^2 + x_2^2) + \omega(x_1\hat{p}_{x_2} - x_2\hat{p}_{x_1}).$$

Hence we have a two-dimensional harmonic oscillator of frequency Ω. The oscillator itself is rotating about the negative x_3-axis with frequency ω. Show that the last term in the Hamilton operator commutes with the harmonic oscillator part. Apply this fact to show that spectrum of the Hamilton operator can be written as

$$E_{n_r,\ell} = (2n_r + |\ell| + 1)\hbar\Omega + \ell\hbar\omega$$

where n_r is the radial quantum number ($n_r = 0, 1, 2, \ldots$) and $\hbar\ell$ is the momentum along the x_3 axis, where $\ell = 0, \pm1, \pm2, \ldots$.

Problem 13. Consider the Hilbert space $L_2(\mathbb{R})$ and the one-dimensional Schrödinger equation (eigenvalue problem)

$$-\frac{\hbar^2}{2m}\frac{d^2u}{dx^2} + c|x|u(x) = Eu(x).$$

The dimension of $\hbar$ is $kg.meter^2.sec^{-1}$, the dimension of m is kg, the dimension of E is $kg.meter^2.sec^{-2}$ and the dimension of the constant c ($c > 0$) is $kg.meter.sec^{-2}$. Show that we can introduce the dimensionless quantities

$$\tilde{x} = \left(\frac{2m}{\hbar^2}c\right)^{1/3}x, \qquad \alpha = \left(\frac{2m}{\hbar^2 c^2}\right)^{1/3}E$$

and together with $\tilde{u}(\tilde{x}(x)) = u(x)$ one obtains

$$\frac{d^2\tilde{u}(\tilde{x})}{d\tilde{x}^2} + (\alpha - |\tilde{x}|)\tilde{u}(\tilde{x}) = 0.$$

The eigenfunctions for bound states satisfy $\tilde{u}(\tilde{x}) \to 0$ as $|\tilde{x}| \to \infty$. Show that since the potential $V(x) = a|x|$ is symmetric ($V(x) = V(-x)$) the eigenfunctions have either add parity or even parity.

Problem 14. (i) Show that the Schrödinger equation

$$\hat{H}u(\mathbf{x}) = \left(-\frac{\hbar^2}{2m}\nabla^2 + V(\mathbf{x}) \right) u(\mathbf{x}) = Eu(\mathbf{x})$$

is satisfied by $u(\mathbf{x}) = a(\mathbf{x})\exp(i\hbar^{-1}S(\mathbf{x}))$ with real function $a(\mathbf{x})$, $S(\mathbf{x})$ if and only if

$$\nabla S \cdot \nabla S + 2mV(\mathbf{x}) - \hbar^2 a^{-1}\nabla^2 a = 2Em$$

and $\nabla \cdot (a^2 \nabla S) = 0$ simultaneously.

(ii) Consider the *Schrödinger equation*

$$i\hbar\frac{\partial\psi}{\partial t} = -\frac{\hbar^2}{2m}\sum_{j=1}^{3}\frac{\partial^2\psi}{\partial q_j^2} + V(\mathbf{q}).$$

Substituting

$$\psi(q,t) = R(q,t)\exp(iS(q,t)/\hbar)$$

where R, S are real, into the Schrödinger equation and separating the real and imaginary parts provides

$$\frac{\partial S}{\partial t} + \frac{(\nabla S)^2}{2m} + V - \frac{\hbar^2}{2m}\frac{(\nabla^2 R)}{R} = 0, \qquad \frac{\partial R^2}{\partial t} + \nabla\left(\frac{1}{m}R^2\nabla S\right) = 0.$$

Let

$$S(q,t) := \hbar\arctan\left(i\frac{\psi^* - \psi}{\psi^* + \psi} \right).$$

Show that

$$p := \nabla S = \frac{1}{|\psi|^2}\Re(\psi^*(-i\hbar\nabla)\psi).$$

(iii) Consider the Banach-Gelfand triple $S(\mathbb{R}^3) \subset L_2(\mathbb{R}^3) \subset S'(\mathbb{R}^3)$ and the Schrödinger equation for a particle scattering in a potential $V(\mathbf{x})$

$$-\frac{\hbar^2}{2m}\nabla^2 u(\mathbf{x}) + V(\mathbf{x})u(\mathbf{x}) = Eu(\mathbf{x}).$$

Show that substituting the ansatz with real $R(\mathbf{x})$, $S(\mathbf{x})$

$$u(\mathbf{x}) = R(\mathbf{x})\exp(iS(\mathbf{x})/\hbar)$$

into the Schrödinger equation provides the *conservation equation*

$$\nabla \cdot (R^2 \nabla S) = 0.$$

Problem 15. Consider the Hilbert space $L_2(\mathbb{R})$ and the Hamilton operator $\hat{H}$ for the one-dimensional *quartic anharmonic oscillator*

$$\hat{H} = -\frac{1}{2m}\frac{d^2}{dx^2} + \frac{1}{2}m\omega^2 x^2 + \gamma x^4$$

with $\gamma \geq 0$. Show that the Hamilton operator $\hat{H}$ is *parity invariant*, i.e.

$$[\hat{H}, S] = 0, \quad Su(x) = u(-x).$$

Show that for $\gamma \geq 0$ the Hamilton operator $\hat{H}$ has bound states. Describe them in terms of translated Gaussian intrinsic states ($r > 0, s > 0$)

$$u(x, a) = \left(\frac{4s}{\pi}\right)^{1/4} \exp(-ra^2)\exp(-2s(x-a)^2) \in L_2(\mathbb{R}).$$

Problem 16. Consider the one-dimensional *nonlinear Schrödinger equation*

$$i\hbar\frac{\partial\psi}{\partial t} + \frac{\hbar^2}{2m}\frac{\partial^2\psi}{\partial x^2} = \lambda|\psi|^2\psi.$$

(i) Show that using the ansatz $\psi(x, t) = u(x)e^{-iEt/\hbar}$ one obtains

$$Eu(x) + \frac{\hbar^2}{2m}\frac{\partial^2 u(x)}{\partial x^2} = \lambda|u(x)|^2 u(x).$$

(ii) Show that

$$u(x) = \frac{A}{\cosh(kx)} \quad A > 0$$

is a solution of this equation with $E = -\hbar^2 k^2/(2m)$, $A^2 = -\hbar^2 k^2/(m\lambda)$. Is $u(x)$ an element of the Hilbert space $L_2(\mathbb{R})$?
(iii) Show that

$$u(x) = B\tanh(kx)$$

is a solution of this equation with $E = \hbar^2 k^2/m$, $B^2 = \hbar^2 k^2/(m\lambda)$.

Problem 17. Consider the equilateral *triangle*. Show that the function

$$u(x_1, x_2) = \sin\left(\frac{2\pi}{\sqrt{3}a}(\sqrt{3}x_1 + x_2)\right) - \sin\left(\frac{2\pi}{\sqrt{3}a}(\sqrt{3}x_1 - x_2)\right) - \sin\left(\frac{4\pi}{\sqrt{3}a}x_2\right)$$

vanishes at the boundary

$$\text{(a) } x_2 = 0, \quad \text{(b) } x_1 - \frac{x_2}{\sqrt{3}} = 0, \quad \text{(c) } x_2 = \sqrt{3}(a - x_1).$$

Hence we consider the domain

$$D := \{\, (x_1, x_2) \, : \, x_2 \geq 0, \, x_1 \geq \frac{x_2}{\sqrt{3}}, \, \sqrt{3}(a - x_1) \geq x_2 \,\}$$

and the Hilbert space $L_2(D)$.

Problem 18. Show that in one-dimensional problems the energy spectrum of the bound state is always non-degenerate. Hint. Suppose that the opposite is true. Let u_1 and u_2 be two linearly independent eigenfunctions with the same energy eigenvalues E, i.e.

$$\frac{d^2 u_1}{dx^2} + \frac{2m}{\hbar^2}(E - V)u_1 = 0, \qquad \frac{d^2 u_2}{dx^2} + \frac{2m}{\hbar^2}(E - V)u_2 = 0.$$

Problem 19. (i) Show that

$$u(x, t) = (A\sin(kx) + B\cos(kx))e^{-ikct}$$

where k is the wave number satisfies the one-dimensional *wave equation*

$$\frac{\partial^2 u}{\partial x^2} - \frac{1}{c^2}\frac{\partial^2 u}{\partial t^2} = 0.$$

(ii) Show that the first boundary condition $u(x = 0, t) = 0$ implies that $B = 0$. Hence we consider now the solution

$$u(x, t) = A\sin(kx)e^{-ikct}.$$

(iii) Let $\ell > 0$ be of dimension length. Show that second boundary condition

$$-T\frac{\partial u(x = \ell, t)}{\partial x} = R_s\frac{\partial u(x = \ell, t)}{\partial t} + K_s u(x = \ell, t)$$

(for given T, R_s, K_s) provides

$$-Tk_n\cos(k_n\ell) = (-ik_n cR_s + K_s)\sin(k_n\ell)$$

where the roots k_n $(n = 0, 1, 2, \dots)$ of this transcendental equation are the eigenvalues of the differential equation.

Problem 20. Let G be the gravitational constant. The *Schrödinger-Newton equations* are given by

$$-\frac{\hbar^2}{2m}\Delta\psi(\mathbf{x}) + U(\mathbf{x})\psi(\mathbf{x}) = \mu\psi(\mathbf{x}), \qquad \Delta\psi(\mathbf{x}) = 4\pi Gm^2|\psi(\mathbf{x})|^2$$

with

$$\int_{\mathbb{R}^3} |\psi(\mathbf{x})|^2 d\mathbf{x} = 1.$$

The second equation is the *Poisson equation*. Show that by multiplying the first equation with $\psi^*(\mathbf{x})$ and integrating one obtains

$$\mu[\psi(\mathbf{x})] = \int_{\mathbb{R}^3} \left(\frac{\hbar^2}{2m} |\nabla\psi(\mathbf{x})|^2 + U(\mathbf{x})|\psi(\mathbf{x})|^2 \right) d\mathbf{x}.$$

Show that energy functional is defined as

$$E[\psi(\mathbf{x})] = \int_{\mathbb{R}^3} \left(\frac{\hbar^2}{2m} |\nabla\psi(\mathbf{x})|^2 + \frac{1}{2}U(\mathbf{x})|\psi(\mathbf{x})|^2 \right) d\mathbf{x}.$$

Problem 21. Consider the one-dimensional *parity operator* $Px = -x$. Hence $P^{-1} = P$. We define $\hat{O}_P$ as $\hat{O}_P u(x) := u(P^{-1}x) = u(-x)$. Find the eigenvalues of $\hat{O}_P$. Do the eigenvalues form a group under multiplication? Let

$$\hat{H} = -\frac{\hbar^2}{2m}\frac{d^2}{dx^2} + V(x)$$

with $V(x) = V(-x)$. Calculate the commutator $[\hat{H}, \hat{O}_P]$. Show that if $u(x)$ is an eigenfunction of $\hat{H}$, then $\hat{O}_P u(x)$ is an eigenfunction with the same eigenvalue.

Problem 22. Consider the Hilbert space $L_2(\mathbb{R})$. Let $a > 0$ and

$$f_a(x) = \begin{cases} x/a^2 + 1/a & \text{for} \quad -a \le x \le 0 \\ -x/a^2 + 1/a & \text{for} \quad 0 \le x \le a \\ 0 & \text{otherwise} \end{cases}$$

(i) Calculate $\int_{-a}^{a} f_a(x)dx$.
(ii) Calculate the derivative of f_a in the sense of generalized functions.
(iii) Calculate the Fourier transform of f_a. Discuss the cases $a \to 0$ and $a \to \infty$.
(iv) Let

$$\hat{H} = -\frac{\hbar^2}{2m}\frac{d^2}{dx^2}$$

be an operator acting in a subspace of the Hilbert space $L_2(\mathbb{R})$. Calculate

$$\langle f_a, \hat{H}f_a \rangle.$$

Problem 23. Calculate the matrix representation of the Hamilton operator (Stark effect of a two-dimensional rotator)

$$\hat{H} = -\frac{\hbar^2}{2\Theta}\frac{d^2}{d\phi^2} - pE\cos(\phi)$$

which acts in the Hilbert space $L_2(\mathbb{S}^2)$. Here p and E are constants. An orthonormal basis in the Hilbert space $L_2(\mathbb{S}^1)$ is given by

$$|m\rangle = \frac{1}{\sqrt{2\pi}}e^{im\phi}, \quad m = 0, \pm 1, \pm 2, \ldots$$

Find the spectrum of the matrix representation.

Problem 24. Let $C^m(\mathbb{R})$ be the m-times continuously differentiable complex valued functions on $\mathbb{R}$. Let $C^m_{(2)}(\mathbb{R})$ be the square integrable and vanishing at infinity functions in $C^m(\mathbb{R})$. Let ψ_1, ψ_2 and their first derivatives $d\psi_1/dx, d\psi_2/dx$ as well as the functions $V\psi_1$ and $V\psi_2$ be from $C^1_{(2)}(\mathbb{R})$. Show that

$$\langle \psi_1(x), -\frac{\hbar^2}{2m}\frac{d^2\psi_2}{dx^2} + V(x)\psi_2(x)\rangle = \langle -\frac{\hbar^2}{2m}\frac{d^2\psi_1}{dx^2} + V(x)\psi_1(x), \psi_2(x)\rangle.$$

Problem 25. Let $n \in \mathbb{Z}$. Consider the Hilbert space $\ell_2(\mathbb{Z})$ and $u(n) \in \ell_2(\mathbb{Z})$. Consider the Hamilton operator

$$(\hat{H}_{\lambda,\theta})u(n) := u(n+1) + u(n-1) + \lambda\cos(2\pi(\alpha n + \theta))u(n)$$

where λ, α, θ are real parameter with λ considered small. Show that the eigenvalue problem $\hat{H}_{\lambda,\theta}u = Eu$ can be written in matrix form

$$\begin{pmatrix} u(n+1) \\ u(n) \end{pmatrix} = \begin{pmatrix} E - \lambda\cos(2\pi(\alpha n + \theta)) & -1 \\ 1 & 1 \end{pmatrix}\begin{pmatrix} u(n) \\ u(n-1) \end{pmatrix}, \quad n \in \mathbb{Z}.$$

Find the eigenvalues of the 2×2 matrix. Consider the *Fourier transform*

$$\hat{u}(k) = \sum_{n\in\mathbb{Z}} u(n)\exp(i2\pi kn), \quad k \in \mathbb{Z}.$$

Show that the sequence $v(n) = \hat{u}(\tau + \alpha n)\exp(in\theta)$ for fixed τ is formally a solution of

$$\hat{H}_{4/\lambda,\tau}v = \frac{2E}{\lambda}v$$

(Aubry-André duality). Note that the $\hat{u}(k)$ ($k \in \mathbb{Z}$) is the Fourier transform of a sequence in the Hilbert space $\ell_2(\mathbb{Z})$. Hence it is defined as an $L_2(\mathbb{S}^1)$

function of the unit circle $\mathbb{S}^1$.

Problem 26. Consider the Hamilton operator

$$\hat{H} = -\frac{\hbar^2}{2m}\frac{d^2}{dx^2} + eEx$$

where e is a charge and E a constant electric field strength. Show that

$$\exp(-i\hat{H}t/\hbar) \equiv \exp\left(-\frac{it}{\hbar}\left(-\frac{\hbar^2}{2m}\frac{d^2}{dx^2} + eEx\right)\right)$$

$$\equiv \exp\left(-\frac{ieEtx}{\hbar}\right)\exp\left(\frac{i\hbar t}{2m}\frac{d^2}{dx^2} + \frac{eEt^2}{2m}\frac{d}{dx} - \frac{it^3(eE)^3}{6m\hbar}\right).$$

Problem 27. A particle is enclosed in a *rectangular box* with impenetrable walls, inside which it can move freely. The Hilbert space is $L_2([0,a] \times [0,b] \times [0,c])$, where $a, b, c > 0$. Find the eigenfunctions and eigenvalues. What can be said about the degeneracy, if any, of the eigenfunctions? The Hamilton operator is

$$\hat{H} = \frac{\hat{p}^2}{2m} \equiv -\frac{\hbar^2}{2m}\left(\frac{\partial^2}{\partial x_1^2} + \frac{\partial^2}{\partial x_2^2} + \frac{\partial^2}{\partial x_3^2}\right).$$

Problem 28. Consider the eigenvalue equation

$$\left(-\frac{\hbar^2}{2m}\frac{d^2}{dx^2} + \gamma x\right)u(x) = Eu(x).$$

We write

$$\left(-\frac{d^2}{dx^2} + \tilde{\gamma}x\right)u(x) = \tilde{E}u(x)$$

with $\tilde{\gamma} = \gamma m/\hbar^2$, $\tilde{E} = Em/\hbar^2$, the boundary condition $u(0) = u(x_0) = 0$ where $x_0 > 0$. With

$$s := (2\gamma)^{1/3}x - e, \qquad e := 2^{1/3}\tilde{\gamma}^{-2/3}\tilde{E}$$

we obtain the *Airy differential equation*

$$\frac{d^2f}{ds^2} - sf(s) = 0, \quad f(s) = u((2\tilde{\gamma})^{-1/3}s + \tilde{E}/\tilde{\gamma})$$

Show that the eigenvalues e_n are given as the roots e_n of the equation

$$\det\begin{pmatrix} Ai(s_{0n}) & Bi(s_{0n}) \\ Ai(-e_n) & Bi(-e_n) \end{pmatrix} = 0, \quad s_{0n} \equiv (2\tilde{\gamma})^{1/3}x_0 - e_n$$

where Ai, Bi are the Airy functions.

Problem 29. Let m be the mass of a particle, μ the chemical potential and g be a positive real constant. The *Gross-Pitaevskii equation* for the superfluid condensate wave $\psi(\mathbf{x}, t)$ is given by

$$i\hbar \frac{\partial \psi}{\partial t} = -\frac{\hbar^2}{2m} \Delta \psi - \mu \psi + g |\psi|^2 \psi.$$

Show that setting

$$\psi(\mathbf{x}, t) = f(\mathbf{x}, t) \exp(i\phi(\mathbf{x}, t))$$

and introducing the scaled variables

$$t' = \frac{\hbar}{2m} t, \quad \bar{\mu} = \frac{2m\mu}{\hbar^2}, \quad \bar{g} = \frac{2mg}{\hbar^2}$$

the Gross-Pitaevski equation can be written as a system of partial differential equation

$$\frac{1}{2} \frac{\partial f^2}{\partial t'} + \nabla \cdot (f^2 \nabla \phi) = 0, \quad f \frac{\partial \phi}{\partial t'} - \Delta f + f(\nabla \phi)^2 = \bar{\mu} f - \bar{g} f^3.$$

Show that in terms of velocity $\mathbf{v}$ and mass density ρ

$$\mathbf{v} = 2\nabla(\phi), \qquad \rho = \frac{f^2}{2}$$

the system can be written as

$$\frac{\partial \rho}{\partial t'} + \nabla \cdot (\rho \mathbf{v}) = 0, \quad \frac{\partial \mathbf{v}}{\partial t'} + (\mathbf{v} \cdot \nabla) \mathbf{v} = -\nabla U$$

with

$$U(\mathbf{x}, t) = -2((\Delta \rho^{1/2} / \rho^{1/2}) + \bar{\mu} - 2\bar{g}\rho).$$

which are the *Euler equation* of continuity and the equation of motion.

Problem 30. Consider the Banach-Gelfand triple $S(\mathbb{R}^2) \subset L_2(\mathbb{R}^2) \subset S'(\mathbb{R}^2)$. Show that the eigenfunctions for a separable integrable system described by the Hamilton operator

$$\hat{H} = \hat{H}_1 + \hat{H}_2 = \frac{p_1^2}{2m} + V_1(x_1) + \frac{p_2^2}{2m} + V_2(x_2)$$

can be expressed in the form $u_{m,n}(x_1, x_2) = \phi_{1,m}(x_1)\phi_{2,n}(x_2)$, where $\phi_{1,m}(x_1)$ are the eigenfunctions of $\hat{H}_1$ and $\phi_{2,m}(x_2)$ are the eigenfunctions of $\hat{H}_2$.

Problem 31. Let $a > 0$ with dimension length. Consider the Hilbert space $L_2([-a, a])$ and the Hamilton operator

$$\hat{H} = \frac{\hat{p}^2}{2m} + V(x), \quad \hat{p} = -i\hbar \frac{d}{dx}$$

with

$$V(x) = \begin{cases} 0 & \text{for} \quad -a \le x \le a \\ \infty & \text{for } x > a \text{ and } x < -a. \end{cases}$$

Note that $V(x) = V(-x)$. Show that the solution of the eigenvalue equation $\hat{H}u(x) = Eu(x)$ is given by

$$u_n(x) = \begin{cases} 0 & \text{for} \quad x \le -a \\ (1/a)^{1/2}\cos((n+1)\pi x)/(2a)) & \text{for } -a < x < a \\ 0 & \text{for} \quad x \ge a \end{cases}$$

for $n = 0, 2, 4, \ldots$

$$u(x) = \begin{cases} 0 & \text{for} \quad x \le -a \\ (1/a)^{1/2}\sin((n+1)\pi x)/(2a)) & \text{for } -a < x < a \\ 0 & \text{for} \quad x \ge a \end{cases}$$

where $n = 1, 3, 5, \ldots$. Show that the energy eigenvalues are

$$E_n(a) = \frac{\pi^2 \hbar^2}{8ma^2}(n+1)^2, \quad n \in \mathbb{N}_0.$$

Show that the functions $u_n(x)$ ($n \in \mathbb{N}_0$) form an orthonormal basis in the Hilbert space $L_2([-a, a])$.

Problem 32. Consider the one-dimensional Schrödinger equation

$$\frac{d^2 u(x_3)}{dx_3^2} + Q^2(x_3)u(x_3) = 0, \quad Q^2(x_3) = \frac{2m}{\hbar^2}(E - V(x_3))$$

where E is the energy of the particle and $V(x_3)$ is a periodic, analytic potential with period a. Assume that $V(x_3) = V(x_3 + a)$. This periodicity implies the periodicity of $Q^2(x_3)$. This implies that the one-dimensional Schrödinger equation is a *Hill equation*. For the *Floquet solution* one has

$$u(x_3 + a) = \lambda(E)u(x_3), \quad u'(x_3 + a) = \lambda(E)u'(x_3)$$

where the Floquet factor depends on E. Show that the quantity defined by

$$\mu := \frac{(u(x_3 + a) + u(x_3 - a))}{2u(x_3)}$$

is a function only of the energy and is independent of the choice of the solution $u(x_3)$.

Problem 33. Show that the bounded solution of the Schrödinger equation for the *Yukawa potential* scattering problem

$$\frac{d^2 u(r)}{dr^2} + \left(k^2 + \lambda \frac{\exp(-r/a)}{r} - \frac{\ell(\ell+1)}{r^2} \right) u(r) = 0, \quad r > 0$$

has the asymptotic property $u(r) = A\sin(kr - \ell\pi/2 + \delta_\ell) + O(1)$.

Problem 34. Consider the Schrödinger equation (eigenvalue equation)

$$\hat{H}u(x_1, x_2) = \frac{1}{2m}(\hat{p} - eA)^2 u(x_1, x_2) = Eu(x_1, x_2)$$

where the *momentum operator*

$$\hat{p} := -i\hbar\nabla$$

and the magnetic vector potential A exist in the plane (x_1, x_2) with $A = \frac{B}{2}(-x_2, x_1)$ and B is the magnetic flux density. Let $z = x_1 + ix_2$, $\bar{z} = x_1 - ix_2$ $(x_1, x_2 \in \mathbb{R}$ with dimension length) and

$$\frac{\partial}{\partial z} := \frac{1}{2}\left(\frac{\partial}{\partial x_1} - i\frac{\partial}{\partial x_2} \right), \quad \frac{\partial}{\partial \bar{z}} := \frac{1}{2}\left(\frac{\partial}{\partial x_1} + i\frac{\partial}{\partial x_2} \right).$$

Introducing the *magnetic length* and the frequency

$$\ell := \sqrt{\frac{2\hbar}{eB}}, \quad \omega := \frac{eB}{m}.$$

Consider the Bose operators

$$\hat{b} := \frac{z}{2\ell} + \ell\frac{\partial}{\partial \bar{z}}, \quad \hat{b}^\dagger := \frac{\bar{z}}{2\ell} - \ell\frac{\partial}{\partial z}$$

with the nonzero constant of dimension length. Show that $[\hat{b}, \hat{b}^\dagger] = 1$ and the Hamilton operator can be written as

$$\hat{H} = \hbar\omega(\hat{b}^\dagger\hat{b} + \hat{b}\hat{b}^\dagger).$$

Problem 35. Let $s, s' = 1/2, 1, 3/2, 2, \ldots$ and $\mu = -s, -s+1, \ldots, +s$, $\nu = -s', -s'+1, \ldots, +s'$. Consider the Bose creation and annihilation operators $b_{\mu,\nu}^\dagger$, $b_{\mu,\nu}$, respectively, with the commutation relations

$$[\hat{b}_{\mu,\nu}, \hat{b}_{\mu',\nu'}^\dagger] = \delta_{\mu,\mu'}\delta_{\nu,\nu'}I, \quad [\hat{b}_{\mu,\nu}, \hat{b}_{\mu',\nu'}] = [\hat{b}_{\mu,\nu}^\dagger, \hat{b}_{\mu',\nu'}^\dagger] = 0.$$

Consider the operators

$$\hat{N} = \sum_{\nu=-s'}^{s'} \sum_{\mu=-s}^{s} b_{\mu,\nu}^{\dagger} b_{\mu,\nu}$$

$$\hat{J}_+ = \sum_{\nu=-s'}^{s'} \sum_{\mu=-s}^{s-1} ((s-\mu)(s+\mu+1))^{1/2} b_{\mu+1,\nu}^{\dagger} b_{\mu,\nu}$$

$$\hat{J}_0 = \sum_{\nu=-s'}^{s'} \sum_{\mu=-s}^{s} \mu b_{\mu,\nu}^{\dagger} b_{\mu,\nu}$$

with $J_- = (J_+)^{\dagger}$. Show that

$$[\hat{J}_0, \hat{J}_+] = \hat{J}_+, \quad [\hat{J}_0, \hat{J}_-] = -\hat{J}_-, \quad [\hat{J}_+, \hat{J}_-] = 2\hat{J}_0.$$

Problem 36. Consider the commutation relation $[\hat{q}, \hat{p}] = i\hbar I$, where I is the identity operator and the Fourier transform

$$|p\rangle = \frac{1}{\sqrt{2\pi}} \int_{-\infty}^{\infty} \exp(iqp/\hbar)|q\rangle dq.$$

Define the unitary operators

$$\hat{U}_\alpha = \exp(i\alpha\hat{p}/\hbar), \qquad \hat{V}_\beta = \exp(i\beta\hat{q}/\hbar)$$

with α dimension *meter* and β dimension $kg \cdot meter \cdot sec^{-1}$. Show that

$$\hat{U}_\alpha \hat{V}_\beta = \exp(i\alpha\beta/\hbar)\hat{V}_\beta \hat{U}_\alpha.$$

Show that there are $n \times n$ unitary matrices U, V such that

$$UV = \omega VU, \quad \omega = e^{2\pi i/n}, \quad U^n = V^n = I_n.$$

Show that the matrices

$$X_{jk} := \frac{1}{\sqrt{n}} U^j V^k, \quad j, k = 0, 1, \ldots, n-1$$

provide a basis in the Hilbert space $M(n, \mathbb{C})$.

Problem 37. (i) Let $\hat{b}^{\dagger}$, $\hat{b}$ be Bose creation and annihilation operators, respectively and $n \in \mathbb{N}$. Show that

$$\exp(\gamma_1 \hat{b}^{\dagger} + \gamma_2 \hat{b}^n) \equiv \exp(\gamma_1 \hat{b}^{\dagger}) \exp\left(\sum_{j=1}^{n} \gamma_2 \gamma_1^j \binom{n}{j} \frac{1}{1+j} \hat{b}^{n-j}\right).$$

(ii) Let $z \in \mathbb{C}$ and $\hat{b}^\dagger$, $\hat{b}$ be Bose creation and annihilation operators, respectively. Show that

$$\exp(z\hat{b} - \bar{z}\hat{b}^\dagger) \equiv \exp(z\hat{b}) \exp(-\bar{z}\hat{b}^\dagger) \exp(-z\bar{z}/2).$$

Note that $[\hat{b}, \hat{b}^\dagger] = I$.

(iii) Let $z \in \mathbb{C}$ and $\gamma \in \mathbb{R}$. Show that

$$e^{i\gamma(z\hat{b} + \bar{z}\hat{b}^\dagger)} = e^{i\gamma\bar{z}\hat{b}^\dagger} e^{i\gamma z\hat{b}} e^{-\gamma^2|z|^2/2} = e^{i\gamma z\hat{b}} e^{i\gamma\bar{z}\hat{b}^\dagger} e^{\gamma^2|z|^2/2}.$$

Problem 38. Let $\hat{b}_1^\dagger$, $\hat{b}_2^\dagger$, $\hat{b}_1$, $\hat{b}_2$ be Bose creation and annihilation operators, respectively. Show that the operators

$$\hat{X}_1 = \frac{1}{2}(\hat{b}_2^\dagger\hat{b}_1 + \hat{b}_1^\dagger\hat{b}_2), \quad \hat{X}_2 = \frac{i}{2}(\hat{b}_2^\dagger\hat{b}_1 - \hat{b}_1^\dagger\hat{b}_2), \quad \hat{X}_3 = \frac{1}{2}(\hat{b}_1^\dagger\hat{b}_1 - \hat{b}_2^\dagger\hat{b}_2)$$

form a basis of the simple Lie algebra $su(2)$. Show that

$$U(\alpha, \beta, \gamma, \delta, \epsilon) = \exp(i(\alpha\hat{b}_1^\dagger\hat{b}_1 + \beta\hat{b}_2^\dagger\hat{b}_2)) \exp(i\epsilon\hat{X}_1) \exp(i(\gamma\hat{b}_1^\dagger\hat{b}_1 + \delta\hat{b}_2^\dagger\hat{b}_2))$$

is a unitary operator with $\alpha, \beta, \gamma, \delta, \epsilon \in \mathbb{R}$.

Problem 39. Consider the Lie algebra $su(1,1)$ with the basis K_0, K_1, K_2 satisfying the commutation relation

$$[K_1, K_2] = -iK_0, \quad [K_2, K_0] = iK_1, \quad [K_0, K_1] = iK_2.$$

(i) Setting $K_\pm = K_1 \pm iK_2$ show that

$$[K_0, K_\pm] = \pm K_\pm, \quad [K_+, K_-] = -2K_0.$$

(ii) Let $\hat{b}^\dagger$, $\hat{b}$ be Bose creation and annihilation operators, respectively. Show that the representation of the Lie algebra is given by

$$K_0 = \frac{1}{4}(\hat{b}^\dagger\hat{b} + \hat{b}\hat{b}^\dagger) \equiv \frac{1}{2}\left(\hat{b}^\dagger\hat{b} + \frac{1}{2}I\right), \quad K_+ = \frac{1}{2}(\hat{b}^\dagger)^2, \quad K_- = \frac{1}{2}\hat{b}^2$$

where I is the identity operator.

(iii) The *Casimir operator* $\hat{C}$ is given by

$$\hat{C} = K_0^2 - K_1^2 - K_2^2.$$

Show that $\hat{C} = -\frac{3}{16}I$.

(iv) Consider the states $\{|m, k\rangle\}$ with $m = 0, 1, 2, \ldots$ and $k > 0$ with (eigenvalue equation)

$$\hat{C}|m, k\rangle = k(k-1)|m, k\rangle.$$

Show that $k = \frac{1}{4}$ or $k = \frac{3}{4}$.

(v) Let $|n\rangle$ $(n = 0, 1, 2, \dots)$ be the number states with $\hat{N}|n\rangle = n|n\rangle$, where $\hat{N} = b^\dagger b$. Show that the states $|m, 1/4\rangle$ corresponds to the number state $|2m\rangle$ (even). Show that the states $|m, 3/4\rangle$ corresponds to the number states $|2m + 1\rangle$ (odd).

(vi) The coherent states of $SU(1,1)$ are defined as

$$|\zeta, k\rangle = \exp(\alpha K_+ - \alpha^* K_-)|0, k\rangle$$

where $\alpha = -\frac{1}{2}\theta e^{-i\phi}$ with θ and ϕ are group parameters with $0 < \theta < \infty$, $0 \leq \phi \leq 2\pi$, $\zeta = -\tanh(\theta/2)e^{-i\phi}$. Show that these states can be expanded in the basis $|m, k\rangle$ as

$$|\zeta, k\rangle = (1 - |\zeta|^2)^k \sum_{m=0}^\infty \left(\frac{\Gamma(m + 2k)}{m!\Gamma(2k)}\right)^{1/2} \zeta^m |m, k\rangle.$$

Problem 40. Consider the Lie algebra $su(1,1)$ with the basis K_+, K_0, K_- and the commutation relations

$$[K_+, K_-] = -2K_0, \quad [K_0, K_+] = K_+, \quad [K_0, K_-] = -K_-.$$

Let $g_1(t)$, $g_2(t)$, $g_3(t)$ be given smooth functions of t and let

$$\hat{H}(t) = g_1(t)K_+ + g_2(t)K_0 + g_3(t)K_-$$

be the Hamilton operator. Consider the evolution equation

$$i\hbar \frac{\partial U(t, 0)}{\partial t} = \hat{H}(t)U(t, 0)$$

with $U(0, 0) = I$, where I is the identity operator. Setting

$$U(t, 0) = e^{f_1(t)K_+} e^{f_2(t)K_0} e^{f_3(t)K_-}$$

where f_1, f_2, f_3 are smooth functions with $f_1(0) = f_2(0) = f_3(0) = 0$. Show that

$$\frac{\partial U}{\partial t} = \left(\left(\frac{df_1}{dt} - f_1\frac{df_2}{dt} + f_1^2 e^{-f_2}\frac{df_3}{dt}\right)K_+ \right.$$
$$\left. + \left(\frac{df_1}{dt} - 2f_1 e^{-f_2}\frac{df_3}{dt}\right)K_0 + e^{-f_2}\frac{df_3}{dt}K_-\right)U(t, 0).$$

Show that f_1, f_2, f_3 satisfy the nonlinear system of differential equations

$$\frac{df_1}{dt} - f_1\frac{df_2}{dt} + f_1^2 e^{-f_2}\frac{df_3}{dt} = -ig_1(t)/\hbar$$

$$\frac{df_2}{dt} - 2f_1 e^{-f_2}\frac{df_3}{dt} = -ig_2(t)/\hbar$$

$$e^{-f_2}\frac{df_3}{dt} = -ig_3(t)/\hbar$$

together with the initial conditions $f_1(0) = f_2(0) = f_3(0) = 0$.

Problem 41. (i) Let $\hat{b}^\dagger$, $\hat{b}$ be Bose creation and annihilation operators, respectively and I the identity operator. Do the operators $(n \in \mathbb{N})$

$$T_1 = 2\hat{b}^\dagger \left(\hat{b}^\dagger \hat{b} + \frac{n}{2} I\right)^{1/2}, \quad T_2 = 2\left(\hat{b}^\dagger \hat{b} + \frac{n}{2} I\right)^{1/2} \hat{b}, \quad T_3 = 2\left(\hat{b}^\dagger \hat{b} + \frac{n}{4} I\right)$$

form a basis of a Lie algebra under the commutator?
(ii) Let $n \in \mathbb{N}$. Do the operators

$$T_1 = x, \quad T_2 = 4x\frac{d^2}{dx^2} + 2n\frac{d}{dx}, \quad T_3 = 2x\frac{d}{dx} + \frac{n}{2}$$

form a basis of a Lie algebra under the commutator? Compare the result with the result from (i).

Problem 42. Let $\hat{b}_1^\dagger$, $\hat{b}_2^\dagger$, $\hat{b}_1$, $\hat{b}_2$ be Bose creation and annihilation operators, respectively. Let $\alpha, \beta \in \mathbb{R}$ and $\alpha\beta \neq 1$. Show that

$$\exp(\alpha\hat{b}_1\hat{b}_2)\exp(\beta\hat{b}_1^\dagger\hat{b}_2^\dagger)|0,0\rangle = \frac{1}{1 - \alpha\beta}\exp(\beta\hat{b}_1^\dagger\hat{b}_2^\dagger/(1 - \alpha\beta))|0,0\rangle.$$

Problem 43. Consider the displacement operator $D(\beta)$ and the squeezing operator $S(\zeta)$

$$D(\beta) = \exp(\beta\hat{b}^\dagger - \overline{\beta}\hat{b}), \quad S(\zeta) = \exp(\zeta\hat{b}^\dagger\hat{b}^\dagger - \overline{\zeta}\hat{b}\hat{b}).$$

Let $|\beta, \zeta\rangle = D(\beta)S(\zeta)|0\rangle$. Show that

$$\overline{n} \equiv \langle\beta, \zeta|\hat{b}^\dagger\hat{b}|\beta, \zeta\rangle = |\beta|^2 + \sinh^2(|\zeta|).$$

Problem 44. Let $\hat{b}_1^\dagger$, $\hat{b}_2^\dagger$, $\hat{b}_1$, $\hat{b}_2$ be Bose creation and annihilation operators, respectively. Note that $\hat{b}_1^\dagger = \hat{b}^\dagger \otimes I$, $\hat{b}_2^\dagger = I \otimes \hat{b}^\dagger$, where I is the identity operator. Consider the Hamilton operator

$$\hat{H} = \hbar\omega_1\hat{b}_1^\dagger\hat{b}_1^\dagger\hat{b}_1\hat{b}_1 + \hbar\omega_2\hat{b}_2^\dagger\hat{b}_2^\dagger\hat{b}_2\hat{b}_2 + \hbar\omega_3\hat{b}_1^\dagger\hat{b}_1\hat{b}_2^\dagger\hat{b}_2$$
$$+ \epsilon_1\hat{b}^\dagger\hat{b} \otimes I + \epsilon_2 I \otimes \hat{b}^\dagger\hat{b} + \hbar\omega_4(\hat{b}_1^\dagger\hat{b}_1^\dagger\hat{b}_2 + \hat{b}_2^\dagger\hat{b}_1\hat{b}_1)$$

which is a many-body two-channel Hamilton operator including scattering interaction. Find the expectation values

$$(\langle n| \otimes \langle n|)\hat{H}(|n\rangle \otimes |n\rangle)), \quad (\langle\beta| \otimes \langle\beta|)\hat{H}(|\beta\rangle \otimes |\beta\rangle)), \quad (\langle\zeta| \otimes \langle\zeta|)\hat{H}(|\zeta\rangle \otimes |\zeta\rangle))$$

where $|n\rangle$ are the number states, $|\beta\rangle$ are the coherent states and $|\zeta\rangle$ are the squeezed states.

Problem 45. Consider the Hamilton operator

$$\hat{H} = \hbar\omega_1(\hat{b}^\dagger\hat{b} + \frac{1}{2}I) + \hbar\omega_2(\hat{b}^\dagger\hat{b})^2.$$

Let $|\beta\rangle$ be a coherent state. Show that

$$|\psi(t)\rangle = e^{-i\hat{H}t/\hbar}|\beta\rangle = \exp(-|\beta|^2/2)\sum_{n=0}^\infty \frac{\beta^n}{\sqrt{n!}}\exp(-i\omega_1 nt - i\omega_2 n^2 t)|n\rangle$$

where $|n\rangle$ are the number states.

Problem 46. (i) Let $m \in \mathbb{N}_0$ and $\zeta = ze^{i\theta}$ $s \geq 0$. Consider the *squeezed state*

$$|\zeta\rangle = \frac{1}{\cosh(s)}\sum_{m=0}^\infty (-1)^m \frac{\sqrt{(2m)!}}{2^m m!}e^{im\theta}\tanh^m(s)|2m\rangle.$$

Show that

$$P_{2m}(s) = |\langle 2m|\zeta\rangle|^2 = \frac{(2m)!}{2^{2m}(m!)^2}\frac{\tanh^{2m}(s)}{\cosh(s)}$$

$$P_{2m+1}(s) = |\langle 2m+1|\zeta\rangle|^2 = 0.$$

(ii) Consider the single-mode *squeezing operator*

$$S(\zeta) = \exp\left(\frac{1}{2}(\zeta(\hat{b}^\dagger)^2 - \zeta^*\hat{b}^2)\right), \quad \zeta \in \mathbb{C}, \quad \zeta = se^{i\theta}$$

with $s \geq 0$. Show that

$$|\zeta\rangle = S(\zeta)|0\rangle = \sum_{m=0}^\infty \frac{(2m)!}{2^{2m}(m!)^2}\frac{\tanh^{2m}(s)}{\cosh(s)}|2m\rangle$$

where s and θ are the squeezing parameter and squeezing angle, respectively.

Problem 47. Fermi creation and annihilation operators $\hat{c}_j^\dagger$, $\hat{c}_j$ $(j = 1,\ldots,n)$ can be represented by matrices via the *Jordan-Wigner transform*. Let $n = 2$. Consider the hermitian Hamilton operator

$$\hat{H} = \hbar\omega_1(\hat{c}_1^\dagger\hat{c}_2 + \hat{c}_2^\dagger\hat{c}_1) + \hbar\omega_2(\hat{c}_1^\dagger\hat{c}_1 + \hat{c}_2^\dagger\hat{c}_2)$$
$$+ \hbar\omega_3(\hat{c}_1^\dagger\hat{c}_2^\dagger + \hat{c}_2\hat{c}_1) + \hbar\omega_4(\hat{c}_1^\dagger + \hat{c}_2^\dagger + \hat{c}_1 + \hat{c}_2).$$

Note that the Hamilton operator $\hat{H}$ does not commute with the number operator $\hat{N} = \hat{c}_1^\dagger \hat{c}_1 + \hat{c}_2^\dagger \hat{c}_2$. The Jordan-Wigner transform is given by

$$\hat{c}_1 = \frac{1}{2}\sigma_- \otimes I_2 \;\Rightarrow\; \hat{c}_1^\dagger = \frac{1}{2}\sigma_+ \otimes I_2$$

$$\hat{c}_2 = \sigma_3 \otimes \frac{1}{2}\sigma_- \;\Rightarrow\; \hat{c}_2^\dagger = \sigma_3 \otimes \frac{1}{2}\sigma_+.$$

Setting $\alpha_j \equiv \hbar\omega_j$ show that the matrix representation is given by

$$\begin{pmatrix} 2\alpha_2 & \alpha_4 & \alpha_4 & -\alpha_3 \\ \alpha_4 & \alpha_2 & -\alpha_1 & \alpha_4 \\ \alpha_4 & -\alpha_1 & \alpha_2 & -\alpha_4 \\ -\alpha_3 & \alpha_4 & -\alpha_4 & 0 \end{pmatrix}.$$

Find the four (real) eigenvalues and the corresponding eigenvectors. Find the matrix representation of $\hat{H}$ with the basis

$$\hat{c}_1^\dagger \hat{c}_2^\dagger |0\rangle, \;\; \hat{c}_2^\dagger |0\rangle, \;\; \hat{c}_1^\dagger |0\rangle, \;\; |0\rangle$$

and the corresponding dual basis

$$\langle 0|\hat{c}_2 \hat{c}_1, \;\; \langle 0|\hat{c}_2, \;\; \langle 0|\hat{c}_1, \;\; \langle 0|.$$

Discuss.

Problem 48. Let $\hat{c}_\uparrow^\dagger$, $\hat{c}_\downarrow^\dagger$, $\hat{c}_\uparrow$, $\hat{c}_\downarrow$ be Fermi creation and annihilation operators with spin up and down, respectively.

(i) Find the matrix representation of the Hamilton operator

$$\hat{H} = \hbar\omega_1(\hat{c}_\uparrow^\dagger + \hat{c}_\uparrow + \hat{c}_\downarrow^\dagger + \hat{c}_\downarrow) + \hbar\omega_2(\hat{c}_\uparrow^\dagger \hat{c}_\uparrow + \hat{c}_\downarrow^\dagger \hat{c}_\downarrow) + \hbar\omega_3(\hat{c}_\uparrow^\dagger \hat{c}_\downarrow + \hat{c}_\downarrow^\dagger \hat{c}_\uparrow)$$
$$+ \hbar\omega_4(\hat{c}_\uparrow^\dagger \hat{c}_\downarrow^\dagger + \hat{c}_\downarrow \hat{c}_\uparrow) + \hbar\omega_5 \hat{c}_\uparrow^\dagger \hat{c}_\uparrow \hat{c}_\downarrow^\dagger \hat{c}_\downarrow$$

with the basis $|0\rangle$, $\hat{c}_\uparrow^\dagger |0\rangle$, $\hat{c}_\downarrow^\dagger |0\rangle$, $\hat{c}_\uparrow^\dagger \hat{c}_\downarrow^\dagger |0\rangle$, i.e. we have a four-dimensional Hilbert space and $\hat{c}_\uparrow |0\rangle = 0|0\rangle$, $\hat{c}_\downarrow |0\rangle = 0|0\rangle$.

(ii) Consider the Hamilton operator

$$\hat{H} = \hbar\omega(\hat{c}_\uparrow^\dagger \hat{c}_\uparrow + \hat{c}_\downarrow^\dagger \hat{c}_\downarrow) + U\hat{c}_\uparrow^\dagger \hat{c}_\downarrow^\dagger \hat{c}_\downarrow \hat{c}_\uparrow.$$

The Hamilton operator can be approximate by

$$\widetilde{H} = \hbar\omega(\hat{c}_\uparrow^\dagger \hat{c}_\uparrow + \hat{c}_\downarrow^\dagger \hat{c}_\downarrow) + \Delta(e^{i\chi}\hat{c}_\downarrow \hat{c}_\uparrow + e^{-i\chi}\hat{c}_\uparrow^\dagger \hat{c}_\downarrow^\dagger)$$

where

$$\Delta e^{i\chi} = U\langle \hat{c}_\uparrow^\dagger \hat{c}_\downarrow^\dagger \rangle$$

is the complex order parameter. Show that the Hamilton operator $\tilde{H}$ can be diagonalized by the transformation

$$\hat{d}_1 = -e^{i\chi}\cos(\phi)\hat{c}_\uparrow + \sin(\phi)\hat{c}_\downarrow^\dagger, \quad \hat{d}_2^\dagger = e^{-i\chi}\cos(\phi)\hat{c}_\downarrow^\dagger + \sin(\phi)\hat{c}_\uparrow$$

where $\tan(2\phi) = -\Delta/(\hbar\omega)$. Show that the diagonal form of the Hamilton operator $\tilde{H}$ is given by

$$\hat{K} = \sqrt{\hbar^2\omega^2 + \Delta^2}(\hat{d}_1^\dagger\hat{d}_1 + \hat{d}_2^\dagger\hat{d}_2).$$

Problem 49. Let $\sigma \in \{\uparrow,\downarrow\}$ and $\hat{c}_{1,\sigma}^\dagger$, $\hat{c}_{2,\sigma}$ be Fermi creation operators with spin σ and $\hat{c}_{1,\sigma}$, $\hat{c}_{2,\sigma}$ be Fermi annihilation operators. Consider the operators

$$\hat{T}_1 = \hat{c}_{1,\uparrow}^\dagger\hat{c}_{2,\uparrow}\hat{c}_{1,\downarrow}^\dagger\hat{c}_{2,\downarrow}, \quad \hat{T}_2 = \hat{c}_{1,\downarrow}^\dagger\hat{c}_{2,\downarrow}\hat{c}_{1,\uparrow}^\dagger\hat{c}_{2,\uparrow}$$

and the basis

$$\hat{c}_{1,\uparrow}^\dagger c_{2,\uparrow}^\dagger|0\rangle, \ \hat{c}_{1,\uparrow}^\dagger c_{2,\downarrow}^\dagger|0\rangle, \ \hat{c}_{1,\downarrow}^\dagger c_{2,\uparrow}^\dagger|0\rangle, \ \hat{c}_{1,\downarrow}^\dagger c_{2,\downarrow}^\dagger|0\rangle, \ \hat{c}_{1,\uparrow}^\dagger c_{1,\downarrow}^\dagger|0\rangle, \ \hat{c}_{2,\uparrow}^\dagger c_{2,\uparrow}^\dagger|0\rangle.$$

Find the matrix representation of $\hat{T}_1$, $\hat{T}_2$ and $\hat{T}_1 + \hat{T}_2$ and the eigenvalues of these matrices.

Problem 50. Let $\hat{c}_1^\dagger$, $\hat{c}_2^\dagger$, $\hat{c}_3^\dagger$, $\hat{c}_4^\dagger$, $\hat{c}_1$, $\hat{c}_2$, $\hat{c}_3$, $\hat{c}_4$ be Fermi creation and annihilation operators. The standard basis with two Fermions is given by

$$|\psi_{12}\rangle = \hat{c}_1^\dagger\hat{c}_2^\dagger|0\rangle \quad |\psi_{13}\rangle = \hat{c}_1^\dagger\hat{c}_3^\dagger|0\rangle \quad |\psi_{14}\rangle = \hat{c}_1^\dagger\hat{c}_4^\dagger|0\rangle$$

$$|\psi_{23}\rangle = \hat{c}_2^\dagger\hat{c}_3^\dagger|0\rangle \quad |\psi_{24}\rangle = \hat{c}_2^\dagger\hat{c}_4^\dagger|0\rangle \quad |\psi_{34}\rangle = \hat{c}_3^\dagger\hat{c}_4^\dagger|0\rangle.$$

The six *Bell states*

$$|\phi_1\rangle = \frac{1}{\sqrt{2}}(\hat{c}_1^\dagger\hat{c}_2^\dagger + \hat{c}_3^\dagger\hat{c}_4^\dagger)|0\rangle, \quad |\phi_2\rangle = \frac{1}{\sqrt{2}}(\hat{c}_1^\dagger\hat{c}_2^\dagger - \hat{c}_3^\dagger\hat{c}_4^\dagger)|0\rangle$$

$$|\phi_3\rangle = \frac{1}{\sqrt{2}}(\hat{c}_2^\dagger\hat{c}_4^\dagger + \hat{c}_3^\dagger\hat{c}_1^\dagger)|0\rangle, \quad |\phi_4\rangle = \frac{1}{\sqrt{2}}(\hat{c}_2^\dagger\hat{c}_4^\dagger - \hat{c}_3^\dagger\hat{c}_1^\dagger)|0\rangle$$

$$|\phi_5\rangle = \frac{1}{\sqrt{2}}(\hat{c}_2^\dagger\hat{c}_3^\dagger + \hat{c}_1^\dagger\hat{c}_4^\dagger)|0\rangle \quad |\phi_6\rangle = \frac{1}{\sqrt{2}}(\hat{c}_2^\dagger\hat{c}_3^\dagger - \hat{c}_1^\dagger\hat{c}_4^\dagger)|0\rangle.$$

Consider the Hamilton operator

$$\begin{aligned}\hat{H} = &\ \hbar\omega_1(\hat{c}_1^\dagger\hat{c}_2 + \hat{c}_2^\dagger\hat{c}_1 + \hat{c}_2^\dagger\hat{c}_3 + \hat{c}_3^\dagger\hat{c}_2 + \hat{c}_3^\dagger\hat{c}_4 + \hat{c}_4^\dagger\hat{c}_3 + \hat{c}_4^\dagger\hat{c}_1 + \hat{c}_1^\dagger\hat{c}_4 \\ &+ \hbar\omega_2(\hat{c}_1^\dagger\hat{c}_1 + \hat{c}_2^\dagger\hat{c}_2 + \hat{c}_3^\dagger\hat{c}_3 + \hat{c}_4^\dagger\hat{c}_4) \\ &+ \hbar\omega_3(\hat{c}_1^\dagger\hat{c}_2^\dagger\hat{c}_3\hat{c}_4 + \hat{c}_4^\dagger\hat{c}_3^\dagger\hat{c}_1\hat{c}_2).\end{aligned}$$

Find the matrix representation with the standard basis and the Bell basis. Then find the eigenvalues and normalized eigenvectors. Show that the Hamilton operators commutes with the number operator

$$\hat{N} = \sum_{j=1}^{4} \hat{c}_j^\dagger \hat{c}_j.$$

Problem 51. Consider the Bose creation and annihilation operators

$$\hat{b}_1^\dagger, \hat{b}_0^\dagger, \hat{b}_{-1}^\dagger, \hat{b}_1, \hat{b}_0, \hat{b}_{-1}$$

the Fermi creation and annihilation operator

$$\hat{c}_{1/2}^\dagger, \hat{c}_{-1/2}^\dagger, \hat{c}_{1/2}, \hat{c}_{-1/2}$$

and the product states

$$|n_1, n_2, n_3, m_1, m_2\rangle = ((\hat{b}_1^\dagger)^{n_1} (\hat{b}_0^\dagger)^{n_2} (\hat{b}_{-1}^\dagger)^{n_3} \otimes (\hat{c}_{1/2}^\dagger)^{m_1} (\hat{c}_{-1/2}^\dagger)^{m_2}) |0\rangle_B \otimes |0\rangle_F$$

where $n_1, n_2, n_3 \in \mathbb{N}_0$, $m_1, m_2 \in \{0, 1\}$. Here $|0\rangle_B$ is the Boson vacuum and $|0\rangle_F$ is the Fermion vacuum. Show that the vector space of these states with $n_1 + n_2 + n_3 + m_1 + m_2 = s$ is isomorphic to the s-th supersymmetric Kronecker product of the representation $\Gamma = D^1 \oplus D^{1/2}$. Show that the rotation operator can be written as ($\alpha \in \mathbb{R}$)

$$R_3(\alpha) = \exp(i\alpha T_3)$$

with

$$T_3 = \sum_{k \in \{+1, 0, -1\}} k \hat{b}_k^\dagger \hat{b}_k \otimes I_F + I_B \otimes \sum_{\ell \in \{+1/2, -1/2\}} \ell \hat{c}_\ell^\dagger \hat{c}_\ell.$$

Problem 52. Let $\hat{b}^\dagger$, $\hat{b}$ be Bose creation and annihilation operators, respectively and $\hat{c}^\dagger$, $\hat{c}$ be Fermi creation and annihilation operators. Consider the Hamilton operator

$$\hat{H} = \hbar\omega_1 \hat{b}^\dagger \hat{b} \otimes I_F + \hbar\omega_2 I_B \otimes \hat{c}^\dagger \hat{c} + \hbar\omega_3 (\hat{b}^\dagger \hat{b}) \otimes (\hat{c}^\dagger + \hat{c})$$
$$+ \hbar\omega_4 (\hat{b}^\dagger + \hat{b}) \otimes \hat{c}^\dagger \hat{c} + \hbar\omega_5 (\hat{b}^\dagger + \hat{b}) \otimes (\hat{c}^\dagger + \hat{c}).$$

(i) Let $|n\rangle_B$ ($n = 0, 1, \dots$) be number states and $|0\rangle_F$, $\hat{c}^\dagger |0\rangle_F$ the two states for the Fermi system. Find

$$(_B\langle n| \otimes {}_F\langle 0|) \hat{H}(|n\rangle_B \otimes |0\rangle_F), \quad (_B\langle n| \otimes {}_F\langle 0|\hat{c}) \hat{H}(|n\rangle_B \otimes \hat{c}^\dagger |0\rangle_F),$$

$$(_B\langle n| \otimes {}_F\langle 0|) \hat{H}(|n\rangle_B \otimes \hat{c}^\dagger |0\rangle_F), \quad (_B\langle n| \otimes {}_F\langle 0|\hat{c}) \hat{H}(|n\rangle_B \otimes |0\rangle_F),$$

(ii) Let $|\beta\rangle$ be a coherent state. Find

$$((\langle\beta|\otimes\langle 0|)\hat{H}(|\beta\rangle\otimes|0\rangle)), \quad ((\langle\beta|\otimes\langle 0|\hat{c})\hat{H}(|\beta\rangle\otimes\hat{c}^\dagger|0\rangle)).$$

(iii) Let $|\zeta\rangle$ be a squeezed state. Find

$$((\langle\zeta|\otimes\langle 0|)\hat{H}(|\zeta\rangle\otimes|0\rangle)), \quad ((\langle\zeta|\otimes\langle 0|\hat{c})\hat{H}(|\zeta\rangle\otimes\hat{c}^\dagger|0\rangle)).$$

Compare the results from (i), (ii) and (iii).

Problem 53. (i) Let A be a nonzero nonnormal $n \times n$ matrix. What can be said about the spectrum of

$$\hat{H} = \hbar\omega_1(A + A^*) + \hbar\omega_2 AA^* ?$$

(ii) Let $\hat{c}^\dagger$, $\hat{c}$ be Fermi creation and annihilation operators, respectively. What can be said about the spectrum of the operator

$$\hat{H} = \hbar\omega_1(\hat{c}^\dagger + \hat{c}) + \hbar\omega_2\hat{c}^\dagger\hat{c} ?$$

(iii) Let $b^\dagger$, b be Bose creation and annihilation operators, respectively. What can be said about the spectrum of the operator

$$\hat{H} = \hbar\omega_1(\hat{b}^\dagger + \hat{b}) + \hbar\omega_2\hat{b}^\dagger\hat{b} ?$$

(iv) Let C be a nonzero $n \times n$ circulant matrix. What can be said about the spectrum of the hermitian matrix

$$\hat{H} = \hbar\omega_1(C + C^*) + \hbar\omega_2(CC^*) ?$$

Bibliography

[1] Adams R. A.
Sobolev Spaces, Academic Press, New York (1975)

[2] Akhiezer N. I. and Glazman I. M.
Theory of Linear Operators in Hilbert Space, Ungar, New York (1961)

[3] Akhiezer N. I. and Glazman I. M.
Theory of Linear Operators in Hilbert Space, vol. 2, Pitman, London (1981)

[4] Balakrishnan A. V.
Applied Functional Analysis, Second edition. Springer Verlag, New York (1981)

[5] Bargmann V.
"On a Hilbert space of analytic functions and an associated integral transform". Part I. Comm. Pure Appl. Math. **14** 187–214, (1961)

[6] Bargmann V.
"On the Representation of the Rotation Group", Rev. Mod. Phys. **34**, 829–845 (1962)

[7] Bargmann V.
"On a Hilbert space of analytic functions and an associated integral transform". Part II. A family of related function spaces applications to distribution theory, Comm. Pure Appl. Math. **20**, 1–101 (1967)

[8] Beauzamy B.
Introduction to Banach spaces and their Geometry, Second edition, North Holland (1985)

[9] Berlinet A. and Thomas C.
Reproducing kernel Hilbert spaces in Probability and Statistics, Kluwer Academic Publishers (2004)

[10] Bhaumik D., Dutta-Roy B. and Ghosh G.
"Classical limit of the hydrogen atom", J. Phys. A : Math. Gen. **19**, 1355–1364 (1986)

[11] Bohm A.
Quantum Mechanics: Foundations and Applications, Third edition, Springer-Verlag, New York (1993)

[12] Bourbaki N.
Topological Vector Spaces, Elements of Mathematics, Springer Verlag, Berlin (1986)

[13] Chan Y. T.
Wavelet Basics, Kluwer Academic Publishers, Boston (1995)

[14] Chui C. K.
An Introduction to Wavelets, Academic Press, Boston (1992)

[15] Ciarlet P. G.
Linear and Nonlinear Functional Analysis with Applications, SIAM, Philadelphia (2013)

[16] Conway J. B.
A Course in Functional Analysis, second edition, Springer-Verlag, New York (1990)

[17] Courant and Hilbert D.
Methods of Mathematical Physics, Vol. 1, Vol. 2, John Wiley, New York (1953)

[18] Davies E. B.
Linear Operators and their Spectra, Cambridge studies in advanced mathematics, Cambridge University Press (2007)

[19] Debnath L. and Mikusiński P.
Introduction to Hilbert Spaces with Applications, Academic Press, Boston (1990)

[20] Dieudonné J.
Foundations of Modern Analysis, Academic Press, New York (1960)

[21] Dirac P. A. M.
The Principles of Quantum Mechanics, 4th revised edition, Clarendon Press, Oxford (1974)

[22] Dunford E. B. and Schwartz J. T.
Linear Operators. Part 1: General Theory, Interscience, New York (1966)

[23] Edmonds A. R.
Angular momentum in Quantum mechanics, Princeton, NJ: Princeton University Press (1957)

[24] Feichtinger H. G.
"Banach Gelfand Triples for Applications on Physics and Engineering", AIP Conference Proceedings **1146**, 189–228 (2009)

[25] Flügge S.
Practical Quantum Mechanics, Springer Verlag, Berlin (1974)

[26] Gelfand I. M. and Shilov G. E.
Generalized functions, II-III, Academic Press, New York (1968)

[27] Gelfand I. and Vilenkin N.
Generalized Functions, vol. 4 Academic Press, New York (1968)

[28] Glimm J. and Jaffe A.
Quantum Physics, Springer Verlag, New York (1981)

[29] Golub G. H. and Van Loan C. F.
Matrix Computations, Third Edition,
Johns Hopkins University Press, Baltimore (1996)

[30] Gottfried K.
Quantum Mechanics, Benjamin, New York (1966)

[31] Haar D. ter
Problems in Quantum Mechanics, 3rd edition, London: Pion (1975)

[32] Halmos P. R.
A Hilbert Space Problem Book, 2nd edition, Springer Verlag, New York
(1982)

[33] Hardy Y. and Steeb W.-H.
Matrix Calculus, Kronecker Product and Tensor Product, World Scientific,
Singapore (2019)

[34] Hernandez E. and Weiss G.
A First Course on Wavelets, CRC Press, Boca Raton (1996)

[35] Hörmander L.
The Analysis of Linear Partial Differential Operators, vol I-III, Springer
Verlag, Berlin, 1983, 1985

[36] Jantcher L.
Distributionen, Walter de Gruyter, Berlin (1971)

[37] Jones D. S.
The Theory of Generalized Functions, Cambridge University Press (1982)

[38] Kaiser G.
A Friendly Guide to Wavelets, Birkhäuser, Boston (1994)

[39] Kato T.
Perturbation Theory of Linear Operators, 2nd edition, Springer, New York
(1980)

[40] Keener J. P.
Principles of Applied Mathematics: Transformation and Approximation, Perseus Books, 2nd revised edition (2000)

[41] Klauder R. and Skagerstam B.
Coherent States: Applications in Physics and Mathematical Physics, World Scientific, Singapore (1985)

[42] Kowalski K. and Steeb W.-H.
Nonlinear Dynamical Systems and Carleman Linearization, World Scientific, Singapore (1994)

[43] Kreyszig E.
Functional Analysis with Applications, John Wiley, New York (1989)

[44] Lax P. D. and Milgram A. N.
"Parabolic Equations" in Contributions to the theory of partial differential equations, Annals of Mathematical Studies **33**, 167–190 (1964)

[45] Lighthill M. J.
Introduction to Fourier analysis and generalized functions, Cambridge University Press, Cambridge (1959)

[46] Littlewood J. E. and Payley R. E.
"Theorems of Fourier Series and Power Series", J. London Math. Soc. **6**, 230–233 (1931)

[47] Mallat S. G.
A Wavelet Tour to Signal Processing, Academic Press (1999)

[48] Mavromatis H. A.
Exercises in Quantum Mechanics, Reidel, Dordrecht (1987)

[49] Mazya Vladimir G.
Sobolev Space, Springer Verlag (1985)

[50] Megginson R.
An Introduction to Banach Space Theory, Springer Verlag, New York (1998)

[51] Mercer J.
"Functions of positive and negative type and their connection with the theory of integral equations", Philosophical Transaction of the Royal Society, **209**, 441–458 (1909)

[52] Messiah A.
Quantum Mechanics, vol. I, North-Holland, Amsterdam (1961)

[53] Miller W.
Symmetry Groups and Their Applications, Academic Press, New York (1972)

[54] Morse P. M. and Feshbach H.
Methods of Theoretical Physics, Part I-II, McGraw-Hill, New York (1998)

[55] Naimark M. A.
Linear Differential Operators, vol. 2, Unger, New York (1968)

[56] Naimark M. A.
Linear Representation of the Lorentz Group, Pergamon Press, London (1964)

[57] Prugovečki E.
Quantum Mechanics in Hilbert Space, Second Edition, Academic Press, New York (1981)

[58] Rademacher H.
"Einige Sätze über Reihen von allgemeinen Orthogonalfunktionen", Math. Ann. **87**, 112–138 (1922)

[59] Reed M. and Simon B.
Methods in Modern Mathematical Physics, Vol. 1: Functional Analysis, 2nd edition, Academic Press, New York (1980)

[60] Reddy B. Daya
Introductory Functional Analysis, Springer Verlag, New York (1990)

[61] Richtmyer R. D.
Principle of Advanced Mathematical Physics, Volume I, Springer Verlag, New York (1978)

[62] Riesz F.
"Sur les systems orthogonaux de fonctions", Comptes rendus del l'Académie des sciences **144**, 615–619 (1907)

[63] Riesz F. and Nagy B. Sz. *Functional Analysis*, Frederick Unger Pub., New York (1965)

[64] Rosenthal J. S.
A First Look at Rigorous Probability Theory, Second Edition, World Scientific, Singapore (2006)

[65] Rudin W.
Functional Analysis, McGraw Hill, New York (1973)

[66] Sakurai J. J.
Modern Quantum Mechanics, Benjamin/Cummings, Menlo Park, CA. (1993)

[67] Schiff L. I.
Quantum Mechanics, New York: McGraw-Hill (1968)

[68] Schwartz L.
Théorie des distributions, Hermann, Paris, 2 vols. (1966)

[69] Stakgold I.
Green's Functions and Boundary Value Problems, 2nd edition, John Wiley, New York (1998)

[70] Steeb W.-H.
Continuous Symmetries, Lie Algebras, Differential Equations and Computer Algebra
World Scientific Publishing, Singapore (1996)

[71] Steeb W.-H.
Hilbert Spaces, Wavelets, Generalized Functions and Quantum Mechanics,
Kluwer Academic Publishers, Dordrecht (1998)

[72] Steeb W.-H.
Theoretical and Mathematical Physics: Problems and Solutions, World Scientific Publishing, Singapore (2019)

[73] Steeb W.-H., Hardy Y., Hardy A. and Stoop R.
*Problems and Solutions in Scientific Computing with C++ and Java Sim-
ulations*, World Scientific Publishing, Singapore (2004)

[74] Stone M. H.
"On one parameter unitary groups in Hilbert space", Annals of Mathematics, **33**, 643–648 (1932)

[75] Titchmarsh E. C.
*Eigenfunction Expansions associated with second-order Differential Equa-
tions*, Part One, second edition Oxford University Press, Oxford (1962)

[76] Titchmarsh E. C.
Introduction to the Theory of Fourier Integrals, Oxford University Press,
London (1962)

[77] Trotter H. F.
"On the product of semi groups of Operators", Proceedings of the Ameri-
can Mathematical Society", **10**, 545–551 (1959)

[78] Vladimirov V. S.
Equations of Mathematical Physics, Marcel Dekker, New York (1971)

[79] von Neumann J.
Mathematical Foundations of Quantum Mechanics, Princeton U.P., Princeton, New Jersey (1955)

[80] Weidmann J.
Linear Operators in Hilbert Spaces, Springer-Verlag, New York (1980)

[81] Weyl H.
The Theory of Groups and Quantum Mechanics, Dover, New York (1931)

[82] Weber H. and Mathis W.
"Analysis and Design of Nonlinear Circuits with a Self-Consistent Carleman Linearization", IEEE Transactions on Circuits and Systems I, Vol. 65, No. 12, pp. 4272–4284 (2018)

[83] Wilansky A.
Modern Methods in Topological Vector Spaces, Dover Publications, New York (2013)

[84] Yosida K.
Functional Analysis, Fifth Edition, Springer Verlag, New York (1978)

[85] Young N.
An Introduction to Hilbert Space, Cambridge University Press (1988)

[86] Zuily C.
Problems in Distributions and Partial Differential Equations, North-Holland, Amsterdam (1988)

Index

Printed in the United States
by Baker & Taylor Publisher Services